Groundwater Hydrology

Groundwater Hydrology
Conceptual and Computational Models

K.R. Rushton
Emeritus Professor of Civil Engineering
University of Birmingham, UK

WILEY

Other Wiley Editorial Offices

John Wiley & Sons Inc., 111 River Street, Hoboken, NJ 07030, USA

Jossey-Bass, 989 Market Street, San Francisco, CA 94103-1741, USA

Wiley-VCH Verlag GmbH, Boschstr. 12, D-69469 Weinheim, Germany

John Wiley & Sons Australia Ltd, 33 Park Road, Milton, Queensland 4064, Australia

John Wiley & Sons (Asia) Pte Ltd, 2 Clementi Loop #02-01, Jin Xing Distripark, Singapore 129809

John Wiley & Sons Canada Ltd, 22 Worcester Road, Etobicoke, Ontario, Canada M9W 1L1

Wiley also publishes its books in a variety of electronic formats. Some content that appears in print may not be available in
electronic books.

British Library Cataloguing in Publication Data

A catalogue record for this book is available from the British Library

ISBN 0-470-85004-3

Typeset in 9/11 pt Times by SNP Best-set Typesetter Ltd., Hong Kong
Printed and bound in Great Britain by Antony Rowe Ltd, Chippenham, Wiltshire
This book is printed on acid-free paper responsibly manufactured from sustainable forestry
in which at least two trees are planted for each one used for paper production.

Contents

Preface

It was more than twenty years ago when visiting India that I was asked the question, 'How should a groundwater investigation be planned?' At that time I had difficulty in giving a convincing answer, but in the intervening years, with involvement in many challenging practical groundwater studies, the important issues have become clearer. The key is *understanding* before *analysis*; this is reflected by the use of the words *conceptual* and *computational* models in the title of this book.

The number of groundwater investigations throughout the world continues to increase. The objectives of these investigations are varied, including meeting regulatory requirements, exploring the consequences of groundwater development and rectifying the results of over-exploitation of groundwater resources. Although hydrogeologists and water resource engineers working on these projects may not carry out the analytical and numerical analysis themselves, it is vital that they understand how to develop comprehensive quantified conceptual models and also appreciate the basis of analytical solutions or numerical methods of modelling groundwater flow. The presentation of the results of groundwater investigations in a form that can be understood by decision makers is another important task. This book is designed to address these issues. The first task in every investigation is to develop conceptual models to explain how water enters, passes through and leaves the aquifer system. These conceptual models are based on the interpretation of field data and other information. Second, techniques and methodologies are required to analyse the variety of flow processes identified during the conceptual model development. A considerable number of computational models are available.

Developments of both conceptual and computational models for groundwater hydrology have continued from early in the twentieth century to the present day. Initially, computational models relied on analytical methods but there is now a greater use of numerical models. Computational models for analysing groundwater problems are the subject of many articles in the literature. However, conceptual models are not discussed as widely. Real groundwater problems are frequently so complex that they can only be analysed when simplifying assumptions are introduced. Imagination and experience are required to identify the key processes which must be included in conceptual and computational models. Furthermore, the selection of appropriate aquifer parameters is not straightforward. A computational model does not need to be complex, despite the availability of model codes which include both saturated and unsaturated flow conditions with the option of a large number of three-dimensional mesh subdivisions and numerous time steps. Simpler, more flexible analytical or numerical models are often suitable for the early stages of modelling; more complex models can be introduced when there is confidence that the important features of the aquifer system have been recognised. In an attempt to indicate how the key processes are identified and aquifer parameters are selected, more than fifty major case studies are included in this book.

Case studies illustrate how crucial insights are gained which lead to a breakthrough in identifying the important flow processes which must be incorporated in the conceptual models. Comparing and contrasting an aquifer system with other field problems has proved to be of immense benefit. A further advantage of case studies is that they can indicate appropriate aquifer parameter values. For example, selecting suitable values for the effective vertical hydraulic conductivities of low permeability strata is notoriously difficult. However, experience gained from other investigations in similar situations often provides suitable first estimates. Due to differences in climate, geology and the way in which groundwater is utilised, the reader may find it difficult to appreciate the significance of some of the case studies. Yet much can be learnt from groundwater studies in other countries. Although the reader may never be involved in studying losses from lined canals or irrigated ricefields, insights gained from these investigations can be transferred to other projects involving surface water–groundwater interaction and recharge estimation.

Groundwater studies involve teamwork, respecting insights from different disciplines and sharing in uncertainties and ambiguities. I am most grateful to the many colleagues with whom I have shared the challenge of a wide diversity of problems. Some of my colleagues can be identified from more than sixty co-authors listed in papers in the References, but there are many more individuals and organisations with whom I have been fortunate to collaborate.

1

Introduction

1.1 GROUNDWATER INVESTIGATIONS – A DETECTIVE STORY

Many of the basic principles of the flow of groundwater through aquifers were established by the early 1970s. In addition, numerous analytical solutions were available. At that time, numerical modelling of groundwater flow was also developing although limited computing power restricted applications. Since the 1970s there have been significant advances in:

- broadening the understanding of important flow processes in a wide variety of aquifer systems,
- developing and applying analytical and numerical techniques to represent these flow processes.

There are many different flow processes in aquifer systems. Inflows occur due to recharge which can result from rainfall, runoff or groundwater–surface water interaction. Outflows occur from natural features such as springs and rivers and through wells and boreholes. The actual flow processes may be complex when, for example, a borehole takes different proportions of the discharge from a number of layers within the aquifer system. Flow paths within an aquifer are frequently complicated with horizontal and vertical components; these flows may be strongly influenced by layering in the aquifer system. Furthermore, the response of an aquifer with substantial seasonal variations in saturated thickness is different from one in which the seasonal water table fluctuations are small compared to the total saturated depth. Additional uncertainties occur when deciding how large an area needs to be studied. Although the purpose of the investigation may relate to the impact of changed abstraction in an area of a few square kilometres, the aquifer response may be influenced by hundreds, if not thousands, of square kilometres of the aquifer system. Recognising and

resolving these issues is the key to *conceptual model* development. The conceptual model must represent all these flow processes; estimates of the magnitudes of all the relevant parameters are also required.

Having developed a conceptual model, *computational models* are used to test whether the flow processes of the conceptual model are plausible and the parameter values are appropriate so that the computational model replicates observed field responses. As a first step, it is often helpful to simplify the physical setting, perhaps eliminating one or two coordinate dimensions or considering a pseudo-steady state. These idealised problems can be studied using analytical models or simple numerical models. Initial studies can be followed by the development of complex regional groundwater models. Insights from computational models often lead to modifications of conceptual models.

Although each groundwater study has its own special features, there has been a growing consensus about the importance of certain flow processes. For example, in many early studies, vertical flows through strata of lower permeability were considered to be sufficiently small to be neglected. Then, a number of investigations in very different hydrogeological settings identified the importance of vertical flow components through low permeability strata. Failure to include these vertical flow components explains why certain earlier resource studies provided unreliable predictions.

The purpose of this book is to describe advances in conceptual and computational modelling. The contents of the book are selected to introduce a wide range of issues which are important in practical groundwater studies. Many of the examples are taken from studies in which the author was involved, although there is cross-referencing to other similar studies. One advantage of using case studies based on the author's experience is that information can be presented about initial misunderstandings and invalid concepts.

Groundwater Hydrology: Conceptual and Computational Models. K.R. Rushton
© 2003 John Wiley & Sons Ltd   ISBN: 0-470-85004-3

There are many similarities between *detective stories* and *groundwater investigations*. Important activities of a good detective include thorough investigations at the scene of the crime, imaginative interpretation of all the available information, and identifying the crucial issues from a wealth of information. In groundwater investigations, this corresponds to the need to become very familiar with all aspects of the study area, to examine imaginatively all field information (not just the 'good' data sets) and to use all this information to hypothesise on the groundwater flow processes. Inevitably false leads will be followed yet information, which appears not to fit, is often a key to identifying crucial aspects of the flow processes. In a detective story there is usually a famous detective who succeeds in identifying vital clues which lead to the naming of the culprit. However, groundwater studies do not rely on a single individual: instead, contributions are made by a number of specialists. Each member of the team must listen to others, respecting their skills and insights; this leads to the breakthroughs that are required during most groundwater investigations.

1.2 CONCEPTUAL MODELS

Conceptual models describe how water enters an aquifer system, flows through the aquifer system and leaves the aquifer system. Conceptual models start with simple sketches although in their final form they may be detailed three-dimensional diagrams. First attempts at conceptual models are often based on simplified geological cross-sections with sketches of flow directions. Then diagrams are prepared for various seasons; more detailed drawings for certain parts of the aquifer system are also helpful. Conceptual models allow other members of the team to assess critically the current thinking and to provide further insights. In addition to conceptual models for the whole aquifer system, specialised models are required for smaller scale processes such as recharge estimation, the performance of pumped boreholes and river–aquifer interaction.

To demonstrate the importance of conceptual models, consider the analysis of pumping tests. Conceptual models are rarely prepared before embarking on the analysis of pumping test data. One example quoted in this book relates to a weathered-fractured aquifer system with pumping from the underlying fractured zone and monitoring in both the weathered and fractured zones (Section 7.4.6). When pumping from the fractured zone stopped, drawdowns continued to increase in shallow piezometers in the weathered zone. This continuing fall was explained away as being a fault

in the construction of the shallow piezometers. Consequently the original analysis was carried out assuming that the fractured zone acts as a confined aquifer. However, preparation of a conceptual sketch of flows through the aquifer system indicates that substantial flows occur from the weathered to the fractured zone. This key feature of the weathered-fractured system was overlooked by moving straight to analysis and ignoring 'inconvenient' data.

1.3 COMPUTATIONAL MODELS

The phrase 'computational models' is used to emphasise that a wide variety of alternative models are available. Darcy's Law is the fundamental computational model; not only is it included in many of the governing equations for groundwater flow but it can also be used for preliminary calculations such as vertical flows through low permeability strata. The diversity of computational models is illustrated by considering radial flow. There are many analytical models for radial flow to pumped boreholes including the Thiem equation for steady-state flow in a confined aquifer, the time-variant Theis equation, classical leaky aquifer theory and many more complex analyses which may include vertical flow components. However, in all of these analytical models, assumptions are introduced so that mathematical expressions can be derived. These assumptions are often discounted or disregarded by those carrying out the analysis. As an alternative, radial flow through an aquifer system can be represented by numerical models based on the finite difference approximation to the differential equations; fewer assumptions are required in deriving the numerical models. On what basis should a computational model be selected? If the objective of the study is to estimate aquifer parameters from a pumping test, the analytical solution which most closely represents the field situation can be compared with field data. Originally type curves were prepared for graphical matching with field information but computer software is now widely available to simplify the procedure. However, the ease of using computer software for matching field data to the analytical solutions often hides the inconsistencies between the assumptions of the analytical model and the real field conditions. Although less convenient to use, numerical radial flow models often represent more closely the actual behaviour of the aquifer system.

Real groundwater problems are three-dimensional in space and time-variant. Due to frequent increases in the power of computers and user-friendly input and output routines, regional groundwater flow problems of

considerable complexity can be represented by numerical models. Yet, if there is only a partial understanding
of the hydrogeology, with limited data and information
about the aquifer, any three-dimensional time-variant
analysis is likely to be tentative. In the early stages of
an investigation, it is preferable to simplify or idealise
the conceptual models so that preliminary analyses can
be achieved using analytical or simpler numerical
models.

Computational models are required for other aspects
of groundwater investigations such as recharge estimation based on a soil moisture balance technique,
runoff-recharge simulation and groundwater–surface
water interaction.

For most of the case studies described in the following chapters, detailed information is not included about
the actual numerical technique or software used in the
analysis. Some of the problems were originally studied
using analogue computers; this was necessary due to
the limited power and accuracy of digital computers at
the time when the analysis was carried out. Today
digital computer software is usually available, which
provides effectively the same results. Changes have also
occurred in the numerical techniques used to solve the
finite difference equations; this is partly due to substantial increases in computer memory and speed of
operation. Developments in software and hardware are
certain to continue. Nevertheless, it is essential to check
that the software does perform what the investigator
requires. With graphical interfaces separating the investigator from the actual code, there are instances where
the software does not perform the task required (see,
for example, Section 12.2.2).

1.4 CASE STUDIES

So that the reader can gain insights into both the
development of conceptual models and the selection
and application of computational models, there are
detailed descriptions of many case studies. Case studies
are taken from eleven different countries with climates
including semi-arid, temperate and humid-tropical. A
wide variety of geological formations are considered.
The objectives of the studies are also varied. Each
chapter contains a number of case studies. For each
case study one or two aspects are highlighted; further
issues are raised in subsequent examples. To obtain full
value from individual case studies, the reader needs to
appreciate the actual situation; this requires imagination, especially when the case study refers to climatic
conditions, hydrogeology or aquifer exploitation which
are outside the experience of the reader. Focusing on

case studies permits explanations of misconceptions
and inadequate assumptions which often occur in the
early stages of a study. The locations of many of the
case studies in England and Wales are indicated in
Figure 1.1.

Another advantage of case studies is that conceptual
and computational models can be examined in the
context of their actual use, rather than in hypothetical
situations. The approach and parameter values from an
existing case study can often be adapted and modified
as the basis for the development of conceptual and
computational models for new projects.

The wide use of case studies means that guidance
is required to identify where information can be
found about the alternative concepts and techniques.
Summary tables are provided in Chapters 5 to 11 to
indicate where different topics are considered. For each
of the major case studies in regional groundwater flow,
objectives of the study and key issues are highlighted.
In addition, there is extensive cross-referencing. The
index is also helpful in finding where specific topics are
considered.

No problem is totally solved; no study is complete.
Nevertheless, individual case studies do make positive
contributions because they move understanding
forward. It is for those who revisit a case study to take
the current understanding and advance it further.

1.5 THE CONTENTS OF THIS BOOK

This book is divided into three parts. Part I: Basic Principles (Chapters 2 to 4) considers three fundamental
topics in groundwater hydrology. The principles of
groundwater flow are presented concisely in Chapter 2,
where reference is also made to numerical methods,
collecting and presenting field data plus a brief consideration of water quality issues. Chapter 3 considers
recharge estimation and includes detailed information
about a soil moisture balance method of recharge estimation for both semi-arid and temperate climates.
Modifications to recharge processes due to the presence
of low permeability drift, including the estimation of
runoff recharge, are described. Surface water–groundwater interaction is considered in Chapter 4; the
surface water bodies include canals, springs, rivers and
lakes, drains and flooded ricefields.

Radial flow towards wells and boreholes is the
subject of Part II: Radial Flow (Chapters 5 to 8). Too
often, radial flow analysis is restricted to the determination of aquifer parameters using analytical solutions.
However, far more can be learnt about the aquifer and
the ability of a well or borehole to collect water by crit-

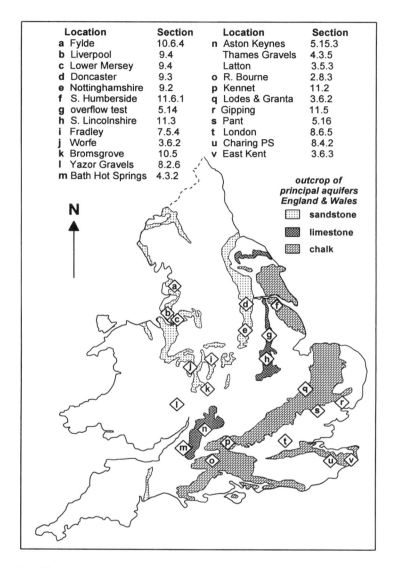

Location	Section		Location	Section
a Fylde	10.6.4		**n** Aston Keynes	5.15.3
b Liverpool	9.4		Thames Gravels	4.3.5
c Lower Mersey	9.4		Latton	3.5.3
d Doncaster	9.3		**o** R. Bourne	2.8.3
e Nottinghamshire	9.2		**p** Kennet	11.2
f S. Humberside	11.6.1		**q** Lodes & Granta	3.6.2
g overflow test	5.14		**r** Gipping	11.5
h S. Lincolnshire	11.3		**s** Pant	5.16
i Fradley	7.5.4		**t** London	8.6.5
j Worfe	3.6.2		**u** Charing PS	8.4.2
k Bromsgrove	10.5		**v** East Kent	3.6.3
l Yazor Gravels	8.2.6			
m Bath Hot Springs	4.3.2			

outcrop of principal aquifers England & Wales

- sandstone
- limestone
- chalk

N

Figure 1.1 Locations of studies in England and Wales reported in this book

ically examining all the field information, developing conceptual models and then selecting suitable analytical or numerical techniques to analyse the data. Chapter 5 is concerned with radial flow problems in which vertical flow components are not significant. Large diameter wells are reviewed in Chapter 6; these are widely used in developing countries. Practical examples include analysing pumping tests in large diameter wells and the use of large diameter wells over extended time periods. Situations where vertical flow components are important are reviewed in Chapter 7.

Analytical and numerical techniques of analysis are described and the significance of aquitard storage in assessing aquifer resources is also discussed. In Chapter 8, step-pumping tests and packer tests are reviewed, and case studies concerning the reliable yield of aquifer systems are introduced. Artificial recharge through injection wells and the hydraulics of horizontal wells are also topics in Chapter 8.

The first three chapters of Part III: Regional Groundwater Flow are concerned with different types of aquifer systems; the discussion is based on twelve major case studies. Chapters 9 refers to aquifer systems where the transmissivity remains effectively constant. In Chapter 10, case studies include situations where vertical flows through lower permeability strata are significant. Chapter 11 focuses on aquifer systems where the hydraulic conductivity varies over the saturated depth. In Chapter 12 a number of techniques and issues related to modelling are examined, including boreholes in multi-aquifer systems, the lateral extent and time period for regional groundwater studies, model refinement and sensitivity analyses together with the use of groundwater models for predications. In the final section of Chapter 12 a brief evaluation is made of conceptual and computational models with a review of achievements and limitations and an indication of the possible direction of future work.

In each part of this book, concepts and techniques are introduced in different ways. In Part I, each chapter is concerned with a specific topic; as ideas are developed they are illustrated by numerical examples. In Part II, the major derivations and techniques are described in Chapter 5 which considers radial flow, and in Chapter 7 where vertical flow components are also included. Chapters 6 and 8 are concerned mainly with applications of these techniques to practical problems. In Part III, the concepts and procedures involved in regional groundwater studies are introduced and developed within the case studies. Each case study is used to present new material. In Chapter 12 many of the insights of the previous three chapters are drawn together. For those with limited experience of conceptual or computational modelling in groundwater hydrology, it will be helpful to quickly read through the material in this book, paying particular attention to the figures. This will provide an overall appreciation of the approach to the wide range of situations which arise in groundwater hydrology. More detailed examination of specific topics can then follow.

1.6 UNITS, NOTATION, JOURNALS

Metric units are used throughout this book, usually in terms of metre-day. These units are convenient for groundwater resource problems because response times due to pumping, recharge, etc. are typically in days or tens of days, with annual groundwater fluctuations in the range of one to fifteen metres. For rainfall and recharge it is preferable to use mm/d. Pumping rates from major boreholes often exceed $1000\,m^3/d$, hence it is advantageous to use units of Ml/d (million litres per day or $1000\,m^3/d$). Hydraulic conductivity for sands lie in the range 1.0 to 10 m/d, while for clays the range is usually 0.0001 to 0.01 m/d.

Since this book relates to several disciplines, it includes a number of conventional notations. Whenever possible, standard symbols are used and they are defined in the text. Occasionally there is duplication, for example K_C is the hydraulic conductivity of clay and also the crop coefficient in recharge estimation. However, the meaning of the symbol is clear from the context. A List of Symbols can be found after the Appendix and before the References.

Some of the journals quoted in the references may not be familiar to all readers.

BHS: British Hydrological Society, contact through Institution of Civil Engineers, 1 Great George Street, Westminster, London SW1P 3AA, UK.

Hydrol. Sci. J.: Hydrological Sciences Journal, IAHS Press, CEH, Wallingford, Oxon OX10 8BB, UK.

IAHS Publ. No.: International Association of Hydrological Sciences, IAHS Press, as above.

J. Inst. Water & Env. Man.: Journal of Institution of Water and Environmental Management, Chartered Institution of Water and Environmental Management, 15 John Street, London WC1N 2EB, UK.

Quart. J. Eng. Geol.: Quarterly Journal of Engineering Geology, Geological Society Publishing House, Unit 7, Brassmill Enterprise Centre, Bath BA1 3JN, UK.

Proc. Inst. Civ. Eng. (London): Proceedings of Institution of Civil Engineering, as above.

PART I: BASIC PRINCIPLES

Part I considers three basic issues in groundwater hydrology. Chapter 2 introduces the principles of groundwater flow, Chapter 3 considers recharge estimation and Chapter 4 examines various forms of groundwater–surface water interaction. In each chapter conceptual models of the flow processes are developed. Specific problems are then analysed using a variety of computational techniques.

Chapter 2 summarises the physical and mathematical background to groundwater flow. One-dimensional Cartesian and radial flow are analysed under both steady state and time-variant conditions. Two-dimensional vertical section formulations (profile models) are studied with an emphasis on boundary conditions. An examination of regional groundwater flow in the x–y plane demonstrates that this formulation implicitly contains information about vertical flow components. A brief introduction to numerical models serves as a preparation for their use in later chapters. The importance of monitoring both groundwater heads and river flows is stressed; case studies are included to illustrate suitable techniques. The chapter concludes with a brief discussion of quality issues with a focus on the analysis of the fresh–saline water interface.

The estimation of groundwater recharge is considered in Chapter 3. Detailed information is provided for potential recharge estimation using soil moisture balance techniques, and the physical basis of this approach is highlighted. Examples include recharge estimates for temperate climates and for semi-arid areas under both rainfed and irrigated regimes. Sufficient information is provided for readers to use this approach for their own setting. Modifications to recharge occur due to the unsaturated zone between the soil and the water table; delays occur before the recharge reaches the water table. Several examples of runoff-recharge are described where runoff subsequently enters the aquifer in more permeable locations.

Chapter 4 considers different forms of interaction between groundwater and surface water. For canals, the importance of the regional groundwater setting and the effect of lining are examined. The study of canals provides useful insights into river–aquifer interaction; the dimensions of the river and any deposits in the riverbed may not be the most important factors in quantifying the interaction. Springs and lakes are also considered. Next, a reappraisal of the theory of horizontal drains shows that it is incorrect to assume that the water table intersects the drain. The importance of time-variant behaviour is stressed. Interceptor drains intended to collect canal seepage are used as an example of conceptual and numerical modelling. Finally, water losses from flooded ricefields are studied; the major cause of losses is water flowing through the bunds.

2

Background to Groundwater Flow

2.1 INTRODUCTION

The purpose of this chapter is to introduce both the basic principles of groundwater flow and the associated analytical and numerical methods of solution which are used in later chapters. These principles and methods of analysis are the basis of conceptual and computational models. To apply them to specific practical problems, appropriate field information and monitoring is required; this is the second topic considered in this chapter. The chapter concludes with a brief introduction of water quality issues.

A clear understanding of fundamental concepts such as groundwater head, groundwater velocity and storage coefficients are a prerequisite for the interpretation of field data, the development of quantified conceptual models and the selection of appropriate analytical or numerical methods of analysis. In Section 2.2 the fundamental parameters of groundwater flow are introduced. Sections 2.3 to 2.6 consider groundwater flow using different co-ordinate systems, namely one-dimensional Cartesian flow, radial flow, flow in a vertical section and regional groundwater flow. Where appropriate, analytical solutions to representative problems are introduced. This is followed by a brief introduction to numerical methods of analysis. The finite difference method is selected because it is the basis of many of the commonly used packages.

The study of groundwater flow requires the use of the language of mathematics; readers who are not familiar with the mathematical approach are encouraged to examine the steps in obtaining the solutions even though the mathematical manipulation is not fully understood. The derivations in this chapter are straightforward; for more rigorous derivations see Bear (1979) and Zheng and Bennett (2002). Awareness of the basic steps in obtaining the mathematical solution, and appreciating the assumptions made, will enable readers to judge whether a particular approach is appropriate for use in practical problems. Furthermore, the same basic steps are followed in obtaining numerical solutions. With computer packages designed to simplify the inputting of data, there is a vital need to appreciate the basic approach to studying groundwater flow, otherwise there is a risk that the physical problem will not be specified adequately, thereby leading to incorrect or misleading conclusions.

A further requirement of a successful groundwater study is the collection and collation of aquifer parameter values and the field monitoring of groundwater flows and heads. The estimation of aquifer parameter values is considered elsewhere in the book, especially in Part II; monitoring aquifer responses is considered in Section 2.8. The fundamental importance of collecting and interpreting field data and other information is stressed with examples of good practice.

The final section of this chapter contains a brief introduction to the study of groundwater quality. Some of the principles are summarised with introductory examples related to saline water–fresh water interaction. A thorough treatment of contaminant transport issues with associated numerical models and case studies is available in *Applied Contaminant Transport Modeling* by Zheng and Bennett (2002).

2.2 BASIC PRINCIPLES OF GROUNDWATER FLOW

This book is concerned with groundwater flow in a wide variety of situations. Although groundwater travels very slowly (one hundred metres per year is a typical average horizontal velocity and one metre per year is a typical vertical velocity), when these velocities are multiplied by the cross-sectional areas through which the flows occur, the quantities of water involved in groundwater flow are substantial. Consequently, the essential feature of an aquifer system is the balance between the inflows, outflows and the quantity of water stored.

Generally it is not possible to measure the groundwater velocities within an aquifer. However, observation piezometers (boreholes) can be constructed to determine the elevation of the water level in the piezometer. This water level provides information about the groundwater head at the open section of the piezometer. Groundwater head gradients can be used to estimate the magnitude and direction of groundwater velocities. A thorough understanding of the concept of groundwater heads is essential to identify and quantify the flow processes within an aquifer system.

2.2.1 Groundwater head

The groundwater head for an elemental volume in an aquifer is the height to which water will rise in a piezometer (or observation well) relative to a consistent datum. Figure 2.1a shows that the groundwater head

$$h = \frac{p}{\rho g} + z \qquad (2.1)$$

where p is the pressure head, ρ is the density of the fluid and z is the height above a datum. All pressures are relative to atmospheric pressure.

In Figure 2.1a a piezometer pipe enters the side of an elemental volume of aquifer. However, an observation well usually consists of a vertical drilled hole with a solid casing in the upper portion of the hole and an open hole or a screened section with slotted casing at the bottom of the solid casing. So that the groundwater head can be identified at a specific location in an aquifer, the open section of the piezometer should extend for no more than one metre. When the borehole has a long open section, unreliable results may occur, as described in Section 5.1.6.

2.2.2 Direction of flow of groundwater

Groundwater flows from a higher to a lower head (or potential). Typical examples of the use of the groundwater head to identify the direction of groundwater

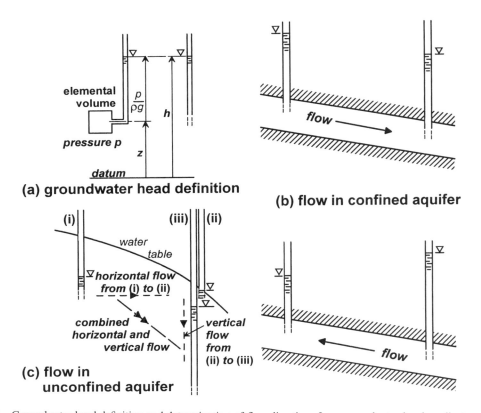

Figure 2.1   Groundwater head definition and determination of flow directions from groundwater head gradients

flow are shown in Figure 2.1b. In this confined aquifer there are two piezometers which can be used to identify the direction of flow. In the upper diagram the flow is from left to right because the lower groundwater head is in the piezometer to the right; this is in the direction of the dip of the strata. For the lower diagram, the groundwater flow is to the left since the water level in the left hand piezometer is lower. Consequently the direction of the flow is up-dip in the aquifer. This flow could be caused by the presence of a pumped well or a spring to the left of the section.

Figure 2.1c refers to an unconfined aquifer in which the elevation of the water table falls to the right. There are three piezometers; their open ends are positioned to identify both horizontal and vertical components of flow. Piezometer (ii) (the numbering of the piezometers can be found at the top of the figure) penetrates just below the water table, hence it provides information about the water table elevation. The other two piezometers have a solid casing within the aquifer with a short open section at the bottom; they respond to the groundwater head at the base of the piezometer. Since the open sections of piezometers (i) and (ii) are at the same elevation they provide information about the *horizontal velocity component* (values of the horizontal hydraulic conductivity are required to quantify the velocity); the horizontal flow is clearly from left to right. Piezometer (iii) is positioned almost directly below piezometer (ii), therefore it provides information about the *vertical velocity component*. Since the groundwater head in piezometer (iii) is below that of piezometer (ii) there is a vertically downward component of flow. The horizontal and vertical velocity components can be combined vectorially to give the magnitude and direction of the flow.

2.2.3 Darcy's Law, hydraulic conductivity and permeability

It was in 1856 that Henry Darcy carried out experiments which led to what we now call Darcy's Law. An understanding of assumptions inherent in Darcy's Law is important; significant steps in the derivation are discussed below.

Groundwater (Darcy) velocity and seepage velocity

Figure 2.2 illustrates flow through a cylinder of aquifer; the groundwater velocity or Darcy velocity is defined as the discharge Q divided by the *total* cross-sectional area of the cylinder A,

$$v = \frac{Q}{A} \tag{2.2}$$

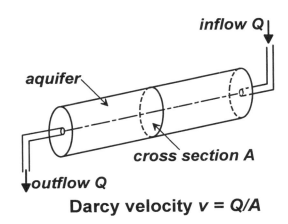

Darcy velocity $v = Q/A$

Figure 2.2 Darcy velocity

Note that the calculation of the groundwater velocity ignores the fact that the aquifer cross-section A contains both solid material and pores; consequently the groundwater velocity is an *artificial velocity* which has no direct physical meaning. Nevertheless, because of its convenience mathematically, the groundwater velocity is frequently used.

An approximation to the actual seepage velocity v_s can be obtained by dividing the Darcy velocity by the effective porosity N.

$$v_s = \frac{v}{N} = \frac{Q}{AN} \tag{2.3}$$

Derivation of Darcy's Law

A mathematical derivation of Darcy's Law is presented in Figure 2.3, and some of the steps are summarised below.

1. In Figure 2.3, the elemental volume of water is inclined at an angle α to the horizontal; consequently the analysis which leads to Darcy's Law applies to flow in any direction and is not restricted to horizontal or vertical flow.

2. If the water pressure on the left-hand face is p, the pressure on the right-hand face, which is a distance δl to the right, will equal

   pressure = p + rate of change of p × distance

   $$= \left(p + \frac{dp}{dl} \delta l \right)$$

3. There are four forces acting on the element. Two forces are due to water pressures on the ends of the

Consider an element with cross section δA and length δl, the element is inclined at an angle α to the horizontal; inflows and outflows only occur at the two ends.

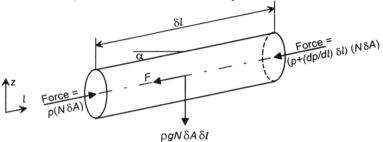

Due to water pressures there are forces on the left hand face and the right hand face. There is a downwards force equal to ρg times the volume of water and there is also a frictional force F resisting the flow of water. Resolving in the direction of l,

$$pN\delta A - (p + \frac{dp}{dl}\delta l)N\,\delta A = (\rho g\,N\,\delta A\,\delta l)\sin\alpha + F \tag{A}$$

where N is the effective porosity. Substituting $\sin\alpha = dz/dl$ and rearranging

$$\frac{F}{N\,\delta A\,\delta l} = -\left(\frac{dp}{dl} + \rho g\frac{dz}{dl}\right) \tag{B}$$

Darcy's experiments showed that for low velocities, the resisting force is proportional to the velocity v; for mathematical convenience this is written as

$$\frac{F}{N\,\delta A\,\delta l} = \frac{\mu}{\overline{K}}\,v \tag{C}$$

where μ is the dynamic viscosity of the flowing fluid and $\overline{K}$ is the intrinsic permeability of the porous material. Combining equations (B) and (C)

$$v = -\frac{\overline{K}}{\mu}\left(\frac{dp}{dl} + \rho g\frac{dz}{dl}\right) \tag{D}$$

Differentiating equation (2.1) and substituting in equation (D)

$$v = -\frac{\overline{K}\rho g}{\mu}\frac{dh}{dl} \qquad \text{or } v = -K\frac{dh}{dl} \tag{E}$$

where the hydraulic conductivity is defined as $\qquad K = \dfrac{\overline{K}\rho g}{\mu}$ \qquad (F)

Figure 2.3   Derivation of Darcy's Law

element, and there is also a force due to the weight of water in the element and a frictional force resisting the flow of water. The water pressure is multiplied by $N\delta A$, where N is the effective porosity.

4. In Eq. C in Figure 2.3, the resisting force per unit volume of water in the element is selected to be proportional to the dynamic viscosity of the flowing fluid and inversely proportional to a parameter called the intrinsic permeability of the porous material. This is consistent with field experiments.

5. Darcy's Law from Eq. (E) is

$$v = -K\frac{dh}{dl} \tag{2.4}$$

where the hydraulic conductivity K is a function of the dynamic viscosity of water and the intrinsic permeability of the porous material.

6. The minus sign in Darcy's Law emphasises that water flows from a higher to a lower groundwater head.

7. Strictly the phrase *hydraulic conductivity* should always be used but in groundwater hydraulics and soil mechanics the word *permeability* is often used as an alternative; in the remainder of this text these words are considered to be interchangeable.

Various authors provide tables of the hydraulic conductivity and storage properties of unconsolidated and consolidated materials. See, for example, McWhorter and Sunada (1977), Bouwer (1978), Todd (1980), Driscoll (1986), Walton (1991) and Fetter (2001).

2.2.4 Definition of storage coefficients

The change in groundwater head with time results in the release of water from storage or water taken into storage. Two mechanisms can apply; the release of water at the water table due to a change in the water table elevation and/or the release of water throughout the full depth of the aquifer due to a change in groundwater head which corresponds to an adjustment of the pressure within the aquifer system. For more detailed discussions on the physical basis of the storage coefficients, see Bear (1979) and Fetter (2001).

Specific yield: to appreciate the meaning of specific yield, consider the decline in water surface elevation in a surface reservoir. This fall in water surface corresponds to a reduction in the volume of water stored due to outflows to a river, withdrawal for water supply, or evaporation. The effect of a fall in the water table elevation in an aquifer is based on similar concepts; see

Figure 2.4a. However, account must be taken of the fact that the aquifer contains both solid material and water, hence it is necessary to multiply the decrease in volume due to the water table fall by the specific yield, S_Y. The specific yield is defined as follows:

The volume of water drained from a unit plan area of aquifer for a unit fall in head is S_Y

Specific storage: a fall in groundwater head within a saturated aquifer also causes a reduction in pressure within the aquifer system (Eq. (2.1)). This fall in pressure means that less water can be contained within a specified volume of the aquifer, and the volume of the soil or rock matrix also changes. Consequently, water is released from storage. The specific storage S_S relates to a unit volume of aquifer; see the lower part of Figure 4b:

The volume of water released from a unit volume of aquifer for a unit fall in head is S_S

Confined storage: the specific yield refers to a unit plan area of aquifer whereas the specific storage relates to a unit volume. Therefore, it is helpful to define a storage coefficient for pressure effects related to a unit plan area. The confined storage coefficient equals

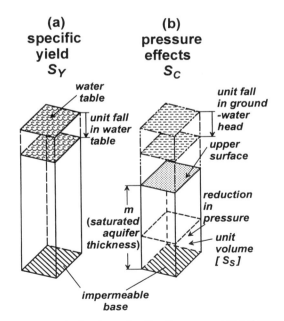

each vertical column is of unit cross-sectional area

Figure 2.4 Diagrammatic explanation of the three storage coefficients

the specific storage multiplied by the saturated thickness m,

$$S_C = S_S \times m$$

Accordingly, the compressive properties of a confined aquifer are quantified as follows:

The volume of water released per unit plan area of aquifer for a unit fall in head is S_C

Note that, as illustrated in Figure 2.4b, the saturated thickness extends from the impermeable base of the aquifer to the upper surface which may be a confining layer or a water table. Therefore, *unconfined* aquifers also have a *confined storage* response. The importance of considering both the unconfined and confined storage properties of a water table aquifer is demonstrated in the pumping test case study in Section 5.1.

Storage properties are summarised in Table 2.1. Note that the specific storage has dimensions L^{-1} all the other storage coefficients are dimensionless. Typical ranges for these three storage coefficients are

$$10^{-6} \leq S_S \leq 10^{-4} \text{ m}^{-1}$$
$$10^{-5} \leq S_C \leq 3 \times 10^{-3}$$
$$10^{-3} \leq S_Y \leq 0.25$$

Although the confined response to a change in groundwater head is effectively instantaneous, the drainage of water as the water table falls can take some time. Hence the concept of delayed yield is introduced for unconfined aquifers; for further information about delayed yield see Section 5.17.

2.2.5 Differential equation describing three-dimensional time-variant groundwater flow

A mathematical derivation for three-dimensional regional groundwater flow is outlined in Figure 2.5;

certain aspects of the derivation are discussed below.

- The two fundamental principles of the governing equation are Darcy's Law and the principle of continuity.
- The magnitudes of the velocities change across the element. Therefore, if the velocity on the left-hand face of the element is v_x, on the right-hand face at distance dx the velocity will become $v_x + \dfrac{\partial v_x}{\partial x} dx$.
- In the flow balance there are four components, net flows in the x, y and z directions and a further component due to the compressibility of the aquifer system which equals the specific storage S_S multiplied by the rate of change of groundwater head with time. These four components must sum to zero leading to Eq. (B).
- The final part of the derivation is to substitute Darcy's Law into the continuity equation; this leads to Eq. (C), the governing equation which is repeated below. The left-hand side of the equation represents flows in the three co-ordinate directions, the right-hand side represents the storage contribution due to the change in groundwater head with time.

$$\frac{\partial}{\partial x}\left(K_x \frac{\partial h}{\partial x}\right) + \frac{\partial}{\partial y}\left(K_y \frac{\partial h}{\partial y}\right) + \frac{\partial}{\partial z}\left(K_z \frac{\partial h}{\partial y}\right) = S_S \frac{\partial h}{\partial t}$$

(2.5)

In most practical situations, there is no need to work in the three co-ordinate dimensions plus the time dimension. In the next four sections of this chapter, analyses are described in which one or more of the dimensions are excluded.

2.3 ONE-DIMENSIONAL CARTESIAN FLOW

This section introduces a one-dimensional mathematical formulation which can be used to understand simple regional groundwater flow when the flow is predominantly in one horizontal direction. Initially steady-state problems are considered, subsequently a

Table 2.1 Summary of storage properties

Type	Symbol	Unit	Dimension	Name	Pressure/drainage
Specific storage	S_S	cube	L^{-1}	confined	pressure
Confined storage	S_C	area	–	confined	pressure
Specific yield	S_Y	area	–	unconfined	drainage
Unconfined storage	$S_Y + S_C$	area	–	unconfined	drainage + pressure

Three-dimensional time-variant groundwater flow (*x, y, z, t*)

(a)   Darcy's Law in direction of principal axes (*x, y, z*)

$$v_x = -K_x \frac{\partial h}{\partial x} \qquad v_y = -K_y \frac{\partial h}{\partial y} \qquad v_z = -K_z \frac{\partial h}{\partial z} \quad (A)$$

(b)   Continuity: considering a three-dimensional element

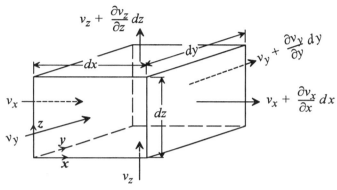

flow entering in the *x* direction in time $\delta t = v_x \, dy \, dz \, \delta t$

flow leaving in the *x* direction in time $\delta t = (v_x + \frac{\partial v_x}{\partial x} dx) \, dy \, dz \, \delta t$

net flow entering in time δt in the *x* direction $= -(\partial v_x/\partial x) dx \, dy \, dz \, \delta t$

net flow entering in time δt in the *y* direction $= -(\partial v_y/\partial y) dx \, dy \, dz \, \delta t$

net flow entering in time δt in the *z* direction $= -(\partial v_z/\partial z) dx \, dy \, dz \, \delta t$

net flow entering in time δt due to increase in head $\delta h = -\delta h \, S_S \, dx \, dy \, dz$

These flows must sum to zero; in the limit $\delta h/\delta t$ tends to $\partial h/\partial t$,

$$-\frac{\partial v_x}{\partial x} - \frac{\partial v_y}{\partial y} - \frac{\partial v_z}{\partial z} = S_S \frac{\partial h}{\partial t} \qquad (B)$$

(c)   Combining equations (A) and (B)

$$\frac{\partial}{\partial x}\left(K_x \frac{\partial h}{\partial x}\right) + \frac{\partial}{\partial y}\left(K_y \frac{\partial h}{\partial y}\right) + \frac{\partial}{\partial z}\left(K_z \frac{\partial h}{\partial z}\right) = S_S \frac{\partial h}{\partial t} \qquad (C)$$

This is the governing equation for three-dimensional groundwater flow.

Figure 2.5   Derivation of governing equation for three-dimensional time-variant groundwater flow

time-variant formulation is examined. Specific problems are selected, to illustrate how numerical values can be obtained from mathematical analyses. Readers with limited mathematical expertise are encouraged to follow the stages of the derivation and then to substitute numerical values in the expressions so that they can appreciate the basis of the approach.

2.3.1 Equation for one-dimensional flow

Using Darcy's Law and the principle of continuity of flow, a differential equation is derived which can be applied to many problems of one-dimensional groundwater flow. The approach is based on a simplified form of the three-dimensional derivation in Figure 2.5.

Consider an element of aquifer of length dx shown in Figure 2.6 which extends upwards from the impermeable base to the water table, a vertical distance m. The vertical recharge at the water table equals q. Continuity of flow can be written as:

inflow at left-hand face + inflow due to recharge
= outflow from right-hand face

In mathematical terms (note that the cross-sectional area perpendicular to v_x is $m \times 1.0$ since the width of the aquifer into the paper is 1.0),

$$v_x \times m \times 1.0 + q \times dx \times 1.0 = \left(v_x + \frac{dv_x}{dx} dx\right) \times m \times 1.0$$

or $\dfrac{d}{dx}(mv_x) = q$ (2.6)

From Darcy's Law

$$v_x = -K_x \frac{dh}{dx} \tag{2.7}$$

Combining Eqs (2.6) and (2.7).

$$\frac{d}{dx}\left(mK_x \frac{dh}{dx}\right) = -q(x) \tag{2.8}$$

In this equation the recharge on the right-hand side is written as $q(x)$, signifying that the recharge may be a function of x.

2.3.2 Aquifer with constant saturated depth and uniform recharge

Figure 2.7a illustrates the first problem to be analysed. An aquifer with an impermeable base and a constant transmissivity T (where $T = K_x \times m$) is in contact with

an impermeable stratum at the left-hand side, $x = 0$. On the right-hand side, $x = L$, the aquifer is in contact with a large lake at an elevation H above datum. The recharge, which is written as q (units L/T), is constant and therefore not a function of x. As with the derivation of the governing equation (Eq. (2.8)), a unit width (into the paper) of the aquifer is considered.

Since the transmissivity, which is the product of the saturated thickness and the hydraulic conductivity, is constant, Eq. (2.8) can be written as

$$\frac{d^2h}{dx^2} = -q/T \tag{2.9}$$

Integrating once,

$$\frac{dh}{dx} = -qx/T + A \tag{2.10}$$

where A is a constant of integration. Integrating again,

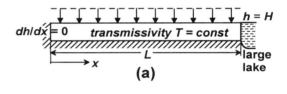

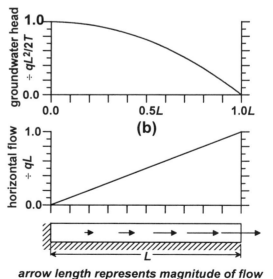

Figure 2.7 One-dimensional flow in an aquifer with constant saturated depth and uniform recharge

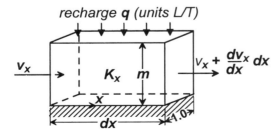

Figure 2.6 One-dimensional groundwater flow

$$h = -qx^2/2T + Ax + B \qquad (2.11)$$

where B is the second constant of integration.

There are *two* constants of integration which can be determined from the *two* boundary conditions.

(i) At $x = 0$ there is no lateral inflow thus $dh/dx = 0$, hence from Eq. (2.10)

$$0 = -q \times 0/T + A \quad \text{thus } A = 0$$

(ii) At $x = L$, $h = H$ and also substituting $A = 0$, Eq. (2.11) becomes

$$H = -q \times L^2/2T + 0 + B \quad \text{or} \quad B = H + qL^2/2T$$

When the values of the two constants of integration are substituted back in Eq. (2.11), the resultant equation for the groundwater head is

$$h = H + \frac{q(L^2 - x^2)}{2T} \qquad (2.12)$$

The flow through the aquifer can also be calculated from the equation

$$Q = -T \frac{dh}{dx} = qx \qquad (2.13)$$

In Figure 2.7b, the distribution of groundwater head above the lake level is plotted as a fraction of $qL^2/2T$. The maximum groundwater head occurs at $x = 0$, the left-hand side. Note also that the slope of the groundwater head on the left-hand side is zero, thereby satisfying the condition of zero inflow. Figure 2.7c shows the variation of flow in the aquifer with x; there is a linear increase from zero at the no-flow boundary to qL per unit width of the aquifer at the lake.

These expressions will be used to estimate the maximum groundwater heads and maximum flows into the lake when the following parameters apply:

transmissivity $T = 250\,\text{m}^2/\text{d}$
constant recharge $q = 0.5\,\text{mm/d}$
length $L = 1\,\text{km}$, $3\,\text{km}$ or $10\,\text{km}$
head in lake $H = 50\,\text{m}$

From Eq. (2.13) the flows into the lake for the three lengths of aquifer are 0.5, 1.5 and $5.0\,\text{m}^3/\text{d}$ per metre

width of aquifer. The maximum calculated groundwater heads from Eq. (2.12) show greater variations:

$$L = 1.0\,\text{km} \quad h_{max} = 50 + 1.0\,\text{m}$$
$$L = 3.0\,\text{km} \quad h_{max} = 50 + 9.0\,\text{m}$$
$$L = 10.0\,\text{km} \quad h_{max} = 50 + 100.0\,\text{m}$$

The maximum groundwater head is far higher for the aquifer length of 10 km since, compared to the 1 km length, the flow to the lake is ten times higher; in addition, some of the water has to travel ten times the distance through the aquifer.

The final issue to be resolved is whether changes in saturated thickness between the lake and the outer boundary invalidate the analysis. Assuming that the base of the aquifer is horizontal at $0.0\,\text{m}$,

for $L = 1.0\,\text{km}$ the saturated thickness varies from $50.0\,\text{m}$ to $51.0\,\text{m}$.
for $L = 3.0\,\text{km}$ the saturated thickness varies from $50.0\,\text{m}$ to $59.0\,\text{m}$.
for $L = 10.0\,\text{km}$ the saturated thickness varies from $50.0\,\text{m}$ to $150.0\,\text{m}$.

The analysis assumes that the transmissivity (and therefore the saturated thickness) remains constant. For $L = 1.0\,\text{km}$ the thickness varies by 2 per cent, therefore the error in maximum groundwater head is small. For $L = 3.0\,\text{km}$ the variation in thickness is 18 per cent so that the assumption of constant saturated depth is just acceptable. However, for $L = 10.0\,\text{km}$ the saturated thickness increases threefold, which means that the calculated maximum groundwater head is likely to be a serious over-estimate. Varying saturated depth is considered in Section 2.3.7.

2.3.3 Definition of transmissivity

In the preceding section, the transmissivity T is defined as the product of the hydraulic conductivity and the saturated thickness. This is a restrictive definition which suggests that the concept of transmissivity is only applicable to homogeneous aquifers. In fact, transmissivity can describe a variety of layered aquifer systems. The definition of transmissivity is the sum, or integral, of the hydraulic conductivities over the saturated depth of the aquifer:

$$T = \sum_{\substack{saturated \\ depth}} K_x \Delta z \quad \text{or} \quad T = \int_{z=base}^{z=top} K_x dz \qquad (2.14)$$

Two examples of calculating transmissivities are shown in Figure 2.8. The vertical scale in each figure represents the height above the effective base of the aquifer; the horizontal scale on the left-hand figures refers to the horizontal hydraulic conductivity, and in the right-hand figures the scale refers to the transmissivity. Figure 2.8a relates to an aquifer with three layers; typical calculations of the aquifer transmissivity with differing saturated thickness are presented below.

(i) water table is 15 m above base of aquifer, $T = 15 \times 10 = 150 \, \text{m}^2/\text{d}$

(ii) water table is 30 m above base of aquifer, $T = 15 \times 10 + 15 \times 1 = 165 \, \text{m}^2/\text{d}$

(iii) water table is 50 m above base of aquifer, $T = 15 \times 10 + 30 \times 1 + 5 \times 20 = 280 \, \text{m}^2/\text{d}$

As shown on the right-hand figure, the increasing transmissivity with saturated depth is represented as a series of inclined straight lines.

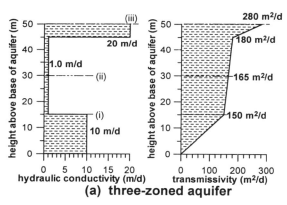

(a) three-zoned aquifer

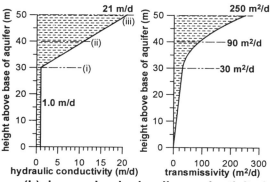

(b) increasing hydraulic conductivity in upper part of aquifer

Figure 2.8 Calculation of transmissivity for aquifers with layers or gradually changing hydraulic conductivity

In Figure 2.8b there is a constant value of hydraulic conductivity up to 30 m, while above 30 m the hydraulic conductivity increases from 1.0 m/d to a maximum value of 21 m/d; this type of variation applies to chalk and limestone aquifers (see Section 8.7 and Chapter 11). The transmissivity is calculated using the second part of Eq. (2.14). The integral represents the shaded area; for this example the integrals can be determined using the areas of rectangles and triangles.

(i) water table is 30 m above base of aquifer, $T = 30 \times 1 = 30 \, \text{m}^2/\text{d}$

(ii) water table is 40 m above base of aquifer, $T = 40 \times 1 + 0.5 \times 10 \times 10 = 90 \, \text{m}^2/\text{d}$

(iii) water table is 50 m above base of aquifer, $T = 50 \times 1 + 0.5 \times 20 \times 20 = 250 \, \text{m}^2/\text{d}$

Up to 30 m there is a linear increase in the transmissivity; above 30 m the transmissivity increase has linear and parabolic components.

2.3.4 Aquifer with constant saturated depth and linear variation in recharge

In Figure 2.9a, the recharge varies linearly from zero at the no-flow boundary to q_m at the lake. This can be expressed mathematically by defining the value of the recharge at any location x by

$$q(x) = q_m x / L \tag{2.15}$$

Therefore the differential equation becomes

$$\frac{d^2 h}{dx^2} = -\frac{q_m}{T} \frac{x}{L} \tag{2.16}$$

Integrating this equation twice and substituting the same boundary conditions as for the constant recharge example, the expression for the drawdown becomes

$$h = H + \frac{q_m \left(L^3 - x^3 \right)}{6TL} \tag{2.17}$$

Note that, because the right-hand side of the governing equation is a function of x, both L and x are raised to the third power in Eq. (2.17). The equation for the horizontal flow in the aquifer is obtained by differentiating Eq. (2.17) to find

$$Q = -T \frac{dh}{dx} = 0.5 q_m x^2 / L \tag{2.18}$$

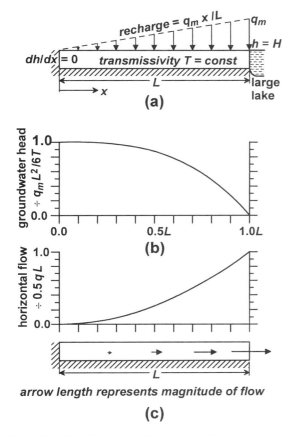

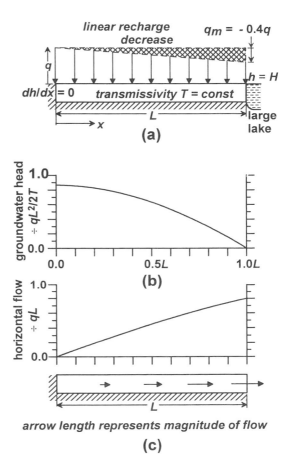

Figure 2.9 One-dimensional flow in an aquifer with constant saturated depth and linear variation in recharge

The groundwater head and horizontal flow are plotted in Figure 2.9b and c; note in particular that the flow at $x = 0.5L$ is only one quarter of the flow at $x = L$.

2.3.5 Aquifer with constant saturated depth and linear decrease in recharge towards lake

Figure 2.10a shows a more realistic variation of recharge in which the recharge is highest on the higher ground distant from the lake, falling linearly to the lake where the recharge is 60 per cent of the maximum value. This situation can be analysed using superposition of the two previous solutions. It is necessary to subtract the recharge represented by the shaded triangle in Figure 2.10a; this can be achieved by setting $q_m = -0.4q$ to give the required recharge distribution. Therefore from Eqs (2.12) and (2.17) the groundwater head equals

Figure 2.10 One-dimensional flow in an aquifer with constant saturated depth and linear recharge decrease; solution obtained by combining examples of Figures 2.7 and 2.9

$$h = H + \frac{q(L^2 - x^2)}{2T} - 0.4\left[\frac{q(L^3 - x^3)}{6TL}\right] \qquad (2.19)$$

with the flow equal to

$$Q = qx - 0.4\left[0.5qx^2/L\right] \qquad (2.20)$$

Variations in groundwater head are plotted in Figure 2.10b; comparison with Figure 2.7b shows a similar shape but with the maximum groundwater head above lake level equal to $0.867qL^2/2T$ compared to $qL^2/2T$ for uniform recharge. Comparisons between Figure 2.7c and 2.10c shows that the maximum flow into the lake equals $0.8qL$ compared to $1.0qL$ for uniform recharge. Superposition is a valid approach in this example because the transmissivity is assumed to remain constant despite different water table elevations. If a varying saturated depth is included in the analysis, superposition cannot be used.

2.3.6 Confined aquifer with varying thickness

If the saturated thickness m is a function of x it is no longer legitimate to work in terms of a constant transmissivity T which, in Eq. (2.9), is transferred to the right-hand side of the equation. Instead it is necessary to use the original equation

$$\frac{d}{dx}\left(mK_x\frac{dh}{dx}\right) = -q(x) \qquad (2.8)$$

There are few mathematical expressions for a variation in saturated thickness for which analytical solutions can be obtained by integration. One possible variation in thickness, illustrated in Figure 2.11, is

$$m = m_0\exp(-\alpha x) \qquad (2.21)$$

The value of α must be selected to give a positive value of the aquifer thickness at the right hand side. With groundwater heads H_1 and H_2 at the left- and right-hand sides, the head at any location x is given by the expression,

$$h = H_1 - (H_1 - H_2)\left(\frac{e^{\alpha x}-1}{e^{\alpha L}-1}\right) \qquad (2.22)$$

with the flow through the aquifer equal to

$$Q = K\alpha m_0(H_1 - H_2)/(e^{\alpha L}-1) \qquad (2.23)$$

A specific example is presented in Figure 2.11. The aquifer is 2000 m in length and the coefficient α is selected so that the aquifer thickness decreases from 20 m to 10 m. The groundwater head profile shows an increasing slope across the aquifer. The flow through the aquifer of $0.0693\,\mathrm{m^3/d}$ per metre width of the aquifer lies between a flow of $0.10\,\mathrm{m^3/d}$ for a constant thickness of 20 m and $0.05\,\mathrm{m^3/d}$ for a constant thickness of 10 m. This example illustrates how analytical solutions can be used to assess the likely influence of features such as decreasing aquifer thickness.

2.3.7 Unconfined aquifer with saturated depth a function of the unknown groundwater head

For unconfined aquifers in which the saturated thickness is a function of the unknown groundwater head, approximate solutions can be obtained using the Dupuit–Forchheimer approach. Figure 2.12a shows a typical example of unconfined flow on a vertical cross-section; lines of equal groundwater head and flowlines are sketched in the figure. In the Dupuit theory, Figure 2.12b, the element through which flow occurs is drawn as a rectangle of length dx in the x direction. Clearly there is a difference between the curved element in Figure 2.12a and the rectangular element of Figure 2.12b. Several studies (Bear (1972), Marino and Luthin (1982)) have considered the assumptions inherent in

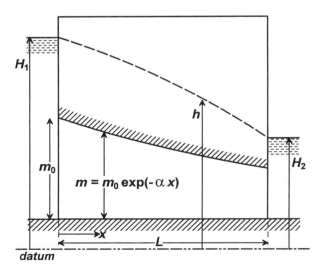

Numerical example:

$L = 2000$ m, $K = 1.0$ m/d,
$H_1 = 30$ m, $H_2 = 20$ m

$m_0 = 20$ m and $\alpha = 0.000347$
so thickness at right = 10 m

groundwater head profile shown by broken line

flow per metre width of
aquifer = 0.0693 m^3/d

compared to 0.10 m^3/d
for constant depth of 20 m

Figure 2.11 One-dimensional flow in confined aquifer with varying depth

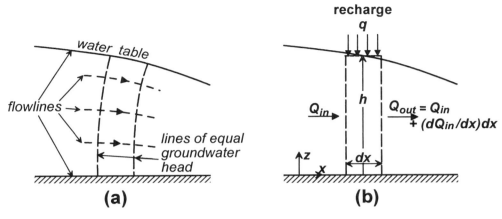

Figure 2.12 Derivation of Dupuit–Forchheimer theory for unconfined aquifer with varying saturated depth: (a) curved flowlines and lines of equal groundwater head, (b) assumptions for Dupuit theory

this Dupuit–Forchheimer approximation and identified situations where the error is sufficiently small for the approach to be acceptable.

In this analysis, the parameter h (which must be measured from the impermeable base) represents *both* the groundwater head in the element *and* the height of the section through which flow occurs. There are three components in the flow balance for the element of Figure 2.12b. From continuity the inflow plus the recharge must equal the outflow (which can be expressed as the inflow plus the rate of change of inflow multiplied by the length of the element dx):

$$Q_{in} + q\,dx = Q_{out} = Q_{in} + \left(\frac{dQ_{in}}{dx}\right)dx \qquad (2.24)$$

hence $\dfrac{dQ_{in}}{dx} = q$

but $Q_{in} = -K\dfrac{dh}{dx}h$ (hydraulic conductivity × head gradient × saturated depth),

thus $\dfrac{d}{dx}\left(K\dfrac{dh}{dx}h\right) = -q$ $\qquad (2.25)$

From the differentiation of products $\dfrac{d[h \cdot h]}{dx} = h\dfrac{dh}{dx} + \dfrac{dh}{dx}h$, Eq. (2.25) becomes

$$\frac{d}{dx}\left(\frac{dh^2}{dx}\right) = -\frac{2q}{K} \qquad (2.26)$$

provided that K is constant. Equation (2.26), the Dupuit–Forchheimer equation, is used to analyse unconfined flow problems by integrating and eliminating the constants of integration by incorporating the boundary conditions. Two examples illustrate the approach.

Example 1: Flow through a rectangular dam

Figure 2.13 represents a rectangular dam with an upstream head H_1 (measured from the impermeable base) and a downstream head H_2. By integrating Eq. (2.26) twice and substituting the conditions that at $x = 0$, $h = H_1$ and at $x = L$, $h = H_2$, the following equation for the variation in groundwater head (and hence the water table elevation) is obtained,

$$h^2 = H_1^2 - \frac{H_1^2 - H_2^2}{L}x \qquad (2.27)$$

The water table elevation, often described as the *Dupuit parabola*, is shown in Figure 2.13 as the full line; the results refer to the case of $H_1 = L$ and $H_2 = 0.2H_1$. Equation (2.27) does not give the correct elevation of the water table since the analysis ignores the presence of the seepage face on the downstream face of the dam between the water table and downstream water level. The broken line in Figure 2.13 indicates the correct water table position determined from a numerical solution (see Section 2.5.1).

Flow through the dam can be calculated from the equation,

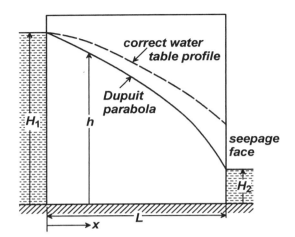

Figure 2.13 Dupuit–Forchheimer solution for flow through a rectangular dam

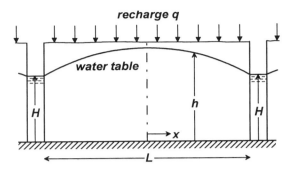

Figure 2.14 Dupuit–Forchheimer solution for flow to parallel ditches

$$Q = -K\frac{dh}{dx}h = \frac{K\left(H_1^2 - H_2^2\right)}{2L} \qquad (2.28)$$

This expression for the total through-flow is correct (Charni 1951, Bear 1972) even though it is derived from an incorrect equation for groundwater heads.

Example 2: Flow to drainage ditches

The Dupuit theory can also be used to estimate the water table elevation between drainage ditches. Figure 2.14 shows two drainage ditches, a distance L apart, penetrating to an impermeable layer; recharge to the aquifer is q and the horizontal hydraulic conductivity K. After integrating Eq. (2.26) and substituting the boundary conditions that at $x = \pm 0.5L$, $h = H$, the resultant equation for the height of the water table above the impermeable base is

$$h^2 = H^2 + q\frac{0.25L^2 - x^2}{K} \qquad (2.29)$$

The curve plotted in Figure 2.14 relates to a typical example in which $q = 0.002\,\text{m/d}$, $K = 0.05\,\text{m/d}$, $L = 20\,\text{m}$ and $H = 2.0\,\text{m}$. A more detailed discussion of drainage issues can be found in Section 4.4.

2.3.8 Time-variant one-dimensional flow

Analytical solutions can be obtained for time-variant flow in a confined aquifer. A representative example is

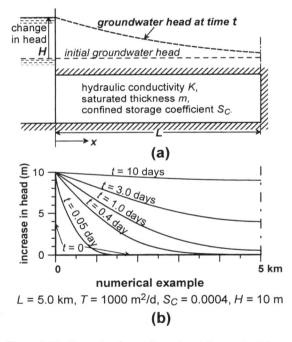

Figure 2.15 Example of one-dimensional time-variant flow in a confined aquifer

presented in Figure 2.15; there is a sudden change in head at the left-hand boundary. Since the thickness and the hydraulic conductivity of the aquifer are constant with flow only occurring in the x direction, Eq. (2.5) multiplied by the saturated thickness m becomes

$$\frac{\partial}{\partial x}\left(mK\frac{\partial h}{\partial x}\right) = S_C\frac{\partial h}{\partial t} \qquad (2.30)$$

In the problem illustrated in Figure 2.15, the initial condition is that the groundwater head is zero everywhere in the aquifer, thus $h = 0$ when $t = 0$ for all x. The boundary conditions include a sudden rise in head at the left-hand boundary with no flow crossing the right-hand boundary, thus for $t > 0$ at $x = 0$, $h = H$ and for $t \geq 0$ at $x = L$, $dh/dx = 0$. A mathematical expression satisfying the differential equation and initial and boundary conditions is

$$h(x,t) = H - \frac{4H}{\pi} \sum_{n=0}^{\infty} \frac{1}{2n+1} \exp\left[-\frac{Kmt}{S_C} \left(\frac{(2n+1)\pi}{2L} \right)^2 \right]$$
$$\sin\left(\frac{(2n+1)\pi x}{2L} \right) \qquad (2.31)$$

with a flow at the left-hand boundary of

$$Q(0,t) = \frac{2KmH}{L} \sum_{n=0}^{\infty} \exp\left[-\frac{Kmt}{S_C} \left(\frac{(2n+1)\pi}{2L} \right)^2 \right] \qquad (2.32)$$

Figure 2.15b presents the changes in groundwater heads with time for a specific problem. The aquifer is confined with $S_C = 0.0004$ and $T = 1000\,\text{m}^2/\text{d}$; groundwater head distributions are plotted for times varying from 0.05 to 10 days. For this confined aquifer more than 90% of the rise in groundwater heads is completed within 10 days.

Other examples of analytical solutions for one-dimensional groundwater flow can be found in Bear (1972, 1979), McWhorter and Sunada (1977) and Marino and Luthin (1982). Analytical solutions have also been obtained for the particular problem of unconfined flow due to time-variant vertical recharge with variations in saturated depth included in the analysis. For example, Rao and Sarma (1981) considered a rectangular aquifer which receives constant recharge over a smaller rectangular area. In another analysis, Rai and Singh (1995) represent recharge which decreases exponentially with time from an initial value of $N_1 + N_0$ to N_1, thereafter the recharge remains constant. A review of analytical solutions to similar problems has been prepared by Rai and Singh (1996).

2.4 RADIAL FLOW

The purpose of this section is to introduce the analysis of both steady state and time-variant radial flow. A limited number of analytical solutions are described, with special reference to boundary conditions. The material in this section is selected to serve as a background to Chapters 5 to 8 in which a wide variety of radial flow situations are examined in detail.

2.4.1 Radial flow in a confined aquifer

First an equation for steady radial flow to a pumped well in a confined aquifer is derived from first principles. Figure 2.16a shows a pumped well with discharge Q which fully penetrates a confined aquifer of constant saturated thickness m and uniform radial hydraulic conductivity K_r. In the plan view (Figure 2.16b), a cylindrical element is drawn at a radial distance r with width dr. The well discharge Q must cross this cylindrical element. From Darcy's Law:

$$Q = \text{cross-sectional area} \times \text{hydraulic conductivity} \times \text{head gradient}$$

so $\quad Q = 2\pi r m \times K_r \times \dfrac{dh}{dr} = 2\pi r T \dfrac{dh}{dr} \qquad (2.33)$

Rearranging $\quad dh = \dfrac{Q}{2\pi T} \dfrac{dr}{r}$,

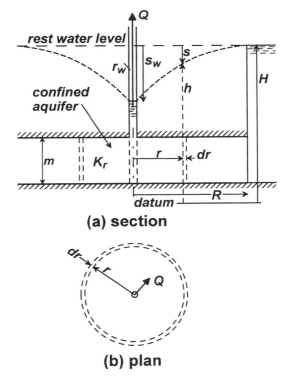

Figure 2.16 Radial flow to a pumped well in a confined aquifer

and integrating $h = \dfrac{Q}{2\pi T}\ln(r) + A$

Note that the integral of $1/r$ is the natural logarithm $\ln(r)$. This logarithmic term is a distinctive feature of radial flow; it implies rapid changes in groundwater head in the vicinity of the well when r is small.

If the groundwater head at radius R is H, Figure 2.16a, the constant of integration A can be evaluated as

$$A = H - \frac{Q}{2\pi T}\ln(R),$$

hence $h = H + \dfrac{Q}{2\pi T}(\ln(r) - \ln(R)) = H - \dfrac{Q}{2\pi T}\ln\!\left(\dfrac{R}{r}\right)$

$$(2.34)$$

An alternative derivation is possible using the *governing differential equation* for radial flow (which is quoted without derivation),

$$\frac{1}{r}\frac{d}{dr}\left(mK_r r\frac{dh}{dr}\right) = \frac{d}{dr}\left(mK_r\frac{dh}{dr}\right) + \frac{m}{r}K_r\frac{dh}{dr} = -q$$

$$(2.35)$$

since Figure 2.16 refers to a confined aquifer, $q = 0$; furthermore the aquifer is of constant thickness hence mK_r is constant. Therefore the equation reduces to

$$\frac{1}{r}\frac{d}{dr}\left(r\frac{dh}{dr}\right) = 0$$

Integrating twice and substituting the boundary conditions that:

1. for any radius, the total flow $Q = 2\pi r T\dfrac{dh}{dr}$ (this expression is based on Darcy's Law, there is no negative sign since the flow Q is in the direction of negative r),
2. at the outer radius R the groundwater head is H,

leads to Eq. (2.34).

Rather than working in terms of h, the groundwater head above a datum, it is often more convenient to work in terms of the drawdown s below the rest water level

$$s = H - h \tag{2.36}$$

A more general form of Eq. (2.34) can be written in terms of drawdowns as

$$s_1 - s_2 = \frac{Q}{2\pi T}\ln\!\left(\frac{r_2}{r_1}\right) \tag{2.37}$$

This is called the Thiem equation; it can be used for preliminary studies of groundwater flow to wells. For example, the equation can be used to determine the drawdown s_w in a pumped well of radius r_w when the drawdown at radius r_1 is s_1:

$$s_w = s_1 + \frac{Q}{2\pi T}\ln\!\left(\frac{r_1}{r_w}\right) \tag{2.38}$$

2.4.2 Radial flow in an unconfined aquifer with recharge

Assuming that the saturated depth remains approximately uniform so that a constant transmissivity can be used, the analysis of steady-state unconfined flow with uniform recharge (see Figure 2.17), uses a simplified form of Eq. (2.35),

$$\frac{1}{r}\frac{d}{dr}\left(r\frac{dh}{dr}\right) = -\frac{q}{T} \tag{2.39}$$

Details of the mathematical integration and substitution of boundary conditions are included in Figure 2.17, and some of the essential features are described below.

- Since Eq. (2.39) is a second-order equation, it is necessary to perform two integrations. The term $\ln(r)$ is introduced during the second integration.
- The two boundary conditions both apply at the outer boundary, $r = R$.
- Substituting $Q = \pi R^2 q$, the final equation in terms of drawdowns can be written as

$$s = \frac{Q}{2\pi T}\left(\ln\frac{R}{r} - \frac{R^2 - r^2}{2R^2}\right) \tag{2.40}$$

The first term in brackets is the same as for confined flow, while the second term is a correction which results from increasing flow towards the well as recharge enters the aquifer. Results for a typical problem in which $Q/2\pi T = 1.0$ with a maximum radius $R = 1000\,\text{m}$, are quoted in Table 2.2. The third column lists the correction term which tends to a maximum of 0.50. Drawdowns at radial distances of 900 m and 300 m are smaller for the unconfined situation because the flow though the aquifer at these radii is smaller than in the confined aquifer for which all the flow enters at the outer

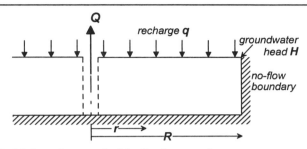

Rewriting Eq. 2.39 in a form suitable for integration,

$$\frac{d}{dr}\left(r\frac{dh}{dr}\right) = -\frac{qr}{T}$$

(a) *First integration and first boundary condition*

$$r\frac{dh}{dr} = \frac{-qr^2}{2T} + A \qquad \text{or} \qquad \frac{dh}{dr} = \frac{-qr}{2T} + \frac{A}{r} \qquad\qquad (A)$$

At the outer boundary, $r = R$, no flow crosses hence $\frac{dh}{dr} = 0$, thus $A = \frac{qR^2}{2T}$

therefore Eq. A becomes $\qquad \frac{dh}{dr} = \frac{q}{2T}\left(\frac{R^2}{r} - r\right)$

(b) *second integration and second boundary condition*

$$h = \frac{q}{2T}\left(R^2 \ln r - \frac{r^2}{2}\right) + B \qquad\qquad (B)$$

at $r = R$, $h = H$, substituting in the above equation

$$H = \frac{q}{2T}\left(R^2 \ln R - \frac{R^2}{2}\right) + B$$

thus $\qquad h = \frac{q}{2T}\left(\frac{R^2 - r^2}{2} + R^2 \ln\frac{r}{R}\right) + H \qquad\qquad (C)$

(c) *for comparison with the confined aquifer response*, Eq. C is converted to drawdowns, also include the substitution that $Q = \pi R^2 q$

$$s = \frac{Q}{2\pi T}\left(\ln\frac{R}{r} - \frac{R^2 - r^2}{2R^2}\right) \qquad\qquad (D)$$

Figure 2.17   Radial flow to a pumped well in an unconfined aquifer with recharge assuming constant saturated thickness

Table 2.2 Comparison of drawdowns in a confined aquifer (Eq. (2.37)) and in an unconfined aquifer with constant recharge (Eq. (2.40)): maximum radius = 1000 m, $Q/2\pi T = 1.0$

Radius r (m)	Drawdown confined (m)	$(R^2 - r^2)/2R^2$	Drawdown unconfined (m)
1000	0.0	–	0.0
900	0.105	0.095	0.01
300	1.10	0.455	0.64
100	2.30	0.495	1.80
30	3.51	0.499	3.01
10	4.61	0.500	4.11
3.0	5.81	0.500	5.31
1.0	6.91	0.500	6.41
0.3	8.11	0.500	7.61

boundary. For radial distances of 100 m or less, the drawdown for the unconfined aquifer with uniform recharge is consistently 0.5 m less than in the confined situation.

2.4.3 Radial flow in an unconfined aquifer with varying saturated depth

Approximate solutions for radial flow in an unconfined aquifer can be obtained using the Dupuit approximation. For the situation illustrated in Figure 2.18, the pumping rate from the unconfined aquifer

$$Q = K \times 2\pi r h \times \frac{dh}{dr} \qquad (2.41)$$

Note that groundwater heads are always measured from the base of the aquifer.

Rearranging $2h\,dh = \dfrac{Q}{\pi K}\dfrac{dr}{r}$

and integrating $h^2 = \dfrac{Q}{\pi K}\ln(r) + A$

Boundary conditions

At $r = R, h = H$ therefore $A = H^2 - (Q/\pi K)\ln(R)$, thus

$$h^2 = H^2 - \frac{Q}{\pi K}\ln\!\left(\frac{R}{r}\right) \qquad (2.42)$$

At $r = r_w, h = h_w$ hence $h_w^2 = H^2 - \dfrac{Q}{\pi K}\ln\!\left(\dfrac{R}{r_w}\right)$

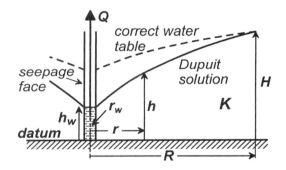

Figure 2.18 Radial flow to a pumped well in an unconfined aquifer with variable saturated thickness

This equation can be rearranged as

$$Q = \frac{\pi K\left(H^2 - h_w^2\right)}{\ln(R/r_w)} \qquad (2.43)$$

As with one-dimensional flow in the x direction, the flow equation, Eq. (2.43), is correct but the expression for the groundwater head, Eq. (2.42), is not, due to the approximations of the Dupuit approach. Comparisons between the flow equations for constant and variable saturated depths allows a judgement to be made about the validity of the constant saturated depth approximation. Equation (2.42) can be rewritten as

$$Q = \frac{\pi K(H - h_w)(H + h_w)}{\ln(R/r_w)}$$

The equation for radial flow in a confined aquifer, Eq. (2.34), can be rearranged as

$$Q = \frac{\pi K (H - h_w) \times 2\,m}{\ln(R/r_w)}$$

The difference between the two equations is the last term in the numerators $(H + h_w)$ and $2m$. If h_w, the elevation of the well water level above the base, is close to the full saturated thickness H (i.e. small drawdowns) then $H + h_w \approx 2m$, where m is the constant saturated depth. Therefore, if $h_w \geq 0.9H$ (the pumped well drawdown is less than 10 per cent of the maximum saturated thickness) the error in the flows will be less than 5 per cent when a constant saturated thickness approach is adopted. Otherwise for $h_w \leq 0.9H$ the errors become significant.

2.4.4 Radial flow in a leaky aquifer

Pumping water from deeper aquifers which are overlain by less permeable strata is a common practice since the yield from the deeper aquifers is often more reliable and the risk of pollution is lower. A typical situation is illustrated in Figure 2.19a. The main aquifer is overlain by a semi-permeable stratum (or aquitard). Above the semi-permeable stratum is an overlying aquifer. The elevation of the water table in the overlying aquifer is assumed to remain unchanged at H above datum. However, pumping from the main aquifer causes a fall in the piezometric head so that the groundwater head

h in the main aquifer is lower than H as indicated by the representative observation piezometer. The difference in groundwater head across the semi-permeable stratum leads to a flow (or leakage) through the semi-permeable stratum.

An element, or control volume, of the semi-permeable stratum adjacent to the observation piezometer is shown in Figure 2.19b. The groundwater heads above and below the semi-permeable stratum are H and h, the thickness of the stratum is m' and the hydraulic conductivity K'. Therefore the vertical velocity (or recharge) through the semi-permeable stratum can be calculated from Darcy's Law as

$$q = -v_z = K' \frac{H - h}{m'} \tag{2.44}$$

Therefore the leakage is equivalent to a recharge varying with radius as shown in Figure 2.19c. The differential equation describing steady-state radial flow with recharge, Eq. (2.35), can be written as

$$\frac{1}{r}\frac{d}{dr}\left(r\frac{dh}{dr}\right) + \frac{q}{Km} = \frac{1}{r}\frac{d}{dr}\left(r\frac{dh}{dr}\right) + \frac{K'(H-h)}{Kmm'} = 0 \tag{2.45}$$

It is convenient to introduce a leakage factor $B = \left(\dfrac{Kmm'}{K'}\right)^{1/2}$ and also to work in terms of the drawdown $s = H - h$. The differential equation then becomes

$$\frac{1}{r}\frac{d}{dr}\left(r\frac{ds}{dr}\right) - \frac{s}{B^2} = 0 \tag{2.46}$$

with boundary conditions

1. as $r \to \infty$, $s = 0$
2. the well is represented as a sink of infinitely small radius hence

$$\lim_{r \to 0} r \frac{ds}{dr} = \frac{-Q}{2\pi T}$$

The solution to Eq. (2.46) with these two boundary conditions is

$$s = \frac{Q}{2\pi T} K_0\left(\frac{r}{B}\right) \tag{2.47}$$

where K_0 is a modified Bessel function of the second kind. Tabulated values of K_0 can be found in Kruseman and de Ridder (1990).

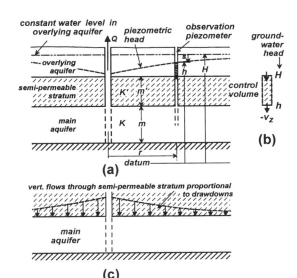

Figure 2.19 Radial flow to a pumped well in a leaky aquifer

Table 2.3 Calculated drawdowns for representative example of a leaky aquifer

r (m)	r/B	K_0	s (m)
10.0	0.01	4.721	4.721
100.0	0.10	2.427	2.427
1000	1.00	0.421	0.421
5000	5.00	0.004	0.004

For a *representative example*, $Q = 1000\pi \, \mathrm{m^3/d}$, $T = 500 \, \mathrm{m^2/d}$, $m' = 20 \, \mathrm{m}$ and $K' = 0.01 \, \mathrm{m/d}$; calculate the drawdowns in the main aquifer at 10 m, 100 m, 1000 m and 5000 m.

$$B = \left(\frac{Kmm'}{K'}\right)^{1/2} = \left(\frac{500 \times 20}{0.01}\right)^{1/2} = 1000 \, \mathrm{m}$$

$$\text{and } \frac{Q}{2\pi T} = 1.0$$

The drawdowns are listed in Table 2.3.

Drawdowns in the main aquifer spread a considerable distance from the pumping well; these drawdowns set up vertical gradients drawing water through the aquitard for radial distances of several kilometres. Smaller values of B would result in more localised effects; B could be smaller due to a lower transmissivity of the main aquifer, a thinner semi-permeable stratum or a higher vertical hydraulic conductivity of the semi-permeable stratum. A consequence of a smaller value of B would be larger drawdowns in the main aquifer.

Time-variant leaky aquifer response is discussed in detail in Section 5.8; the impact of the storage properties of a semi-permeable stratum are considered in Section 7.5.

2.4.5 Time-variant radial flow

The governing equations for one-dimensional time-variant radial flow, derived using Darcy's Law and continuity, becomes

$$\frac{\partial}{\partial r}\left(mK_r \frac{\partial s}{\partial r}\right) + \frac{m}{r}K_r \frac{\partial s}{\partial r} = (S_C + S_Y)\frac{\partial s}{\partial t} + q \qquad (2.48)$$

The application of this equation to a wide range of field problems is considered in detail in Chapters 5 and 6. For completeness, both the confined storage coefficient and the specific yield are quoted in Eq. (2.48). As explained in Section 5.17, the release of water from

unconfined storage is not instantaneous due to the mechanism of delayed yield; consequently the confined storage coefficient may be important in the early stages of pumping.

2.4.6 Time-variant radial flow including vertical components

When a practical problem requires the consideration of combined radial and vertical flow components, the governing equation must include both vertical and radial co-ordinates,

$$\frac{\partial}{\partial r}\left(K_r \frac{\partial s}{\partial r}\right) + \frac{K_r}{r}\frac{\partial s}{\partial r} + \frac{\partial}{\partial z}\left(K_z \frac{\partial s}{\partial z}\right) = S_S \frac{\partial s}{\partial t} \qquad (2.49)$$

This equation refers to an element within the aquifer, therefore the relevant storage coefficient is the specific storage. Lateral, upper and lower boundary conditions must be specified in a manner similar to the two-dimensional vertical section which is considered in Section 2.5. The application of Eq. (2.49) to practical problems is examined in Section 7.2.

2.5 TWO-DIMENSIONAL VERTICAL SECTION (PROFILE MODEL)

The essential flow processes of certain groundwater problems can be represented by considering a vertical section in the x-z plane, provided that the flow in the y direction is sufficiently small to be neglected. The time-variant form of the governing differential equation in the vertical plane is the same as that for three-dimension flow (Eq. (2.5)), but with the flow in the y direction eliminated:

$$\frac{\partial}{\partial x}\left(K_r \frac{\partial h}{\partial x}\right) + \frac{\partial}{\partial z}\left(K_z \frac{\partial h}{\partial z}\right) = S_S \frac{\partial h}{\partial t} \qquad (2.50)$$

2.5.1 Steady-state conditions, rectangular dam

For a steady-state analysis of flow in a rectangular dam (see Figure 2.20), it is necessary to identify the differential equation and five boundary conditions. The steps in specifying the problem and preparing a mathematical description are as follows.

1. *Define the co-ordinate axes*: the co-ordinate axes are x and z; it is preferable to measure z from the base of the dam.
2. *Specify the differential equation describing flow within the dam*: flow through the dam is steady state

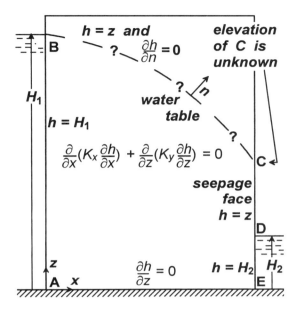

Figure 2.20 Mathematical specification of flow through a rectangular dam in the *x-z* plane

and occurs in the *x-z* plane; the flow is described by Eq. (2.50) but with the time-variant term on the right-hand side set to zero.

3. *Boundary* AB: this is the upstream face; the groundwater head is constant on this face since at **A** the pressure is H_1 but $z = 0$ while at **B** the pressure is zero (atmospheric) but $z = H_1$, therefore $h = H_1$ for the whole boundary.

4. *Boundary* DE: this is the downstream face where $h = H_2$.

5. *Boundary* AE: on this impermeable base there is no flow in the *z* direction, therefore from Darcy's Law $\partial h/\partial z = 0$.

6. *Boundary* CD: on the vertical face of the dam above the downstream water level, water seeps out into the atmosphere. Since the pressure is atmospheric,

$$h = \frac{p}{\rho g} + z, \text{ but } p = 0 \text{ so } h = z$$

The location of the bottom of the seepage face is known, but the elevation of the top of the seepage face is unknown as discussed under (7) below.

7. *Boundary* BC: this boundary, which connects the top of the upstream face **B** to an unknown location **C** on the seepage face, is the water table (often called

the free surface). It is the upper boundary of the saturated flow regime. To appreciate conditions on this boundary, consider a particle of water entering the aquifer at **B** and moving to the right. The particle moves laterally and downward; as it moves through the aquifer, the downwards movement is strongly influenced by the particles of water beneath it. As the particle moves through the dam, it is 'unaware' of the location of the downstream water level **D** (although the Dupuit assumption assumes that it will move directly to **D**). Eventually this particle, which left from the top of the upstream face, emerges from the aquifer above **D** at location **C**. This location **C** cannot be explicitly defined since it depends on the movement of all the particles within the aquifer system. The series of question marks along the water table boundary in Figure 2.20 highlight the unknown location of the water table.

This simplified description of the flow of a particle along the water table is presented to give insights into the conditions on this boundary. Since the location of the water table boundary is unknown, there will be two boundary conditions; for other boundaries the position is known, hence only one boundary condition is required. The two conditions are:

(i) the pressure everywhere is atmospheric hence $h = z$,

(ii) no flow crosses this boundary hence $\partial h/\partial n = 0$ where *n* is the direction normal to the boundary.

Note that this analysis assumes a sudden change across the water table from fully saturated conditions to zero flow. However, in practice there is a capillary fringe above the saturated water table. Bouwer (1978) examines the likely errors which occur when ignoring lateral flow through the capillary fringe. He introduces the concept of a fictitious saturated capillary fringe which produces the same flow as the actual capillary fringe with its varying unsaturated hydraulic conductivity. For coarse sands the equivalent additional depth of saturated flow is about 0.2 m while for fine sands it is about 0.6 m. For many practical situations these equivalent depths of flow are small compared to the total saturated thickness.

Currently there is no exact solution for the location of the water table and the groundwater head distribution within the dam, therefore numerical solution techniques must be used. This problem of flow through a dam has been solved by a variety of techniques including finite difference (Rushton and Redshaw 1979), finite element (Neuman and Witherspoon 1970, Kazda 1978) and the boundary integral equation method (Liggett and Liu 1983).

Results obtained from a numerical solution for steady flow through a rectangular dam are included in Figures 2.21 and 2.22. The dam is square with an upstream water level 10 m above the impermeable base and a downstream water level of 2.0 m. The location of the water table and equipotential lines within the dam

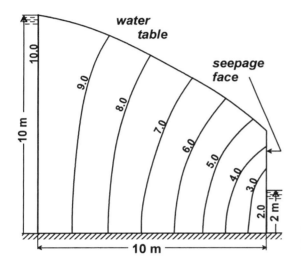

Figure 2.21 Location of water table and lines of equal groundwater head in rectangular dam

solution from finite difference grid 1.0 m x 1.0 m; flow arrows centred in middle of squares

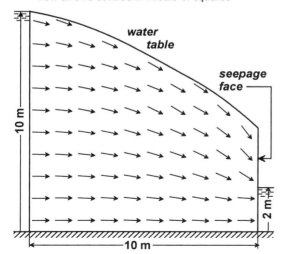

Figure 2.22   Flow directions through the rectangular dam

are drawn in Figure 2.21. Further insights into the flow through the dam are provided by the flow direction arrows of Figure 2.22; the way in which these directions are calculated is explained in Figure 2.23. The importance of flows leaving the dam at the seepage face is indicated by the flow arrows; from the detailed numerical results it is estimated that, for this particular problem, more than 45 per cent of the total flow through the dam leaves from the seepage face.

From the date of publication of the above references, it is clear that the methods of solution for problems such as the flow through dams have been available for several decades. More recent contributions relate to the use of these techniques for field problems where the location of the unknown water table is important. Examples reported elsewhere in this book include:

- seepage losses from canals (Section 4.2),
- the analysis of water table control using tile drainage (Section 4.4),
- water losses through bunds of flooded ricefields (the location of three separate water tables has to be determined) (Section 4.5),
- pumping from unconfined aquifers (Section 7.2).

Many of these problems also involve time-variant water table movement, which is considered in the following section.

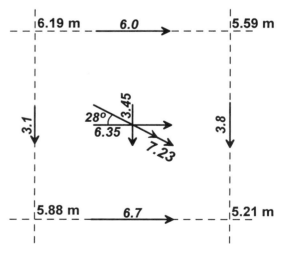

Figure 2.23   Calculation of magnitude and direction of flows from groundwater heads on a square grid. **Methodology**: grid 1.0 m by 1.0 m, $K = 10$ m/d. Values of heads at intersection of grid lines. Calculated flows (m^3/d) in italics. Flows on grid lines calculated from head differences. Flows at centre, average of flows on grid lines. Resultant flow and angle from component flows at centre

2.5.2 Time-variant moving water table

A number of workers, including Kirkham and Gaskell (1951), Polubarinova-Kochina (1962), Todsen (1971), Bear (1972) and Rushton and Redshaw (1979), have considered conditions at a moving water table. Mathematically, the moving boundary can be described in a number of alternative ways; in this discussion the conditions are derived using a graphical construction so that the physical significance of the boundary condition is apparent.

Figure 2.24, which refers to the *x-z* plane, illustrates how a particle on the water table moves from location **A** to location **A'** during a time increment δt. In this analysis the symbol *H* is used to describe the elevation of the water table. *H* is a function of *x* alone; *h* is the groundwater head within two-dimensional space. There are three components of the total movement, two component velocities and a third component due to the recharge *q*. Note that the movements are expressed in terms of the Darcy velocities at the water table hence, to obtain actual velocities, they must be divided by the specific yield.

In the vertically upwards direction the movement is $v_z \, \delta t / S_Y$
In the horizontal direction to the right the movement is $v_x \, \delta t / S_Y$
Due to the recharge there is an upward movement of $q \, \delta t / S_Y$.

The signs of each of these terms is important; in particular the vertical velocity and the movement due to the recharge are both positive upwards.

In numerical studies the information required is usually the *vertical displacement* of the water table, AB

(see Figure 2.24). The water table at **A** is at an angle of α below the horizontal, therefore using the construction shown in Figure 2.24, a mathematical expression can be written to determine the vertical movement at node **A**.

The *vertical rise* in the water table δH during time step δt is given by the expression

$$\delta H = -AB + \frac{q\delta t}{S_Y} = \frac{\delta t}{S_Y}(v_z + v_x \tan\alpha + q) \quad (2.51)$$

where the distance AB is determined from the geometric construction of Figure 2.24. From Darcy's Law

$$v_x = -K_x \frac{\partial h}{\partial x} \quad v_z = -K_z \frac{\partial h}{\partial z},$$

Also, provided that the change in slope of the water table is small,

$$\frac{\partial h}{\partial x} = \left[\frac{\partial h}{\partial z} - 1\right]\tan\alpha$$

Therefore Eq. (2.51) can be written as

$$\delta H = \frac{\delta t}{S_Y}\left(-K_z \frac{\partial h}{\partial z} + K_x\left(1 - \frac{\partial h}{\partial z}\right)\tan^2\alpha + q\right) \quad (2.52)$$

Practical examples of time-variant moving water tables can be found in Section 4.4.5 which considers water table control using horizontal tile drains. Numerical values for a specific time step are presented to illustrate how the three components on the right-hand side of Eq. (2.52) are combined to calculate the water table movement.

The moving water table is also important in radial flow problems. Equation (2.52) can be converted to (r, z) co-ordinates; for radial flow problems it is convenient to write

$$\tan\alpha = -\partial H / \partial r$$

so that the radial form of Eq. (2.51) becomes

$$\delta H = -\frac{\delta t}{S_Y}\left(K_z \frac{\partial h}{\partial z} - K_r \frac{\partial h}{\partial r}\frac{\partial H}{\partial r} - q\right) \quad (2.53)$$

The numerical analysis of moving water table problems in radial flow is considered in Section 7.2.

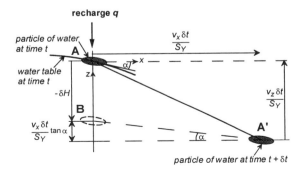

Figure 2.24 Geometric representation of a moving water table

2.6 REGIONAL GROUNDWATER FLOW

Many regional groundwater flow studies are based on the equation

$$\frac{\partial}{\partial x}\left(T_x\frac{\partial h}{\partial x}\right)+\frac{\partial}{\partial y}\left(T_y\frac{\partial h}{\partial y}\right)=S\frac{\partial h}{\partial t}-q \qquad (2.54a)$$

In this equation the groundwater head h is a function of the co-ordinates x, y and time t, i.e. $h(x, y, t)$. When developing this equation, it is often stated that vertical flows are so small that they can be neglected, hence only the horizontal co-ordinates x and y are included in the analysis. However, in practical regional groundwater problems vertical flows are important. Recharge is a vertical inflow, also boreholes and groundwater-fed rivers draw water both horizontally and vertically from within the aquifer. This issue was considered by Connorton (1985) in a paper entitled 'Does the regional groundwater-flow equation model vertical flow?' His answer is yes! Connorton's approach is summarised below, and this is followed by an arithmetic calculation which demonstrates that vertical flow components are automatically included when using the regional groundwater equation. Finally, situations where the vertical flow components must be represented explicitly are identified.

2.6.1 Analysis of Connorton (1985)

A summary of the mathematical basis for Connorton's affirmative answer can be found in Figure 2.25; the essential features of his argument are illustrated in Figure 2.26 and considered below.

The differential equation for three-dimensional time-variant groundwater flow has previously been derived as

$$\frac{\partial}{\partial x}\left(K_x\frac{\partial h}{\partial x}\right)+\frac{\partial}{\partial y}\left(K_y\frac{\partial h}{\partial y}\right)+\frac{\partial}{\partial z}\left(K_z\frac{\partial h}{\partial z}\right)=S_S\frac{\partial h}{\partial t} \qquad (2.25)$$

in which the groundwater head is a function of the three co-ordinate directions and time, $h(x, y, z, t)$. The equation can be described in words with reference to the unit cube of Figure 2.26a. The first term of Eq. (2.25) describes the change in flux (velocity multiplied by unit cross-sectional area) in the x direction as water passes from the left-hand to the right-hand side of the cube. In Figure 2.26a this is described as *difference in flux, x direction* (in the following discussion a shorthand is used of *flux diff in x*). There are also flux differences in the y and z directions. A further component of the flow

balance is water taken into storage as the pressure changes. The pressure change component equals the change in groundwater head multiplied by the specific storage ($S_S\partial h/\partial t$). These four components sum to zero (this is equivalent to Eq. (A) in Figure 2.25):

flux diff in x + flux diff in y + flux diff in z
+ specific storage × head change with time = 0

Figure 2.26b refers to an element of the aquifer of unit plan area but extending over the full saturated depth. Therefore the difference in fluxes in the x and y directions are multiplied by the saturated depth to give the difference in flows (*flow diff in x* etc.); the hydraulic conductivities K_x and K_y are replaced by the transmissivities T_x and T_y. Also, because the full depth of the section is considered, the specific storage is replaced by the confined storage coefficient. Therefore the equivalent equation written in words for the full saturated depth is:

flow diff in x + flow diff in y + flux diff in z
+ confined storage × head change with time = 0

Note that the term *flux diff in z* is retained since the cross-sectional area in the z direction is the same in Figure 2.26a and b. The above equation is the same as Eq. (B) in Figure 2.25.

Assuming that the base of the aquifer is impermeable, the velocity and hence the flux at the base is zero. At the top of the element there are two physical processes which provide the vertical flux, the recharge $-q$ and the flux component due to the specific yield $S_Y\partial h/\partial t$. Therefore *flux diff in z* (vertically upwards is positive) is

$$S_Y\frac{\partial h}{\partial t}-q$$

Consequently the flow balance for the full depth of the aquifer (Figure 2.26b), becomes

$$\frac{\partial}{\partial x}\left(T_x\frac{\partial h}{\partial x}\right)+\frac{\partial}{\partial y}\left(T_y\frac{\partial h}{\partial y}\right)=S_C\frac{\partial h}{\partial t}+S_Y\frac{\partial h}{\partial t}-q \qquad (2.54b)$$

This is the standard two-dimensional regional groundwater flow equation. From this outline derivation, it is clear that vertical components of flow are a fundamental part of this equation. A similar analysis is described by Pinder (2002); he emphasises that in the presence of significant vertical gradients, the areal

This summary derivation, based on Connorton (1985), retains his notation. The continuity equation in terms of the specific discharge vector is

$$\int_{z_1}^{z_2} -\nabla \bullet q \; dz = \int_{z_1}^{z_2} S_S \frac{\partial \phi}{\partial t} dz \tag{A}$$

where $\nabla \equiv (\partial / \partial x, \; \partial / \partial y, \; \partial / \partial z)$

$q = (q_x, \; q_y, \; q_z)$; the specific discharge vector in the x, y, z directions,

z_1 and z_2 are the elevations of the lower and upper boundaries,

$\phi = z + p(x, y, z, t)/\rho g$, in which p is the pressure of the water,

S_S is the specific storage coefficient.

This equation can be divided into horizontal and vertical components and, after differentiating under the integral sign,

$$\int_{z_1}^{z_2} -\nabla \bullet q \, dz = -\frac{\partial}{\partial x} \int_{z_1}^{z_2} q_x \, dz - \frac{\partial}{\partial y} \int_{z_1}^{z_2} q_y \, dz + A_1 + A_2 \tag{B}$$

where A_1 relates to the lower boundary $z = z_1(x, y)$

$$A_1 = -q_{x|z_1}\left(\frac{\partial z_1}{\partial x}\right) + q_{y|z_1}\left(\frac{\partial z_1}{\partial y}\right) + q_{z|z_1}$$

and A_2 relates to the upper boundary $z = z_2(x, y)$

$$A_2 = -q_{x|z_2}\left(\frac{\partial z_2}{\partial x}\right) + q_{y|z_2}\left(\frac{\partial z_2}{\partial y}\right) - q_{z|z_2}$$

The potential function ϕ in Eq. A is three-dimensional in space whereas the regional groundwater head h (often called the average head) is two-dimensional in space,

$$\phi(x, y, z, t) = h(x, y, t) + \varepsilon(x, y, z, t) \tag{C}$$

where $\varepsilon(x, y, z, t)$ is an error term;

errors Δq_x, Δq_y can then be calculated for the specific discharges,

$$\Delta q_x = -K_x \frac{\partial \varepsilon}{\partial x} \text{ and } \Delta q_y = -K_y \frac{\partial \varepsilon}{\partial y} \tag{D}$$

Different situations on the lower and upper boundaries of the aquifer system, such as confined, unconfined, leaky etc. result in different errors. Connorton (1985) studies a range of problems giving analytical expressions for the errors in groundwater head and specific discharges. He concludes that *for straightforward regional-flow problems the vertical flow components, as given by the average-head solution of the conventional governing equation, is remarkably accurate.*

Figure 2.25   Derivation of two-dimensional regional groundwater flow equation from flow components in three dimensions

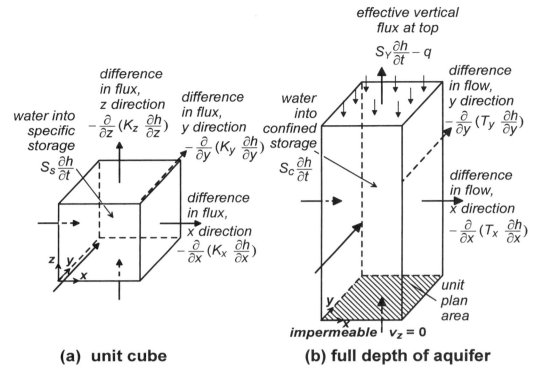

(a) unit cube   **(b) full depth of aquifer**

Figure 2.26   Diagram illustrating how vertical flow components are implicitly contained in the two-dimensional regional groundwater flow formulation

two-dimensional form of the groundwater flow equation is not appropriate.

2.6.2 Illustrative numerical examples of one-dimensional flow

An example is presented in Figure 2.27 to demonstrate how vertical flows can be deduced for one-dimensional steady-state flow in which the groundwater heads are known on a 1.0 km grid. Details of the aquifer system are given in Figure 2.27a. Note in particular,

1. the aquifer has a slight slope to the right, elevations of the base of the aquifer (45, 44, 43 and 42 m) are given at 1 km intervals,
2. the lower 10 m of the aquifer has a hydraulic conductivity of 120 m/d (shown shaded in Figure 2.27), the upper part has a maximum saturated thickness of 20.1 m with a hydraulic conductivity of 15 m/d,
3. the recharge rate to the aquifer is 0.598 mm/d,
4. groundwater heads (74.9, 74.0, 72.7 and 72.1 m) are known at the 1 km spacing.

Flows are calculated for a unit width into the paper using Darcy's Law, first for the lower stratum and subsequently for the upper stratum. The flow from A to B *in the lower stratum,*

$$Flow_{AB} = K \times depth \times head\ gradient$$
$$= 120 \times 10.0 \times (0.9/1000) = 1.080\ m^3/d/m$$

Also $Flow_{BC} = 1.560\ m^3/d/m$, $Flow_{CD} = 0.720\ m^3/d/m$.

These flows are recorded in Figure 2.27b.

The flow in the lower stratum between B and C is 0.48 m³/d/m higher than the flow from A to B; this change in flow can only occur if there is a vertical inflow to the lower layer of 0.48 m³/d/m. This inflow is shown in the figure as a vertical arrow but in practice the vertical flow occurs between the mid-point of AB and the mid-point of BC. An examination of the flow difference $Flow_{CD} - Flow_{BC}$ shows that there must be an upwards flow of 0.84 m³/d/m.

In calculating the flows *in the upper stratum* account must be taken of the varying saturated depth.

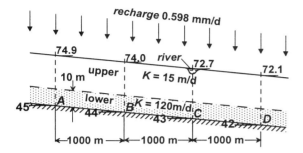

(a) properties of aquifer (*all elevations in* m)

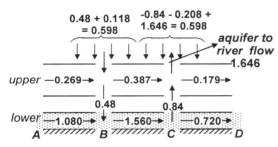

(b) component flows (*all flows in* m³/d/m)

Figure 2.27   Numerical example of estimation of horizontal and vertical flows from a one-dimensional solution: (a) details of the problem, (b) identification of horizontal and vertical flow components

$Flow_{AB} = K \times depth \times head\ gradient$
$= 15 \times [(19.9 + 20.0)/2] \times (0.9/1000)$
$= 0.269\ \text{m}^3/\text{d/m}.$

also $Flow_{BC} = 0.387\,\text{m}^3/\text{d/m}$, $Flow_{CD} = 0.179\,\text{m}^3/\text{d/m}$
Thus $Flow_{BC} - Flow_{AB} = 0.118\,\text{m}^3/\text{d/m}$ and $Flow_{CD} - Flow_{BC} = -0.208\,\text{m}^3/\text{d/m}$

Considering the total inflow to the aquifer system between the mid-point of AB and the mid-point of BC, this must equal the differences $Flow_{BC} - Flow_{AB}$ in *both* the lower *and* upper strata:

Total inflow = 0.480 + 0.188 = 0.598 m³/d/m which, when distributed over a length of 1.0 km by 1.0 m, is equivalent to the recharge of 0.598 mm/d.

These components are shown in Figure 2.27b.
For the inflow between the mid-point of BC and the mid-point of CD, account must be taken of the flow from the aquifer to the river at C. The presence of the river can be identified from the decreased gradient

of the length CD. The calculation takes a different form:

River flow = $Flow_{BC} - Flow_{CD}$ in both the lower and upper strata + recharge so river flow = 0.840 + 0.208 + 0.598 = 1.646 m³/d/m;

the river flow is indicated by the bold arrow *aquifer to river flow* on the right of Figure 2.27b.
This example demonstrates that the one-dimensional groundwater head distribution of Figure 2.27a implicitly contains information about both horizontal and vertical flow components as indicated in Figure 2.27b. The interpretation of the flows, which is based on Darcy's Law, provides estimates of the flow from aquifer to river and the change in horizontal flows from one side of the river to the other. In this example, a two-layered system with different hydraulic conductivities in each layer is introduced to highlight the flow processes. The same type of analysis can be carried out for homogeneous aquifers with an arbitrary division between upper and lower layers (Rushton 1986a).

2.6.3 When vertical flows must be included explicitly

The detailed analysis of Connorton (1985), summarised in Figure 2.25, provides expressions which permit the assessment of errors in the two-dimensional formulation for confined, unconfined and leaky aquifers. These errors are also examined by Rushton and Rathod (1985) using numerical experiments. They compare detailed analyses in the x-z plane (Eq. (2.50)) with one-dimensional solutions in terms of the x co-ordinate alone using Eq. (2.8) or (2.30); both confined and unconfined examples were considered. For homogeneous isotropic aquifers ($K_z = K_x$), even with saturated thickness equal to 20 per cent of the horizontal extent of the aquifer, errors in the vertical flux are no more than 5 per cent. However, when the aquifer contains a horizontal zone of lower hydraulic conductivity, errors become significant. In one example, the thickness of the lower conductivity zone is 4 per cent of the total saturated thickness with a vertical hydraulic conductivity in this zone 1 per cent of the remainder of the aquifer. This results in discontinuities in heads and flows across the low conductivity zone, a feature which cannot be represented when working in terms of a groundwater head which is only a function of x. Therefore, analyses using Eq. (2.54b) are unreliable if there are low conductivity zones with vertical hydraulic conductivities 1 per cent or less than the remainder of the aquifer.

In Part III of this book, where practical examples of regional groundwater flow are considered, Chapters 9 and 11 use Eq. (2.54b) with the groundwater heads as functions of x, y and t. However, Chapter 10 is concerned with regional groundwater problems in which the aquifer system consists of multiple layers. Consequently the groundwater heads are functions of x, y, z and t.

2.7 NUMERICAL ANALYSIS

Numerical analysis is a powerful tool when solving groundwater problems; detailed information about numerical techniques can be found in Smith (1985), Anderson and Woessner (1992) and Kresic (1997). The purpose of this section is to introduce numerical techniques and demonstrate how numerical solutions can be obtained for some of the problems considered earlier in this chapter. This discussion will be based on the finite difference method, since it provides insights into the principles on which MODFLOW is based. Reference to numerical solutions of practical problems can be found throughout the book. Alternatively the finite element approximation can be used; information about the finite element technique including comparisons with finite difference solutions is presented by Pinder (2002).

2.7.1 One-dimensional flow in *x*-direction

The finite difference equations for one-dimensional flow with recharge are derived using two approaches. First, the equations are deduced using a *lumping* argument. Figure 2.28 is a vertical cross-section of an aquifer; it is divided into a mesh with a spacing of Δx (an unequal spacing can be included without diffi-

culty). The unknown heads at three nodes are h_{-1}, h_0 and h_1. The product of the saturated thickness and hydraulic conductivity is defined at the mid-point between pairs of nodes. Between nodes −1 and 0 the product is written as $(mK)_{-1,0}$, between nodes 0 and 1 the appropriate value is $(mK)_{0,1}$.

From Darcy's Law the flow from node −1 to node 0 can be written as

$$Q_{-1,0} = -(mK)_{-1,0}\frac{h_0 - h_{-1}}{\Delta x} \qquad (2.55a)$$

also

$$Q_{1,0} = -(mK)_{1,0}\frac{h_1 - h_0}{\Delta x} \qquad (2.55b)$$

From continuity, with the recharge intensity at node 0 written as q_0,

$$Q_{-1,0} + q_0\Delta x = Q_{1,0} \qquad (2.56)$$

Substituting Eqs (2.55) in Eq. (2.56) and rearranging

$$\frac{(mK)_{-1,0}}{\Delta x}\left[\frac{h_{-1} - h_0}{\Delta x}\right] + \frac{(mK)_{1,0}}{\Delta x}\left[\frac{h_1 - h_0}{\Delta x}\right] = -q_0 \quad (2.57)$$

An *alternative derivation* starts from the differential equation

$$\frac{d}{dx}\left(mK_x\frac{dh}{dx}\right) = -q(x) \qquad (2.8)$$

The curve of Figure 2.29 represents the variation of groundwater head with the x co-ordinate for part of the aquifer. Three nodal points, a distance Δx apart, are defined as −1, 0 and 1; in addition the mid-points between these nodes are defined as P and Q. Using the simplest finite difference expression, the gradients of the groundwater heads at locations P and Q can be written as

$$\left(\frac{dh}{dx}\right)_P = \frac{h_0 - h_{-1}}{\Delta x} \qquad \left(\frac{dh}{dx}\right)_Q = \frac{h_1 - h_0}{\Delta x} \qquad (2.58)$$

In addition, using the same finite difference approximation, Eq. (2.8) becomes

$$\frac{d}{dx}\left(mK_x\frac{dh}{dx}\right) = \frac{(mK)_{0,1}(dh/dx)_Q - (mK)_{-1,0}(dh/dx)_P}{\Delta x}$$
$$= -q_0 \qquad (2.59)$$

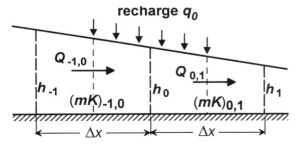

Figure 2.28 Lumping approach for finite difference equation in one-dimensional steady state flow with recharge

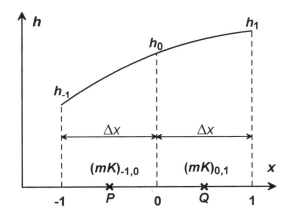

Figure 2.29 Derivation of finite difference equations. Note that the products of the saturated thickness and horizontal hydraulic conductivity are defined at the mid-points between nodes

Note that (mK) is defined at the mid-points between the nodes.

Combining Eq. (2.58) and (2.59) leads to Eq. (2.57); this confirms that the finite difference approach leads to the same algebraic equations as the lumping argument. A third way of deriving the same equation using Taylor's series (Rushton and Redshaw 1979) allows estimates to be made of the mathematical error (truncation error) in the finite difference equation.

This, the simplest finite difference equation, is node centred with the recharge (and if time-variant the storage) identified at the node but the saturated thickness and hydraulic conductivity are defined between nodes. Certain codes use this *node-centred* formulation while others such as MODFLOW use a *block-centred* approach but, for the numerical computation, the saturated thickness and hydraulic conductivity are transferred to inter-nodal values.

2.7.2 Example

The example of Figure 2.30 is introduced to illustrate the versatility of the finite difference approach. Details of the problem are shown in Figure 2.30a, the aquifer is 4.0 km in length, with no inflow or outflow on the left-hand boundary but with a large deep lake having a water surface elevation H to the right. The thickness decreases linearly from 20.0 m adjacent to the lake to 10.0 m at the no-flow boundary while the recharge increases linearly from 0.5 mm/d at the lake to 1.0 mm/d at the no-flow boundary. The horizontal hydraulic conductivity is 30.0 m/d everywhere.

Equation (2.57) is rewritten as

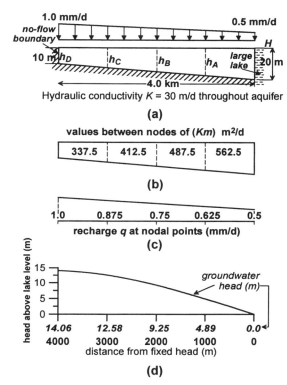

Figure 2.30 Example of finite difference solution of one-dimensional flow with varying saturated depth and varying recharge: (a) description of the problem, (b) values of (Km) between nodal points, (c) values of recharge at nodal points, (d) calculated groundwater heads

$$(mK)_{-1,0}h_{-1} - [(mK)_{-1,0} + (mK)_{0,1}]h_0 + (mK)_{0,1}h_1$$
$$= -q_0\Delta x^2$$

$$(2.60)$$

The aquifer is divided into four mesh intervals, consequently $\Delta x = 1000$ m; there are four unknown heads h_A, h_B, h_C and h_D. Since the product (Km) is required between nodes, the values are presented in Figure 2.30b, and Figure 2.30c lists values of the recharge at nodal points.

When Eq. (2.60) is written with h_0 equal in turn to h_A, h_B, h_C and h_D, the following four equations are derived:

$$562.5H - 1050h_A + 487.5h_B = -625$$
$$487.5h_A - 900h_B + 412.5h_C = -750$$
$$412.5h_B - 750h_C + 337.5h_D = -875$$
$$337.5h_C - 675h_D + 337.5h_C = -1000$$

For the fourth equation when $h_0 = h_D$ there is no node to the left of h_D. However, a no-flow boundary condition applies at D; this can be represented by introducing a fictitious node to the left of D such that the groundwater head h_{-1} is equal to h_C.

There are four equations with four unknowns. These equations lead to a tridiagonal matrix; the four unknown heads are

$$h_A = H + 4.8889 \text{ m}$$
$$h_B = H + 9.2479 \text{ m}$$
$$h_C = H + 12.5812 \text{ m}$$
$$h_D = H + 14.0627 \text{ m}$$

These heads are plotted in Figure 2.30d.

2.7.3 Radial flow

The governing equation for steady-state radial flow through an aquifer in terms of drawdown, derived from Eq. (2.35), is

$$\frac{1}{r}\frac{d}{dr}\left(mK_r r \frac{ds}{dr}\right) = q \tag{2.61}$$

It is helpful to use a logarithmic variable in the radial direction (see Figure 2.31b),

$$a = \ln(r)$$

Equation. (2.61) can be rewritten as

$$\frac{\partial}{\partial a}\left(mK_r \frac{\partial s}{\partial a}\right) = qr^2 \tag{2.62}$$

Equation (2.62) is similar to Eq. (2.8) apart from the hydraulic conductivity equalling K_r instead of K_x and q is multiplied by r^2. Consequently similar approaches can be used for the derivation of the finite difference equations, the inclusion of boundary conditions and the solution of the simultaneous equations. When there is an outer no-flow boundary in a radial flow problem, a convenient method of representing the no-flow condition is to set to zero the hydraulic conductivity for the mesh interval beyond the outer boundary. More detailed information about the finite difference method for solving radial flow problems can be found in Section 5.5.

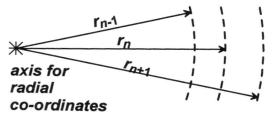

axis for radial co-ordinates

(a) radial increments

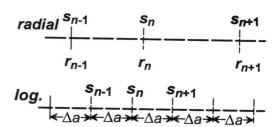

(b) radial and logarithmic increments

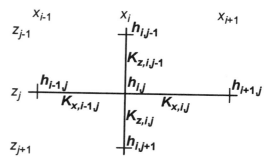

(c) vertical section

Figure 2.31   Finite difference mesh arrangements: (a) radial flow in plan, (b) logarithmic mesh spacing for radial flow, (c) vertical section in *x*-*z* plane

2.7.4 Vertical section

In the *x*-*z* plane the steady-state form of Eq. (2.50) is

$$\frac{\partial}{\partial x}\left(K_x \frac{\partial h}{\partial x}\right) + \frac{\partial}{\partial z}\left(K_z \frac{\partial h}{\partial z}\right) = 0 \tag{2.50a}$$

This equation is written in finite difference form for the two-dimensional mesh of Figure 2.31c. Note that nodes are numbered in the x and z directions by *i* and *j*, with $K_{x,i,j}$ and $K_{z,i,j}$ to the right and below node *i, j*

$$\frac{2}{x_{i+1}-x_{i-1}}\left[\frac{K_{x,i-1,j}\left(h_{i-1,j}-h_{i,j}\right)}{x_i-x_{i-1}}+\frac{K_{x,i,j}\left(h_{i+1,j}-h_{i,j}\right)}{x_{i+1}-x_i}\right]$$
$$+\frac{2}{z_{j+1}-z_{j-1}}\left[\frac{K_{z,i,j-1}\left(h_{i,j-1}-h_{i,j}\right)}{z_j-z_{j-1}}+\frac{K_{z,i,j}\left(h_{i,j+1}-h_{i,j}\right)}{z_{j+1}-z_j}\right]$$
$$=0$$

(2.63)

This approach can be adapted for flow in the *r-z* plane (see, for example, Section 6.3 of Anderson and Woessner (1992)).

2.8 MONITORING AND ADDITIONAL INDIRECT EVIDENCE

2.8.1 Introduction

The key to reliable groundwater analysis is the availability of data and additional indirect evidence on which the analysis can be based. In most regional groundwater studies, monitoring is not limited to the current aquifer conditions; information is also required for previous decades. Consequently there are two sources of information: data derived from appropriate monitoring and indirect evidence such as when wells had to be deepened, changes in agricultural practice due to the availability of groundwater, the size of culverts to take away stormwater, etc. Before any realistic groundwater studies can be carried out, those working on the project must visit the study area at times of different climatic conditions (summer and winter, dry and wet season).

Skill and ingenuity are required in collecting and analysing existing monitoring data and in identifying useful indirect evidence. This is not a straightforward process due to factors such as the large areal scale of most aquifer studies compared to significant small physical features such as springs and wells. Long timescales may also be involved; for high storage aquifers it is often decades before the impact of changes in aquifer inflows or outflows become apparent. Irregular and unpredictable changes in climatic conditions add further complexity. An understanding of groundwater flow is the purpose of most groundwater studies, yet groundwater flows within aquifers cannot be measured. Nor is the determination of the groundwater head at specific locations within an aquifer straightforward. The main properties control-

ling the groundwater flow in an aquifer system, hydraulic conductivities and storage coefficients, have to be estimated from field or laboratory experiments. The motivation for all groundwater monitoring is to gather information which will allow the development of quantified conceptual models.

References which provide information about monitoring associated with groundwater studies include Driscoll (1986), Brassington (1988, 1992) and Neilson (1991). A UNESCO report, in the series *Studies and Reports in Hydrology* (van Lannen 1998), considers groundwater monitoring in (semi-) arid regions. Monitoring of surface water flows is a further important aspect of most groundwater investigations; important references concerned with flow monitoring include Replogle and Bos (1982), Herschy (1978, 1985) and Shaw (1994).

The wide availability of high-speed computers permits the recording of information and screening for unreliable values, together with the presentation of data in various forms. However, it is crucial to ensure that an experienced worker keeps a careful check on all the data handling. An example of potentially unreliable estimates is when computer software is used to draw groundwater contours. A number of the machine-drawn contours are likely to be physically unrealistic, with water tables above the ground surface; these errors should be identified by an experienced worker. Collecting all historical data, sometimes from paper records, entering the data in a database, verifying the consistency and presenting the information should be carried out before commencing detailed groundwater studies. This is a substantial task.

The foregoing paragraphs will fill many with dismay since, for their study area, only limited historical data are available and little money is to hand for new monitoring. Yet it is surprising how much indirect information can be found. When collected and interpreted, the indirect information can form the basis of a groundwater investigation leading to the development of preliminary conceptual models. Those who use groundwater, even for domestic purposes, will have experiences which can be converted to numerical data. The response of the whole catchment to different types of rainfall events and the reliability of groundwater supplies during drought periods are typical examples of information which can be converted to data. When planning new monitoring, there are many monitoring locations which are *desirable* but few monitoring locations which are *essential*. During the conceptual model development, additional data requirements can be identified.

2.8.2 Monitoring groundwater heads

Monitoring of groundwater heads may refer to a *pumping test* or to *long-term aquifer response*. When monitoring in a pumped borehole or an observation well during a pumping test, rapid changes in water levels often occur. Frequently water levels are measured using some form of dipper which consists of a graduated tape and a device which indicates when contact is made with the water surface. Times at which readings should be taken are predetermined; the time interval between readings increases as the rate of change in water level decreases. This procedure is usually acceptable for the pumping phase but it may not be suitable for the early stages of recovery. Often recovery of water levels in the pumped and nearby observation wells is rapid; this rapid recovery occurs primarily because the well losses are no longer operative (see Section 8.2). In the first minute, up to 80 per cent of the recovery can occur. Consequently, before the pump is switched off, the dipper should be raised to, say, 70 per cent of the total pumped drawdown and the time for the water level to reach this elevation recorded. The dipper is then raised to 50 per cent of the total drawdown and the time of recovery noted. After five minutes it is usually acceptable to return to reading water levels at specified times. Changing atmospheric pressures can lead to discrepancies in measurements of drawdowns; corrections for barometric effects in several aquifer systems are described by Rasmussen and Crawford (1997). When pressure transducers with data loggers are used to record groundwater heads, time increments between readings must be small at the start of both pumping and recovery.

Much of the monitoring of water levels in observation wells or boreholes is of limited value, either because of the borehole design, or because the location of the monitoring depth is not selected to answer specific questions. Boreholes open over a considerable depth of the aquifer are supposed to measure depth-averaged responses, but they are more likely to behave as a column of non-aquifer which disturbs the flow within the aquifer system. This is apparent from Figure 5.1 which shows that, when a single open observation borehole is used to monitor a pumping test in an unconfined aquifer, the response is different to that observed when separate piezometers are used. Although differences as large as those identified in Figure 5.1 may not be common, an open borehole does *not* integrate the heads on a vertical section. Therefore, whenever possible, groundwater heads should be mon-itored using boreholes or piezometers which have an open section of no more than 2.0 m.

Six different hydrological situations can be identified which require monitoring of groundwater heads (WMO 1989).

1. Each aquifer unit requires separate observation piezometers whenever there may be differences in groundwater head or chemical composition.
2. An attempt should be made to monitor each aquifer zone that has a distinctive response.
3. Normally more observation locations are required in unconfined aquifers than for confined aquifers.
4. Observation piezometers should be located close to water-courses or other locations of aquifer–surface water interaction.
5. The spacing of observation wells should be less than the distances over which significant changes in hydrogeological conditions occur.
6. Observations should be made on either side of possible major discontinuities in the aquifer system.

Packer testing is a method of determining groundwater heads within open boreholes (Brassington and Walthall 1985, Price and Williams 1993); Driscoll (1986) describes the installation of packers in boreholes. Once they are inflated, two packers isolate a section of the borehole (see Figure 2.38a). The measured groundwater head corresponds to the section of aquifer between the packers. Price and Williams (1993) describe packer testing in a borehole penetrating 71.3 m into a sandstone aquifer, with a solid casing extending to 31.2 m. As indicated in Figure 2.32, the groundwater heads, deduced from packer testing immediately below the solid casing, were up to 0.45 m higher than the groundwater head in the open borehole whereas at depths greater than 54 m all the heads were less than the open borehole groundwater head (the phrase-averaged head is deliberately not used). These results show that there is a distinctive head gradient vertically downwards, which results from recharged groundwater moving downwards to be pumped from the aquifer several kilometres from this experimental site. For further information about packer testing see Sections 2.8.5 and 8.3.

Specially designed observation boreholes or piezometers, with a small section open to the aquifer, should be used whenever possible. Nevertheless, with the scarcity of good observation sites, any field information can be used provided that the validity of the measurements is examined carefully. Even measure-

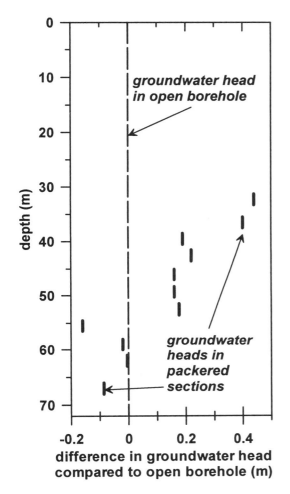

Figure 2.32 Comparison of groundwater heads in an open borehole with those measured between packers. Reproduced by permission of CIWEM from Price and Williams (1993)

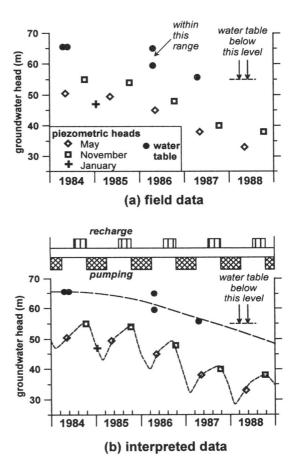

Figure 2.33 Measured groundwater heads at water table and at depth within an alluvial aquifer: (a) actual field data, (b) probable variation in groundwater head when account is taken of periods of recharge and heavy pumping

ments from pumping wells can be used provided that 'pumped' and 'non-pumped' readings are available (see Figure 9.18). Other information can be gleaned from the original drilling records.

Two practical examples of regional groundwater head monitoring are introduced below.

Example where frequency of measurement is hardly sufficient

The first field example is selected to demonstrate that the frequency of measurement is often critical. Figure

2.33a shows field data available for an alluvial aquifer in Gujarat, India. The solid circles represent measurements in shallow wells; there is ambiguity in the records in early 1986 and the only information in 1988 is that the wells previously used for monitoring are dry. Information about the piezometric heads in the deeper aquifer is also limited; there are two readings each year, the diamonds show the pre-monsoon head while the squares represent the post-monsoon reading. There is one additional reading indicated by the plus sign; this is a reading for January 1985.

When further information about this area is included, the interpretation of Figure 2.33b can be developed.

1. Total abstraction from the aquifer system has increased with time, therefore the water table and the piezometric heads should each show a steady decline; the decline is estimated to be about four metres each year.
2. The water table is more than 20 m below the ground surface of 90 m above Ordnance Datum (AOD).
3. The piezometric head for January 1985 is below the two previous and the two following readings. This arises because pumping from the deeper aquifers occurs between late October and March (see the small bar charts between Figures 2.33a and b which indicate the duration of recharge and pumping). This causes a decline in piezometric heads from late October to January, continuing until March; recovery commences as soon as pumping stops (see Section 10.2.3).

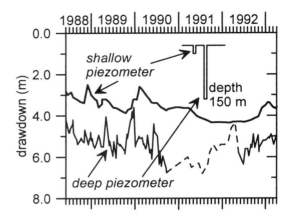

Figure 2.34 Differing responses of a shallow and a deep piezometer in a sandstone aquifer

With the above information it is possible to draw tentative lines to show the behaviour of both the water table and the deeper piezometric heads. Figure 2.33b is more informative than Figure 2.33a because tentative lines, based on additional information, have been drawn to represent annual variations.

Different responses of shallow and deep piezometers

Observation wells are often used to determine whether over-exploitation of an aquifer is taking place. In the Bromsgrove sandstone aquifer in the UK, observation boreholes, which are open to the aquifer over depths of up to 200 m, did not provide convincing evidence. Consequently, two piezometers were constructed towards the centre of the aquifer unit at New Road, 2.3 km from the nearest production borehole; the Bromsgrove aquifer is discussed in Section 10.5 and the location of the New Road observation site shown in Figure 10.22. One piezometer penetrates no more than 5 m below the water table, the second borehole penetrates to 150 m with solid casing extending to within 3.0 m of the bottom of the borehole.

Data from these two piezometers covering a five-year period are plotted in Figure 2.34. For the deep piezometer, readings at weekly intervals are available apart from late 1990 to early 1992; the period with limited readings is indicated by a broken line. There is a distinct difference between the response of the shallow and deep piezometers.

1. The shallow piezometer depth is selected to monitor changes in water table elevation; when annual fluc-

tuations arise they are less than 1 m. Maximum groundwater heads occur when there is significant recharge during the winter.
2. The water table fluctuations reflect the balance between inflows to, and outflows from the aquifer system. The years 1990–92 were drought years with recharge well below average. This leads to a decline of about 1.2 m which, for a specific yield of 0.15, is equivalent to a net loss from the aquifer system of 180 mm of water. However there are signs of recovery in late 1992 which suggests that the inflows to and outflows from the aquifer are approximately in balance.
3. The response of the deeper piezometer is more complex; rapid fluctuations of more than one metre frequently occur. The primary cause of these fluctuations is changes in abstraction; the effects of changes in abstraction spread rapidly through the deeper aquifer zones (see Section 10.5).
4. There are two occasions during the record in Figure 2.34 when the difference between the readings in shallow and deep piezometers is less than 0.3 m; this occurs when the abstractions from neighbouring production boreholes are reduced. In late 1989 the abstraction from the nearest pumping station was almost halved; consequently there was little difficulty in meeting this reduced demand leading to a rise in the deep groundwater heads.
5. The overall trend of the deeper piezometer is similar to, but 2.0–2.5 m lower than, the shallow piezometer. If data were only available from the deep observation piezometer, interpretation of the hydrographs would be more difficult.

This example demonstrates the value of having a pair of piezometers at different depths. In many situations, two piezometers at the same location but at different depths provides far more useful information than having a number of open observation boreholes distributed throughout the aquifer.

2.8.3 Surface water monitoring

Field information about groundwater–surface water interaction is as important as groundwater head data, but there are many practical difficulties in obtaining appropriate information. Often, gauging stations are not positioned at locations where rivers cross aquifer boundaries; they are more likely to be positioned immediately upstream of the confluence of rivers. Furthermore, the flow measured at a gauging station is the sum of many different contributions; identifying the groundwater components is not straightforward. The information needed for a groundwater investigation is the variation in time and the distribution along a river of the transfer of water between an aquifer and river. Frequently groundwater leaves the aquifer from a number of small springs, hence measurements are required of flows in headwater streams. Along streams and rivers there is usually a complex interaction with groundwater. Locations of gaining and losing sections may change due to different patterns of rainfall and groundwater abstraction. Further down the catchment, interaction may occur between the aquifer and the river. Changes in flow in a reach due to the groundwater component may be difficult to identify if they are smaller than the measuring tolerances of the river flows. Nevertheless, it is essential to make estimates at all locations where groundwater–surface water interaction takes place. From this information, provision can be made in mathematical models for this interaction to occur.

Flow measuring techniques

A variety of flow-measuring techniques are available; for detailed information see Herschy (1985) and Shaw (1994).

1. *Velocity–area measurements*: accurate estimates of river flow can be obtained by selecting a cross-section with a straight and uniform approach. Velocity readings are taken at a number of depths on a series of vertical lines across the river; each velocity is multiplied by the fraction of the total cross-section area represented by that reading. For large, deep rivers the velocity needs to be measured at many locations on the cross-section; for smaller rivers, five verticals with three readings per vertical are often sufficient. If the depth of flow is small so that only a single depth reading is possible, the velocity meter should be set at 0.6 of the depth from the surface.

2. *Stage–discharge measurements*: for continuous records of river flow, a suitable cross section is selected (or a measuring structure constructed as discussed below) with frequent measurements of the water level in the river. There are many devices used to record, store and transmit the data. The water level (or river stage) is converted to flows using a stage–discharge relationship which is often in the form of a rating curve. The relationship between stage and discharge is determined from a range of discharge estimates at different river water levels.

3. *Measuring structures (flumes and weirs)*: if a measuring structure such as a flume or weir is constructed in the river bed, the stage–discharge relationship is based either on mathematical analysis or extensive model testing. There are many designs of flumes. A significant advantage of a flume is that the overall head loss across the flume is small so that flumes are suitable for rivers (or canals) with shallow hydraulic gradients. Sharp-crested or thin-plate weirs provide accurate flow estimates but atmospheric pressure must occur under the nappe of the flow. This requires a significant difference in the upstream and downstream water surfaces.

4. *Dilution gauging*: a chemical solution or tracer is added to the stream or river either at a constant rate or as a known volume. The downstream concentration, when complete mixing has occurred, is measured; from direct calculations (Shaw 1994) the total discharge can be estimated.

5. *Ultrasonic method*: this is an advanced form of the velocity–area method. Acoustic pulses are projected through the water from transmitters on one side of the river to sensors on the other side; the time taken for the transmission of the pulses can be used to estimate the flow.

6. *Other methods*: surface velocities estimated by timing the movement of a float, multiplied by cross-sectional areas with corrections for the slower velocities at the bed and sides of the channel, can give a first approximation of the flows. Recommended correction factors are 0.7 for a river with a depth of 1 m and 0.8 for a river 6 m or more in depth. The

size of culverts and bridges can provide an estimate of high flows; the lack of flood damage following a normal monsoon season in Bangladesh provided evidence of limited runoff and hence significant recharge to an aquifer system.

7. *Signatures*: visual inspections of intermittent rivers can provide valuable information; visual observations are made of whether or not the river is flowing. For example, Figure 2.37 (which is considered in detail later in this section) shows how the observations for the end of September 1994 are presented as filled diamonds for locations with flow and open diamonds for locations with no flow. When this information is combined with observations for other months, a picture of the intermittent nature of the river is developed. Therefore, zero flows in a particular section of a river or at a spring are important data.

A helpful case study by Bhutta *et al.* (1997) describes the use of various flow-measuring methods to estimate all the inflows and outflows to a canal in Pakistan. These flow measurements were an essential part of the estimation of losses from the canal to groundwater. A contrasting example of unreliable results arose when a gauging structure was constructed at a location where a river, which had traversed limestone outcrop, started to flow across unconsolidated deposits including clay and permeable river gravels. Even though zero flow was noted at the gauging structure, there were significant flows through the gravels. Therefore the true river flow was not zero.

When investigating groundwater contributions to river flows, flow naturalisation should be carried out to correct for inflows to the river from water reclamation works. Corrections are also required for outflows such as withdrawals for water supply or irrigation. The river flow has many components; for groundwater studies it is necessary to identify the groundwater component, which is often called the baseflow. An important study by Nutbrown and Downing (1976) considers the structure of baseflow recession curves, demonstrating that the recession is not a single exponential relationship but the superposition of many distinct exponential terms which arise from the dynamics of groundwater flow. Various techniques are available for separating the baseflow from the total hydrograph; for example in the UK there is a five-day turning point method (IOH 1992). However, in other climatic conditions and with different aquifers and topography, different approaches should be used. Because of the difficulty involved in baseflow separation, there are advantages in consider-

ing the total catchment response rather than concentrating only on the groundwater components.

Example of a non-perennial river

Imaginative presentation of results can help to focus on the important hydrogeological issues. This is demonstrated by a case study concerning a groundwater-fed river in a chalk catchment in south-Western England (information is presented courtesy of the Environment Agency). Three diagrams have been prepared to summarise the field responses. Figure 2.35 refers to conditions in January at the start of the recharge season, Figure 2.36 presents results for February and September while Figure 2.37 displays the river's 'signatures' from 1992 to 1996. Distance are quoted above a gauging station which is located just above the confluence with a larger river; note that Figure 2.37 only

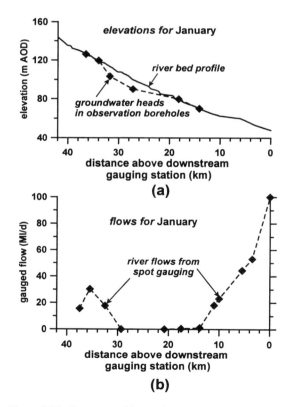

Figure 2.35 Response of intermittent river during January: (a) groundwater heads along the length of the river relative to river bed profile, (b) gauged river flows

includes information 10 km or more above the gauging station.

Three alternative situations can be identified:

1. the river flows in the upper reaches but is dry for a reach extending for about 16 km above the perennial head, or
2. the river flows only from the perennial head which is about 12 km above the gauging station, or
3. the river flows over its whole length.

Condition (1) is represented by Figure 2.35. The elevation of the riverbed can be identified in Figure 2.35a; there is a fall of 100 m in just over 40 km. The figure also shows the groundwater heads in six observation boreholes positioned close to the river; the field values of groundwater heads are joined by a broken line. Figure 2.35b shows the flows in the river determined by current meter gauging. Flows occur in the top 10 km of the river, there is then a river reach of almost 16 km where the flows are small or zero. From about 14 km above the gauging station, river flows increase significantly. This river is groundwater fed, consequently river flows increase when the groundwater head is at or slightly above river bed level and decrease when the groundwater head is below river bed level.

Conditions (2) and (3), field readings for February and September, are included in Figure 2.36. Groundwater heads in February, indicated by **x**, are at or slightly above the riverbed level. Consequently the flows increase downstream. There is one location where there is an apparent reduction in river flow, but this could be due to the difficulty in using current meters to measure flows of 300–400 Ml/d (refer to right-hand scale on left axis in Figure 2.36b). River flows are very different in September (use the left-hand scale). From the upper reaches to 14 km above the gauging station, groundwater heads (indicated by the symbol ■) are below riverbed level; consequently there is no flow from the aquifer to the river. From 14 km to the gauging station, river flows increase steadily.

A fuller picture of the changing conditions can be obtained from the river 'signatures' in Figure 2.37. These results are derived from visual observation of whether the river is flowing or not; surveys are carried out each month (Environment Agency 2002, Figure 4.3.2). The vertical scale is the distance from the downstream gauging station, the horizontal scale represents time. Locations where the river is flowing are indicated by shading; if there is no flow the diagram remains white. In the note to the figure, four particular times

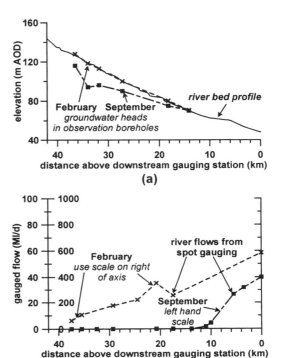

Figure 2.36 Response of intermittent river during February and September; note the different vertical scales for gauged flows in diagram (b)

are considered but a detailed examination indicates how the river starts flowing in various sections and subsequently stops flowing.

Why does the river behave in this way? There are a number of reasons for the cessation of flow, and all play a part in the observed response.

1. The gradient of the riverbed is not constant; at about 13 km above the gauging station where the slope becomes shallower, perennial conditions occur.
2. Along the river section there are different types of chalk with different hydraulic properties.
3. Groundwater abstraction for water supply occurs in the catchment; inevitably this influences the river flows.
4. To either side of the River Bourne there are two major river systems which are 20 m or more below River Bourne elevations. Consequently, groundwater flow often occurs from the River Bourne surface water catchment to adjacent catchments.

Diagram showing where flow occurs in an intermittent river at different times of the year

1. headwaters dry, flow over short section at 34 km and flow over lowermost 13 km,
2. flowing over most of the length of the river,
3. flow in the upper sections and lower sections; a month later flow for whole river,
4. only flowing over lowest 12 km.

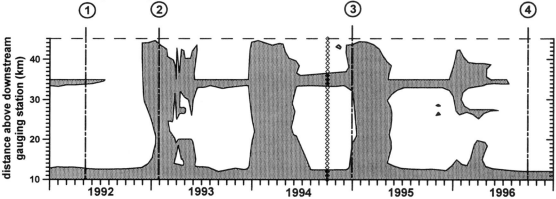

Figure 2.37 Diagram showing which sections of an intermittent river are flowing (shading indicates that river is flowing). Vertical scale represents the distance upstream from the gauging station (the lower perennial section is not included in this figure), the horizontal scale represents time

2.8.4 Monitoring borehole discharge

In this section, reference is made to monitoring flows during pumping tests, measuring the discharge from existing groundwater abstraction sites, and estimating historical abstraction rates.

For pumping tests, an integrating flow meter is frequently used but the weir tank and orifice methods (Driscoll 1986) are also suitable. Discharge rates from major abstraction sites can usually be obtained from integrating flow meters, but the readings should be carefully scrutinised since inconsistencies often occur due to faulty meters and uncertainties about the units of measurement.

When there are a large number of smaller wells or boreholes used for irrigation or domestic supply, estimating the historical and current groundwater discharge is much more difficult. A successful approach is to select a number of wells for a detailed study of usage, then scale up these results to represent the whole area. Alternatively, when the pumps use mains electricity, the consumption of electricity should be monitored and converted to pumping returns from information about the power requirement of pumps. When groundwater is utilised for irrigation, reasonable estimate of the consumptive use of the crops can be made from cropping patterns, water requirements of the crops and fre-

quency of irrigation together with any changes in cropping patterns with time (Kavalanekar *et al.* 1992). If water is used for domestic consumption, a few households can be monitored in detail and the findings scaled up according to the population. Actual measurements are essential since water usage differs between households (Russac *et al.* 1991). Estimates of industrial usage can be obtained by summing the consumption of the individual processes.

2.8.5 Monitoring groundwater quality

Monitoring of groundwater quality is essential for groundwater resource management but it is equally important as an investigation technique. Many examples of the use of hydrochemistry (or aqueous geochemistry) in the development of conceptual models can be found in Lloyd and Heathcote (1985) and Domenico and Schwartz (1997). This discussion will concentrate on methods of collecting samples which are representative of the chemical composition of groundwater at specific locations within the aquifer system.

The difference between *non-point* and *point* sampling is considered by Otto (1998). Non-point samples are obtained when pumping from open boreholes or fully screened observation wells. This form of sampling can be used to examine broad changes in chemical composition with time. However, for point samples, a piezometer must be used with a short screened length. The piezometer should be of a small diameter so that the quantity of water which needs to be withdrawn to purge the piezometer is limited. Point sampling is necessary to monitor features such as plumes from waste disposal sites.

Despite the desirability of using special purpose piezometers, most chemical sampling occurs in boreholes open or screened over most of their depth. A number of alternative techniques of sampling including bailers, syringe samplers and various types of pumps, are described and evaluated by Otto (1998). Lerner and Teutsch (1995) provide recommendations for level-determined sampling in boreholes.

Detailed field studies by Price and Williams (1993) highlight the difficulties in obtaining reliable information about the chemical composition of pore water in an aquifer. They describe a field experiment in which alternative methods were used to identify the chemical composition of the water. A borehole was constructed in the Sherwood Sandstone aquifer in north-east England to a depth of 71.7 m below ground level; the drilling method was rotary air-flush with coring. Solid casing extends to 31.2 m; the remainder of the borehole was uncased. Chemical samples were obtained using a variety of techniques including packer testing and depth sampling (Figure 2.38a and b). Sulphate concentration was used to compare the sampling methods.

In Figure 2.38c and d four sets of values of the estimated sulphate concentration are plotted.

1. Core water samples at 1 m intervals; these are plotted as a broken continuous line in Figure 2.38c and d. The results show higher values over the upper 25 m with two peak readings followed by a steady decrease to a depth of 50 m with a low constant value below 50 m. These are the true values of sulphate concentration in the aquifer.
2. Values were obtained from packer tests using double packers (see Figure 2.38a and Section 8.3). The packer testing was carried out immediately after borehole completion and geophysical logging. Results are plotted in Figure 2.38c with vertical lines indicating the extent of the packered section. Between 32 and 50 m the agreement between the pore water and packer test samples is good, but below 50 m the low pore water sulphate concentration of about 30 mg/l is not identified. The groundwater heads measured in the packered sections

(Figure 2.32) show that there is a downward groundwater head gradient. Accordingly, as soon as the borehole was completed, water of higher sulphate concentration in the upper part of the open section of the borehole travelled down the borehole and entered the aquifer at depths between 50 and 70 m, displacing the lower-concentration water. Consequently, packer testing between 50 and 70 m was unable to identify the true low concentrations at these depths.
3. Values of concentration determined by depth sampling (see Figure 2.38b), while the borehole was pumped are plotted as triangles in Figure 2.38d; they fail to represent the actual chemical composition.
4. Depth sample values under non-pumping conditions, the circles in Figure 2.38d, are also inaccurate.

This example highlights the unreliability of chemical samples taken from open boreholes and the need to use special purpose piezometers to monitor water quality changes within an aquifer. For an example of reliable information concerning sea-water intrusion at a location in China, Wu *et al.* (1993) present readings from numerous observation piezometers at different depths and with screens less than 0.5 m in length.

2.8.6 Data handling and presentation

Data storing, processing and presentation have been revolutionised by the availability of powerful computer hardware and software. Unfortunately, there is a tendency to allow the power of the software to dominate the data handling and presentation, with the result that graphical outputs are often cluttered with too much information. To indicate how data handling and presentation can be improved, a number of suggestions are given below.

- Beware of allowing a computer routine to reject data because it is not consistent with other data according to some statistical procedure. Often it is the anomalies which lead to valuable insights. Unless there are clear reasons for rejecting data, it should be retained.
- When data are summed over a day or week or month they should be presented as bar charts (rainfall is an example). For groundwater head or river flow hydrographs (i.e. point readings in time) these readings can be joined by lines. However, if any data are missing, the gap in readings should be represented as a broken line (see Figure 2.34).

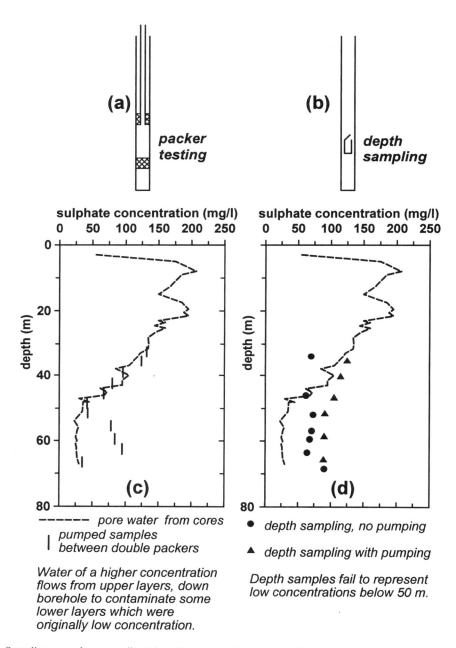

Figure 2.38   Sampling groundwater quality: (a) packer testing, (b) depth sampling, (c) sulphate concentration with depth determined from core samples and packer testing, (d) depth sampling compared with core sampling under pumping and non-pumping conditions. Reproduced by permission of CIWEM from Price and Williams (1993)

- Much confusion is caused by allowing the software to choose scales for graphs; in any report there should be no more than three scales for groundwater heads and no more than five scales for river flows.
- The number of lines or data sets included in a single diagram should be restricted to ensure that the significance of the results is clear. For example, Figures 2.35 and 2.36 could be combined as one diagram, but it would make the interpretation far more difficult.
- There are usually two reasons for presenting hydrographs; one is to show the full range of data availability, the second is to select specific results to aid interpretation. For the latter purpose the number of years plotted should be restricted. Seven years is often sufficient to identify detailed variations although, when considering long-term trends, longer time series are needed.
- The number of colours used should be limited since many people are partially colour-blind; furthermore, despite being prepared in colour, the graphs may be photocopied without colour.
- Before the wide availability of computer hardware and software, data were often drawn on tracing paper and the different sheets representing different properties were overlain to identify how the aquifer properties influenced the aquifer response. The same type of procedure can be followed using computer overlays, but the number of parameters on a single diagram must be restricted, otherwise trends may not be apparent.
- Great care must be taken when preparing groundwater head contour plots, which should not only depend on groundwater heads measured in observation boreholes. Other constraints, such as the ground surface topography, the elevation of springs and the impact of pumping, can all influence groundwater heads. A recommendation is that at least one diagram based on hand-drawn contours should be prepared before using computer-generated contours.

2.9 INTRODUCTION TO QUALITY ISSUES

2.9.1 Introduction

The focus of this book is the *quantity* of water flowing through aquifer systems; however, the issue of the *quality* of the groundwater is also very important. Historically, studies of the impact of groundwater quality started with the examination of the interaction of naturally occurring saline water and freshwater both at the coast and where saline water is present at depth. However, disturbances to the groundwater regime caused by man have led to the contamination of aquifers; typical examples include the impact of underground waste disposal or underground storage facilities. Contamination has also occurred due to agricultural practices such as the use of fertilisers and pesticides.

Quality issues are associated with many of the case studies considered in subsequent chapters; in most examples the aim has been to understand the flow processes to ensure that any spread of a contaminant is controlled. Typical examples are listed in Table 2.4.

In this section a preliminary treatment is presented of saline water–freshwater situations both at the coast and inland where upconing of saline water can occur. Monitoring the movement of saline water is also considered. For the more complex topics of hydrodynamic dispersion and contaminant transport, there is a brief introduction together with reference to sources of further information. An adequate understanding of the *quantity* of water passing through an aquifer system is a prerequisite of a study of *quality* issues.

2.9.2 Fresh and saline groundwater

This and the following two sections consider aquifers in which both fresh and saline water are present. First, an abrupt interface is assumed to occur between the fresh and saline zones; the continuity of pressure at the interface is used to develop a method of analysis. By including the Dupuit assumption of constant groundwater head on each vertical, the Ghyben–Herzberg approximation is derived. Two specific situations are examined, conditions in the vicinity of the coast where fresh water from inland flows out to the sea (Section 2.9.3) and upconing of saline water towards boreholes pumping from fresh water layers (Section 2.9.4). Difficulties in monitoring the movement of an interface are considered in Section 2.9.5. The abrupt interface approach ignores the diffusion process; reference is made to comparisons between results determined from the abrupt interface approach and those which include the effect of dispersion.

Equilibrium of pressures at the interface

In Figure 2.39, a curved line is drawn representing an interface between freshwater and saline water. The height of this interface above a specified datum is η; the freshwater has a density ρ_f and the saline water a density ρ_s. The groundwater heads within the fresh and saline water are written using Eq. (2.1) as

Table 2.4 Quality issues considered in other sections of this book

Type	Location	Section	Details
Lateral ingress of saline water	Lower Mersey	9.4.5	Boreholes near Mersey Estuary have drawn in significant quantities of saline water
	Limestone, W. India.	11.4	Saline water in coastal region cannot be displaced by artificial recharge
Ingress of deep saline water	Lower Mersey	9.4.3	Saline water at depth drawn in when boreholes drilled too deep, prevented by backfilling boreholes
	SCARPS	10.6.3	Saline and brackish water pumped out during water table control, difficulty in saline water disposal
Contamination from landfills	Southern Lincs. Lmst.	11.3.1	Contaminant crosses the supposedly impermeable Marholm–Tinwell Fault from landfill to south
Source protection studies	Notts Sandstone	9.2.6	Regional model used to trace areas of aquifer which feed pumping stations, importance of rivers
	Southern Lincs. Lmst.	11.3.8	Source protection zones change significantly between summer and winter
Thermal effects	San Luis Potosi	10.4.3	Higher temperature water from deep flows contributes to groundwater resources
	Bath Hot Springs	4.3.2	Thermal water from depth supplies springs, restricting other outflows, maintains temperatures

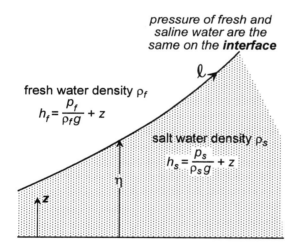

Figure 2.39 Diagram showing location of freshwater–salt water interface and definition of groundwater heads in the fresh and saline zones

$$h_f = \frac{p_f}{\rho_f g} + z \text{ and } h_s = \frac{p_s}{\rho_s g} + z \qquad (2.64)$$

At the interface $z = \eta$ the pressures of the fresh and saline water are equal, $p_f = p_s$, hence

$$\rho_f(h_{f,i} - \eta) = \rho_s(h_{s,i} - \eta) \qquad (2.65)$$

where the subscript i signifies that these are the groundwater heads on the interface.

This equation can be solved to give

$$\eta = \frac{\rho_s}{\Delta\rho} h_{s,i} - \frac{\rho_f}{\Delta\rho} h_{f,i} \text{ where } \Delta\rho = \rho_s - \rho_f \qquad (2.66)$$

The next step in the analysis is to examine the velocities in the saline and fresh zones.

Defining ℓ as a co-ordinate along the interface

$$\frac{\partial\eta}{\partial\ell} = \frac{\rho_s}{\Delta\rho} \frac{\partial h_{s,i}}{\partial\ell} - \frac{\rho_f}{\Delta\rho} \frac{\partial h_{f,i}}{\partial\ell} \qquad (2.67)$$

From Darcy's Law,

$$\frac{\partial h}{\partial\ell} = -\frac{v}{K}$$

therefore with the appropriate subscripts Eq. (2.67) becomes

$$\frac{\partial\eta}{\partial\ell} = \frac{\rho_f}{\Delta\rho} \frac{v_{f,i}}{K_f} - \frac{\rho_s}{\Delta\rho} \frac{v_{s,i}}{K_s} \qquad (2.68)$$

From Eq. (2.68), if the saline water is static, i.e. $v_{s,i} = 0$, there can still be movement in the freshwater zone with

$$\frac{\partial \eta}{\partial \ell} = \frac{\rho_f}{\Delta\rho}\frac{v_{f,i}}{K_f} \qquad (2.69)$$

This equation shows that the elevation of the saline interface must increase in the direction of freshwater flow.

Ghyben–Herzberg approach

In the light of Eqs (2.67) and (2.69), the incremental change in interface elevation for static saline water can be written as

$$\Delta\eta = -\frac{\rho_f}{\Delta\rho}\Delta h_{f,i} \qquad (2.70)$$

Using the Dupuit approach (Section 2.3.7), in which the groundwater head on a vertical line is assumed to be constant,

$$\Delta\eta = -\frac{\rho_f}{\Delta\rho}\Delta h \qquad (2.71)$$

For a water table aquifer, h also equals the water table elevation. If the datum is selected as the sea-water surface (sea level), and $\bar{z}$ is defined as the distance vertically *downward* to the interface (see Figure 2.40), then

$$\bar{z} = \frac{\rho_f}{\Delta\rho}h \qquad (2.72)$$

This is the Ghyben–Herzberg approach, which is based on four features; the balance of pressures at the interface, the assumption of zero flow in the saline water body, constant hydraulic conductivity throughout the aquifer, and the Dupuit assumptions. This approach is suitable for obtaining a first approximation to the location of the saline interface. Because the ratio of densities of sea-water to freshwater is approximately 1.025 : 1.000, $\rho_f/\Delta\rho = 40$. Consequently, the depth to the interface below sea level is forty times the height of the water table above sea level.

2.9.3 Conditions at the coast

Analytical solutions based on the Ghyben–Herzberg approximation

Figure 2.40 shows a conceptual model which can be used to consider the formation of a saline wedge beneath freshwater which flows to the coast. Assuming

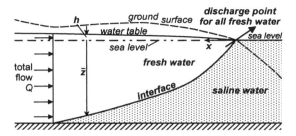

Figure 2.40 Conditions at the coast using the Ghyben–Herzberg approximation

that there is a uniform flow to the coast of Q per unit width of aquifer, this flow equals the hydraulic conductivity multiplied by the saturated depth of freshwater and the head gradient (there is no minus sign since in this formulation x is measured to the left in the opposite direction to the flow),

$$Q = K(\bar{z} + h)\frac{\partial h}{\partial x} \qquad (2.73)$$

Substituting from Eq. (2.72), integrating and eliminating the constant of integration using the boundary condition that $h = 0$ at $x = 0$,

$$h = \left[\frac{2Q}{K}\frac{\Delta\rho}{\rho_f + \Delta\rho}x\right]^{1/2} \approx \left[\frac{2Q\Delta\rho}{K\rho}x\right]^{1/2} \qquad (2.74a)$$

$$\text{and } \bar{z} \approx \left[\frac{2\rho Q}{\Delta\rho K}x\right]^{1/2} \qquad (2.74b)$$

Examination of Figure 2.40 identifies one unsatisfactory feature of this analysis, namely that freshwater leaves the aquifer at the point where the interface and the water table intersect. Clearly this is incorrect, since all the freshwater flow cannot exit at a point.

In an alternative approximate solution, Glover (1959) presents an expression for the interface which makes allowance for horizontal and vertical components of flow,

$$\bar{z} = \left[\frac{2\rho Q}{\Delta\rho K}x + \left(\frac{\rho Q}{\Delta\rho K}\right)^2\right]^{1/2} \qquad (2.75)$$

Glover's expression leads to an interface slightly lower than the Ghyben–Herzberg relationship in the vicinity of the coast; however, his analysis does not satisfy the boundary conditions of the two-dimensional formulation introduced in the following sub-section. Limitations of the Ghyben–Herzberg expression are discussed by Bear and Dagan (1964).

Inouchi *et al.* (1985) developed a time-variant analysis of a coastal aquifer in plan (*x*, *y*) allowing for the unsteady motion of the fresh–saline interface. A finite element program is used to predict the thickness and location of the interface for a specific problem. Reasonable agreement between field and modelled results is obtained. An analytical solution based on the Dupuit approximation is described by Isaacs and Hunt (1986); expressions are quoted which can be used to examine the impact of changes in the lateral freshwater flow towards the coast on interface elevations.

Two-dimensional formulation

A two-dimensional steady-state formulation of a coastal saline interface, Figure 2.41, will be discussed with reference to a specific problem. There are important differences from the idealised situation of Figure 2.40 since the water table and the interface do not intersect at the coastline. In addition, there is a shallow cliff exposed above sea level, and a seepage face EF, forms on this cliff. This means that the interface intersects the submerged ground surface at D, a considerable distance offshore. A further feature of this example is recharge to the water table in addition to a lateral specified inflow to the aquifer.

The mathematical description is based on Figure 2.41. It is assumed that the saline water is static, therefore it is only necessary to solve for flow in the fresh-water zone. The equation describing flow in the freshwater zone is Eq. (2.50) but with the storage term set to zero,

$$\frac{\partial}{\partial x}\left(K_x \frac{\partial h}{\partial x} \right) + \frac{\partial}{\partial z}\left(K_z \frac{\partial h}{\partial z} \right) = 0 \tag{2.50a}$$

A careful examination of Figure 2.41 shows that there are six boundary conditions:

1. On **AB** there is a known inflow (which may be zero), represented as a specified value of $\partial h/\partial x$.
2. The lower boundary **BC** is an impermeable boundary, hence $\partial h/\partial z = 0$.
3. Boundary **CD** is the saline interface. The locations of C and D are unknown but C does lie on the horizontal impermeable base and D is located on the ground surface beneath the sea. Two conditions apply on this boundary, the first, derived from Eqs (2.64)–(2.66), is that

$$h = z + \frac{\rho_s}{\rho_f}(z_s - z) \tag{2.76}$$

   where z_s is the elevation of sea level above datum; also, there is no flow from the saline zone into the fresh zone, hence $\partial h/\partial n = 0$ where n is the direction normal to the interface.

4. The boundary **DE** is similar to a seepage face but account must be taken of the contact with sea-water hence

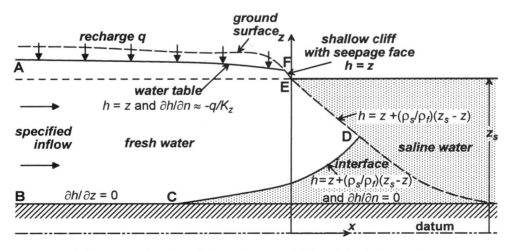

Figure 2.41   More realistic representation of conditions at the coast with formulation in the *x-z* plane

$$h = z + \frac{\rho_s}{\rho_f}(z_s - z)$$

5. Boundary **EF** is a seepage face open to the atmosphere; the condition is derived in Section 2.5.1 as $h = z$.
6. For boundary **FA**, the water table, the position is unknown; hence there are two boundary conditions,

$$h = z \text{ and } \partial h / \partial n = q / K_z \tag{2.77}$$

where n is direction normal to the boundary; the second condition assumes that the slope of the water table is shallow (see Eq. (4.16a)).

Specific problems can be solved using a numerical approach; iterative techniques are required to determine the location of the interface and the water table.

For the typical field situation illustrated in Figure 2.41, due to the shallow slope of the ground surface beneath the sea and the presence of a small cliff with a seepage face above the beach, the freshwater can extend almost 200 m from the coast. This result was confirmed in a field situation by drilling observation boreholes beyond the coastline.

Effect of ignoring dispersion

An important question is whether it is acceptable to ignore dispersion and use the sharp interface approximation. Volker and Rushton (1982) compare steady-state solutions obtained considering both a miscible fluid model, which includes dispersion and density differences, with the two-dimensional sharp interface solution similar to that of Figure 2.41. For higher dispersion coefficients, circulation occurs within the saline region hence the assumption of stationary conditions in the saline zone is not acceptable. However, for low dispersion coefficients the sharp interface is close to the contour for a concentration equal to 90 per cent of the saline water concentration. For a given freshwater discharge, the agreement between the two solutions improves with a decrease in dispersion coefficients. For a given dispersion coefficient the interface becomes more diffuse with decreasing freshwater discharges.

Time-variant numerical model solutions for the sharp interface problem require careful formulation and program preparation (see, for example, Chapter 13 of Bear and Verruijt 1987). When miscible fluid numerical models with dispersion and density differences are used to investigate time-variant behaviour, the results depend critically on the assumed flux and concentra-tion boundary conditions. An example of a three-dimensional density dependent model is described by Huyakorn *et al.* (1987).

2.9.4 Upconing

Freshwater layers or lenses above saline water bodies are frequently exploited, but there is a risk of drawing saline water into the pumped boreholes. If a partially penetrating pumped borehole is positioned in a fresh groundwater layer, the underlying saline water tends to rise in the form of a cone due to the upward potential gradient caused by pumping (Figure 2.42).

Solution using the Ghyben–Herzberg approach

The Ghyben–Herzberg approach suggests that the rise in the interface will be $\rho_f / \Delta \rho$ times the drawdown of the water table or piezometric surface. Despite the significant vertical flows in the vicinity of the borehole, which are not consistent with the Ghyben–Herzberg assumptions, an analysis based on the Ghyben–Herzberg approach does provide a first approximation to the upconing process. Laboratory and field investigations supported by theoretical studies suggest that, if a critical elevation of the interface is exceeded, the interface

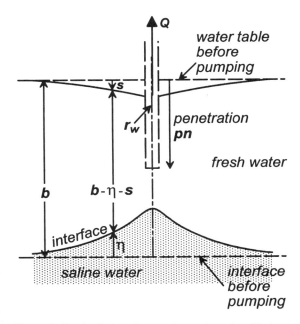

Figure 2.42 Analysis of upconing using the Ghyben–Herzberg approach

is not stable and saline water enters the borehole. McWhorter and Sunada (1977) and Motz (1992) suggest that the interface appears to be stable provided it does not rise more than one third of the distance from the undisturbed interface to the bottom of the borehole. This assumption is used in the following analysis.

The steady discharge from a borehole, which causes upconing of an interface, is proportional to the saturated depth between the water table and the interface (see Figure 2.42) multiplied by the hydraulic gradient (McWhorter and Sunada 1977). Based on the Dupuit assumption,

$$Q = 2\pi r K(b - s - \eta)\frac{ds}{dr} \tag{2.78}$$

Integration of Eq. (2.78), using the relationships of Eq. (2.72) and introducing the conditions that in the borehole of radius r_w, the drawdown is s_w, while there is no drawdown at a large radius R, leads to the equation

$$Q = \frac{\pi K}{\ln(R/r_w)}\left(2bs_w - (1 + \rho/\Delta\rho)s_w^2\right) \tag{2.79}$$

Using the experimental finding that for a well penetration of pn, the critical upconing is $(b - pn)/3$ or

$$s_w = (\Delta\rho/\rho_f)(b - pn)/3 \tag{2.80}$$

the maximum safe discharge is

$$Q_m = \frac{\pi K b^2}{\ln(R/r_w)}$$
$$\times \left\{\frac{\Delta\rho}{\rho_f}\left(\frac{1 - pn/b}{3}\right)\left[2 - \left(1 + \frac{\rho}{\Delta\rho}\right)\frac{\Delta\rho}{\rho}\left(\frac{1 - pn/b}{3}\right)\right]\right\} \tag{2.81}$$

Upconing in an aquifer overlain by a leaky confining bed is considered by Motz (1992).

Solution in the radial-vertical plane

Due to the significant vertical velocities in the vicinity of a pumped borehole, analyses should include both horizontal and vertical components of flow. At higher pumping rates, significant quantities of saline water are drawn into the well. Using resistance network analogues, Bennett *et al.* (1968) analysed a number of possible upconing situations. These results are replotted by Reilly *et al.* (1987) for a dimensionless maximum permissible discharge function.

Informative comparisons of the sharp interface and density-dependent solute transport approaches as applied to upconing problems are reported by Reilly and Goodman (1986) and Reilly *et al.* (1987). Comparisons are made between analyses based on the two approaches for a number of different scenarios; in many cases the sharp interface location is close to the 50 per cent isochlor. However, when the discharge rates are close to the maximum for a stable interface, the interface locations differ under the pumped borehole. Nevertheless, due to dispersion there is always some saline water entering the borehole when the density-dependent solute transport approach is used. The upconing of salt water at a field site in Massachusetts is simulated to gain practical insights into methods of analysis.

Recommendations are presented for a practical method of design based primarily on the sharp interface approach.

2.9.5 Monitoring the movement of a saline interface

Unreliable information from open boreholes

Difficulties in identifying the true location of a saline interface in the field are demonstrated by a case study in the Lower Mersey Sherwood Sandstone aquifer in north-west England (Rushton 1980). Analyses of the quality of pumped water indicated a small increase in the salinity. Therefore two exploratory boreholes were drilled, borehole I at 40 m and borehole II at 300 m from the pumped borehole (Figure 2.43a). Cores were taken during drilling, and the pore water was tested for salinity. When borehole I reached a depth of 232 m below ground level (BGL) an increase in the conductivity of the drilling fluid was measured; subsequently pore water analysis showed a change in chloride ion concentration from 150 to 12000 mg/l over a distance of 1.0 m. Drilling continued to a total depth of 320 m. For borehole II, at a distance of 300 m from production borehole, a sharp saline interface was identified at 260 m BGL. From this information, the location of the saline water was sketched as shown in Figure 2.43a.

After cleaning out the boreholes, conductivity measurements were made in the open boreholes; the findings were unexpected. In borehole I a saline interface, AA in Figure 2.43a, was identified at a depth of 74 m BGL, a rise of almost 160 m. For borehole II the rise in the interface to BB at 230 m BGL was less dramatic. If the saline interface in the aquifer had risen as high as indicated in borehole I, the salinity of the pumped

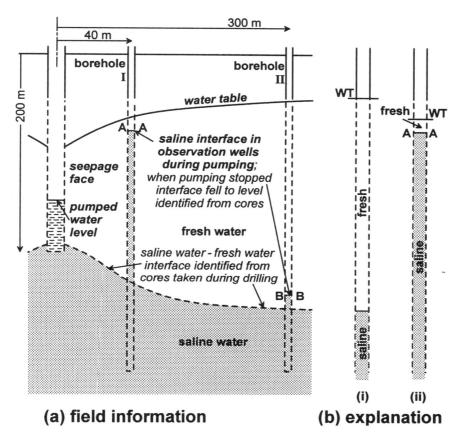

Figure 2.43 Identification of saline water–freshwater interface from cores, also misleading information gained from location of interface in observation boreholes. Reprinted from *Journal of Hydrology* **48**, Rushton, Differing positions of saline interfaces is aquifers and observation bokeholes, 185–9, copyright (1980) with permission of Elsevier Science

water would be very high, but the quality from the pumped borehole showed no change.

Later the borehole pump was stopped, and the position of the interface in the boreholes was monitored. After several hours the interface elevations in boreholes I and II fell to approximately the locations identified from the core samples. Why were spurious results for the interface location obtained in boreholes I and II? The reason is that the open boreholes, which penetrate through the aquifer, act as columns of liquid which respond according to the laws of hydrostatics. Figure 2.43b illustrates two columns, each containing both saline and freshwater; the hydrostatic pressure at the bottom of each column is the same. For column (i) the water table elevation corresponds to that some distance

from the pumped borehole; the vertical extent of the saline water is small compared to the freshwater. Column (ii) represents borehole I with the water table, WT, lower than in column (i). Since the hydrostatic pressure at the base of each column is the same but the water table elevation is lower than in column (i), the vertical extent of the saline water must be much larger with a higher interface, AA.

Open observation boreholes intended to monitor the elevation of saline interfaces in aquifers often provide misleading information. For accurate monitoring of the movement of a saline interface, a number of independent piezometers with short open sections must be provided within the region through which the interface might move.

*Effect zone of zones of lower hydraulic
conductivity on upconing*

The Lower Mersey study area provides further information about the risk of drawing saline water into boreholes which are located in freshwater zones but underlain by saline water. Over much of the study area there is saline water at depth (see Figure 9.14). When boreholes were drilled, many penetrated to the saline zone. However, backfilling the drilled hole with about 10 m of impermeable material usually meant that little saline water was drawn into the borehole. Figure 9.17 shows that a low conductivity layer between freshwater at the bottom of the pumped borehole and the saline water can largely prevent the upward movement of saline water since the vertical velocity upwards through the low conductivity zone is small.

This finding suggests that many of the foregoing methods of analysis may not be suitable for analysing the likelihood of upconing in sandstone and alluvial aquifers since the aquifers contain layers of lower hydraulic conductivity. In the Ghyben–Herzberg approach, no account is taken of the nature of the aquifer. In numerical model studies, Chandler and McWhorter (1975) did represent anisotropic conditions and found that higher discharge rates are permissible with anisotropic aquifers. However, the presence of a zone of lower hydraulic conductivity between the borehole and the saline water leads to a conceptual model different to that for anisotropic conditions. A numerical model solution with a layer of lower hydraulic conductivity between the bottom of the borehole and the saline water (Section 9.4) indicates that the upward velocity of the saline water is reduced to one-hundredth of the velocity when there is no low conductivity layer.

2.9.6 Hydrodynamic dispersion and contaminant transport

As indicated in the introduction to this section, there is a growing body of work concerning hydrodynamic dispersion and contaminant transport. The basic principles are presented by Bear (1972), Freeze and Cherry (1979), Bear (1979), Bear and Verruijt (1987), Pinder (2002) and Zheng and Bennett (2002). These texts provide a detailed discussion of hydrodynamic dispersion theories, the parameters of dispersion, the governing equations with boundary conditions and a number of analytical solutions. Other topics considered include heat and mass transport and the movement of water injected into aquifers. Bear and Verruijt

(1987) and Zheng and Bennett (2002) describe numerical models with computer codes for many dispersion and contaminant transport problems.

An informative review published by the National Research Council in 1984 includes a survey of the location and magnitude of groundwater contamination, reviews the scientific understanding of contaminant transport and chemistry, considers some of the problems associated with waste-disposal methods and provides a number of specific examples of groundwater contamination. Although there has been considerable progress in understanding and modelling groundwater contamination since the preparation of these reports, the conclusions and recommendations of each chapter provide a realistic assessment of the challenges associated with groundwater contamination.

A book presenting a British perspective of groundwater hydrology (Downing and Wilkinson 1991) contain a series of chapters which focus on topics related to groundwater contamination including landfill disposal of wastes, urban and industrial groundwater pollution, rural and agricultural pollution. In addition there is consideration of bacterial activity in aquifers, advection, adsorption, diffusion, dispersion and the transport of colloids in fractured rocks and radionuclide migration in aquifers.

The proceedings of two conferences provide insights into the wide range of topics and techniques associated with contaminant transport. Jousma *et al.* (1989) includes papers which address the topic of groundwater contamination and the use of models in decision making while Bear and Corapcioglu (1991) have edited papers under the title *Transport Processes in Porous Media*. No summary can do justice to the very different insights gained from these papers. Collecting appropriate field data, developing conceptual and numerical models and using the field information is the general framework within which the individual contributions lie. A book by Canter *et al.* (1987) reviews the rapidly expanding body of knowledge on groundwater pollution sources and their evaluation and control.

A textbook, *Groundwater Contamination: Transport and Remediation* by Bedient *et al.* (1994) provides a broad coverage of the varied aspects of groundwater contamination. To quote from the introduction, 'Our book has been written to address the scientific and engineering aspects of subsurface contamination and remediation in groundwater.' The numerous sources and types of contamination are described, and contaminant transport mechanisms, sorption and other chemical reactions are considered together with biological issues and conditions in the unsaturated zone.

Issues related to non-aqueous-phase liquids are reviewed and groundwater remediation is introduced. Reference is made to analytical and numerical solutions. Another helpful text containing a wealth of practical information is *Contaminant Hydrogeology* by Fetter (1999). *Applied Contaminant Transport Modeling* by Zheng and Bennett (2002) provides a thorough description of the basic principles and theories of contaminant transport in groundwater. Detailed explanations are included of numerical techniques for solving transport equations, together with step-by-step guidance on the development and use of field-scale models. The final part of the book examines the simulation of flow and transport under variable water density and variable saturation. The use of the simulation–optimisation approach in remediation system design is also discussed.

Processes involved in dispersion and contaminant transport

There are several processes involved in the migration and transport of solutes in groundwater.

Advection: this process is the movement of the solute with the flowing water; it depends on the seepage velocity, which can be approximated by the Darcy velocity divided by porosity.

Diffusion and dispersion: these two processes have similar effects. Molecular diffusion occurs as solute moves from locations with higher concentrations to locations with lower concentrations, while mechanical dispersion is a mixing process due to velocity variations arising from the inhomogeneity of the aquifer material on a macroscopic scale. Due to diffusion and dispersion, sudden changes in concentration become more diffuse away from the location of the change.

Adsorption involves the transfer of contaminants from the soluble phase onto the soil/rock matrix.

Biodegradation occurs when certain organic solutes are degraded by microbes to carbon dioxide and water; other chemical reactions can occur to solutes in groundwater.

Non-aqueous-phase liquids do not mix with water; either they are of a lower density than water and float just above the water table or they are denser than water and can sink to the bottom of an aquifer.

The analysis of contaminant transport problems may be complicated due to *unsaturated flow* conditions.

Great care is required in developing conceptual models for contaminant transport problems. Although the contaminant may be contained within a relatively small area, the total flow field which can influence the movement of the contaminant must be considered. It is inadvisable to use integrating parameters such as the transmissivity; usually it is necessary to represent the detailed three-dimensional time-variant nature of flow processes. For fieldwork, piezometers with a small open section should be used. Identification of suitable parameter values including dispersion coefficients is difficult. Furthermore, there may be added complications in obtaining reliable numerical model solutions due to features such as numerical dispersion.

An example of radial flow with dispersion

A simplified radial flow example is introduced here to illustrate issues which often arise when studying solute transport. Consider a confined aquifer of constant thickness and hydraulic conductivity, Figure 2.44a. A borehole pumps at a constant rate from the aquifer. At a specified radial distance encircling the borehole, a contaminant suddenly enters the aquifer. This example could represent pumping from the centre of an island surrounded by sea-water.

- *velocity of flow in the aquifer*: there is a constant discharge from the borehole of Q; hence groundwater flows with increasing velocity from the outer boundary, a. The Darcy velocity v at a radius r can be calculated directly from the continuity equation,

$$Q = 2\pi r m v \qquad (2.82)$$

However, the seepage velocity v_s is required for the solute transport calculation, $v_s = v/N$, where N is the porosity, hence

$$v_s = Q/(2\pi r m N) \qquad (2.83)$$

- *dispersion processes with the aquifer*: the equation for radially-symmetrical horizontal dispersion in a porous medium is

$$\frac{\partial}{\partial r}\left(D\frac{\partial C}{\partial r}\right) + \frac{D}{r}\frac{\partial C}{\partial r} + v_s\frac{\partial C}{\partial r} = \frac{\partial C}{\partial t} \qquad (2.84)$$

where C is the concentration of the solute and D is the dispersion coefficient.

The initial condition is that at $t = 0$ the concentration of the contaminant is everywhere zero with the

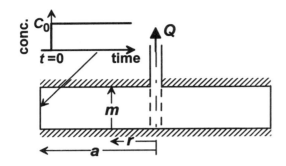

(a) details of the problem

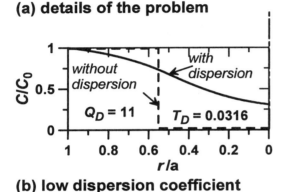

(b) low dispersion coefficient

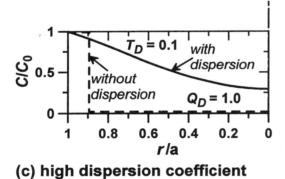

(c) high dispersion coefficient

Figure 2.44 Radial flow with dispersion, contaminant enters from outer boundary

boundary condition that at $t \geq 0$ the concentration at the outer boundary suddenly increases to C_0. It is assumed that the dispersion coefficient is independent of the velocity (see later comment). Non-dimensional parameters are introduced:

$$Q_D = Q / (2\pi DmN) \text{ and } T_D = tD / a^2$$

- *analytical solutions* can be obtained using Laplace transforms for specific values of Q_D (Al-Niami and Rushton 1978). To illustrate the significance of dispersion for flow towards a pumped borehole, two sets of results are presented in Figures 2.44b and c. In each of the plots, the concentration from the outer boundary $r/a = 1.0$ to the borehole where $r/a = 0.0$ is plotted as a continuous line. Non-dimensional times T_D are chosen so that in each example the concentration at the borehole is close to $C/C_0 = 0.3$. In addition, results when there is no dispersion are shown by a broken line; there is an abrupt change in concentration from C_0 to zero.
- *substitution of specific parameter values*: the results for Figure 2.44b refer to a low dispersion coefficient; typical parameter values are $Q = 627\,\mathrm{m^3/d}$, $a = 500\,\mathrm{m}$, $m = 70\,\mathrm{m}$, $D = 0.68\,\mathrm{m^2/d}$, $N = 0.19$; hence $Q_D = 11.0$.

At a non-dimensional time $T_D = 0.0316$, which is equivalent to 11 618 days or 31.8 years, the concentration at the borehole is close to 30 per cent of the concentration at the outer boundary. However, when dispersion is ignored, the contaminant only moves from 500 m to 276 m as indicated by the broken line.

A higher dispersion coefficient applies for Figure 2.44c, with the parameter values $Q = 627\,\mathrm{m^3/d}$, $a = 500\,\mathrm{m}$, $m = 70\,\mathrm{m}$, $D = 7.5\,\mathrm{m^2/d}$, $N = 0.19$; this is equivalent to $Q_D = 1.0$. For a non-dimensional time $T_D = 0.1$, which is equivalent to 3333 days or 9.1 years, the concentration in the borehole is again 30 per cent of that at the outer boundary; the contaminant moves more rapidly to the pumped borehole due to the higher dispersion coefficient. Without dispersion, the contaminant would only move from 500 m to 425 m in 9.1 years.

This example demonstrates the importance of considering dispersion in the direction of groundwater flow. In practice the dispersion coefficient is likely to be a function of the velocity; solutions for a problem with a velocity-dependent dispersion coefficient can be obtained using numerical techniques.

2.10 CONCLUDING REMARKS

In this chapter, three aspects are considered: the basic principles of groundwater flow together with analytical and numerical methods of solution, field monitoring of aquifer responses and an introduction to quality issues.

The presentation of the basic principles of groundwater flow is intended to encourage readers, who are not very confident with mathematical procedures to examine the derivations so that the physical significance can be appreciated. Numerical examples supplement the derivations. Some of the information on the basic principles of groundwater flow in Sections 2.2 to 2.7 is also presented in other texts on groundwater hydrology including McWhorter and Sunada (1977), Bouwer (1978), Bear (1979), Todd (1980), Kresic (1997) and Fetter (2001).

In Section 2.6 an important issue is considered in detail. The question addressed is whether *vertical* flows are included in the two-dimensional time-variant flow equation for regional groundwater flow when the groundwater head *h* is a function of the *horizontal* co-ordinates *x* and *y*. Starting with the three-dimensional time-variant equation for groundwater flow, the fluxes in the vertical direction are represented in terms of boundary conditions on the lower and upper boundaries (for example, no flow across the lower boundary with recharge and water released from unconfined storage on the upper water table boundary). This leads to the conventional two-dimensional time-variant regional groundwater flow equation. Therefore this equation implicitly represents vertical flow components. Numerical examples are introduced to show how vertical flow components can be calculated from the results of a solution in *x*, *y*, *t* co-ordinates. This finding is of great importance in planning and interpreting regional groundwater flow solutions.

The section on numerical analysis is based on the finite difference method which is the approximation used in MODFLOW. Further information about the finite difference method, the finite element method and the boundary integral equation method can be found in Anderson and Woessner (1992), Pinder and Gray (1977) and Liggett and Liu (1983).

In Section 2.8, examples are given of the monitoring of groundwater flows and heads. Monitoring of groundwater flows is often undervalued. However, information about flows between the aquifer and surface water bodies is vitally important in developing quantified conceptual models. Furthermore, adequate agreement between field and modelled groundwater flows provides convincing evidence about the validity of the numerical model. Advice is also given about the monitoring of groundwater heads, including the location of piezometers and the frequency and duration of readings. References quoted in Section 2.8 should be consulted for further information about monitoring techniques.

The discussion on quality issues is mainly concerned with the density differences arising when saline water and fresh water occur in the same aquifer. These conditions occur in several of the case studies described in the following chapters. The summary of contaminant transport issues only serves to highlight the complexities involved when contaminants enter the aquifer system; both theoretical and practical information about these issues can be found in Zheng and Bennett (2002).

The next two chapters in Part I relate to further basic issues in groundwater hydrology. Chapter 3 is concerned with the estimation of groundwater recharge. Most of the chapter relates to recharge estimation using a soil moisture balance technique; sufficient information is provided for the technique to be used directly in different climatic conditions. In Chapter 4, interaction between an aquifer and surface water bodies is discussed and information about the interaction between canals and an aquifer system is used to refine the techniques of representing river–aquifer interaction.

3

Recharge due to Precipitation or Irrigation

3.1 INTRODUCTION

This chapter is concerned primarily with the estimation of recharge due to precipitation or irrigation. Other sources of recharge include river–aquifer interaction, water losses from canals and from flooded ricefields (see Sections 4.2, 4.3.3 and 4.5). For many regional groundwater resource studies, insufficient attention is paid to the estimation of recharge. However, the case studies in this chapter demonstrate that realistic recharge estimates can be made using a soil moisture balance technique when account is taken of actual field conditions, especially vegetation growth and soil properties. The primary aim of this chapter is to provide sufficient detailed information to allow the reader to estimate recharge in their own field situations.

The first part of this chapter is concerned with the estimation of *potential rainfall recharge*; this is defined as water which can move vertically downwards from the base of the soil zone. Three representative field situations are described to highlight the practical issues involved in the estimation of potential recharge. In Section 3.2 a number of alternative methods of recharge estimation are outlined. The conceptual basis of a soil moisture balance technique for recharge estimation is introduced in Section 3.3 with practical details of the method in Section 3.4. In many field situations, the flow of water between the base of the soil zone and the aquifer water table is restricted by low-permeability material in the unsaturated zone so that the *actual recharge* to the main aquifer system may be less than the potential recharge; this is the subject of Section 3.5. A further complication in recharge estimation is that recharge may enter the aquifer system at locations other than where the rainfall occurs. This *runoff-recharge* occurs due to water flowing across less permeable regions to infiltrate into the more permeable parts of the aquifer system; see Section 3.6.

3.1.1 Representative field situations

To appreciate the physical processes which can influence the timing and magnitude of recharge, three representative field situations are introduced. Detailed calculations for each of these case studies can be found in Sections 3.4.9, 3.4.10 and 3.4.11.

Temperate climate of the UK with a crop of winter wheat

In the UK, summer and winter rainfall are generally of similar magnitudes although individual months may exhibit substantial differences. Apart from very dry years, a period of more than thirty days without rainfall is unusual. Annual rainfall, except for high ground, is in the range 500–900 mm. Average daily temperatures vary from 0°C to 20°C although there are occasionally a few days with higher temperatures. Potential evapotranspiration for cereal crops lies in the range 250–550 mm per annum.

In eastern England where winter wheat is grown on a sandy loam soil, the seed is sown before the end of September within four to six weeks of the previous harvest. The crop germinates and becomes established before the temperature of the soil falls during November inhibiting further growth; it is not until March that crop growth resumes. Until the spring the crop provides partial ground cover, consequently crop transpiration and bare soil evaporation occur simultaneously. Also, during the winter periods of rainfall, the soil reaches

Groundwater Hydrology: Conceptual and Computational Models. K.R. Rushton
© 2003 John Wiley & Sons Ltd ISBN: 0-470-85004-3

field capacity and becomes free-draining so that excess water passes through the soil zone to become recharge. From March through to July, crop growth (including root growth) and maturing occurs; moisture is taken from the soil to meet the transpiration demand of the crop. If the crop roots are unable to collect sufficient water, the crop is stressed, with the result that evapotranspiration is at a rate below the potential. After harvest, the time taken for the soil moisture deficit to reduce to zero determines the start of the recharge period. In summary, crop growth occurs in autumn, spring and summer, and most recharge occurs in winter.

Semi-arid climate of north-eastern Nigeria, rain-fed crop of millet

Although the same physical processes occur with a rain-fed crop in Nigeria, the timing of the various stages is totally different to that in the temperate climate of the UK. During the dry season, which lasts for about eight months, no crops are grown. When the rainy season commences in late June, millet seeds are sown; hopefully this is followed by further rainy days resulting in germination of the seeds. The first twenty days are critical to the establishment of the crop. Some of the rainfall is intercepted by shallow roots, but rainfall also reduces the deficit of moisture in the soil. As the roots grow, they are able to collect water from deeper within the soil. Between July and August there are typically thirty to forty days with rainfall; on several occasions the daily rainfall exceeds 30 mm. Some runoff occurs on days with heavy rainfall. When the soil moistures content reaches field capacity, excess rainfall passes through the soil to become potential recharge. Normally recharge does not occur until at least fifty days after the first rainfall. Harvest occurs towards the end of the rainy season, in late September or early October. For a year with low rainfall, free drainage conditions may not be achieved, hence no recharge occurs.

Semi-arid climate of western India, rainfed crop of millet followed by an irrigated crop of oilseed

The third case study considers inputs of water from both rainfall and irrigation. The timing of the first crop during the monsoon season is similar to the Nigerian situation. If there is insufficient water during the monsoon season, there may be limited irrigation. On the other hand, during periods of intense rainfall there may be sufficient water to bring the soil to field capacity; any excess water moves via the free-draining soil to become recharge. After the harvest of the first crop the

soil is bare for a number of weeks. When irrigation water is available, the land is ploughed and prepared for the second crop. Sowing of the second crop takes place in late October or early November; irrigation water is applied to ensure that there is sufficient water in the soil to meet the demands of the crop. Harvest occurs towards the end of February. The soil then remains bare during the period March to mid-June when the temperatures are high and the humidity is low.

3.1.2 Symbols used in this chapter

Dimensions: L = *length*, T = *time*

A_B	fractional area of bare soil
A_C	fractional area of crop
AE	actual evapotranspiration [L/T]
ES	bare soil evaporation [L/T]
ET_C	crop evapotranspiration [L/T]
ET_0	reference crop evapotranspiration [L/T]
In	infiltration [L/T]
K_C	crop coefficient
K_E	evaporation coefficient
K_S	water stress coefficient for evapotranspiration
K'_S	soil stress coefficient for evaporation
p	RAW as a fraction of TAW (or REW/TEW)
PE	potential evapotranspiration [L/T]
Pr	precipitation [L/T]
RAW	readily available water [L]
REW	readily evaporable water [L]
RO	runoff [L/T]
SMD'	soil moisture deficit at the start of the day [L]
SMD	soil moisture deficit at the end of the day [L]
TAW	total available water [L]
TEW	total evaporable water [L]
Z_E	depth of surface layer subject to drying by evaporation [L]
Z_r	rooting depth [L]
θ_{FC}	moisture content at field capacity [L³/L³]
θ_{WP}	moisture content at wilting point [L³/L³]
Ψ	total potential [L]
ψ	matric potential [L]

3.2 BRIEF REVIEW OF ALTERNATIVE METHODS OF ESTIMATING RECHARGE

This section provides a brief summary of a number of alternative methods of estimating recharge. Further information can be found in three publications of the International Association of Hydrologists. *Groundwater Recharge* (Lerner *et al.* 1990) considers a wide range of methods of estimating recharge in a variety of climates, while *Recharge of Phreatic Aquifers in (Semi-) Arid Areas* (Simmers 1997) focuses on detailed reviews of the principles and practicalities of assessing recharge from precipitation, from intermittent flow and from permanent water bodies. The theme issue on *Groundwater Recharge* in the *Hydrogeology Journal* (Scanlon and Cook 2002) provides extensive reviews of various aspects of groundwater recharge. These three volumes contain extensive lists of references which should be consulted for further information.

3.2.1 Methods based on field measurements

Of the variety of techniques available for estimating recharge, certain methods are based on field studies such as the water table fluctuation method, lysimeter and tracer experiments. In the water table fluctuation method the rise in water table during the recharge season is multiplied by the specific yield to give a direct estimate of the recharge. However, as explained by Healy and Cook (2002), the method is only valid over short time periods of a few days so that the other components of the water balance (groundwater flows to rivers, abstraction from boreholes, flow out of the catchment, etc.) are small compared to the recharge. A further issue is the difficulty in obtaining realistic estimates of the specific yield.

Lysimeters, which provide a direct means of measuring recharge, consist of enclosed blocks of soil with or without vegetation. Commonly used methods of measuring the quantity of water recharging the aquifer include the provision of drainage pipes at depth which collect the water draining from the soil zone, determining the change in weight of the whole lysimeter, or removing water so that the water table within the lysimeter remains the same as the water table in the adjacent parts of the aquifer. Lysimeters covering areas of up to $100\,m^2$ in England (Kitching *et al.* 1977) and Cyprus (Kitching *et al.* 1980) provided daily recharge estimates. However, lysimeter estimates of the recharge refer to a specific location. This is illustrated by two lysimeters only 50 m apart with recharge measurements during the same year of 159 and 114 mm (Kitching *et al.* 1977).

Workers in many countries have carried out tracer studies using a wide variety of chemicals to identify flow pathways in the unsaturated zone (Simmers 1997). Observations of the movement of a tracer with changes in the moisture content allow an estimate to be made of the recharge rate. Tracers can be divided into three groups:

1. *Historical* tracers such as pollution events, the change of farming practice or nuclear testing in the 1950s and 1960s,

2. *Environmental* tracers such as chloride, nitrate or the stable isotopes of water, and

3. *Artificial* tracers which are applied by the investigator at or below the soil surface.

Many of the field experiments, in which injected tritium is used as an artificial tracer, have been carried out in locations where there is no crop. However in a field study in Western India, Sharma and Gupta (1987) studied the movement of tritium and change in moisture content under a non-irrigated crop; they obtained eight readings over a period of twenty-six months. Their results showed a complex pattern of the movement of tracer together with significant changes in moisture content. Interpretation of the results indicated that there were periods with negative recharge.

3.2.2 Estimates using properties of unsaturated soil

The interaction in the unsaturated zone between water and air in the pores of a permeable medium is complex. In saturated conditions, the flow can be estimated from Darcy's Law in terms of the head gradient and the saturated hydraulic conductivity. For unsaturated flow, Darcy's Law still applies. However the hydraulic conductivity does not remain constant but depends on the matric potential (or negative pressure), which in turn depends on the volumetric moisture content. The interaction between these three parameters,

volumetric moisture content θ,
matric potential ψ which for unsaturated flow is always a suction (negative),
hydraulic conductivity K,

can be determined from field experiments. Typical results for loamy fine sand are included in Figure 3.1. The curves are derived from information in Simmers (1997); the matric potential is negative. The scales of the axes are important; the volumetric moisture content lies within the range is 0.05 to 0.4 but the matric potential covers the range 1.0 to 5000 cm of water while the hydraulic conductivity has a maximum saturated value of about 65 cm/d but falls to 2.5×10^{-6} cm/d.

In saturated flow the groundwater head is defined by

$$h = \frac{p}{\rho g} + z \tag{2.1}$$

but in unsaturated flow the total potential Ψ is determined from the expression

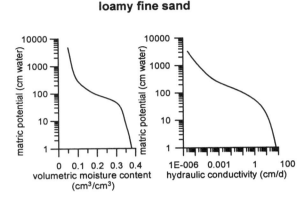

Figure 3.1 Relationships between matric potential and both volumetric water content and hydraulic conductivity for a loamy fine sand

$$\Psi = \psi + z \tag{3.1}$$

Darcy's Law for vertical unsaturated flow is written as

$$v = -K \frac{\partial \Psi}{\partial z} \tag{3.2}$$

Numerical models based on Eq. (3.2) together with continuity equations are available, but solution of the equations is not straightforward. Insights into the movement of water in the unsaturated zone have been gained from numerical solutions (see, for example, Watson 1986) but the techniques are not a suitable basis for routine recharge estimates.

An understanding of the movement of water in the unsaturated zone can be obtained from plots of field readings of the variation of total potential and moisture content with depth below ground surface. Figure 3.2 contains field information for a soil above a chalk aquifer in late May during the early stages of the main growing season (information from Wellings and Bell 1980). The variation of the total potential with depth can be interpreted using Figure 3.3 which shows four alternative distributions (figure developed from Wellings and Bell 1982). In Figure 3.3a the total potential becomes more negative (decreases) with depth which indicates a downwards vertical flow (flux). For Figure 3.3b, the flow is upwards whereas in Figure 3.3c the flux is upwards near the soil surface and downwards below the zero flux plane (ZFP). In Figure 3.3d there are two zero flux planes. Conditions are similar

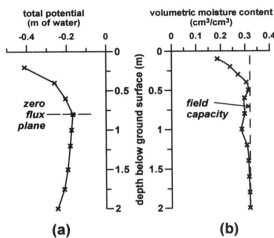

(a)   **(b)**

Figure 3.2 Variation of total potential and volumetric moisture content with depth for a soil above a chalk aquifer. Reproduced by permission of Geological Society, London from Wellings and Bell (1980)

to in Figure 3.3c but recent irrigation or heavy rainfall results in a downwards flow in the upper few centimetres of the soil zone. The total potential distribution for the soil above the chalk aquifer, Figure 3.2a, is of the same form as Figure 3.3c, indicating that there is a generally upwards flow above 0.8 m but below this depth the flow is downwards.

Field measurement of the matric potential (and hence the estimation of the total potential) requires careful long-term experimental work. Nevertheless, the identification of a zero flux plane provides detailed information about the changing directions of flows in the soil. Conversion of the gradient of the total potential to actual flows requires information about the variation of the unsaturated hydraulic conductivity with matric potential (equivalent to Figure 3.1).

The experimental and theoretical approaches described above are helpful in understanding the flux in the unsaturated zone and can lead to recharge estimates. However, none of the methods is suitable for routine estimation of recharge over long time periods. For regular prediction of recharge, a technique is required which is based on meteorological and field data available in most locations. Strictly the analysis should be based on interactive water and energy balances. Nevertheless, the estimation of recharge based on a water (soil moisture) balance technique, which incorporates insights gained from detailed studies such as those described above, provides a

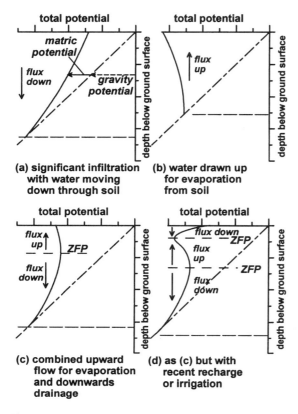

Figure 3.3 Four possible variations of total potential with depth, *ZFP* is zero flux plane. Reproduced by permission of Geological Society, London, from Wellings and Bell (1982)

practical methodology for recharge estimation in many situations.

3.3 CONCEPTUAL MODELS FOR THE SOIL MOISTURE BALANCE TECHNIQUE

3.3.1 Introduction

Any method used for routine recharge estimation must be based on information which is widely available. Insights into the form of a suitable methodology can be gained from a related problem, the estimation of crop water requirements for irrigated agriculture. FAO Irrigation and Drainage Paper 56, *Crop Evapotranspiration* (FAO 1998) provides a detailed methodology which enables the practitioner to calculate the water requirement for a specific crop. Several of the techniques, together with information about crop and soil properties in FAO (1998), have been used to develop the

recharge estimation methodologies presented in Sections 3.3 and 3.4. The approach is based on daily soil moisture balance calculations with coefficients introduced to represent both runoff and reduced evapotranspiration (or evaporation) when there is limited moisture availability. Inflows and outflows to the soil-plant-atmosphere system are illustrated in Figure 3.4 (which is developed from Hillel 1982, 1998). The essential features of the system are summarised below.

1. At the soil surface, some of the water infiltrates *In*, some runs off and some is ponded in hollows.
2. Of the water which enters the soil system, some is drawn towards the roots of the plant and is extracted by the roots. Where there are no roots present, water may be drawn to the soil surface and evaporated from the bare soil. Under certain conditions water passes vertically downwards from the bottom of the soil zone as recharge.
3. Water is transpired by the plants; this water is extracted from the soil zone by the roots.

The key to the estimation of recharge is to represent conditions in the soil zone; this can be achieved by carrying out a daily soil moisture balance. However, Figure 3.4 does not represent many of the complexities in the soil zone. For example, unsaturated conditions occur in the soil; how does this influence the quantity of water extracted by the roots and the quantity of water leaving the base of the root zone as recharge?

In this section, conceptual models for the various processes indicated in Figure 3.4 are described; the *quantification* of these processes is considered in Section 3.4. A new technique is introduced in Section 3.3.9 which takes account of the storage of moisture close to the soil surface following significant rainfall. This moisture is available to meet evaporative demands during the next and subsequent days. One further preliminary point concerns the use of the terms *evaporation*, *transpiration* and *evapotranspiration*. There is no consistency in the literature about the use of these words. In the following discussion the usual conventions will be followed, but the phrase *total evaporation* will be used when referring to the removal of water from the soil whether by evaporation, transpiration or a combination of the two.

3.3.2 Representation of moisture conditions in the soil

The purpose of this sub-section is to consider how the moisture content of a soil varies with depth, to represent the soil moisture as an equivalent depth of water

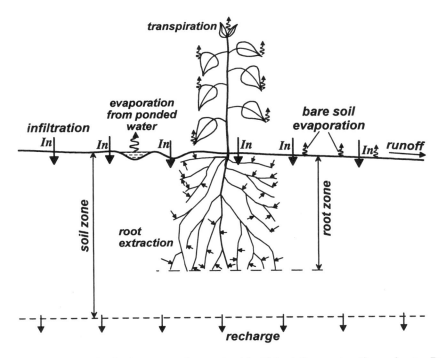

Figure 3.4   Conceptual diagram of soil–plant–atmosphere system identifying inflows and outflows of water. Reproduced from *Introduction to Soil Physics*, Hillel, copyright (1982) with permission from Elsevier Science.

and then to introduce the concept of a soil moisture deficit. Figure 3.5 illustrates the representation of soil moisture conditions; the upper diagram use a true depth scale, the lower ones use an equivalent depth scale. Diagram (i) in Figure 3.5a contains a typical distribution of moisture content with depth. The figure also shows the *field capacity* θ_{FC} which is defined as the amount of water that a well-drained soil can hold against gravitational forces, or the amount of water remaining when downward drainage has markedly decreased. Another important limit is the *permanent wilting point* θ_{WP} which is the soil moisture content below which plant roots cannot extract moisture. In diagram (i) the moisture content above wilting point (the usable moisture content for the plants) is shown shaded. This usable moisture is transferred to diagram (ii) and indicated by the same shading. The unshaded area between the wilting point and field capacity is the volume of water which must be supplied to bring the soil profile up to field capacity. In Figure 3.5b, an equivalent depth scale is introduced. The moisture content between the wilting point and field capacity is shown in terms of a depth of water. Since $(\theta_{FC} - \theta_{WP}) = 0.3$, the equivalent depth of water above 2.0 m is 0.3 ×

2.0 m = 0.6 m. Computationally, it is convenient to work in terms of a soil moisture deficit *SMD* which is the equivalent depth of water required to bring the soil up to field capacity. It is represented in diagram (ii) by the unshaded area.

Since a clear appreciation of the approach of working in terms of an equivalent depth of water is critical for successful soil moisture balance calculations, the stages involved in developing this approach are summarised below.

1. Identify the distribution of volumetric moisture content with depth; diagram (i).
2. Calculate the difference between the actual soil moisture distribution and field capacity which is plotted as shown in diagram (ii). Note that this does not infer that at a certain depth the moisture content suddenly changes between wilting point and field capacity. The equivalent soil moisture content distribution can represent a wide range of moisture content variations with depth.
3. Convert the soil moisture content above the wilting point, diagram (ii), to an equivalent *depth* of water by dividing by the difference of $(\theta_{FC} - \theta_{WP})$; this is

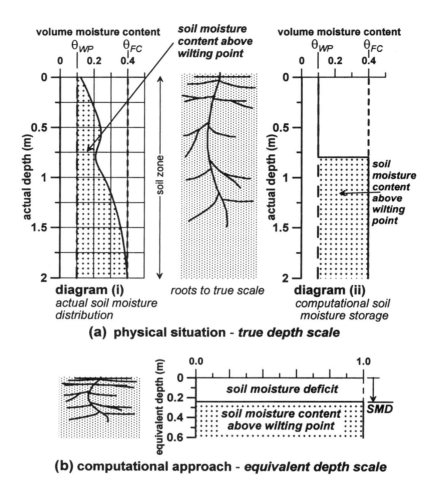

Figure 3.5   Idealisations involved in relating actual moisture content and rooting depth with soil moisture deficit and equivalent depth

shown on the right of Figure 3.5b. Define the soil moisture deficit as the equivalent depth of water required to bring the profile up to field capacity.

4. The soil moisture deficit is used to determine whether evaporation from bare soil or transpiration from the crops will occur at the potential rate or at a reduced rate. When developing expressions to determine when reduced evaporation takes place, account must be taken of the moisture-holding properties of the soil.

The physical significance of a soil moisture deficit is demonstrated by Figure 3.6, which relates to Central India. The graph contains field results for the total soil moisture to a depth of 2.5 m; the variation of total soil

moisture with depth was determined from Neutron probe measurements. Sowing of a wheat crop occurs towards the end of the monsoon season in early October; the crop is grown in a black cotton soil. As the crop grows, moisture is drawn from the soil so that the total soil moisture to a depth of 2.5 m decreases. There are some apparently erratic results; these can occur due to rainfall or perhaps inaccurate field readings. The rate of decline in soil moisture decreases as the crop ripens and harvest occurs. Monsoon rainfall occurs in June resulting in a rapid increase in the total soil moisture.

Although this figure refers to the total soil moisture to a depth of 2.5 m, it can be converted to a soil moisture deficit. In July–August the soil is assumed to reach

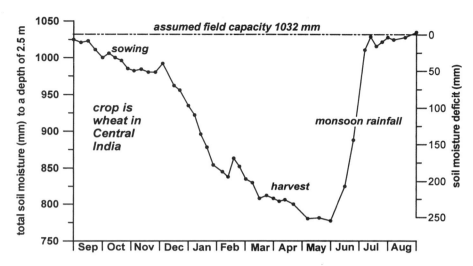

Figure 3.6 Field results from India for total soil moisture to a depth of 2.5 m; soil moisture deficit axis is drawn to the right

field capacity; this is represented by a total soil moisture content of 1032 mm. This value is defined as *zero soil moisture deficit*; a soil moisture deficit axis is drawn on the right-hand side of Figure 3.6. A maximum soil moisture deficit of about 250 mm is consistent with a rooting depth of 1.5 m and $(\theta_{FC} - \theta_{WP}) = 0.17$.

3.3.3 Bare soil evaporation

The physical processes of evaporation from bare soil depends on conditions in the atmosphere, namely a continual supply of heat (radiated or advected energy) at the soil surface, while the vapour pressure in the atmosphere over the evaporating surface must remain lower than the vapour pressure of the soil surface. The part played by the soil is that there must be a continual supply of water from or through the interior of the soil body to the site of evaporation (at what rate can the soil transfer water to the site of the evaporation?).

There are three stages in the evaporation process.

I. An initial constant rate which occurs early in the process when the soil is moist and sufficiently conductive to supply water to the site of evaporation at a rate limited by and controlled by meteorological conditions. This is the *weather controlled* stage.

II. An intermediate falling rate during which the evaporation rate falls progressively below the potential rate. The evaporation is limited by the rate at which the gradually drying soil profile

can deliver water to the evaporation zone. This is the *soil profile controlled* stage.

III. A *residual slow rate* stage when the surface zone has become so desiccated that conduction through the zone virtually ceases.

The end of the first stage and beginning of the second can occur abruptly; irregular surfaces and shrinkage cracks modify the processes.

These stages are illustrated in the four sketches of Figure 3.7. Stage I is represented by diagrams (a) and (b) in which the soil moisture deficit is zero or small (note that the figures represent equivalent deficits of soil moisture, not actual depths). (c) refers to stage II when there is sufficient soil moisture for evaporation to continue but at a reduced rate, whereas in (d) the soil moisture is so depleted that the soil is unable to transmit water to the soil surface for evaporation. These different stages of the evaporation process can be represented as shown in (e), the horizontal scale refers to the soil moisture deficit while the vertical scale is a coefficient K_S' which describes the actual evaporation as a fraction of the potential evaporation. Thus

$$\text{actual evaporation} = K_S' \times \text{potential evaporation}$$

$$(3.3)$$

The straight inclined line of Figure 3.7e, used to represent stage II, is an approximation but it is supported by field evidence (Hillel 1998). Two specific soil moisture deficits are defined. The soil moisture deficit at

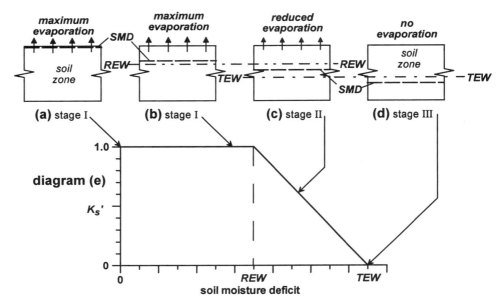

Figure 3.7   Reducing evaporation from a soil due to increasing soil moisture deficits

which evaporation ceases is defined as TEW, the total evaporable water. The corresponding value beyond which evaporation occurs at a reduced rate is REW, the readily evaporable water. The quantification of these parameters is considered in Section 3.4.

3.3.4 Crop transpiration

Crop transpiration involves more complex physical processes than bare soil evaporation. After summarising some of the physical processes, the reference crop evapotranspiration is defined. The representation of different transpiration rates for seasonal crops is considered, and the effect of limited soil water on crop transpiration is explored.

Summary of physical processes

The *soil–plant–atmosphere continuum* is concerned with the movement of water from the soil, through plants, to the leaves and out to the atmosphere (Figure 3.4). Only a small fraction (generally less than 1 per cent) of the water absorbed by plants is used for photosynthesis; most is lost as a vapour by transpiration. Roots perform a number of essential functions including absorption and conveyance of water and nutrients from the soil and acting as anchorage for the plant's

superstructure. Soil water suction increases as soil wetness decreases, consequently the plant–water suction required to extract water from the soil must increase correspondingly. Once the rate of extraction drops below the rate of transpiration (either due to the high evaporative demand from the atmosphere, low soil conductivity and/or because the root system is too sparse) the plant is under stress. Unless it adjusts its root–water suction or root density so as to increase the rate of soil–water uptake, the plant may be unable to grow normally.

Reference crop evapotranspiration

It is helpful to define a reference crop so that the behaviour of other crops can be related to this reference. Grass is selected as the reference crop since it continues to grow throughout the year. The reference crop evapotranspiration ET_0 can be estimated using the FAO version of the Penman–Monteith equation (FAO 1998).

Crop coefficients

Different crops require different amounts of water depending on the date of sowing, their rate of growth, date of harvest, etc. Figure 3.8 is a conceptual diagram

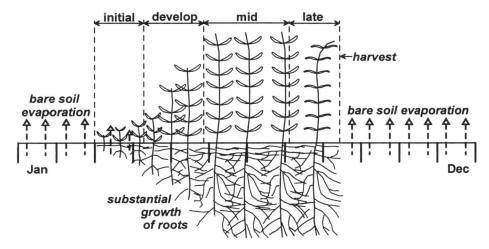

Figure 3.8 Conceptual diagram of crop and root development for a spring-sown crop in the northern hemisphere. Reproduced by permission of the Food and Agriculture Organisation of the United Nations, FAO (1998)

of a crop which is sown in March, develops during April and May, experiences main growth during June, July and August and is harvested in early September. From late September to February the ground is without crops. The figure also shows how the roots and the crop develop. This growth can be related to the reference evapotranspiration of grass. During the development stage of crop growth and the mid stage when the crop matures, the transpiration is higher than for grass. During the late stages when the crop is ripening, little water is required so that the evapotranspiration is less than for grass. The multiplying factor is the crop coefficient K_C, and the potential crop evapotranspiration is estimated as

$$ET_C = K_C ET_0 \tag{3.4}$$

The crop coefficient, which changes with the different growth stages, is an aggregate of the physical and physiological differences between crops. Quantitative information about crop coefficients is presented in Section 3.4.1.

Crops under stress due to limited water availability

It is the meteorological conditions rather than soil or plant conditions which exercise the greatest influence on the transpiration rate as long as there is sufficient moisture in the soil. However, as soil wetness diminishes, even though not completely depleted, the actual evapotranspiration begins to fall below the potential rate either because the soil cannot supply water fast enough and/or because the roots can no longer extract water fast enough to meet the meteorological demand. Plants with growing roots try to reach into moist regions of the soil rather than depend entirely on the conduction of water over considerable distances in the soil against a steadily increasing hydraulic resistance. Usually there are more shallow than deep roots. As reported by Hillel (1998), many field experiments have been carried out and numerical models devised to represent crops under stress. However, for a soil moisture balance calculation a direct, straightforward method of representing the effect of limited soil moisture is required.

Figure 3.9 is a diagrammatic representation of the extraction of water by roots from the soil zone; it is of the same form as Figure 3.7 which refers to the evaporative processes. Again there are four sketches which illustrate,

(a) soil moisture deficit close to zero, transpiration at the potential rate,
(b) the soil wet enough for transpiration at the potential rate,
(c) a reduced rate of transpiration due to limitations in the soil or roots to meet the potential evapotranspiration demand and
(d) the soil moisture so low that no water is collected by the roots.

Condition (d) is more likely to arise in arid zones following a long period with no rainfall.

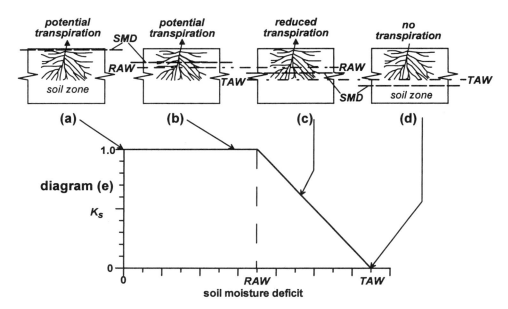

Figure 3.9   Reducing transpiration due to increasing soil moisture deficit

Diagram (e) is similar to that for evaporation; the coefficient K_S defines the reduction in the actual transpiration. The limiting soil moisture conditions are the readily available water (RAW) and the total available water (TAW). If the soil moisture deficit is less than RAW, evapotranspiration will occur at the potential rate. However, if the soil moisture deficit is between TAW and RAW, transpiration occurs at a reduced rate. Since TAW corresponds to the wilting point, transpiration ceases when the soil moisture deficit exceeds TAW.

3.3.5 Nature of the soil

The nature of the soil, and in particular its moisture-holding properties, are crucial in determining TAW and RAW. It is well known that a lack of rainfall or irrigation causes stress to crops in sandy soils more rapidly than in loams or clays. This occurs because the difference between field capacity and wilting point for a sand is approximately half that for a loam or clay. The depth of roots is also important since the total available water is directly proportional to the rooting depth. In the initial stages when the roots are shallow (see Figure 3.8), TAW is small and crops are susceptible to stress.

3.3.6 Runoff

For all but the most permeable soils, runoff occurs during periods of heavy rainfall due to the intensity of the rainfall and the restricted infiltration capacity of the soil. Precipitation which becomes runoff is a loss to the soil moisture balance. There are many methods of estimating runoff but they are not usually designed to give an approximate value of the runoff for a single day over an area of, say, 1 km². The approach chosen for this study assumes that the runoff RO is a function of two of the parameters used in the soil moisture balance, namely the daily rainfall intensity and the current soil moisture deficit. An illustrative example is contained in Section 3.4.4.

3.3.7 Occurrence of recharge

It is assumed that recharge from the soil zone can only occur when the soil is at field capacity. Field capacity is defined as *the amount of water that a well-drained soil can hold against gravitational forces*; once the field capacity is exceeded, excess water moves by free drainage through the soil zone. Field capacity is equivalent to zero soil moisture deficit.

The manner in which recharge can occur is illustrated by the three sketches of a soil profile in Figure

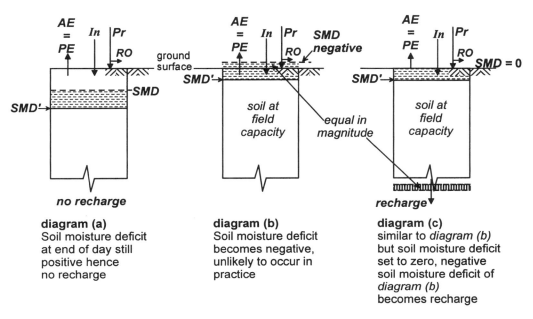

Figure 3.10 Diagram illustrating the soil moisture conditions necessary for recharge to occur

3.10. *SMD'* is the soil moisture deficit at the start of a day, *SMD* is the value at the end of the day. In Figure 3.10a the precipitation *Pr* is high, there is some runoff *RO* hence the infiltration to the soil zone *In = Pr − RO* is almost 90 per cent of the precipitation. The infiltration exceeds the actual evapotranspiration *AE*; the excess water is used to reduce the soil moisture deficit, with the reduction in soil moisture deficit shown shaded in Figure 3.10a. Since the soil moisture does not reach field capacity, no recharge occurs. In Figure 3.10b, the precipitation, runoff, infiltration and actual evapotranspiration are identical to Figure 3.10a; consequently, the excess water goes to reduce the soil moisture deficit. Because the soil moisture deficit at the start of the day *SMD'* is small, the calculated soil moisture deficit at the end of the day is negative. However, when the soil moisture deficit becomes zero, the soil moisture reaches field capacity so that the soil is free-draining. Consequently, the condition of water above the soil surface, as sketched in Figure 3.10b, is unlikely to occur; instead the excess water drains through the soil to become recharge as shown in Figure 3.10c.

3.3.8 Combining crops and bare soil

As indicated in Figure 3.8, the area of the soil covered by crops and the area with bare soil changes through-

out the year. Figure 3.11a indicates how the proportion of bare soil and crop for a spring-sown crop in England varies at four different times of a water year. During autumn and winter, bare soil conditions apply with no crop cover; during spring there is a partial crop cover which becomes a total crop cover in summer. The figure starts with autumn and winter conditions since these are the periods when recharge is more likely to occur. A winter-sown crop is illustrated in Figure 3.11b. Since the crop is sown soon after harvest, there is always a proportion of the area covered by the crop although, during autumn and winter, bare soil still predominates. In climates where there is a distinct rainy season followed by a dry season, the distribution of crop and bare soil is different since crop growth and recharge both occur during the rainy season. How to include the changing proportions of bare soil and crop in the calculation of the total evaporation is considered in Section 3.4.5.

3.3.9 Water storage near the soil surface

The soil moisture deficit concept provides no direct information about the distribution of soil moisture with depth. Although soil stress coefficients take account of the reduced evapotranspiration when the soil moisture deficit exceeds the Readily Available

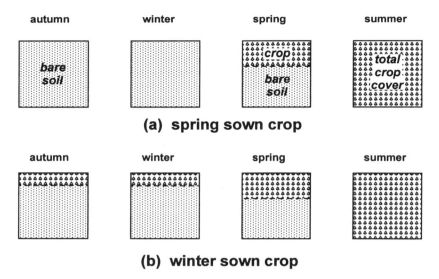

Figure 3.11   Seasonal changes in proportions of crop and bare soil

Water, there is an important situation which a single soil moisture balance fails to represent. This occurs on days following significant rainfall when the *SMD* is greater than RAW (or REW). On the day when the rainfall occurs, evapotranspiration is at the potential rate. According to the conventional soil moisture balance model, any excess water (precipitation less runoff and actual evapotranspiration) reduces the soil moisture deficit. Consequently, none of this excess water is available to the shallow roots during the following days. In practice, however, heavy rainfall leads to an increase in the soil moisture in the upper part of the soil profile. Accordingly, there may be sufficient soil moisture for the shallow roots to collect enough water for evapotranspiration at the potential rate on day(s) following the heavy rainfall. This phenomenon is described as *near surface soil storage*.

Field observations show that, for less permeable soils, some of the excess water remains close to the soil surface so that evapotranspiration occurs on the following days. For instance, with a field in a temperate climate in early winter when there is still a large soil moisture deficit, there are different responses following heavy rainfall between sandy and clay soils. For a sandy soil, little water is retained close to the soil surface so that on the day following heavy rainfall the soil surface appears to be 'dry' so that it is possible to walk over the ground. Yet, if the soil has a substantial clay content, it may be several days before the soil becomes sufficiently dry to walk across the surface without

causing damage. With the sandy soil, most of the excess water following rainfall moves downward to increase the moisture content in the lower part of the soil profile, whereas for a clay soil a proportion of the excess water is held near the soil surface for evapotranspiration during the following days. For similar reasons, vegetation in a clay soil, which is under stress, shows signs of recovery for several days following heavy rainfall.

Near surface soil storage is illustrated in Figure 3.12 (the upper part of this figure is similar to Figure 3.5). The soil moisture distribution at the start of the day is indicated by the chain-dotted line in Figure 3.12a. During the day with rainfall (or irrigation) the infiltration is sizeable. Some of the infiltration is used for evapotranspiration, and there is also a redistribution so that the soil moisture content at the end of the day is as shown by the full line. The increase in moisture content during the day is indicated by the shaded area (comment i). Some of this increased moisture is available on the following day for transpiration by shallow roots (or evaporation from bare soil). Therefore a proportion of the additional soil moisture is retained near the soil surface. This is represented in the computational model of Figure 3.12b by the pentagon shape in the soil column (comment ii), while the remainder of the water is used to reduce the soil moisture deficit (comment iii). The proportion of the excess soil water retained near the soil surface depends on the vertical hydraulic conductivity of the soil. For permeable sandy

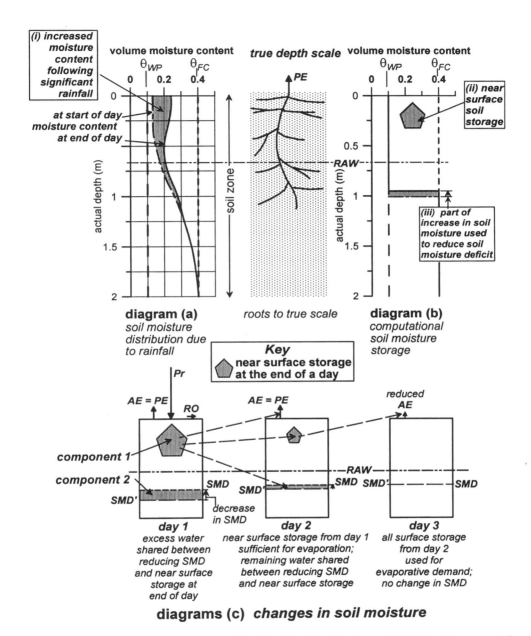

Figure 3.12 Conceptual diagram of water stored near the soil surface: (a) soil moisture distribution, (b) representation of soil moisture for computational purposes including near surface storage, (c) computational model of near surface storage meeting evapotranspiration demand

soils, only a small proportion of the water is retained near the soil surface whereas for soils with a higher clay content more water is retained. In Figure 3.12b the quantity retained as near surface storage is indicative of a moderately draining soil.

The implication of near surface storage is demonstrated by the daily soil moisture balances in Figure 3.12c; this diagram refers to equivalent depths of water. On *day 1* there is substantial rainfall which leads to runoff and evapotranspiration at the potential rate. At the end of *day 1* some of the excess water is retained close to the soil surface (*component 1*) with a smaller proportion of the excess water available to reduce the soil moisture deficit (*component 2*). Some of the near surface storage may be water ponded above the soil surface but most of the water is held as an increase in the moisture content in the upper part of the soil profile. Consequently at the start of *day 2* there is sufficient water above the RAW line in the location of extensive roots to meet the full evaporative demand. The remaining near surface moisture is distributed between near surface storage available for the following day and a further reduction in the soil moisture deficit. At the start of *day 3*, there is still sufficient water retained from *day 2* to meet part of the evaporative demand. There is no water retained at the end of *day 3*.

In locations with a dry season followed by a rainy season, near surface storage is important for the early stages of crop growth. At the start of the rainy season, when there is a substantial soil moisture deficit, farmers ensure that moisture from the early rains is retained near the soil surface so that the seed can germinate and start to grow. Therefore Figure 3.12 also represents conditions early in the rainy season.

To summarise, the retention of some recent infiltration as a higher moisture content close to the soil surface is an important factor in estimating the magnitude of the evaporation or transpiration when the soil moisture deficit is greater than RAW/REW. Consequently, the conventional conceptual model of a single store soil moisture balance is modified to allow for near surface storage. Examples of the actual representation of near surface storage can be found in Section 3.4.7.

3.4 QUANTIFIED CONCEPTUAL AND COMPUTATIONAL MODELS FOR THE SOIL MOISTURE BALANCE TECHNIQUE

This section describes a quantified conceptual soil moisture balance model for recharge estimation. Helpful insights into parameter values can be gained from FAO Irrigation and Drainage Paper 56 (1998) *Crop Evapotranspiration; Guidelines for Computing Crop Water Requirements*. The main focus of that document is irrigated agriculture. For rain-fed crops and periods when crops are not growing, such as the dry season in semi-arid or arid areas or winter in temperate climates, additional procedures are required. Numerical information for certain crops and soils, derived largely from FAO (1998), is included in the fol-lowing tables; for other crops and soils it is necessary to consult agricultural departments and soil surveys.

The basic conceptual models are described in Section 3.3. This section is concerned with the selection of suitable parameter values and the computational model for the calculation of a daily soil moisture balance which includes the estimation of actual total evaporation and hence recharge.

3.4.1 Crops and crop coefficients

Estimation of the potential evapotranspiration of a specific crop is based on a reference evapotranspiration ET_0; the reference crop closely resembles an extensive surface of green grass of uniform height, actively growing, completely shading the ground and with adequate water. This reference surface has an assumed crop height of 0.12 m, a fixed surface resistance of 70 s/m and an albedo of 0.23. ET_0 can be calculated from meteorological data; the FAO report recommends a form of the Penman–Monteith equation since it can provide consistent estimates in most regions and climates.

The evapotranspiration of a particular crop ET_C depends on a crop coefficient K_C which varies during the growing and harvest stages

$$ET_C = K_C ET_0 \qquad (3.5)$$

The crop coefficient represents three primary characteristics that distinguish the crop from reference grass. Factors determining the numerical values of the crop coefficients include the following.

- *Crop type* including crop height, aerodynamic properties, leaf and stomata properties and the closeness of spacing of the plants. For most deciduous trees, stomata are only on the lower side of the leaf; large leaf resistances lead to relatively smaller values of K_C. The spacing of trees, providing say 70 per cent ground cover, may cause the value of K_C to be less than 1.0 if the cultivation is without ground cover.
- *Climate*: the crop coefficients refer to standard climatic conditions; the coefficients tend to increase with increasing wind and decreasing humidity. Corrections can be made for differing climatic conditions (for further information see Figure 21 in FAO 56).
- *Crop growth stages*: as the crop develops, crop heights and leaf areas change. The growing period is divided into four stages: initial stage, crop development, mid-season and late season. These stages and the crop coefficients are presented in Figure 3.13.

Table 3.1 lists crop coefficients and the duration of crop development for various crops. Note that there are five stages for the crop., initial, development, mid-season, late season and finally harvest; they can be defined using four time durations and three values of K_C.

The note relating to the first column of numbers in Table 3.1 is important because it highlights an issue which is not always addressed clearly in FAO 56. As indicated in Section 3.3.8, the approach adopted in this presentation is to identify the separate areas of crop and bare soil whereas the single crop coefficient approach in FAO 56 allows for a reduced area of crop in the initial stages of growth by introducing a reduced crop coefficient. However, in Table 3.1, the coefficient for the initial stage is close to 1.0 since small plants transpire at a rate close to that of grass. The different assumptions are illustrated on the left of Figure 3.13.

3.4.2 Reduced transpiration when crops are under water stress

An introduction to the situation when crops are under stress can be found in Section 3.3.4 and in Figure 3.9 which contains sketches of the relationship between the transpiration and the soil moisture deficit. In Figure 3.9e, a relationship is presented for the variation of the soil stress coefficient K_S with the soil moisture deficit. This coefficient can be used to calculate the actual crop evapotranspiration from the equation

$$ET_C = K_S K_C ET_0 \qquad (3.6)$$

Note that this equation only applies for days when there is no precipitation. To calculate K_S, numerical values of TAW and RAW must be defined.

There are four important factors involved in the estimation of the soil stress coefficient K_S:

1. the soil moisture deficit
2. the depth of the roots of the crop

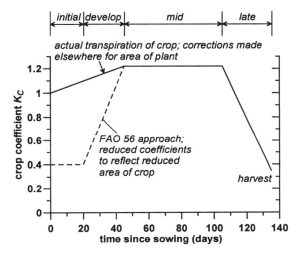

Figure 3.13 General form of variation of crop coefficient during the growing season; the early part of the graph is different from FAO 56 since the current method directly represents the smaller area of the young crop rather than including it as a reduced crop coefficient

Table 3.1 Crop coefficients and duration of crop development stages (days). Three crop coefficients are sufficient to describe the variation of the crop evapotranspiration; the meanings of *initial, develop, mid* and *late* are shown in Figure 3.13. Reproduced by permission of the Food and Agriculture Organisation of the United Nations, FAO (1998)

	K_C initial	K_C mid	K_C end	Initial (days)	Develop (days)	Mid (days)	Late (days)	Total (days)
Spring wheat	1.00[†]	1.15	0.30	20	25	60	30	135
Winter wheat	1.00[†]	1.15	0.30	30	>140	50	30	>250
Potatoes	1.00[†]	1.15	0.75	30	35	50	30	145
Lettuce	0.70[†]	1.0	0.95	25	35	30	10	100
Sugar beet	0.70[†]	1.20	0.70	30	45	90	15	180
Rapeseed	1.00[†]	1.15	0.35	25	35	55	30	145
Apples	0.60	0.95	0.75	20	70	90	30	210
Conifers	1.00	1.00	1.00					365
Grapes	0.30	0.85	0.45	20	50	90	20	180
Grass	1.00	1.00	1.00					365

Note: † the values quoted for K_C initial in FAO 56 are much lower than the values used in this table. The values of FAO 56 appear to directly reflect the small area covered by the young crop whereas it appears reasonable that, if account is taken of the area of the small plant (see Section 3.3.8), crop coefficients approaching 1.0 should be used.

3. the moisture holding properties of the soil
4. the behaviour of the crop under water stress.

The *soil moisture deficit* (*SMD*) (or the root zone deple-tion) is described in detail in Section 3.3.2.

In assessing whether the roots are able to draw water from the soil, information is required about the *depths of the roots*. The depth of the roots gradually increases following sowing, as illustrated in Figure 3.8. During main crop development a linear increase in the depth of the roots is assumed; during the mid and late season (and harvest) the roots remain at an approximately constant depth. A number of maximum root depths for different crops are listed in Table 3.2.

Different soils have different *moisture holding prop-erties*; it is widely known that plants can withstand water shortages more easily if the soil is a clay rather than a sand. Soil water availability is concerned with the capacity of a soil to retain water which can be extracted by the roots of plants. As the water uptake progresses, the remaining water is held to the soil par-ticles with greater force, lowering its potential energy and making it more difficult for the plant to extract the moisture. Eventually a point is reached when the crop can no longer extract the remaining water; this is defined as the wilting point, which is the moisture content when the plant ceases to grow.

The *Total Available Water* for a plant (in mm) is

$$\text{TAW} = 1000(\theta_{FC} - \theta_{WP})Z_r \qquad (3.7)$$

where   TAW = total available water in mm,
θ_{FC} = the moisture content at field capacity,
θ_{WP} = the moisture content at wilting point,
Z_r = the rooting depth (m).

Table 3.3 (which is derived from FAO 1998) lists the moisture contents at field capacity and wilting point for a number of soils. Hence for spring wheat with a rooting depth of 1.5 m in a sandy loam with (θ_{FC} − θ_{WP}) = 0.133, the total available water is TAW = 1500 × 0.133 = 200 mm. The contrast between sand and other soil types should be noted.

The *behaviour of a crop under water stress* can be described using the water stress coefficient K_S which depends on the *SMD* as defined in Figure 3.9. The fraction of the TAW that a crop can extract without suffering water stress, the *Readily Available Water* (RAW in mm) is calculated from RAW = *p*TAW, in which *p* is the average fraction of TAW that can be depleted before the moisture content falls below the threshold value. Values of *p* (Table 3.2) vary from 0.30 for lettuce to 0.70 for conifer trees.

The relevant equation for the sloping part of the diagram for the water stress coefficient is:

$$K_S = \frac{\text{TAW} - SMD}{\text{TAW} - RAW} \quad \text{when RAW} < SMD < \text{TAW}$$

$$(3.8)$$

This relationship, which is presented in Figure 3.14, shows how the water stress coefficient varies with the soil moisture deficit (root zone depletion). The darker lines refers to a sandy loam and the lighter broken lines to a sand.

3.4.3 Reduced evaporation due to limited soil water availability

The potential evaporation for a soil, *ES*, can be deduced from the reference crop evapotranspiration using the equation

Table 3.2   Rooting depths and depletion factor *p* for various crops. Reproduced by permission of the Food and Agriculture Organisation of the United Nations, FAO (1998)

Crop	Maximum root depth (m)	*p* (for RAW)
Spring wheat	1.0–1.5	0.55
Winter wheat	1.5–1.8	0.55
Potato	0.4–0.6	0.35
Lettuce	0.3–0.5	0.3
Sugar beet	0.7–1.2	0.55
Rapeseed	1.0–1.5	0.6
Apples	1.0–2.0	0.5
Conifers	1.0–1.5	0.7
Grapes	1.0–2.0	0.35
Grass	0.5–1.0	0.5

Table 3.3   Moisture contents at field capacity and wilting point (m^3/m^3). Reproduced by permission of the Food and Agriculture Organisation of the United Nations, FAO (1998)

Soil type	θ_{FC}	θ_{WP}	(θ_{FC}−θ_{WP})
Sand	0.07–0.17	0.02–0.07	0.05–0.11
Loamy sand	0.11–0.19	0.03–0.10	0.06–0.12
Sandy loam	0.18–0.28	0.06–0.16	0.11–0.15
Loam	0.20–0.30	0.07–0.17	0.13–0.18
Silt loam	0.22–0.36	0.09–0.21	0.13–0.19
Silt	0.28–0.36	0.12–0.22	0.16–0.20
Silt clay loam	0.30–0.37	0.17–0.24	0.13–0.18
Silty clay	0.30–0.42	0.17–0.29	0.13–0.19
Clay	0.32–0.40	0.20–0.24	0.12–0.20

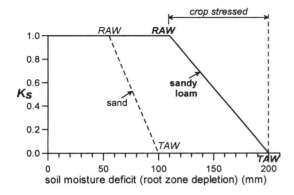

Figure 3.14 Soil stress coefficients for fully grown spring wheat

Table 3.4 Selected values of readily evaporable water REW and total evaporable water TEW for $Z_E = 100$ mm. Reproduced by permission of the Food and Agriculture Organisation of the United Nations, FAO (1998)

Soil type	REW (mm)	TEW (mm)
Sand	2–7	6–12
Loamy sand	4–8	9–16
Sandy loam	6–10	15–22
Loam	8–10	16–23
Silt loam	8–11	18–25
Silt	8–11	22–26
Silt clay loam	8–11	22–27
Silty clay	8–12	22–28
Clay	8–12	22–29

$$ES = K_E ET_0 \qquad (3.9)$$

where K_E is the evaporation coefficient which is set at 1.10 for temperate climates and 1.05 for semi-arid climates.

As explained in Section 3.3.3, there is a limit to the depth from which evaporation from the soil can occur; there is also a reducing efficiency of evaporation when there is insufficient water available. The parameters introduced to represent the reducing rate of evaporation are the *Total Evaporable Water* (TEW) and the *Readily Evaporable Water* (REW). TEW is estimated from

$$TEW = 1000(\theta_{FC} - 0.5\theta_{WP})Z_E \quad (mm) \qquad (3.10)$$

where Z_E is the depth of the surface soil layer that is subject to drying by evaporation; it lies within the range 0.10–0.25 m. The coefficient 0.5 is introduced before θ_{WP} since evaporation can dry the soil to mid-way between the wilting point and oven dry.

Values of REW are estimated from field studies. Ranges of values of REW and TEW for $Z_E = 0.10$ m are given in Table 3.4.

The alternative soil stress coefficient for actual soil evaporation K_S' is introduced in Section 3.3.3 and Figure 3.7. To illustrate the manner in which REW and TEW relate to RAW and TAW, Figure 3.15 has been prepared. The soil stress coefficient can be calculated as

$$K_S' = \frac{TEW - SMD}{TEW - REW} \quad \text{when } REW < SMD < TEW \qquad (3.11)$$

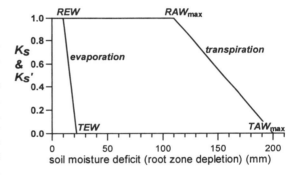

Figure 3.15 Comparison of soil stress coefficients for evaporation and transpiration for a sandy loam

3.4.4 Runoff

Runoff following rainfall may be substantial when the soil is wet and the rainfall intensity is high. Quantative information can be gained from field observations of the occurrence of runoff and how this depends on soil conditions. In addition, the sizes of drains, ditches and culverts can provide estimates of the maximum intensities of runoff. When the soil is dry, the soil moisture deficit is high; runoff is less likely to occur. All available information is used to develop a matrix such as that presented in Table 3.5 which relates to a cultivated sandy soil overlying a sandstone aquifer.

The table shows that for dry soil conditions with a soil moisture deficit greater than 60 mm, no runoff occurs if the rainfall intensity is less than 20 mm/d. However, when the soil approaches field capacity with the soil moisture deficit less than 10 mm, substantial

Table 3.5 Runoff for different rainfall intensity (*Pr* in mm/d) and current soil moisture deficit

Rainfall intensity → SMD' ↓	0–10 mm/d	10–20 mm/d	20–30 mm/d	30–50 mm/d	50+ mm/d
0–10 mm	0.00	0.2 (Pr − 10)	0.15 Pr	0.30 Pr	0.50 Pr
10–30 mm	0.00	0.00	0.10 Pr	0.20 Pr	0.45 Pr
30–60 mm	0.00	0.00	0.05 Pr	0.10 Pr	0.40 Pr
60+ mm	0.00	0.00	0.02 Pr	0.05 Pr	0.30 Pr

runoff occurs, especially with higher rainfall intensities. For soils with a high clay content, larger runoff coefficients are appropriate. Inevitably there are uncertainties about the magnitude of the coefficients in Table 3.5. However, a sensitivity analysis can be used to show how modifications in these coefficients influence the recharge estimates.

3.4.5 Combining the influence of crop transpiration and bare soil evaporation

When using a soil moisture balance technique to estimate recharge, both the crop and bare soil must be considered (see Figures 3.8 and 3.11 which show the changing areas between crops and bare soil). For example, in the UK and other temperate climates most recharge occurs in winter when either the soil is bare or crops cover only a small proportion of the area. The combination of crop response and bare soil evaporation is equally important in regions where there is a rainy season followed by a dry season. Any change in the soil moisture deficit during the dry season when the soil is bare will have an impact on the conditions at the start of the rainy season when the crops are sown.

Typical parameter values for a crop of winter wheat in a sandy loam soil in England are listed in Table 3.6. For winter wheat, sowing occurs soon after harvest; the growth of the crop and roots is similar to that shown in Figure 3.8 but the *initial* period with a small area covered by plants starts in early October and continues through to March (Figure 3.11b). Note that Table 3.6 contains information about:

- the crop and evaporation coefficients K_C and K_E which are used to estimate both the potential evapotranspiration for the crop and the bare soil evaporation from daily values of the reference crop evapotranspiration,
- percentage areas of crop and bare soil, A_C and A_B,
- values of TAW and RAW which depend on the depth of roots and the nature of the soil,

- values of TEW and REW which are constant throughout the year.

Values quoted are for the first day of the month; a linear variation is usually assumed between monthly values. Table 3.6 can be used as a template for studies in other locations with different crops and soils.

Since the relationships between the soil stress factor K_S or K'_S and the soil moisture deficit are similar (see Figure 3.15), it is acceptable to combine the coefficients for bare soil and the crop in Table 3.6, weighting the parameters according to the relative areas. The result of combining parameters is shown in Table 3.7. The comments on the right-hand side of Table 3.7 show that the bare soil parameters dominate during the early and late months.

The suitability of this approach of using weighted values of the coefficients was tested by comparing daily recharge estimates using the parameter values in Table 3.7 with those calculated with the crops and bare soil represented individually. The difference in the estimated monthly recharge between the two approaches was never more than 1 per cent.

3.4.6 Soil moisture balances

Since the estimation of the soil stress coefficient and hence the actual total evaporation depends on the current soil moisture deficit, a daily calculation is required to track the changes in soil moisture deficit. A number of possible situations can occur, some of which are illustrated in Figures 3.10, 3.16 and 3.17. The input to the soil moisture balance is the infiltration which depends on precipitation (and/or irrigation) and runoff; the main output is the actual total evaporation. The relationship between the soil moisture deficit at the start of the day, *SMD'* and RAW/REW or TAW/TEW determines the equation for estimating the actual total evaporation. Four situations can be identified.

1. ***SMD'* < RAW/REW:** When the soil moisture deficit at the start of the day is less than RAW/REW, the

Table 3.6 Parameters for winter wheat in sandy-loam soil: values refer to conditions at start of month, use linear variations during month

Start of month	K_C	% area of crop A_C	% area of bare soil A_B	K_E	RAW (mm)	TAW (mm)	REW (mm)	TEW (mm)	Comments
Jan	1.00	0.15	0.85	1.10	22	40	10	22	Assuming 15% crop area hence 85% bare soil; crop is almost dormant
Feb	1.00	0.15	0.85	1.10	22	40	10	22	Conditions the same as January
Mar	1.00	0.25	0.75	1.10	33	60	10	22	The crop is just beginning to grow, covering a larger area with deepening roots
Apr	1.00	0.40	0.60	1.10	44	80	10	22	Further growth and increasing area
May	1.15	0.90	0.10	1.10	88	160	10	22	Continuing growth, greater area with roots reaching down a lot further
Jun	1.15	1.00	0.00	1.10	110	200	10	22	Full growth
Jul	1.15	1.00	0.00	1.10	110	200	10	22	Full growth
Aug	0.35	1.00	0.00	1.10	110	200	10	22	Crop ready to harvest, little evapotranspiration and no bare soil
Sep	1.00	0.00	1.00	1.10	–	–	10	22	Field ploughed, only bare soil evaporation possible
Oct	1.00	0.10	0.90	1.10	10	22	10	22	Small newly planted crop, mainly bare soil
Nov	1.00	0.10	0.90	1.10	16	30	10	22	Newly planted crop beginning to become established, roots slightly deeper
Dec	1.0	0.15	0.85	1.10	22	40	10	22	Slightly more growth during past month but moving to dormant period

Table 3.7 Parameter values determined as weighted averages of values for crop and bare soil for winter wheat in a sandy-loam soil: values refer to conditions at start of the month, normally use linear variations during month

Start of month	Combined K_C	RAW/REW (mm)	TAW/TEW (mm)	Comments
Jan	1.09	12	25	Bare soil is dominant
Feb	1.09	12	25	Bare soil is dominant
Mar	1.07	16	32	Area of crop increases slightly
Apr	1.06	23	45	Further increase in crop area
May	1.14	80	146	Crops cover most of the area
Jun	1.15	110	200	Crops are dominant
Jul	1.15	110	200	Crops are dominant
Aug	0.35	110	200	Ripening of crops for harvest
Sep	1.10	10	22	Bare soil
Oct	1.09	10	22	Winter crop sown, covers only 10% of area
Nov	1.09	11	22	Bare soil still dominant
Dec	1.09	12	25	Bare soil still dominant

soil stress coefficient, which is given by the upper horizontal line of Figure 3.15, equals 1.0. Consequently, whatever the magnitude of the rainfall, the actual total evaporation AE equals the potential total evaporation PE (combined K_C of Table 3.7 multiplied by ET_0).

2. **RAW/REW < *SMD'* < TAW/TEW**: The situation when the soil moisture deficit at the start of the day is between the readily and total available water is illustrated in Figure 3.16. When infiltration is greater than the potential total evaporation (Example (a)), the shallow roots intercept part of

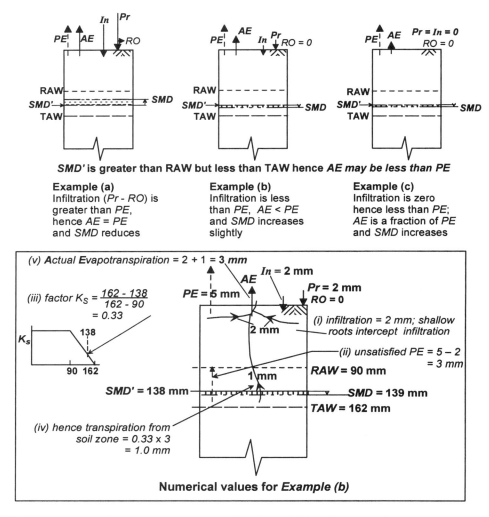

Figure 3.16   Three examples of water balances when the soil moisture deficit at the start of the day is between RAW and TAW

the infiltrating water so that they meet the potential total evaporation demand, hence $AE = PE$; any excess water leads to a reduction in the SMD. However, when $In < PE$ (Example (b)), the infiltration is transpired by the shallow roots but the remaining transpiration demand $(PE - In)$ is only met at the reduced rate, the factor being K_S. This results in a slight increase in the SMD. In Example (c), there is no precipitation or infiltration, in the absence of any rainfall, $AE = K_S PE$; this leads to an increase in the SMD.

To illustrate how the calculations are performed, numerical values for Example (b) are included in the box in the lower part of Figure 3.16. The five steps in performing the calculation are explained. The actual evapotranspiration equals 2.0 mm from the interception of the infiltration by the shallow roots and 1.0 mm from the soil moisture store which corresponds to $K_S = 0.33$ and an unsatisfied potential evapotranspiration of $(5.0–2.0)$ mm. Note that the arrows representing precipitation, PE, etc. are drawn to an exaggerated scale.

3. **SMD' > TAW/TEW**: After harvest another set of conditions occur if the soil moisture deficit is greater than the total evaporable water TEW (Figure 3.17). In Example (a) infiltration is greater than potential

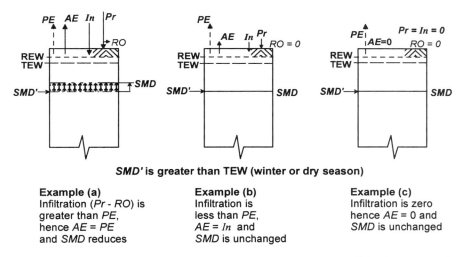

SMD' is greater than TEW (winter or dry season)

Example (a)
Infiltration (*Pr - RO*) is
greater than *PE*,
hence *AE = PE*
and *SMD* reduces

Example (b)
Infiltration is
less than *PE*,
AE = In and
SMD is unchanged

Example (c)
Infiltration is zero
hence *AE = 0* and
SMD is unchanged

Figure 3.17 Three examples of water balances when the soil moisture deficit at the start of the day is greater than TAW/TEW

evaporation, hence the total evaporation is at the potential evaporation rate, the excess water reduces the soil moisture deficit. For Example (b), the infiltration is less than the potential evaporation for that day; the infiltration is evaporated from the shallow soil, but further water cannot be drawn from deeper within the soil, hence the soil moisture deficit remains unchanged and $AE = In$. For Example (c) there is no precipitation or infiltration, hence there is no evaporation and no change in the soil moisture deficit. Note that in temperate climates with significant winter rainfall, the condition of the soil moisture deficit being greater than Total Evaporable Water continues until late autumn or early winter. For locations with a rainy season followed by a dry season, the soil moisture deficit is likely to remain greater than TEW throughout the dry season.

4. **SMD ≤ 0**: The situation when the soil moisture deficit at the end of the day is calculated as zero or negative has already been discussed in Section 3.3.7; the water balance conditions are illustrated in Figure 3.10. It is assumed that only when the soil is at field capacity can water drain freely through the soil zone to become recharge. Consequently, recharge occurs when there is a negative soil moisture deficit. This negative deficit becomes recharge from the base of the soil zone with the soil moisture deficit at the end of that day equal to zero.

3.4.7 Soil moisture balances with near surface soil storage

Water storage near the soil surface is introduced in Section 3.3.9 and illustrated in Figure 3.12; the computational procedure is shown in (c) of that figure. Near surface storage is included only when $SMD' >$ RAW/REW. The computational procedure is explored using the following idealised example; detailed calculations are presented in Table 3.8. On day 1 there is 15mm of rainfall with no rainfall on the succeeding days (runoff is ignored to simplify the calculations). The near surface storage is quantified using a storage retention fraction, *FRACSTOR*. In the following discussion, two storage retention fractions, *FRACSTOR* = 0.0 and 0.75, are examined. Two alternative times of the year are considered, one in the summer with *PE* of 4.0mm/d and another in winter with *PE* = 1.0mm/d.

In the calculations, the near surface store *SURFSTOR* receives on the first day a quantity equal to *FRACSTOR**(*In* − *PE*); this is available for the following day. In subsequent days the same value of *FRACSTOR* is used to recalculate the fraction of water remaining in the shallow zone. The quantity of water transferred deeper into the soil zone to reduce the soil moisture deficit [to *SMD*] is also calculated.

Results in the upper part of Table 3.8 show how the computation proceeds. Because the rainfall on Day 1

Table 3.8 Use of coefficient *FRACSTOR* to represent near-surface soil storage; the tabulated calculations include the water held in the near-surface store *SURFSTOR*. Two sets of results are presented, one with PE = 4.0 mm/d, the second with PE = 1.0 mm/d; values of *FRACSTOR* are 0.0 and 0.75

Day 1 rain 15 mm		*FRACSTOR* = 0.0			*FRACSTOR* = 0.75		
Day	*PE*	*AE*	*SURFSTOR*	to *SMD*	*AE*	*SURFSTOR*	to *SMD*
1	4.0	4.0	0.0	11.0	4.0	8.3	2.7
2	4.0	0.0	0.0	0.0	4.0	3.2	1.1
2	4.0	0.0	0.0	0.0	3.2*	0.0	0.0
Total		*4.0*		*11.0*	*11.2*		*3.8*

* ignoring any total evaporation at the reduced rate from the deeper soil zone

Day 1 rain 15 mm		*FRACSTOR* = 0.0			*FRACSTOR* = 0.75		
Day	*PE*	*AE*	*SURFSTOR*	to *SMD*	*AE*	*SURFSTOR*	to *SMD*
1	1.0	1.0	0.0	14.0	1.0	10.5	3.5
2	1.0	0.0	0.0	0.0	1.0	7.1	2.4
3	0.0	0.0	0.0	0.0	1.0	4.6	1.5
4	0.0	0.0	0.0	0.0	1.0	2.7	0.9
5	0.0	0.0	0.0	0.0	1.0	1.3	0.4
6	0.0	0.0	0.0	0.0	1.0	0.2*	0.1
7	0.0	0.0	0.0	0.0	0.2*	0.0	0.0
Total		*1.0*		*14.0*	*6.2*		*8.8*

of 15.0 mm/d exceeds the *PE* of 4.0 mm/d, total evaporation occurs at the potential rate; this leaves excess water of 11.0 mm. When *FRACSTOR* = 0.0 (very permeable soil) all of the 11.0 mm goes to reduce the soil moisture deficit. However, with *FRACSTOR* = 0.75, 8.3 mm is stored near the soil surface with the remaining 2.7 mm reducing the soil moisture deficit. On Day 2, with *FRACSTOR* = 0.0 there is no water available for evapotranspiration (the same condition holds on Day 3). However with *FRACSTOR* = 0.75, 4.0 mm of the 8.3 mm supplies *AE*; of the remaining 4.3 mm, 3.2 mm is retained in near surface soil storage with 1.1 mm to reduce the *SMD*. On the third day, all of the 3.2 mm in the near soil store is lost as total evaporation. With *FRACSTOR* = 0.0 the total evaporation sums to 4.0 mm/d whereas with *FRACSTOR* = 0.75 the total evaporation following the rainfall event equals 11.2 mm.

For the second part of the table the rainfall is again 15 mm on day 1 but the potential evaporation is 1.0 mm/d; the sum of the total evaporation values are 1.0 and 6.2 mm respectively. The introduction of a near surface soil store results in increased total evaporation and consequently a reduced estimated recharge.

Different values of *FRACSTOR* have been explored; appropriate values for coarse sand, sandy loam and silty clay are typically 0.0, 0.45 and 0.75 respectively.

3.4.8 Algorithms for soil moisture balance

Although there are several alternative conditions for daily soil moisture balances, they can all be represented by the following algorithms (note that TAW and RAW are used to represent the area-weighted averages of TAW/TEW and RAW/REW).

The *reference crop potential total evaporation* (for grass) is always multiplied by the crop coefficient K_C to give the potential total evaporation, ET_C (written as *PE* in the algorithms).

When estimating the *actual total evaporation*, all possible conditions can be represented by the following algorithms; note that the water entering the soil zone is the infiltration *In* where $In = Pr - RO$ (*RO* is calculated from a relationship such as Table 3.5).

1. For *SMD′* < RAW or for *In* ≥ *PE*, then *AE* = *PE*.
2. For TAW ≥ *SMD′* ≥ RAW and *In* < *PE*, then *AE* = *In* + *KS* (*PE* − *In*)

where K_S is calculated using the soil moisture deficit for the previous day.

$$K_S = \frac{TAW - SMD'}{TAW - RAW}$$

3. For $SMD' \geq TAW$ and $In < PE$, then $AE = In$ (this condition can occur after harvest).

Soil moisture balance and ***estimating recharge***: The method of estimating recharge is illustrated in Figure 3.10; the equation for the daily soil moisture balance is as follows:

$$SMD = SMD' - In + AE$$

where SMD' is the soil moisture deficit at the end of the previous day. A suitable starting value needs to be chosen, for the UK if the calculation starts in January, a value of zero is often acceptable. For study areas with a rainy followed by a dry season the calculation should commence with a dummy year to provide appropriate initial conditions.

If $SMD < 0.0$, then $RECH = -SMD$ and $SMD = 0.0$.

Near surface soil storage can be included in the computational routines by modifying the algorithms as shown below; these changes are only included when $SMD' > RAW$.

After $In = Pr - RO$ add the two following statements

$In = In + SURFSTOR$ [this is the quantity from the previous time step]
If $In > PE$, then $SURFSTOR$ [for current time step] $= FRACSTOR * (In - PE)$ otherwise $SURFSTOR = 0.0$

Also the line $SMD = SMD' - In + AE$ is changed to

$$SMD = SMD' - In + SURFSTOR + AE$$

3.4.9 Annual soil moisture balance for a temperate climate

The results of an annual soil moisture balance calculation for winter wheat in a chalk catchment (location h of Figure 1.1) are presented graphically in Figure 3.18. There are four separate graphs.

(a) The upper graph shows the rainfall; the recharge is represented by shading part of the rainfall bar. Any runoff is plotted below the axis.

(b) The second graph displays the effective crop coefficient calculated by weighting the coefficients for the crop and for bare soil.
(c) The third graph shows the potential total evaporation (seven day average values are used) with the calculated actual total evaporation shaded.
(d) The lower graph shows the Readily Available and Total Available depths of water and the calculated daily soil moisture deficit. They are plotted below the zero line to emphasise that the SMD is a deficit.

Note that near surface soil storage is not included in this analysis to make it easier to interpret the results. Some of the important conditions, indicated by numbers in Figure 3.18, are discussed below.

① Runoff is indicated as negative rainfall. During the first 120 days of the year there are many days with significant runoff, the occasional high rainfall during the last sixty-five days also leads to substantial runoff. For the summer months there is no runoff, this arises because there is a substantial soil moisture deficit, also the daily rainfall never exceeds 10 mm/d. However, on day 230 the rainfall is 25.9 mm, hence runoff does occur.
② Shortly after day 150, the soil moisture deficit becomes greater than (i.e. moves below) RAW/REW. Prior to this time, the actual total evaporation always equals the potential value, but for the following 180 days the actual total evaporation is less than the potential unless the precipitation exceeds the potential total evaporation. Even though this period is more than a hundred days before recharge commences, it is essential to represent the reduced evapotranspiration (and hence the change in SMD) adequately, since the time when SMD becomes zero determines when recharge commences.
③ At day 218, the SMD is greater than TAW/TEW; this occurs because harvest has occurred and bare soil evaporation becomes the dominant mechanism so that TAW/TEW reduces towards 22 mm. After day 218 and until day 325 when SMD becomes less than TAW/TEW, evaporation only occurs on days when there is rainfall.
④ On day 332 the SMD becomes zero hence recharge occurs (part of the rainfall is shaded black). On several further days before the end of the year recharge occurs but the amounts are frequently small.
⑤ Significant recharge occurs early in the year up to day 72. There is then a period with zero recharge although the soil moisture deficit never exceeds

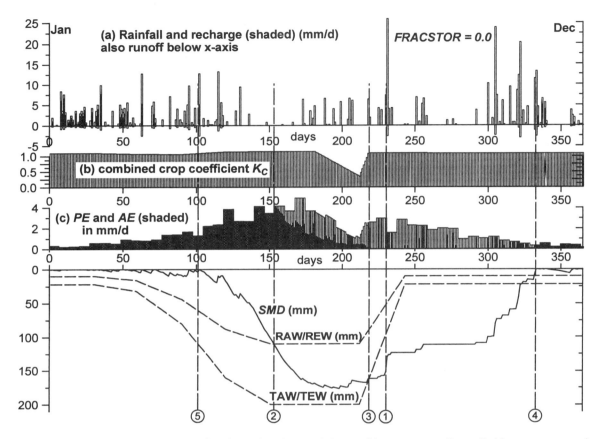

Figure 3.18 Daily soil moisture balance for winter wheat in a sandy loam soil in a temperate climate; limiting parameters and coefficients taken from Table 3.7

10 mm. On day 102 the *SMD* at the start of day is 3.7 mm; the rainfall is 12.7 mm, leading to a runoff of 2.6 mm. With the actual total evaporation of 1.3 mm, the excess water, which equals 5.1 mm, becomes recharge.

3.4.10 Estimating recharge in a semi-arid region of Nigeria

The soil moisture balance approach is equally well suited to semi-arid Nigerian conditions provided that realistic choices are made for the conceptual model and parameter values. The fundamental difference between temperate and semi-arid regions is that in semi-arid regions recharge occurs during the main growing season whereas in temperate climates recharge occurs during the winter when crop growth is minimal. The inclusion of near surface soil storage is a crucial feature of this semi-arid study area. A typical rainy season is

selected to illustrate the water balance calculations, and sufficient information is provided below and in Table 3.9 to indicate how the calculation can be reproduced for other field situations.

Details of the crop and soil are as follows; this discussion will also refer to Figure 3.19 which shows how the parameters change during the rainy season from 24 June to the end of October.

1. The crop is millet, the seed is sown on 24 June, the first rainfall day; germination and the start of crop growth is 4 July. The final stage of the crop cycle is the ripening which starts on 20 September with harvest on 2 October. The maximum crop coefficient is increased to 1.10 to allow for the aridity and the effect of wind.
2. The soil is sandy with $(\theta_{FC} - \theta_{WP}) = 0.10$.
3. The maximum rooting depth is 1.3 m; the depth from which evaporation can occur is $Z_E = 0.25$ m.

Table 3.9 Table of significant times and conditions for millet in eastern Nigeria

Date	K_C	% area of crop A_C	% area of bare soil A_B	K_E	RAW (mm)	TAW (mm)	REW (mm)	TEW (mm)	combined K_C	RAW/REW (mm)	TAW/TEW (mm)	Comments
1 Jan	–	0.00	1.00	1.05	–	–	17	29	1.05	17	29	Bare soil, towards middle of dry season
24 Jun	–	0.00	1.00	1.05	–	–	17	29	1.05	17	29	First rains, crop sown
4 Jul	1.10	0.08	0.92	1.05	20	36	17	29	1.05	18	30	Germination – very small plants
14 Jul	1.10	0.30	0.70	1.05	48	80	17	29	1.07	26	45	Roots and crop growing
8 Aug	1.10	1.00	0.00	1.05	78	130	17	29	1.10	78	130	End of development stage
20 Sep	1.10	1.00	0.00	1.05	78	130	17	29	1.10	78	130	End of mid stage
21 Sep	1.05	1.00	0.00	1.05	78	130	17	29	1.05	78	130	Ripening of crop ready for harvest
28 Sep	0.77	1.00	0.00	1.05	78	130	17	29	0.77	78	130	Crop effectively dying prior to harvest
2 Oct	0.60	1.00	0.00	1.05	78	130	17	29	0.60	78	130	Harvest
3 Oct	–	0.00	1.00	1.05	–	–	17	29	1.05	17	29	Day after harvest, no transpiration, possible evaporation
31 Oct	–	0.00	1.00	1.05	–	–	17	29	1.05	17	29	Early in dry season, only bare soil evaporation

4. The period of main growth starts on 13 August. Using all the preceding information together with data from FAO (1998), the combined crop coefficient and values of RAW/REW and TAW/TEW in Figure 3.19b and d are derived.

5. After a number of trial solutions, the coefficient for near surface storage was set at *FRACSTOR* = 0.45.

Rainfall and the total potential evaporation in Figures 3.19a and c are obtained from meteorological data. The soil moisture balance continues on a day by day basis, the initial value of the soil moisture deficit of 72 mm being obtained after running the simulation for a year. Detailed results for the soil moisture balance during the rainy season are presented in Figure 3.19. Of the large amount of information in this figure, certain issues are highlighted.

Comment (i): for successful initial growth of a plant, sufficient water must be available; this can be assessed by examining the actual total evaporation. Examination of Figure 3.19, where *comment (i)* is written, shows that from germination on 4 July there are ten days when there is no rainfall during the first sixteen days of crop growth. It is because of near surface soil storage that water can be collected by the shallow roots; consequently there is some actual evaporation (the shaded area of Figure 3.18c) on all but one of these days. Detailed calculations for 4–19 July can be found in Table 3.10. As the soil moisture deficit decreases and the root growth leads to increases in RAW and TAW, the roots are able to collect some water from the soil store. If no allowance is made for near surface soil storage in the simulation (*FRACSTOR* = 0.0) the soil moisture balance calculation predicts

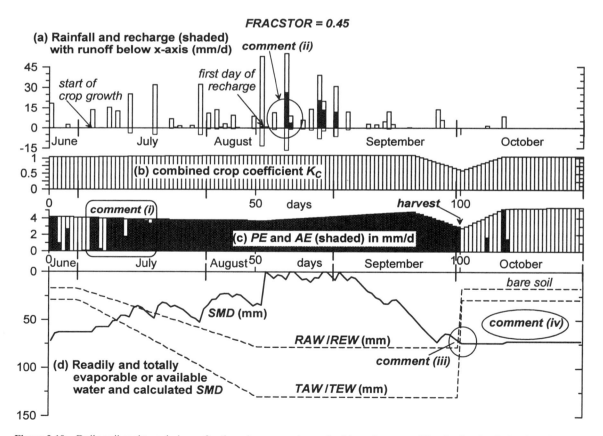

Figure 3.19 Daily soil moisture balance for the rainy season in semi-arid north-eastern Nigeria; for detailed calculations see Table 3.10

Table 3.10 Daily soil moisture balance for the Nigerian catchment from 20 August showing the effect of near surface soil storage, *NSSS*; all quantities are mm/d. Due to round-off of numbers there is often a lack of balance of 0.1 mm/d. The terms in brackets (columns 7 to 9) show how the excess water is distributed between near-surface storage and water to reduce the soil moisture deficit

Day	Start of day SMD'	NSSS'	Pr	RO	AE	$\begin{bmatrix} Pr + NSSS' \\ -RO- AE \end{bmatrix}$	to NSSS	to SMD	New SMD	Comments
04/7	62.9	0.0	13.7	0.0	4.1	9.6	4.3	5.3	57.6	Significant rainfall
05/7	57.6	4.3	0.0	0.0	4.0	0.3	0.1	0.2	57.5	Water from *NSSS* provides *AE*
06/7	57.5	0.1	0.2	0.0	0.3	0.0	0.0	0.0	57.5	Small *AE*
07/7	57.5	0.0	0.0	0.0	0.0	0.0	0.0	0.0	57.5	*AE* is zero
08/7	57.5	0.0	15.5	0.0	4.0	11.5	5.2	6.3	51.2	Rainfall so *AE* = *PE*
09/7	51.2	5.2	0.0	0.0	4.0	1.2	0.5	0.7	50.5	*AE* due to *NSSS*
10/7	50.5	0.5	12.6	0.0	4.0	9.1	4.1	5.0	45.5	Rainfall so *AE* = *PE*
11/7	45.5	4.1	0.0	0.0	4.0	0.1	0.1	0.1	45.4	*AE* due to *NSSS*
12/7	45.4	0.1	0.0	0.0	1.8*	−1.7	0.0	0.0	47.1	Reduced *AE* since TAW > SMD' > RAW
13/7	47.1	0.0	25.4	3.8	3.9	17.7	8.0	9.7	37.4	Rainfall so *AE* = *PE*
14/7	37.4	8.0	0.0	0.0	3.9	4.1	1.8	2.3	35.2	*AE* due to *NSSS*
15/7	35.2	1.8	0.0	0.0	3.9†	−2.1	0.0	0.0	37.2	*AE* = *PE* since SMD' < RAW
16/7	37.2	0.0	0.0	0.0	3.9*	−3.9	0.0	0.0	41.1	Reduced *AE* since TAW > SMD' > RAW
17/7	41.1	0.0	0.0	0.0	3.7*	−3.7	0.0	0.0	44.9	Reduced *AE* since TAW > SMD' > RAW
18/7	44.9	0.0	0.0	0.0	3.4*	−3.4	0.0	0.0	48.3	Reduced *AE* since TAW > SMD' > RAW
19/7	48.3	0.0	32.0	4.8	3.9*	23.3	10.5	12.8	35.4	Rainfall so *AE* = *PE*

NSSS' and *NSSS* near surface soil storage at beginning and end of day.
* TAW > SMD' > RAW   † SMD' < RAW

many more days with zero actual evapotranspiration. This suggests that the value of *FRACSTOR* = 0.45 is appropriate for this simulation. The ability of the soil moisture balance model to represent the early stages of crop growth confirms the plausibility of the model (Carter *et al.* 2002).

Comment (ii): the first day on which recharge is predicted to occur is 14 August, which is when the *SMD* first becomes zero. During this year there are six days when recharge is estimated to occur, the total recharge is 79 mm.

Comment (iii): as the harvest date is approached, the *SMD* becomes close to but slightly above RAW, therefore the crop is able to transpire at the potential rate throughout the growing season resulting in a good crop yield.

Comment (iv): at harvest on 2 October the soil moisture deficit is 74 mm. There is a small decrease following rainfall on 13 October but for the remainder of the dry season the soil moisture deficit remains constant. In practice there may be a small downward drainage out of the soil zone during the dry season; this could be represented by a small increase in the *SMD*.

3.4.11 Estimating recharge due to precipitation and irrigation in a semi-arid region of India

This third example of potential recharge estimation relates to western India. An investigation in a similar area in east Africa is described by Taylor and Howard (1996); their water balance estimates were supported by stable isotope tracer experiments. Full details of the first rain-fed crop and the second irrigated crop can be found in Table 3.11; the inputs and outputs of the soil moisture balance calculation are plotted in Figure 3.20. The near surface storage factor is set to 0.55 to reflect soil preparation which ensures that there is substantial moisture retention near the soil surface. These results refer to 1993, a year of high rainfall totalling 699.5 mm; most of this rainfall occurred during five days in July.

Monsoon rainfall commenced towards the end of June. The first crop of millet was sown in early July. During July and early August the crop, including the roots, grew with an increase in the area covered by the crop. From 12 August to 21 September the crop continued to develop but the roots did not extend further;

Table 3.11 Parameter values for rain-fed millet and irrigated sunflower oil in India

Date	K_C	% area of crop A_C	% area of bare soil A_B	K_E	RAW (mm)	TAW (mm)	REW (mm)	TEW (mm)	combined K_C	RAW/ REW (mm)	TAW/ TEW (mm)	Comments
1 Jun	–	0.00	1.00	1.05	9	19	9	19	1.05	9	19	Towards end of dry season, bare soil
4 Jul	1.00	0.00	1.00	1.05	9	19	9	19	1.05	9	19	Monsoon millet crop sown
16 Jul	1.00	0.33	0.67	1.05	22	40	9	19	1.03	13	27	Crop and roots growing
29 Jul	1.00	0.67	0.33	1.05	77	140	9	19	1.01	58	100	Continuing growth
12 Aug	1.00	1.00	0.00	1.05	132	240	9	19	1.00	132	240	Reached maximum growth of crop and roots
21 Sep	1.00	1.00	0.00	1.05	132	240	9	19	1.00	132	240	Moving towards harvest
16 Oct	0.30	1.00	0.00	1.05	132	240	9	19	0.30	132	240	Harvest
17 Oct	–	0.00	1.00	1.05	–	–	9	19	1.05	9	19	No growth following harvest
1 Nov	1.00	0.00	1.00	1.05	9	19	9	19	1.05	9	19	Sowing sunflower oil seeds
4 Nov	1.01	0.05	0.95	1.05	14	30	9	19	1.05	10	20	Crop becoming established
7 Dec	1.03	0.50	0.50	1.05	50	111	9	19	1.04	30	65	Mid growth
30 Dec	1.05	1.00	0.00	1.05	86	192	9	19	1.05	86	192	Reached maximum growth of crop and roots
13 Feb	1.05	1.00	0.00	1.05	86	192	9	19	1.05	86	192	Moving towards harvest
10 Mar	0.35	1.00	0.00	1.05	86	192	9	19	0.35	86	192	Harvest
11 Mar	–	0.00	1.00	1.05	9	19	9	19	1.05	9	19	No crops until late June

Irrigation consists of eleven water applications each of 50.8 mm (2 inches); *FRACSTOR* = 0.55.

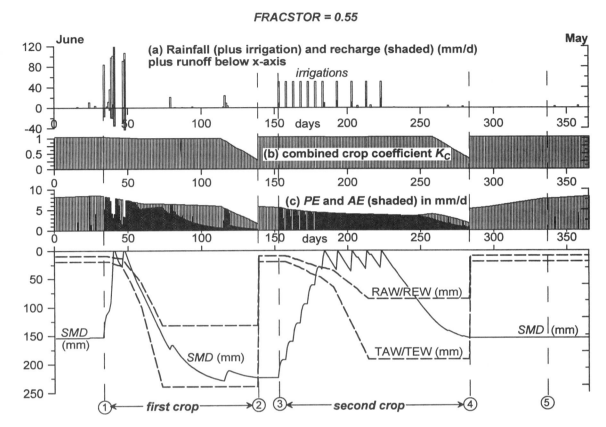

Figure 3.20 Daily soil moisture balance and recharge estimation due to rainfall and irrigation in western India

ripening then occurred and the crop was harvested on 16 October. This process and the corresponding rainfall, soil moisture deficit and recharge are shown between lines ① and ② on Figure 3.20. Very heavy rainfall occurred soon after planting; although some of the water was lost by runoff, there was sufficient infiltration to quickly reduce the soil moisture deficit to zero. Consequently, the heavy rainfall in July resulted in a recharge of more than 200 mm (note that an accurate estimate of recharge requires a good estimate of the magnitude of the runoff). From 19 July there was only sporadic rainfall; this meant that after day 55 the soil moisture deficit is always greater than the Readily Available Water. Therefore the actual evapotranspiration is significantly lower than the potential rate apart from two periods when rainfall occurs. Consequently the millet crop was under severe stress with a resultant reduction in the quality and yield of the crop.

A second crop of sunflower oilseed was planted on 1 November at the time of the first irrigation of 50.8 mm; this is indicated by line ③. Harvest is indicated by line ④. On eleven occasions there is an application of irrigation water of 50.8 mm (2 inch); applications initially occur at five-day intervals increasing later to ten-day intervals. During the first five applications, water moves down to reduce the soil moisture deficit. Since $SMD' >$ TAW the actual evapotranspiration is small on the days before the next irrigation. For the main stages of crop growth the actual evapotranspiration is at the potential rate because $SMD' <$ RAW. Consequently the crop remains healthy. The final five irrigations are higher than required; accordingly the soil moisture deficit becomes zero and recharge occurs totalling 67 mm.

When harvest of the second crop occurs the soil moisture deficit is 154.3 mm. Throughout the dry season, typified by the line ⑤, the soil moisture deficit remains greater than the Total Evaporable Water of 19 mm so that no water is drawn up to the soil surface for evaporation. Therefore, it is only on days with rainfall that any evaporation occurs. The soil moisture deficit remains at 154.3 mm; this is the starting value for the calculation on 1 June.

This example is chosen to illustrate how the soil water balance method can be used to estimate the potential recharge due to the combined inputs of rainfall and irrigation. This computational model can also be used to develop more efficient irrigation scheduling.

3.4.12 Bypass recharge

Bypass recharge is a process in which some of the rainfall less runoff bypasses the soil moisture balance to move rapidly to the aquifer. Recovery in groundwater levels (even though there is a soil moisture deficit) indicates the occurrence of bypass recharge (Mander and Greenfield 1978, Rushton *et al*. 1989). Many workers have observed that there is a recovery in groundwater hydrographs in chalk and limestone aquifers following heavy summer rainfall. A number of physical explanations have been proposed including a *bypass mechanism* whereby water travels through some form of preferential pathways to the water table (Hendrickx and Walker 1997). An alternative mechanism of bypass recharge for the outcrop of the South Humberside Chalk is that surface runoff on the steep-sided hills occurs following heavy rainfall; this runoff quickly moves to lower ground and then through stream beds into the aquifer (Rushton and Ward 1979). Whatever the physical explanation, various workers introduce bypass recharge which equals a fraction of the precipitation (or precipitation minus potential total evaporation) in excess of a threshold value (Smith *et al*. 1970). The adequacy of the predictions can be assessed by comparisons with the timing of hydrograph recoveries.

This feature can be included in recharge calculations by taking a proportion of the rainfall, or the effective rainfall ($Pr - PE$), and adding it directly to the potential recharge. Whenever some of the precipitation is included as bypass recharge, the bypass contribution must be subtracted from the infiltration available for the soil moisture balance.

3.4.13 Estimating catchment-wide recharge distributions

The preceding sections have shown that many factors influence potential recharge. Ideally, a separate daily water balance should be carried out whenever the crop type, soil type, rainfall or potential total evaporation changes. However, it is important to be pragmatic and not make the procedure of catchment-wide recharge estimation too computationally complex. One of the keys to successful recharge estimation is the ability to check that the individual calculations are realistic by comparing with field evidence such as whether the estimated actual total evaporation is consistent with the experience of farmers.

Whenever detailed field information is available, calculations should be carried out for 1 km squares. A good estimate of the rainfall is usually the most important factor. Any method which interpolates linearly between rain gauges should be treated with caution since the topography often causes rain shadows or areas of enhanced rainfall. Areas with different crop/soil combinations must be identified; satellite imagery can be useful for this task. If significant land-use changes have occurred during the study period, this should be included in the recharge estimation.

In many situations, detailed field information is not available. Nevertheless, by performing a number of recharge estimations with different crop types, soil types, runoff parameters etc. it is possible to explore the sensitivity of recharge estimates to these parameter values.

3.5 ESTIMATING RECHARGE WHEN DRIFT IS PRESENT

3.5.1 Introduction

For many aquifer systems lenses, layers or zones of low permeability material between the base of the soil zone and the permeable aquifer restrict recharge. In this chapter the term *Drift* is used to describe the material from the base of the soil zone to the top of the main aquifer. When Drift is present, the *actual recharge* to the main aquifer system may be less than the *potential recharge* from the base of the soil zone.

It is not possible to measure directly the vertical flow through low permeability strata above the main aquifer. Nevertheless, indirect field evidence, which can often be obtained from walk-over surveys at different times of the year, can be invaluable in quantifying the likely vertical flow since ground conditions following heavy rainfall provide an indication of the ability of the Drift to transmit water. The provision of tile drains and drainage ditches is another important source of information. If the drains or ditches run for several days following a period of heavy rainfall, this is a clear sign that the Drift has a limited *effective* vertical permeability. Auger holes, trial pits and trial boreholes are also useful in determining the nature of low permeability strata and the likely effective vertical permeability.

Even though there are extensive low permeability zones within Drift deposits, this does not necessarily mean that the Drift restricts the recharge. In the Doncaster area (Section 9.3), there are areas of Drift, containing extensive clay lenses, which were expected to restrict recharge. However, during periods of heavy and continuous winter rainfall, no runoff or ponding is observed; this indicates that the clay lenses do not limit the recharge. Although the route followed by the recharging water is unknown, the actual recharge to the sandstone aquifer equals the potential recharge leaving the soil zone. The validity of this assumption is confirmed by the accuracy of the regional groundwater model of this area (see, for example, Figure 9.10).

There are *two alternative approaches* to the representation of reduced recharge due to the presence of Drift. One approach is to *factor* the potential recharge to represent the impact of the Drift. When this procedure is followed, control over the estimation remains with the investigator. The alternative is to use *Darcy's Law* to calculate how much water passes through the Drift by multiplying an estimated effective vertical hydraulic conductivity by the vertical hydraulic gradient. When the effective vertical hydraulic conductivity is low (i.e. less than 0.0001 m/d), this approach is often reliable but, with higher estimates of vertical hydraulic conductivity, there is a risk that the calculated recharge through the Drift is higher than the potential recharge.

3.5.2 Recharge factors

Recharge factors reflect the nature of the Drift. When data are available about the thickness and lithology of the Drift together with information about the runoff, which is an indicator of rejected recharge, factors can be introduced to estimate the actual recharge to the underlying aquifer as a fraction of the potential recharge. Table 3.12 contains typical recharge factors

for a representative location. When the Drift is all sand the recharge factor is 1.0. However, for sandy clay Drift more than 10 m thick containing substantial clay lenses, an estimated 3 per cent of the potential recharge enters the aquifer system.

The recharge factor approach was used successfully in a study of recharge to Drift covered areas of the Lower Mersey Sandstone aquifer; for detailed information see Section 9.4.2 and in particular Table 9.3. Where the Drift has a high sand content the recharge factor is 1.0 so that the actual recharge equals the potential recharge. For the lower permeability sandy clay drift, a recharge factor of 0.02 was identified based partly on grading curves, but supported by a sensitivity analysis using a numerical model. Furthermore, in the Lower Mersey study, the influence of the hydraulic gradient across the Drift is represented by a further multiplying factor as illustrated in Figure 9.15f. This additional multiplying factor equals 1.0 when the main aquifer water table is below the Drift. The other limit is a factor of zero when the groundwater head in the main aquifer is at the same elevation as the water table in the soil zone. The inclusion of this additional multiplying factor for the Lower Mersey Sandstone aquifer study resulted in a gradual increase in actual recharge of almost 25 per cent following the lowering of the main aquifer water table due to long-term heavy pumping.

For unpumped unconfined aquifers overlain by low-permeability Drift, the actual recharge can be approximated by a specified long-term average recharge. For instance, in the Gipping catchment (see Section 11.5) recharge through the overlying boulder clay is set at 24 mm/yr (Jackson and Rushton 1987). This value was first estimated from the nature of the boulder clay and the long-term potential recharge but subsequently refined during numerical model development. For the Southern Lincolnshire Limestone investigation (Section 11.3) recharge through boulder clay is estimated at 18 mm/yr. For both of these catchments, much of the rejected recharge becomes runoff and subsequently enters the main aquifer system in Drift-free areas.

Table 3.12 Typical recharge factors for a representative location

Thickness/Drift type	Sand	Clayey-sand	Sandy clay	Clay
0 to 3 m	1.00	0.90	0.20	0.02
3 to 10 m	1.00	0.80	0.10	0.01
>10 m	1.00	0.60	0.03	0.00

3.5.3 Recharge through Drift estimated using Darcy's Law

In the second approach, recharge through the Drift is calculated from the product of the vertical hydraulic gradient and the effective vertical hydraulic conductivity. This calculation is illustrated by an alluvial aquifer in India. Due to extensive canal irrigation and in particular leaking canals and watercourses, waterlogging was observed at the ground surface. The aquifer system, Figure 3.21a, consists of an upper sand zone with a perched water table overlying a low permeability zone 8.0 m thick which contains extensive clay lenses. The low permeability zone is underlain by a sand layer and a permeable aquifer zone. Wells into the underlying aquifer show that, due to heavy pumping, the main aquifer water table is a few metres below the base of the low permeability zone. The presence of a perched water table indicates that the quantity of water from the leaking canals and watercourses is greater than the quantity which can move vertically through the low permeability zone. Using the parameter values of Figure 3.21a the vertical gradient across the low permeability zone is 8.6/8.0 (approximately atmospheric conditions apply beneath the low perme-

ability zone) hence the vertical flow with a vertical hydraulic conductivity of 0.0005 m/d is 0.00054 m/d (0.54 mm/d). This vertical flow is lower than the canal and watercourse losses, which are equivalent to at least 2 mm/d.

Reliable estimates of the effective vertical hydraulic conductivity of the Drift are required for this method of actual recharge estimation. However, there is no field technique available by which a direct estimate can be made. For zones with predominantly low permeability material, vertical hydraulic conductivity values obtained from laboratory tests may provide a realistic estimate of the effective vertical hydraulic conductivity. Numerical model studies have shown that if the Drift consists of inter-fingered lenses and layers of different materials, the effective vertical hydraulic conductivity is typically double the value when all the low-conductivity layers are combined as a single continuous layer (see Section 4.2.3 and Figure 4.5).

A further feature is illustrated by Figure 3.21b in which the groundwater (piezometric) head is above the top of the underlying aquifer. Under these conditions the vertical hydraulic gradient is 3.6/8.0, hence the estimated vertical flow with $K_z = 0.0005$ m/d is 0.00023 m/d. As more water is pumped from the underlying aquifer, the groundwater head is likely to fall, thereby increasing the flow through the Drift.

A second case study is introduced to demonstrate that vertical recharge can occur through strata which are often considered to be effectively impermeable. The study area is a limestone aquifer in the Cotswolds, UK. Latton pumping station, in the confined part of the Great Oolite limestone aquifer, provides a peak supply of up to 30 Ml/d but this results in a fall in piezometric head over several km² of more than 25 m. The high yield at Latton cannot be fully explained by flows from the adjacent unconfined area. However, for a region of about 16 km² adjacent to Latton pumping station, the Great Oolite is overlain by the Forest Marble and Cornbrash (Figure 3.22a). Elsewhere in the confined region the uppermost stratum is the very low permeability Oxford Clay; Figure 3.22b. Since the Forest Marble consists of a variable sequence of intercalated limestone and clays, with a predominance of clays in the upper part of the formation (Rushton *et al.* 1992), it is usually considered to be effectively impermeable. Nevertheless, it is important to consider whether water can be drawn from the overlying Cornbrash minor aquifer through the Forest Marble into the limestone. The physical situation is similar to that in Figure 3.21a with the Forest Marble acting as a low permeability zone and the Great Oolite limestone as the underlying

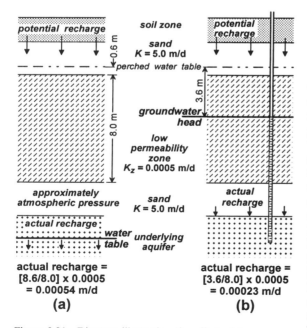

Figure 3.21 Diagram illustrating the effect of low permeability drift in limiting recharge: (a) water table in underlying aquifer, (b) underlying aquifer under a confining pressure

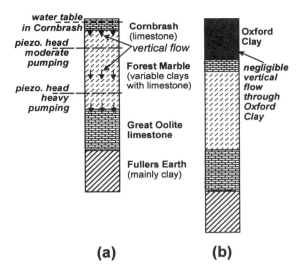

Figure 3.22 Vertical lithology associated with Great Oolite aquifer in the Cotswolds: (a) recharge from the Cornbrash through the Forest Marble clays into the Great Oolite, (b) no recharge due to overlying Oxford Clay

aquifer. Preliminary calculations indicate that the effective vertical hydraulic conductivity of the 25 m thickness of Forest Marble is 0.0005 m/d. With the piezometric head 20 m below the water table in the Cornbrash, 6.4 Ml/d of the abstraction from Latton can be supplied by vertical flows. It is essential to check whether this quantity of water can be supplied by the overlying Cornbrash aquifer. For the vertical gradient of 20/25, the maximum vertical flow is 0.0004 m/d or 146 mm/yr; this is less than the average potential recharge to the Cornbrash. When this mechanism of vertical flow through the Forest Marble is incorporated in a mathematical model of the Great Oolite aquifer system (Rushton *et al.* 1992), the model response in the vicinity of Latton is similar to that observed in the field.

A valuable and detailed study of a regional till aquitard is described by Gerber and Howard (2000). Vertical flow through the till is a major cause of recharge, but estimates of the bulk vertical hydraulic conductivity of the aquitard, derived from a variety of techniques, vary from 10^{-12} to 10^{-5} m/s (approximately 10^{-7} to 1.0 m/d). To resolve this uncertainty, a seven-layer MODFLOW model was developed. In a sensitivity analysis, the bulk vertical hydraulic conductivity of the till is varied with comparisons between model results and field values for both river–aquifer flow

and groundwater heads. Acceptable simulations are obtained with the vertical hydraulic conductivity in the range 5×10^{-10} to 5×10^{-9} m/s (0.00005 to 0.0005 m/d approximately).

With the wide availability of multi-layered regional groundwater models, there is an increasing tendency to use an upper layer of the model to represent low-permeability drift. This means that the actual recharge is determined by the model rather than from calculations such as those described above. On occasions, unrealistic recharge estimates have resulted from this procedure. There are risks in allowing the regional groundwater model to calculate the actual recharge from the product of the vertical hydraulic conductivity and the vertical hydraulic gradient. The following warnings are all derived from case studies in which the recharge estimates proved to be unreliable.

1. The assumed effective vertical permeability may result in a calculated vertical flow (i.e. a recharge) which is not consistent with the magnitude of the potential recharge; therefore it is essential to check that the flow through the Drift is less than the potential recharge estimated from a soil moisture balance.
2. Although in the pre-pumping condition (or historical pumping condition) the vertical flow through the Drift may be less than the potential recharge, extensive pumping can lead to far more water being drawn through the low permeability layer than is provided by the potential recharge.
3. If the low-permeability drift is partly unsaturated, this leads to lower values of the vertical hydraulic conductivity and hence reduced recharge.
4. When the water table is below the low permeability layer (Figure 3.21a), the low-permeability layer becomes 'disconnected' so that the flow does not increase as the groundwater head in the underlying aquifer is lowered further.

In many regional aquifer studies, the recharge is estimated using a combination of the methods outlined in Sections 3.4 and 3.5. Table 3.13 summarises a number of investigations in which recharge through Drift proved to be important.

3.5.4 Delay in recharge reaching the water table

For drift-covered aquifers there may be a delay between the potential recharge leaving the base of the soil zone and a response at the permanent water table. This delay can be identified from a study of the groundwater head hydrographs (which need to be based on weekly

Table 3.13   Examples of the importance of vertical flows through low permeability strata

Example	Nature of drift	Main methods of estimation
Lower Mersey Sandstone, UK	Boulder clay, reducing groundwater head in underlying sandstone	Actual recharge = percentage of potential recharge depending on nature and thickness of drift; increase in recharge due to falling groundwater heads (Section 9.4)
Gipping Chalk, UK	Boulder clay with underlying gravels above low transmissivity chalk	Steady vertical flow through unsaturated zone of 24 mm/yr (Sections 3.5 and 11.5)
Southern Lincolnshire Limestone, UK	Boulder clay	Steady vertical flow through unsaturated zone of 18 mm/yr (Sections 3.5 and 11.3)
Cotswolds Confined Gt. Oolite Lmst. UK	Cornbrash and Forest Marble formations	Vertical flow (leakage) due to difference between head in Great Oolite and water table in Cornbrash (Section 3.5)
Sherwood Sandstone in Notts and Doncaster, UK	Mercia Mudstone, Colwick Formation, 25-foot clays, more permeable alluvium and clays	Vertical flow downwards (leakage) or outflows (upwards) depending on vertical head gradients (Sections 9.2 and 9.3)
Mehsana alluvial, India	Alluvial clays	Leaking canals cause perched water table, deeper water tables in underlying aquifers (Section 10.2)
Vanathavillu, Sri Lanka	Clays, silts and sands above a limestone aquifer	Vertical hydraulic conductivity estimate from leaky aquifer analysis which ignores aquitard storage; estimated K_z and hence recharge too high (Section 10.3)
Barind, Bangladesh	Barind Clay	Study of stratigraphy and paleo-climate shows that K_z is sufficiently high to transmit potential recharge (Section 9.5)
Illinois, USA	Some shales but mostly glacial drift	From the vertical gradient and vertical hydraulic conductivity, vertical flows estimated to be 1330–500 000 US gal/sq mi (0.013–4.9 mm/d) (Walton 1970)
Ontario, Canada	Northern till	Sensitivity analysis with a multi-layer MODFLOW model, 0.00004–0.0004 m/d (Gerber and Howard 2000)
Cardiff Bay, UK	Estuarine alluvium overlying river gravels	K_z in the range 0.0001 to 0.001 m/d; if too high a value of K_z, drains and sewers unable to control water table (Crompton and Heathcote 1994)
Madras aquifer, India	Alluvial aquifer, main aquifer overlain by clays	Partial dewatering of main aquifer due to abstractions (Krishnasamy and Sakthivadivel 1986)

readings) and daily recharge estimates. Due to the highly complex nature of the flow through Drift, it is not possible to develop precise mathematical models for the flow processes since the actual mechanisms by which water moves from beneath the soil zone through the unsaturated zone to the water table are not well understood. The actual movement of water particles is slow (as demonstrated by the use of tritium as a tracer or the movement of nitrates) at typically one metre per year vertically. However, some form of 'piston flow' mechanism occurs whereby, when water enters the upper part of the unsaturated zone, a similar quantity of water crosses the water table. For more permeable unsaturated zones the transfer is rapid, especially where the unsaturated zone contains vertical fissures or other higher permeability pathways. In other situations, there is a measurable delay between potential recharge occurring and the response in observation well hydrographs at the water table. Consequently, the delays must be identified from field evidence.

Estimated delays observed in the Nottinghamshire Sherwood Sandstone aquifer between the onset of potential recharge and changes in observation well hydrographs are used to illustrate the methodology. The magnitude of the delay depends on the unsaturated thickness. Consideration of the vertical movement of water through a column of low permeability aquifer material (Senarath and Rushton 1984) shows that a pulse of one-day duration entering the top of the column leaves the bottom of the column as a smoothed distribution over a period of many days. Since the regional groundwater model for the Nottinghamshire Sherwood Sandstone aquifer system uses a monthly time step, average monthly recharge values are calculated with the recharge for month M distributed over months M, $M + 1$, $M + 2$, $M + 3$ etc. For a drift thickness of up to 10 m, 90 per cent of that month's recharge enters the aquifer during the current month with 10 per cent in the following month. However, with a Drift thickness in excess of 50 m, only 10 per cent of the recharge enters during the current month with 15 per cent, 20 per cent, 30 per cent, 20 per cent and 5 per cent entering in the next five months. The delay factors for the Nottinghamshire Sherwood Sandstone aquifer are listed in Table 9.2.

A further step is required to determine the actual recharge to the aquifer during the month Mn. For a drift thickness of up to 10 m, it is the current month Mn and previous month $Mn - 1$ which provide recharge. For Drift with a thickness in excess of 50 m, it is the current and the five previous months which contribute to the recharge. This procedure is summarised in Table 3.14.

3.6 DELAYED RUNOFF AND RUNOFF RECHARGE

3.6.1 Delayed runoff from minor aquifers in the Drift

Water which does not pass through the Drift to become recharge leaves the drift as *delayed runoff*; this runoff may become recharge when the streams cross more permeable parts of the aquifer, or alternatively it will add to the total stream or river flow. Field observations indicate that delayed runoff leaves the sub-surface system as seepages and springs from the drift forming small streams which continue to flow for several weeks after periods of heavy rainfall. Seepages and spring flows tend to be highest in wet winters; there may be little or no spring-fed runoff in dry winters. In wet summers significant delayed runoff also occurs.

The drift may be alluvium consisting of clays, sands and gravels which act as minor aquifers storing water and then slowly releasing water as delayed runoff. Boulder clay (till) is another form of drift; it usually consists of lenses or layers of moderate to low permeability materials. These strata allow a small downward flow to the underlying aquifer. In addition water is stored within permeable zones (sand lenses or minor aquifers); this stored water is released over a period of time as delayed runoff.

Rarely is there sufficient information about the Drift to develop a precise understanding of the delayed runoff processes, hence a simplified conceptual model is devised to represent the important responses; Figure 3.23. In the schematic diagram Figure 3.23, potential recharge enters the Drift from the soil zone. The storage properties of the various minor aquifers are represented as a storage reservoir. Some water leaves this store by a steady vertical downwards flow to become recharge at the underlying water table (see Section 3.4) but most leaves the store to become delayed runoff (sometimes called interflow). The

Table 3.14 Estimation of actual recharge for month Mn from recharge during the current and previous months for the Nottinghamshire Sherwood sandstone aquifer; values derived from Table 9.2

Thickness of unsaturated zone	Month					
	$Mn - 5$	$Mn - 4$	$Mn - 3$	$Mn - 2$	$Mn - 1$	Mn
0 to 10 m	–	–	–	–	0.10 $R(Mn - 1)$	0.90 $R(Mn)$
10 to 30 m	–	–	–	0.20 $R(Mn - 2)$	0.50 $R(Mn - 1)$	0.30 $R(Mn)$
30 to 50 m	–	0.05 $R(Mn - 4)$	0.10 $R(Mn - 3)$	0.35 $R(Mn - 2)$	0.30 $R(Mn - 1)$	0.20 $R(Mn)$
50 m +	0.05 $R(Mn - 5)$	0.20 $R(Mn - 4)$	0.30 $R(Mn - 3)$	0.20 $R(Mn - 2)$	0.15 $R(Mn - 1)$	0.10 $R(Mn)$

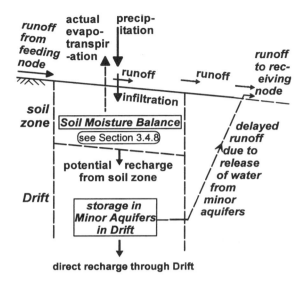

Figure 3.23   Conceptual model of flow processes at the soil surface, within the soil and in the boulder clay drift together with delayed runoff from the minor aquifers. Reprinted from *Journal of Hydrology* **211**, Bradbury and Rushton, Estimating runoff-recharge in the Southern Lincolnshire Limestone catchment, 86–99, copyright (1998) with permission from Elsevier Science

Table 3.15   Example showing release from storage in the Drift; the only inflow to the drift storage is 20 mm on day 1, the release coefficient is 0.15 day^{-1}

Day	In Store (mm)	Release (mm/d)	Day	In Store (mm)	Release (mm/d)
1	20.000	3.000	9	5.450	0.817
2	17.000	2.550	10	4.632	0.695
3	14.450	2.168	11	3.973	0.591
4	12.283	1.842	12	3.347	0.502
5	10.440	1.566	13	2.845	0.427
6	9.874	1.331	14	2.418	0.363
7	7.543	1.131	15	2.055	0.308
8	6.412	0.962	16	1.747	0.262

inclined upwards arrow represents the release of water from Drift to become delayed runoff. In practice this flow will reach the ground surface as springs or seepages some distance from the point of infiltration. These outlets are usually at a lower elevation than the storage zones in the Drift, hence the flow is not physically upwards as indicated by the delayed runoff arrow in Figure 3.23. Farmers often construct drainage ditches to collect and dispose of this water.

The computational technique which represents delayed runoff involves releasing each day a proportion of the water currently stored in the sub-soil; this is equivalent to a release of water which decreases exponentially. Table 3.15 refers to an idealised example in which 20 mm of water enters the drift storage on day 1 with no further inflows on subsequent days. The water is released according to a delay coefficient of 0.15 day^{-1}. This means that on the first day, 15 per cent of the water in storage (20 mm) is released, hence the release is 3.0 mm/d leaving 17 mm in store at the start of day two. On day 2 the release is $0.15 \times 17.00 = 2.55$ mm/d. By day 15 about 10 per cent of the water is still stored and the release on that day is just over one tenth of the release on day 1.

Similar responses can be observed when areas are drained by tile drains; however, the delay in releasing water is much shorter than when water is stored in and released from minor aquifers.

Detailed studies of storage of water and delayed runoff from the Drift were possible in the Southern Lincolnshire Limestone aquifer, UK, due to the availability of measurements from a stream flow gauge located on the margins of the boulder clay. These flow records were invaluable in identifying and quantifying the water released from the strata beneath the soil zone but above the main aquifer. The stream flow measurements showed responses due both to runoff directly following rainfall (i.e. water which did not enter the soil zone) and also water stored and slowly released from minor aquifers. The computational model of Figure 3.23 represents these processes. Water moves from below the soil zone into the drift store, some water (0.05 mm/d) drains from this store to move vertically to the underlying aquifer. Water is also released to become interflow or runoff into ditches and streams. For boulder clay the release is 8 per cent each day, hence the release factor is 0.08 d^{-1}.

There is a small boulder clay catchment at Easton Woods (the location is shown in Figure 11.6) covering an area of about 6 km^2 for which continuous flow measurements are available. In Figure 3.24 a comparison is made between the measured runoff in 1984 (unbroken line) and the runoff determined by summing the calculated surface runoff and the water released from minor aquifers for the Easton Wood catchment (broken line). Agreement between field and modelled flows at Easton Wood is acceptable. At the end of January there is an obvious mismatch; also there is a slight over-estimate of runoff in the summer. However, detailed differences are not significant since the runoff

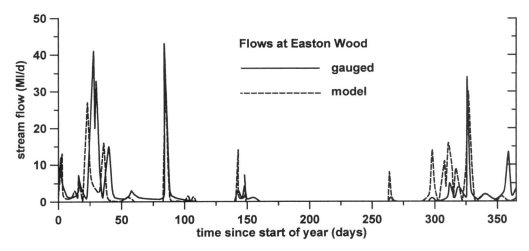

Figure 3.24   Runoff from Easton Wood; comparison between gauged and model results. Reprinted from *Journal of Hydrology* **211**, Bradbury and Rushton, Estimating runoff-recharge in the Southern Lincolnshire Limestore catchment, UK, pp. 86–99, copyright (1998) with permission from Elsevier Science

estimates are part of the runoff recharge input to a regional groundwater model which has a time step of 15 days. More detailed information concerning the estimation of delayed runoff can be found in Bradbury and Rushton (1998).

3.6.2 Runoff recharge

Runoff recharge is an important input to many aquifer systems. Runoff recharge is water which flows from less permeable areas of the catchment in streams, drainage channels, ditches etc. onto the permeable aquifer so that this water can subsequently recharge the aquifer system. Many aquifer units in the UK have some inflows due to runoff recharge. The first example of runoff recharge was in the Worfe catchment of the Severn River basin (Miles and Rushton 1983); other examples include the River Pant catchment (Senarath and Rushton 1984), the Gipping catchment (Jackson and Rushton 1987), the Lodes and River Granta (Rushton and Fawthrop 1991) and the Southern Lincolnshire Limestone (Bradbury and Rushton 1998). Detailed descriptions of runoff recharge for the Southern Lincolnshire Limestone can be found in Section 11.3 and, for the Gipping catchment, in Section 11.5. Information from these sections provides insights which can be used to develop conceptual and computational models of runoff recharge for other aquifer systems.

Stone *et al.* (2001) describe a methodology for estimating recharge due to runoff in the Great Basin region of south-west USA. Their first task was to form a conceptual understanding of the primary areas of precipitation, runoff and infiltration within the basin. Precipitation–elevation relationships are derived. Using a digital elevation model combined with GIS techniques and an empirical surface runoff model, the runoff recharge is estimated and added to the potential recharge in 'receiving' sub-basins.

Essential features of the routing approach for estimating runoff recharge are summarised below.

1. Identify the area of the surface water catchment which provides runoff. This runoff subsequently becomes recharge as the ditches, drainage channels, streams etc. cross onto the aquifer outcrop.
2. Divide the catchment area into cells (1.0 km^2 is usually adequate); identify the route taken by the direct or delayed runoff as it moves down-gradient until the water reaches the margins of the aquifer. Topographical maps and digital terrain information can be useful but field observations are essential. Figure 3.25 is a representative simplified diagram of this process.
3. Perform a daily soil moisture balance for each cell; also calculate the storage in minor aquifers and, hence, the delayed runoff.
4. Route the runoff down the streams to the margins of the low permeability strata. It may be appropriate

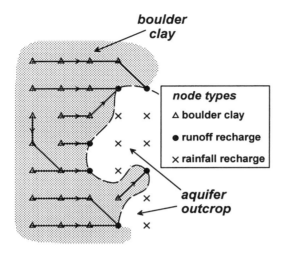

**boulder
clay**

node types
△ boulder clay
● runoff recharge
× rainfall recharge

**aquifer
outcrop**

Figure 3.25 Estimation of runoff-recharge to the margins of the aquifer at locations indicated by solid circles; runoff-recharge occurs due to direct and delayed runoff from cells on the overlying boulder clay

to incorporate some delay as the water moves across the catchment but the delay will not be critical if the time step of the regional groundwater model is 15 days. This procedure provides estimates of the runoff recharge at a number of locations on the margins of the low permeability strata which are indicated in Figure 3.25 by the solid circles. Conventional rainfall recharge occurs at the locations indicated by crosses.

5. Runoff recharge may be rejected by the aquifer when water tables are high; preferably this rejection is determined by a regional groundwater model of the area.

3.6.3 Incorporating several recharge processes

The East Kent Chalk aquifer system is an example of when alternative methods are required to represent precipitation recharge and the impact of different kinds of drift. One distinctive feature of this aquifer system is that the valleys are deeply incised below the interfluves; differences in elevation can be 100 m with the distance between adjacent valleys as little as 2 to 3 km. There is a distinctive contrast between the transmissivities of the interfluves and the valleys; there is also a substantial variation in transmissivity between high and low water tables (Cross *et al.* 1995). Extensive drift deposits occur on the high ground, also alluvium is present in

many of the valleys. Recharge estimation is based on the following principles.

- In areas where Head (reworked material equivalent to a clayey sand of not more than 2–3 m thickness with an effective vertical hydraulic conductivity greater than 0.005 m/d) or alluvium (which again has an effective vertical hydraulic conductivity greater than 0.005 m/d and usually lies directly above the Chalk) is present, the actual recharge equals the potential recharge.
- For Chalk with a thin soil cover, where observation well hydrographs shows small recoveries following heavy summer rainfall, a bypass recharge is included as 15 per cent, of the precipitation with a threshold of 5 mm/d (see Section 3.4.12).
- There are also extensive deposits of Clay-with-Flints covering most of the hills. Due to their low permeability there is only a small drainage through the Clay-with-Flints. In some locations, farmers have dug holes through the deposits to the underlying Chalk to obtain material to make the soil more workable and to provide a route for drainage. Due to the low permeability, runoff coefficients are higher than those quoted in Table 3.5. A small vertical flow of 0.05 mm/d is assumed to pass through the Clay-with-Flints but the extensive drainage intercepts most potential recharge (negative soil moisture deficit). The runoff, which consists of direct runoff and rejected recharge, is routed down-gradient to the margins of the Clay-with-Flints. Since the regional groundwater model uses a 1-km square grid on the interfluves, this runoff often becomes recharge at adjacent nodes. However, where the Clay-with-Flints deposits are extensive, the runoff may be transferred several kilometres before it is added to the rainfall recharge at a drift free node.

3.7 CONCLUDING REMARKS

The main purpose of this chapter is to provide detailed information about the soil moisture balance method, which is a straightforward technique for estimating recharge based on readily available field information. However, the chapter also contains a brief review of alternative methods of recharge estimation including techniques based on field experiments using lysimeters or identifying the movement of tracers in the soil zone with time. Alternatively, techniques based on soil physics can be used either by taking extensive measurements in the soil profile or using a numerical solution for flow in the unsaturated zone. These alternative methods provide valuable insights into the recharge processes; for further information see Lerner *et al.* (1990), Simmers (1997) and Scanlon and Cook (2002). These methods are not suitable for the routine recharge estimation required for regional groundwater studies.

To obtain reliable recharge estimates using the soil moisture balance technique, it is essential to ensure that the approach is consistent with physical insights gained from the alternative methods referred to above. Furthermore, the calculation must represent soil moisture conditions throughout the year. Because recharge only occurs on a few days each year, and the recharge is of a small magnitude compared to the precipitation and actual evapotranspiration, care is required in developing quantified conceptual models which accurately represent the physical processes. Of particular importance is the need to perform a daily calculation since a larger time step may fail to identify certain days when recharge occurs.

Soil moisture balance techniques have been used for more than forty years. However, questions have been raised about certain concepts such as the root constant. The soil moisture balance technique described in this chapter is based on a reappraisal of the conceptual model and the field properties which influence soil moisture conditions. Therefore the calculation requires information about specific soil properties and about the crop including its water requirements, root development and behaviour under stress. In addition, bare soil evaporation is included plus information about the relative areas covered by crop and bare soil throughout the year. Some crop and soil properties are quoted in the tables in this chapter, but more local information is required for detailed studies. One new feature introduced in this chapter is the concept of near surface soil storage. When there is substantial rainfall at a time when the soil moisture deficit is greater than the readily available water (or readily evaporable water), the conventional single soil store approach assumes that the excess water above the potential evapotranspiration goes to make up the soil moisture deficit and is not available for transpiration or evaporation on the following days. To represent the response observed in the field that water is retained near the soil surface for evapotranspiration on the following days, near-surface soil storage is introduced.

Frequently there is extensive drift cover between the soil zone and the saturated aquifer. This can limit the recharge. The representation of this drift in a numerical model as an upper layer with a low vertical hydraulic conductivity is not recommended because flow through drift as calculated by the model may not be consistent with the recharge. There are examples where the calculated recharge has been in excess of the actual recharge. When the vertical hydraulic conductivity is less than 0.0002 m/d (0.2 mm/d) it may be appropriate to include this as an upper layer in the numerical model, but careful checks must be made to ensure that the actual recharge rate is not exceeded.

Drift with a low hydraulic conductivity often has the further property of storing water and slowly releasing it as delayed runoff. This may subsequently infiltrate into the aquifer system as the rivers, streams and drainage ditches pass it on to more permeable strata. Conceptual and computational models which can represent these processes are described in the text. However more field studies are required to provide information to allow the development of conceptual models of runoff recharge for a greater range of physical situations.

4

Interaction between Surface Water and Groundwater

4.1 INTRODUCTION

Significant interaction between groundwater and surface water bodies occurs in many practical situations. Surface water–groundwater interaction is important in irrigation schemes where losses occur from canals or from flooded ricefields or when sub-surface drains are used to control the water table. A second group of problems is concerned with aquifer–river interaction and groundwater flow to springs. A further natural surface water–groundwater interaction occurs when there is contact between an aquifer and a lake; an overall water balance including the effect of evaporation from the surface water body is necessary to quantify this process. Similar conditions arise with wetlands where the volume of the surface water is smaller so that changes in groundwater–surface water interaction may have a more critical effect on the surface water conditions. A detailed review of interactions between groundwater and surface water by Sophocleous (2002) emphasises the hydrogeological framework, groundwater resources implications and the impact on ecosystems.

When studying surface water–groundwater interaction, the first step is to identify the important mechanisms and develop reliable conceptual models. Earlier methods of analysis for many of these problems involve major assumptions introduced to fit available mathematical solutions. For instance, with horizontal sub-surface drains it is often assumed that the water table intersects the drain, whereas in practice the water table usually lies above the drain. When analytical methods were the only means of studying these problems, such simplifying assumptions could be justified.

However, with the wide availability today of numerical methods of analysis, fewer simplifying assumptions need to be made. Consequently, for each of the surface water–groundwater interactions discussed in this chapter, there is a thorough reappraisal of the physical situation; this is followed by the development of appropriate conceptual models.

The first topic to be considered is the interaction between canals and aquifers. Different approaches are required when the groundwater head is above the canal water level or significantly below the canal water level. Both analytical and numerical methods are used; the analyses relate to vertical cross-sections with steady state conditions. The next topic is the interaction of aquifers with springs, rivers and lakes. Some of the concepts developed for canals are directly applicable. However, for springs, rivers and lakes it is necessary to consider time-variant responses. Coefficients describing the interaction with groundwater should be determined from field information. For tile drains a thorough reappraisal is necessary since conventional drainage theory is often based on assumptions introduced for mathematical convenience. Additional features, which are not included in most current methods of analysis, include time-variant responses and the limited flow capacity of drains. The final section is concerned with the vitally important problem of the loss of water from flooded ricefields to the underlying aquifer. Significant losses can occur through the bunds between ricefields. The study of this problem requires the development of valid conceptual models and appropriate numerical methods of analysis.

Groundwater Hydrology: Conceptual and Computational Models. K.R. Rushton
© 2003 John Wiley & Sons Ltd   ISBN: 0-470-85004-3

4.2 CANALS

4.2.1 Introduction

Canals often lose water to the underlying aquifer; this has been recognised for many years. In 1883, following the construction of major canals in the Indian sub-continent, measurements of flows indicated major losses of up to 30 per cent (Ahmad 1974). These canal losses are the main cause of serious water-logging which occurs in the vicinity of many canals. The importance of these conveyance losses in canals has been confirmed by Bos and Nugteren (1982), who collated field results from a large number of irrigation schemes. Conveyance and distribution losses typically account for at least 50 per cent of the water entering the canal system. For design purposes, seepage losses are often assumed to be proportional to the wetted perimeter of the canal; Ahmad (1974) quotes values of $0.2\,\mathrm{m^3/d}$ per square metre of wetted perimeter for main canals and 0.05–$0.1\,\mathrm{m/d}$ for water-courses. Since losses due to evaporation from the water surface of the canal are rarely more than $0.007\,\mathrm{m/d}$ ($7\,\mathrm{mm/d}$), they can be ignored compared to the seepage losses. Note that in the following discussion of *quantitative* information about canal losses the losses are expressed per unit wetted area of the canal base and sides; however, for the *numerical solutions* discussed later in this section the estimated losses are quoted as total losses per metre length of canal.

Field studies have been carried out by a number of investigators. Worstell (1976) considered losses from canals in Idaho, USA; for medium clay loams, losses are typically 0.15–$0.45\,\mathrm{m/d}$, for pervious soils losses are in the range 0.45–$0.6\,\mathrm{m/d}$ while for gravels losses are 0.75–$1.0\,\mathrm{m/d}$. In the Ismailia Canal, Egypt, seepage meters were used to identify losses of 0.6–$2.4\,\mathrm{m/d}$ (Pontin *et al.* 1979). In the Kadulla irrigation scheme, Sri Lanka, ponding tests for an unlined canal passing through an aquifer of relatively low hydraulic conductivity, indicated losses of $0.13\,\mathrm{m/d}$ (Holmes *et al.* 1981). A lined section of the canal had losses as high as $0.12\,\mathrm{m/d}$; these losses were reduced by carefully filling the cracks. Field measurements in South India for main canals showed losses of $0.37\,\mathrm{m/d}$ and $0.11\,\mathrm{m/d}$ for unlined and lined canals respectively; the corresponding figures for small distributaries were 0.06 and $0.045\,\mathrm{m/d}$ (Wachyan and Rushton 1987).

In parallel with actual field measurements, physical, analytical and numerical models have been developed to estimate canal losses. Physical model studies have been carried out by Bhagavantam (1975) and Abdel-Gawad *et al.* (1993); information about model studies can be found in Section 4.2.4 and Figure 4.6. There have also been a number of analytical and numerical model studies. Analytical solutions for two-dimensional flow in a vertical section have been derived for a number of different idealisations (Harr 1962, Bouwer 1969, Sharma and Chawla 1979); approximations are made in the representation of the physical features and/or in the mathematical analysis. Numerical models using resistance networks (Bouwer 1969), finite difference models (Wachyan and Rushton 1987) and finite element techniques (Rastogi and Prasad 1992) have also provided valuable insights. Rastogi and Prasad show that variations in the hydraulic conductivity of the canal lining have less influence on the seepage losses than the aquifer hydraulic conductivity or changes in the depth of water in the canal.

4.2.2 Classification of interaction of canals with groundwater

Bouwer (1969, 1978) classified flows from trapezoidal channels to groundwater as indicated in Figure 4.1. Conditions A, A′ and B depend primarily on the nature of the stratum underlying the aquifer through which the canal passes, and Condition C occurs due to the low permeability of the channel lining.

Condition A: the aquifer through which the canal passes is underlain by a layer which has a far higher hydraulic conductivity than the aquifer. When the regional water table is within the aquifer, the conditions are as sketched in the upper left of Figure 4.1. However, if the regional water table lies in the underlying high hydraulic conductivity layer, this layer acts as a drain with a flow pattern as indicated under Condition A′.

Condition B: the aquifer through which the canal passes is underlain by a layer which has a much lower hydraulic conductivity than the aquifer, hence the underlying layer is taken to be impermeable.

Condition C: around the sides and bed of the channel there is a low permeability layer which may occur due to deposits of silt or lining of the channel. If the effective hydraulic conductivity of this layer significantly reduces flow from the channel, unsaturated conditions may occur beneath the canal with effectively vertical flow to the underlying water table.

Figure 4.1 is important because it illustrates the two alternative situations of the regional water table merging with the water surface in the canal (Conditions A and B) or the canal perched above the regional water table (Conditions A′ and C). Note, however, that for the first two examples, the conditions at the base of the aquifer are totally different. For the final two examples, saturated conditions occur for Condition A′ while

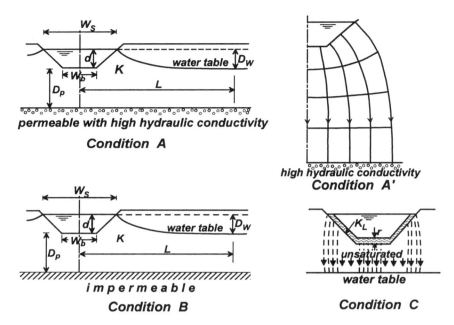

Figure 4.1 Classification of alternative conditions for canals: Bouwer, *Groundwater Hydrology*, 1978, McGraw-Hill. Reproduced with permission of The McGraw-Hill Companies

unsaturated conditions may occur for Condition C. Many practical field situations do not fit precisely into any of these categories. For instance there may be a lower permeability layer beneath the aquifer through which substantial vertical flows can occur. Alternatively, the low-permeability layer around the perimeter of the channel may reduce the outflow but saturated conditions may still occur in the aquifer. These and other possibilities are discussed in Sections 4.2.3 and 4.2.4.

For Condition A', where the flow from the canal is predominantly vertical, a number of analytical solutions are available (Harr 1962). These analytical solutions are applicable when the regional water table is a considerable distance below the canal water level.

Numerical solutions for Conditions A and B, using a resistance network analogue, are described by Bouwer (1969); typical results are reproduced in Figure 4.2. The channel is of a trapezoidal shape with side slopes of 1:1. The loss from the canal is defined as q_L per unit plan area of the water surface; hence the total loss from a unit length of canal $Q_c = q_L W_s$ where W_s is the width of the water surface. Sets of curves are presented for different ratios of d/W_b (depth of water in channel divided by width of base of channel); the curves for $d/W_b = 0.25$ are redrawn as in Figure 4.2. The horizontal axis of the graph is D_w/W_b (the depth to the regional water table below the canal water level

at the outer boundary D_w divided by the width of the base of the canal W_b). The full lines in Figure 4.2 refer to Condition A where there is a high permeability layer beneath the aquifer, individual lines are drawn for different ratios of D_p/W_b (ratio of the depth from base of canal to underlying highly permeable layer D_p divided by the width of the base of the canal W_b). For Condition B the results are indicated by broken lines.

When the aquifer is of infinite depth $D_p/W_b = \infty$, the two sets of curves merge, this condition is represented by a thicker line on Figure 4.2. Another limiting condition occurs when the water table for Condition A is at the same elevation as the top of the underlying high permeability zone; this is Condition A' of Bouwer's classification which is represented by the chain-dotted line. Note that for D_p/W_b greater than 3.0, Condition A' is effectively a horizontal line, this means that the loss from the canal remains constant at $Q_c/KW_s \approx 1.55$. This constant loss occurs because the flow is predominantly vertical with a unit vertical hydraulic gradient.

One important feature of these results is that they were obtained from a resistance network model having a limited lateral extent; the groundwater head was set at a fixed value on the outer vertical boundary at a distance $10W_b$ from the channel centre-line. Consequently, this lateral boundary condition is equivalent to a fully penetrating vertical channel at a distance from the

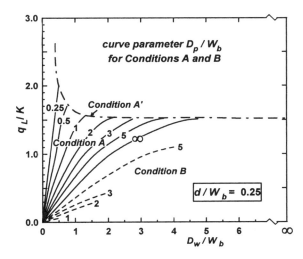

Figure 4.2 Typical design graph for canal losses: from Bouwer, *Groundwater Hydrology*, 1978, McGraw-Hill. Reproduced with permission of The McGraw-Hill Companies

canal centre-line of only $10W_b$ (ten times the width of the base of the canal), the channel is filled with water at an elevation well below the canal water level. This artificial lateral boundary is likely to have a substantial impact on the results.

4.2.3 More detailed consideration of boundary conditions and canal dimensions

Although Bouwer's analysis was an important step forward in understanding losses from canals, a number of assumptions were introduced which may not coincide with true field conditions. In particular the assumptions that the underlying layer is either highly permeable (Figure 4.3a), or totally impermeable with a fully penetrating vertical channel close to the canal (Figure 4.3b), are not consistent with normal field conditions. Guidance is given by the United States Bureau of Reclamation, that an effectively impermeable underlying boundary (or barrier layer) occurs when the hydraulic conductivity of the layer is less than one-fifth of the weighted hydraulic conductivity of the upper layer (USBR 1978). Furthermore, Ahmadi (1999) proposes a technique of using the vertical gradient between pairs of piezometers to identify the location of this 'impermeable layer'. However, regional groundwater studies reported in Chapter 10 indicate that substantial vertical flows can occur through low permeability layers even when the hydraulic conductivity is three orders of magnitude less than the main aquifer.

A numerical model study of the impact of a low permeability zone (neither highly permeable nor totally impermeable) at some depth below the canal indicates that losses from the canal are far higher than the calculated losses assuming that the underlying boundary is impermeable (Wachyan and Rushton 1987). These higher losses are often associated with a second aquifer beneath the low permeability zone from which deep borewells pump water. Due to these questions about the effect of underlying low permeability zones, further analyses were carried out with an alternative condition introduced on the lower boundary, Figure 4.3c, in which there is a low conductivity layer of thickness T_C and hydraulic conductivity K_C. The pressure beneath the low conductivity layer is atmospheric.

Representative results for these three problems, obtained using a finite difference vertical section model (see Section 2.5), are quoted in Figure 4.3. For each problem the width of the canal water surface is 4.0 m, the depth of water in the canal is 1.0 m and the side slopes of the canal are 45°. For example (a), Condition A′, with the highly permeable layer at a depth of 11.0 m below the canal water surface, the total loss from a one metre length of the canal is $Q_C = 6.80 K$ which, when divided by the width of the water surface, is equivalent to $q_L = 1.70 K$.

Figure 4.3b refers to *Condition B* with an impermeable base 11.0 m below the canal water surface. Water lost from the canal flows to a distant fixed head boundary which is equivalent to a deep channel fully penetrating the aquifer at 102 m from the canal centre-line with the water level in the drainage channel 6.0 m below the canal water level, i.e. $D_w = 6.0$ m. From the numerical model results, the total loss from the canal is calculated to be $Q_C = 0.93 K$. The hypothesis of a fixed head boundary is unlikely to occur in practice since it is equivalent to a drainage channel only 100 m from the canal with the water level in the drainage channel 6 m below the canal water surface level!

In example (c) there is no drainage channel, but at a distance of 102 m from the canal centre-line there is a condition of zero horizontal flow (this would occur with a series of canals 204 m apart). Consequently water can only leave the system by vertical flow through the low permeability layer which is 2.0 m thick and has a vertical permeability of 0.0021 times the aquifer permeability. The factor of 0.0021 is selected so that the water table elevation at 102 m from the canal centre-line is approximately 6 m lower than the elevation of the canal water surface (i.e. similar to example (b)). In this case the total loss is $Q_C = 1.91 K$.

The main reason for the higher loss for example (c) can be identified from the flow patterns from the canal

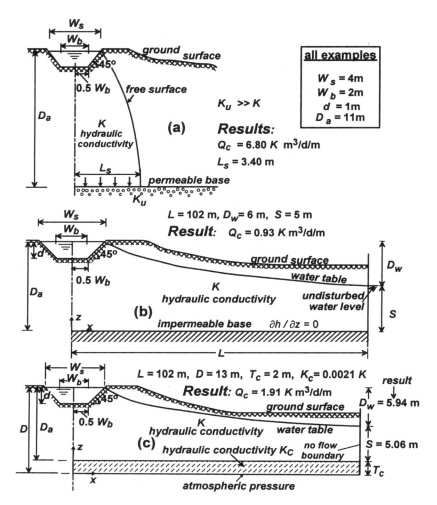

Figure 4.3 Three alternative problems with numerical values of parameters and loss from canal Q_c (Wachyan and Rushton 1987). Reprinted from *Journal of Hydrology* with permission from Elsevier Science. Copyright 1987

periphery to the outlet, either to the lateral channel or through the low permeability underlying layer. Diagram (a) in Figure 4.4a, which refers to the example with the impermeable base, shows that the flow lines are close together as they converge to the outlet which is a 5.0 m deep column of water. However, when the outlet for the flow is vertically through the underlying low conductivity layer (Figure 4.4b), the flowlines in the main aquifer are further apart, allowing an easier passage of water. Therefore the horizontal flow to the 5.0 m high column of water of the lateral channel is less than one-half of the vertical flow through the horizontal low permeability layer which extends for 102 m either side of the canal centre-line.

Further examples are discussed by Wachyan and Rushton (1987). Increased spacing between adjacent canals or increases in the vertical hydraulic conductivity of the low permeability layer all lead to greater losses. In example (c) of Figure 4.3, the groundwater head at the base of the low permeability zone is atmospheric; if the groundwater head is above atmospheric, the loss from the canal is reduced.

Breaks in underlying low permeability zone

Further detailed work has been carried out to explore the difference between the approximation of a uniform

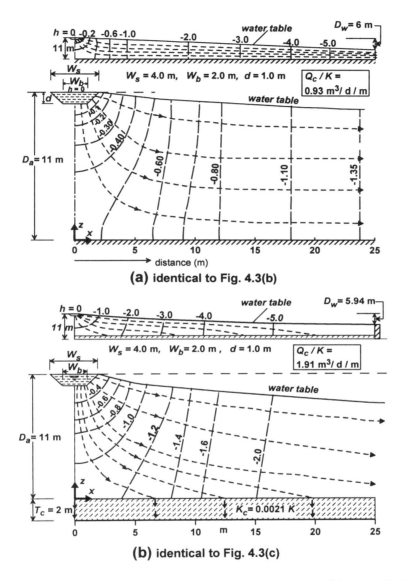

Figure 4.4   Flow lines and equipotentials for examples of losses from canals: (a) canal with impermeable base, (b) canal with low permeability base (Wachyan and Rushton 1987). Reprinted from *Journal of Hydrology* with permission from Elsevier Science. Copyright 1987

underlying low conductivity zone and the more realistic circumstance of discontinuities in the zone. Figure 4.5 shows three possible arrangements for sand and clay layers (each 1.0 m thick) forming the low permeability layer; there is approximately the same cross-sectional area of sand (hydraulic conductivity = 10 m/d) and clay (hydraulic conductivity = 0.002 m/d) for each example. The centre-line of the canal is represented by

$x = 0$, the low permeability zone extends for 100 m from the canal centre-line (dimensions are similar to the problem of Figure 4.3c).

For *Example A* there is a continuous clay thickness of at least 2.0 m across the whole section; the quantity of water passing through the low permeability zone is 0.457 m³/d/m. The bar chart shows the flow through each mesh increment; note that there are smaller mesh

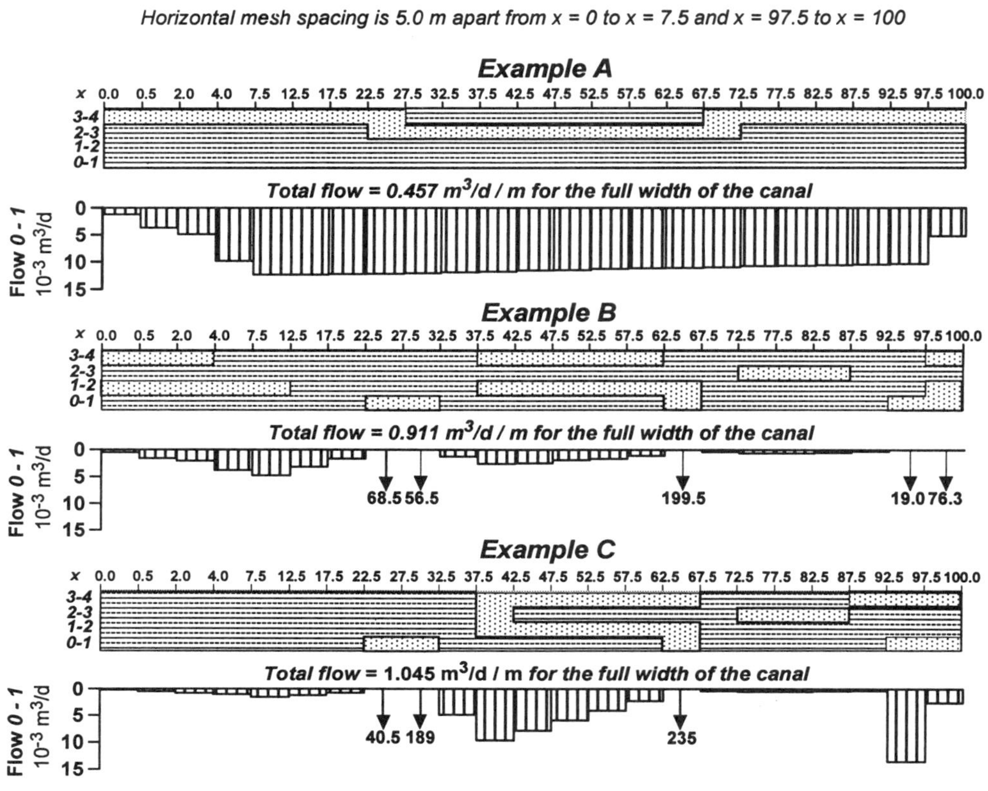

Horizontal mesh spacing is 5.0 m apart from x = 0 to x = 7.5 and x = 97.5 to x = 100

Figure 4.5   Effect of breaks due to sand layers in underlying clay layer; the geometry is similar to the problem of Figure 4.3c

increments at either end of the section. In *Example B* the sand is distributed so that there are several isolated sand lenses; the flow is nearly double the first example at 0.911 m³/d/m. For *Example C* the sand is arranged so that there is a direct route through the clay; the resultant flow of 1.045 m³/d/m is only slightly higher than in *Example B*. This finding, that a continuous high permeability route through the clay layer only doubles the flow, indicates that discontinuous clay layers respond in a similar manner to equivalent continuous clay layers.

Width of water surface

In most methods of estimating losses from canals, the loss from the canal is assumed to be proportional to the width of the water surface or to the wetted perimeter of the canal. For Condition A′ this is approximately true since the width of the column of vertical flow beneath the canal is directly influenced by the width of the canal. However, for the situation where

there is an underlying low permeability layer, the width of the canal has a smaller effect. The results of Table 4.1 refer to the same overall geometry as Figure 4.3c but with the hydraulic conductivity of the underlying layer of $0.002\,K$. Three widths of canal water surface are considered: 3.0, 4.0 and 8.0 m. The losses are quoted in the fourth column of Table 4.1. With the canal width of 8.0 m and depth of water of 2.0 m the loss of $Q_C/K = 1.95$ is only 8 per cent greater than for a canal 3.0 m wide and 1.0 m deep where the value is $Q_C/K = 1.80$. This small difference occurs because the dominant effect is the hydraulic condition at the outlet of the system, namely the vertical gradient across the low conductivity layer. For the three examples in Table 4.1, vertical gradients across the zone of low hydraulic conductivity are similar.

4.2.4 Effect of lining of canals

The large losses of water from canals can be reduced if the hydraulic conductivity around the periphery of the

Table 4.1   Effect of canal dimensions on losses from canal: $D_a = 11$ m, $T_C = 2$ m, $L = 102$ m, $K_C = 0.002K$

Dimensions of unlined canal (m)			Loss from unit length of canal Q_c/K (m²)	Maximum depth to water level D_w (m)
W_s	W_b	d		
3.0	1.0	1.0	1.80	6.0
4.0	2.0	1.0	1.86	5.55
8.0	4.0	2.0	1.95	5.22

canal is lower than the remainder of the aquifer. This can occur naturally by the deposition of silt or by the provision of a lining. Field information indicates that linings often have a limited impact in reducing losses although the lining may improve the hydraulic efficiency of the canal.

Water-logging in the vicinity of lined canals is one indicator of losses due to the movement of water through canal linings. Other evidence is provided by water seeping back into a canal when the canal is emptied, or by damage to the lining due to back pressures when the canal is drained for maintenance. Losses from a lined section of an irrigation canal in Sri Lanka of 0.12 m/d have already been quoted; when the lining was examined carefully, many cracks and joints filled with silt were identified. When these were made good, the losses reduced to 0.07 m/d (Holmes *et al.* 1981).

Insights into losses from lined canals are provided by Abdel-Gawad *et al.* (1993), they describe a series of sand tank experiments designed to investigate losses from lined canals. Figure 4.6 provides details of the experimental set-up. The canal was constructed from permeable no fines concrete, the lining consisted of plastic sheets with pin-pricks providing a route for water to escape from the canal. Figure 4.6a shows the water table elevations for unlined and lined canals.

The full line indicates the water table elevation for an unlined canal with the water in the outlet channel at 0.95 m above the base (or 1.15 m below the canal water level); the water table intersects the water surface in the canal. The broken line shows the water table for a lined canal with the holes in the polythene lining sheet equal to 0.006 per cent of the total area (this is a device for representing the resistance to flow due to lining); the water level in outlet channel is set at 1.54 m below the canal water level so that the water table is below the base of the canal.

Figure 4.6b plots the canal discharge divided by the hydraulic conductivity of the soil for different depths of the outlet water level. Two sets of results are

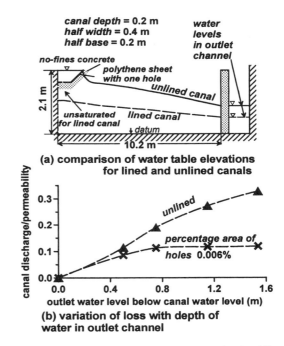

(a) comparison of water table elevations for lined and unlined canals

(b) variation of loss with depth of water in outlet channel

Figure 4.6   Details of sand tank studies for unlined and lined canals; results deduced from Abdel-Gawad *et al.* (1993)

included; the line with triangles refers to an unlined canal and the line with crosses to a lined canal with the area of holes of 0.006 per cent. For both the unlined and lined canals the loss increases for greater differences between the canal water level and the water level in the outlet channel. For the lined canal, once the channel water level is 0.8 m or more below the canal water level, the loss becomes effectively constant. This occurs because the canal is perched (or disconnected) with the vertical hydraulic gradient beneath the canal close to unity (i.e. due to gravity). This is Condition A′ of Bouwer's classification.

To learn more about imperfect linings, numerical solutions have been obtained. A typical example is shown in Figure 4.7; there are two breaks on the side of the canal and two on the bed. The loss from the canal is 0.412 m³/d/m length of the canal where the hydraulic conductivity of the aquifer through which the canal passes is 1.0 m/d. This loss is 94 per cent of the value for an unlined canal. Further results for alternative arrangements of breaks in the canal lining are listed in Table 4.2. In these simulations the canal is not perched, so that the regional water table intersects the canal.

In a second numerical modelling investigation, the lining is represented as a layer 0.1 m thick with a hydraulic conductivity K_L. Six values of the ratio K_L/K_A (K_A is the hydraulic conductivity of the aquifer) are considered; results are quoted in Table 4.3 for three alternative sets of conditions on the bed and/or sides of the canal. With the effective permeability of the lining at 10 per cent of the remainder of the aquifer, the loss is 93.6 per cent of the unlined case, a reduction of only 6.4 per cent. When the sides are permeable and

the bed impermeable, the loss is 84 per cent of the unlined canal while with impermeable sides and permeable bed the loss is 89.8 per cent of the unlined canal. This indicates that more of the water is lost through the bed of a canal. Even if the effective hydraulic conductivity of the lining is 5 per cent of the remainder of the saturated aquifer, the losses are still 80 per cent of the unlined situation. These results from mathematical models support the field information that substantial losses can occur from canals that are lined.

4.2.5 Perched canals

Condition A′, where the canal is perched above the regional groundwater table, occurs in many practical situations. Insights into the perched condition can be gained from the sand tank model experiments (Abdel-Gawad *et al.* 1993). The line with crosses in Figure 4.6b shows that the regional water table can be below the bed of a lined canal so that the canal is perched. Field results, obtained from the Chesma Right

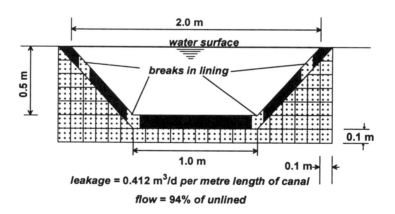

Figure 4.7 Numerical solution for canal with breaks in canal lining

Table 4.2 Effect of different breaks in the canal lining (see Figure 4.7 for positions of the breaks); units of the loss are m³/d per metre length of canal

Example	[Number of breaks in lining]		[Flow]	
	[Bed]	[Sides]	Loss m³/d/m	% unlined
1	2	2	0.412	94.0
2	4	0	0.404	92.0
3	2	0	0.382	87.1
4	0	2	0.370	84.3

Table 4.3 Effect of a lining 0.1 m thick on losses from a canal with canal dimensions as shown in Figure 4.7; saturated conditions beneath the canal. Results are expressed as a percentage of the loss with no lining

K_L/K_A	Bed and sides permeable	Bed permeable; sides impermeable	Sides permeable; bed impermeable
1.0	100.0	100.0	100.0
0.30	98.6	91.4	95.0
0.20	97.4	89.2	93.2
0.15	95.5	86.6	91.6
0.10	93.6	84.0	89.8
0.05	80.3	–	–

Bank Canal in Pakistan (which are presented and discussed in Section 4.4.7 and Figure 4.27) demonstrate that a major canal can be perched above the regional water table.

The following study of water loss from a lined perched canal is based on the idealisation that a canal has impermeable vertical sides and a permeable lined bed perched above the regional water table; the geometry is defined in the left-hand diagram of Figure 4.8a. Alternative conditions of saturated or unsaturated flow beneath the canal are presented in Figure 4.8a, which shows the pressure distributions above the canal lining, through the canal lining and beneath the canal lining; the canal lining has a thickness t_L and a vertical permeability K_L. Taking the water surface as datum, the pressure increases linearly to the base of the canal. However, the changing pressures within the lining depend on whether or not there is sufficient flow to achieve saturated conditions within the zone extending from directly below the lining to the underlying regional water table. If unsaturated conditions apply, there is a tension (pressure less than atmospheric) beneath the lining which results in a greater hydraulic gradient across the lining.

A series of calculations to estimate the flow through the canal lining is described below. The first step in the calculation is to assume a moisture content in the unsaturated or saturated soil beneath the canal. Relationships between the moisture content, matric potential and hydraulic conductivity are used to calculate the matric potential and hydraulic conductivity for the specified moisture content. Subsequently the vertical flux of water through the lining is calculated. From this information, the hydraulic conductivity of the canal lining corresponding to this flux can be determined.

A number of different relationships between the three fundamental parameters of unsaturated flow are available. The empirical equations of Gardener (1958) are used in this calculation; for other relationships see Chapter 2 of Simmers (1997).

The effective saturation $\quad S_e = \dfrac{\theta - \theta_r}{\theta_s - \theta_r}$ (4.1)

the matric potential $\quad \Psi_m = -a[(1.0 - S_e)/S_e]^b$ (4.2)

the hydraulic conductivity $\quad K_u = K_s e / (e + |\Psi_m|^f)$ (4.3)

The meaning of the symbols and the numerical values, [ ] in cm-sec units, for this calculation are listed below:

θ: water content [0.200],
θ_r and θ_s: residual and saturated water contents [0.075, 0.287],
a, b, e, f: constants for a particular soil [36.935, 0.252, 1175000, 4.74],
K_s: the saturated hydraulic conductivity [944 × 10^{-5} cm/sec]
also the depth of water in the canal, $d = 100$ cm, the thickness of the lining, $t_L = 10$ cm.

The method of calculation will be illustrated by considering a moisture content of $\theta = 0.200$:

Step 1: With $\theta = 0.200$, $S_e = 0.5896$,
Step 2: hence $\Psi_m = -33.71$ cm,
Step 3: and therefore the unsaturated hydraulic conductivity $K_u = 0.000596$ cm/sec.
Step 4: Since a unit hydraulic gradient is assumed below the lining, the vertical flux (quantity flowing per unit area) $q = 0.000596$ cm/sec.
Step 5: The head difference cross the lining equals

$$\Delta h = d + t - \Psi_m = 100 + 10 - (-33.71) = 143.71 \text{ cm}$$

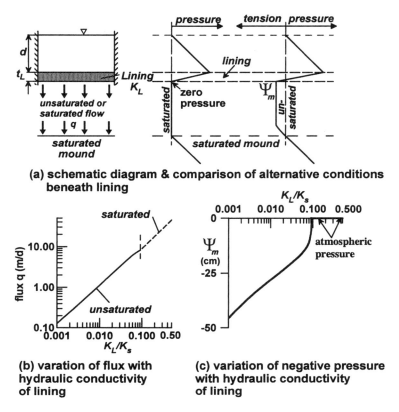

(a) schematic diagram & comparison of alternative conditions beneath lining

(b) varation of flux with hydraulic conductivity of lining

(c) variation of negative pressure with hydraulic conductivity of lining

Figure 4.8   Conditions beneath an idealised lined canal with unsaturated conditions

Writing Darcy's Law as

$$K_L = q t_L / \Delta h$$
$$= 0.000596 \times 10 / 143.71$$
$$= 4.15 \times 10^{-5} \text{ cm/sec. or } 0.0358 \text{ m/d}$$

This is the lining hydraulic conductivity which results in a moisture content of the soil below the lining of 0.20.

A series of calculations have been performed for the unsaturated conditions with different values of the moisture content. The particular case when the flow is just saturated, $\theta = 0.287$ and $\Psi_m = 0$, occurs with a lining permeability of 0.741 m/d (85.8×10^{-5} cm/sec) compared to the saturated hydraulic conductivity of the soil of 8.16 m/d (944×10^{-5} cm/sec). Calculations can also be performed for flow under saturated conditions. Since the pressure under the lining is atmospheric, the head difference becomes

$$\Delta h = d + t_L$$

The results of certain calculations are listed in Table 4.4, and more extensive results are plotted in Figure 4.8. In considering these results, it must be stressed that they refer only to the case of a perched canal with the water moving vertically to the underlying regional groundwater table. Considering first Table 4.4, results are included for linings varying from 0.135 per cent to 50 per cent of the saturated hydraulic conductivity. Lines 1 to 4 of Table 4.4 refer to unsaturated conditions; lines 5 to 7 refer to saturated conditions. Even for very low lining hydraulic conductivities, losses still occur from the canal.

Figure 4.8b relates the flux through the lining to the lining hydraulic conductivity. For saturated conditions (the broken line) the flux increases linearly with increasing lining hydraulic conductivity. However, for unsaturated conditions, a slightly non-linear relationship

Table 4.4 Calculations for different canal bed lining permeabilities for unsaturated flow beneath a perched canal, $K_S = 944 \times 10^{-5}$ cm/sec, (8.16 m/d)

Line	K_L m/d	K_L/K_S	θ	Ψ_m cm	q m/d
1	0.011	0.00135	0.150	−43.0	0.17
2	0.036	0.0044	0.200	−33.7	0.52
3	0.098	0.012	0.240	−26.9	1.34
4	0.46	0.057	0.280	−15.8	5.81
5	0.74	0.091	0.287	0.0	8.16
6	1.73	0.21	0.287	0.0	19.0
7	4.08	0.5	0.287	0.0	44.8

holds. In Figure 4.8c, the matric potential in centimetres of water is plotted against the ratio of lining hydraulic conductivity to saturated hydraulic conductivity. For higher lining hydraulic conductivities, saturated conditions occur; the pressure is atmospheric which leads to the horizontal line for $K_L/K_S > 0.091$. With lower values of the lining hydraulic conductivity, the pressures become more negative but never fall below −50 cm for a lining hydraulic conductivity 0.1 per cent of the saturated aquifer conductivity.

4.2.6 Estimation of losses from canals

There are several approaches to the field estimation of losses from canals.

Ponding test: a section of the canal is isolated, then filled with water and the decline in the water level in the canal is measured. Ponding tests to estimate canal losses are described by Holmes *et al.* (1981). Limitations of ponding tests include:

- because the water is stationary, a ponding test may not represent real field conditions,
- losses which are difficult to quantify may occur at the end of the section and at off-takes,
- during a ponding test silt can settle and reduce the seepage losses,
- it is not possible, under most circumstances, to carry out ponding tests in large canals.

Nevertheless, important information about canal losses can be gained from ponding tests.

Water balance for an operating canal: if all the inflows and outflows for a canal system are quantified, an estimate can be made of losses from a canal. The canal should be maintained under steady flow conditions. A study carried out by the International Water-logging and Salinity Institute, Lahore, Pakistan, (Bhutta *et al.* 1997) demonstrates that, with careful fieldwork, a reasonable estimate of the canal losses can be obtained. A section of the Lower Gugera Branch Canal, 32.3 km long, was studied with discharge measurements taken upstream and downstream of the test section and at nineteen off-takes. During the canal closure period, canal cross-sections were measured; the average surface width is 32.4 m and the average wetted perimeter is 34.1 m. During operation of the canal, flow measurements were made using broad crested weirs, rectangular orifices and current meters on twelve successive days. The off-take discharges are approximately 16 per cent of the incoming canal discharge while evaporation accounted for no more than 0.05 per cent of the incoming discharge. The canal loss averaged 3.2 m³/s or 6.2 per cent of the incoming discharge (losses on individual days ranged from 2.8 to 3.5 m³/s); this is equivalent to 0.25 m/d per unit area of wetted perimeter.

A thorough *field investigation* of canal losses using multiple piezometers, water chemistry and isotopes is described by Harvey and Sibray (2001).

Other techniques of estimating seepage losses from canals include seepage meters in the bed of the canal (Lerner *et al.* 1990, p. 180). Tracers have also been used; they are injected and the tracer movement monitored in a number of piezometers (Lerner *et al.* 1990, p. 187). A summary of techniques for predicting losses from canals can be found in Chapter 4 of Simmers (1997); methods including Darcy's Law calculations, design tables and mathematical models.

4.2.7 Artificial recharge using spreading techniques

Losses from canals can be used to artificially recharge aquifers. Requirements for successful artificial recharge from spreading channels include permeable surface soils, an unsaturated zone which does not contain extensive low permeability layers and an aquifer with a sufficiently high transmissivity to move water away from the location of the spreading channel. Bouwer (2002) provides practical information and valuable insights about artificial recharge from surface sources.

Artificial recharge schemes may cover substantial areas; typical examples include percolation tanks (reservoirs) in India. Mehta and Jain (1997) describe a study of a tank with a catchment area of 82 ha. From water balance calculations the estimated recharge was 121 Ml during 1990. A careful examination of observation well hydrographs suggests that the effect of the recharge from the tank spreads for more than 1 km from the source after several months. Further evidence of the effectiveness of the artificial recharge was the two- or three-fold increase in the yield of dug wells in the zone of influence of the percolation tank. A much smaller scheme is described by Abu-Zreig *et al.* (2000) in which ditches filled with sand are used to recharge the soil zone to overcome the limited infiltration capacity of the near surface soil.

As with all artificial recharge methods, clogging is a serious problem. Clogging of spreading channels can occur due to the deposition of silt and for biological and chemical reasons. Bouwer (2002) provides further information about the reasons for clogging and practical methods of alleviating the effect of clogging. Improving the quality of the recharging water can minimise the impact of clogging. Furthermore, if the recharging water is of potable standards, the water can be abstracted and used directly for public supply. When estimating the effectiveness of an artificial recharge scheme, allowance must be made for the time periods when recharge cannot take place due to the remedial action required in improving the soil surface conditions to increase the infiltration capacity. The quantity injected in a year for a sandy loam is unlikely to exceed 30 m; for clean medium sands the infiltration may approach 300 m in a year. This assumes that high-quality water is available for recharge throughout the year (a condition that is rarely met in practice).

In Gujarat, western India, a pilot study at three sites examined artificial recharge by injection wells in an alluvial aquifer (described in Section 8.6), injection wells in a Miliolite Limestone aquifer (Section 11.4) and spreading channels in an alluvial aquifer (for information about this aquifer, see Section 10.2.2). For the 'common recharge zone' where the soil is predominantly sandy, a spreading channel 2.0 m deep was constructed with a plan area 100 m by 6.0 m and sides

sloping at 45°. The initial infiltration rate was 400 m³/d (equivalent to 0.67 m/d over the plan area of the channel) falling to 100 m³/d (equivalent to 0.17 m/d). Since this spreading channel uses excess monsoon runoff, it is likely to operate for 30 days in a typical year, hence the annual recharge is about (0.17 × 600 × 30) = 3000 m³. For a crop water requirement of 500 mm, this recharge could support an irrigated crop covering only 0.6 ha (100 m by 60 m). With these restricted infiltration rates, artificial recharge using spreading channels is appropriate for meeting domestic needs, but is rarely able to support extensive dry season irrigation.

Comparisons of construction and running costs for the alternative artificial recharge methodologies indicate that the injection well in the alluvial aquifer is the most expensive. Expressing the construction and annual running costs for the injection well in the alluvial aquifer as [100, 100] the corresponding costs for the spreading channel are [9, 10], for the injection well in a limestone aquifer [6, 21] and for a percolation tank [2, 7] (Rushton and Phadtare 1989). Note that these costs relate to specific projects.

4.3 SPRINGS, RIVERS, LAKES AND WETLANDS

4.3.1 Introduction

Natural outflows from aquifers occur through springs, rivers, lakes and wetlands. Alternatively rivers, lakes and wetlands can act as a source of recharge to an aquifer. Winter (1999) emphasises that surface water bodies are integral parts of groundwater systems even if a surface water body is separated from the groundwater system by an unsaturated zone. Groundwater flows to springs and rivers often change with the time of year. For instance, in chalk and limestone aquifers, high water tables during the recharge season result in high transmissivities leading to flows to groundwater-fed rivers, which may be an order of magnitude greater than the groundwater contributions to these rivers during dry conditions. Furthermore, flows in the upper reaches of rivers are often intermittent. If pumping occurs in the vicinity of a river, the effect on river flows can be large, especially during dry periods. In semi-arid areas, with intense short rainfall events, rivers are dry for much of the year. The issue of aquifer recharge due to intermittent flow is considered in Section 4.3.4.

To understand the interaction between aquifers and springs, rivers, lakes or wetlands, detailed fieldwork is required (see, for example, Sophocleous *et al.* 1988) followed by the development of realistic conceptual models of the flow processes. In particular, all the properties that influence the flows must be identified; they include the geometry, elevation and physical setting, the nature of any bed deposits and the relevant aquifer properties.

The coefficients used to measure groundwater flows from springs or between aquifer and rivers, lakes or wetlands should be deduced from field measurements of flows and groundwater heads. However, when there is only limited historical information about groundwater components of surface flows, the estimation of the surface water–groundwater interaction coefficients may rely on comparisons with other similar catchments.

4.3.2 Spring–aquifer interaction

Springs occur when groundwater heads are above the elevation of the spring outlet; Figure 4.9 illustrates different situations which result in spring flows.

a) Certain springs occur at the intersection of the groundwater table with the ground surface. Figure 4.9a shows a regional water table intersecting the sides of a valley. Springs occur at the water table with seepages in the valley below the water table. A typical example of this type of spring is in the valley of the River Alt in Liverpool (Rushton *et al.* 1988). In urban areas, many of these springs are now intercepted by culverts.

b) Springs also occur at the interface between higher and lower permeability material. In Figure 4.9b, water in a permeable sandstone overlying a low permeability mudstone issues from the valley side at the interface of the sandstone and the low permeability layer.

c) In a detailed study of a limestone aquifer, Smith (1979) showed that springs occur at different elevations on the sides of a valley issuing from major fissured zones. The higher level limestone springs (*HLSp*) flow for about two months when the water table is towards its maximum; the medium level springs (*MLSp*) operate for up to six months and the lower level springs (*LLSp*) are perennial.

d) Another form of spring flow can occur from abandoned artesian boreholes. The relationship between outflow and excess groundwater head is different from a normal spring (see Section 11.3.4 and Johnson *et al.* 1999).

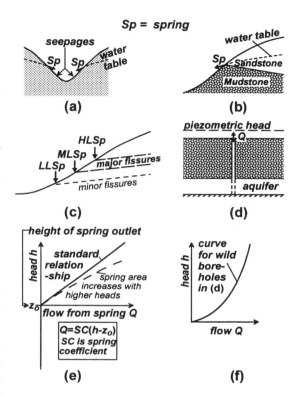

Figure 4.9 Examples of the occurrence of springs: (a) spring on valley sides, (b) spring forming at interface of high and low permeability strata, (c) springs associated with major fissure zone (HLSp = high level spring, MLSp = medium level spring, LLSp = low level spring), (d) overflowing uncontrolled (wild) borehole, (e) relationship between excess groundwater head and spring flow, (f) non-linear relationship for wild borehole

For situations (a) to (c), Figure 4.9e illustrates the dependence of spring flow (horizontal axis) on the difference between the groundwater head and spring elevation (vertical axis). It is usually acceptable to assume a linear relationship indicated by the unbroken line and the equation in Figure 4.9e. However, when the area of the seepages increases with increasing groundwater head, the relationship indicated by the broken line may be more appropriate. For situation (d), the non-laminar flow up the pipe of abandoned boreholes results in a non-linear relationship between the flow and the excess head, with the discharge proportional to the square root of the excess head. This relationship is shown in Figure 4.9f and discussed further in Section 11.3.4.

Case study: Bath Hot Springs

An example of complex spring flow is provided by the Hot Springs of Bath in England (Figure 4.10). Archaeological evidence indicates that the springs have been known to man for about seven thousand years. In Roman times there was a temple with hot water fed to various baths. After Roman times the springs fell into disrepair but were subsequently renovated and have been used extensively for the past eight hundred years or more. In 1977, due to the discovery of the presence of pathogenic amoebae, the springs were closed to the public; this closure was followed by an integrated multi-disciplinary research study which included drilling boreholes. Important hydrogeological findings from the research studies are summarised below. These findings are illustrated by the schematic cross-section of Figure 4.10 which is based on a number of diagrams in Kellaway (1991).

Locations of springs and maintenance of spring flows

- From the three boreholes (Kingsmead, Hot Bath Street and Stall Street) drilled in the 1980s into the underlying Carboniferous Limestone, artesian heads of 27–28 m AOD were measured compared to ground levels of 20–26 m AOD. Reductions in these piezometric heads were identified depending on the outflows from the springs and boreholes.
- The spring feeding the King's Bath is associated with a complex sub-vertical fault zone; the Cross Bath spring is supplied by a smaller fault. The funnels through which the thermal water enters into the springs are filled with gravel.

- Strata above the Carboniferous Limestone act as a confining layer preventing the hot water from leaving the aquifer over an extended area. The importance of this overlying layer is confirmed by observations at the Batheaston coal mine which is about 4 km from the Hot Springs. Pumping from this coalmine had an effect on the flows from the Hot Springs. An historical account reports that, 'the pit engine of Batheaston was sufficient to empty the pit without pumping on Sundays . . . its effect on the Cross Bath was that it took two hours more to fill on Saturdays than it did on Mondays from the engine's rest on the Sabbath'. One day of non-operation of drainage from the coal mine resulted in higher flows from the hot springs even though the distance between the two is 4 km.
- Temperatures of the water leaving the springs are indicated in Figure 4.10; when more water is taken from the system, both the piezometric heads and the temperatures fall. The total heat output from the system appears to remain reasonably constant. The heat output is the product of the flow and the temperature above the natural groundwater temperature of 10°C and equals about 45 500 Mcal/day.

Origin of the water

The thermal water has spent a substantial proportion of its travelling time in Carboniferous Limestone strata. Recent estimates of the age of the thermal water based on radioactivity and chemical evidence range from a few hundred to 8000 years BP. It is difficult to identify the source of the thermal water. The outflow from the springs is less than 2 Ml/d, yet the recharge to

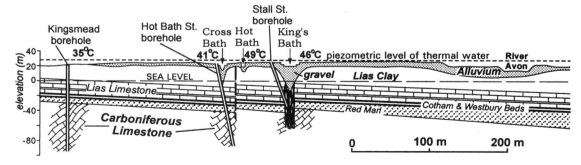

Figure 4.10   Schematic cross-section showing hot springs of Bath (Kellaway 1991, Copyright Bath City Council)

the extensive Carboniferous Limestone outcrop which could feed through to the Bath Hot Springs is more than one hundred times the output of the springs.

The reader is encouraged to examine this fascinating multi-disciplinary study as described in Kellaway (1991).

4.3.3 River–aquifer interaction: basic theory and modelling

Insights into the basic principles of river–aquifer inter-action can be gained from the flow mechanisms for canals described in Section 4.2.

(a) When a canal is perched above the water table mound (Figures 4.3a and 4.8), the loss from the canal approaches a constant value. The loss is approximately proportional to the width of the canal and is influenced by the vertical hydraulic conductivity and thickness of any lining or deposits in the bed of the canal.

(b) For canals in which the regional water table inter-sects the sides of the canal (Figures 4.3b and c, 4.4a and b, 4.6), the loss from the canal to the aquifer depends primarily on the difference between the water level in the canal and the regional ground-water heads.

(c) When the regional water table intersects the sides of the canal the loss is *not* directly proportional to the width of the canal, or to the vertical hydraulic conductivity and thickness of any lining or deposits in the bed of the canal (see, for instance, the results in Tables 4.1–4.3). The loss depends principally on the properties of the aquifer and to a lesser extent on the dimensions and lining of the canal. Field measurements are required to estimate the canal losses (Section 4.2.6).

Flow processes which commonly occur in river–aquifer interaction are illustrated in Figure 4.11a; a diagram quantifying the interaction is presented in Figure 4.11b. Four possible situations must be considered:

(i) The regional water table is higher than the surface of the river, hence water flows from the aquifer to the river; this condition is represented by the line *AB* in Figure 4.11b.

(ii) The regional water table is at the same elevation as the water surface of the river; there is no flow between the aquifer and river (location *B* in Figure 4.11b).

(iii) When the regional water table is slightly below the

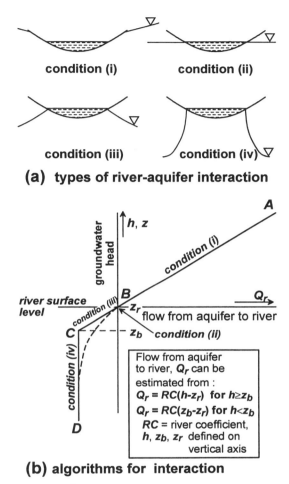

(a) types of river-aquifer interaction

(b) algorithms for interaction

Figure 4.11   River–aquifer interaction diagrams

water level in the river but still in contact with the sides of the river, flow occurs from the river into the aquifer. The line *BC* indicates that the loss from the river increases as the head difference increases; this is similar to the canal conditions described in (b) and (c) above.

(iv) If the regional water table is significantly below the river water level comment (a), the loss under a unit vertical gradient is approximately constant as indicated by line *CD* in Figure 4.11b.

Field information about river–aquifer interaction is examined by Rushton and Tomlinson (1979), who proposed a number of alternative relationships to the straight lines of Figure 4.11b. When water flows to a river from the aquifer, the flow may not be linearly proportional to the increase in head difference as defined by the line **BA**. For instance, the increasing seepage area resulting from a higher water table on the valley sides may lead to a higher than linear increase in flow. Nevertheless, numerical modelling studies using alternative non-linear relationships result in calculated flows very similar to those obtained using the linear relationship of the line **AB**. It is when the river loses water to the aquifer that great care must be taken in defining the relationship. The important principle is that there is a maximum loss under unit vertical hydraulic gradient as indicated by the lower part of the line **CD**. The broken line from **B** to **D** in Figure 4.11b more closely represents the actual losses but the two straight lines, **BCD**, are a satisfactory approximation.

Equations for calculating Q_r, the flow from aquifer to river, are quoted in Figure 4.11b. The river coefficient RC can be defined as the flow from aquifer to river when the groundwater head in the vicinity of the river is 1.0 m higher than the river surface level. Consequently RC equals the slope of the line **AB**. The second equation in Figure 4.11b defines the loss from the river when the water table is below the river surface level. If location **C** is 1.0 m below river level, the maximum loss from the river equals the river coefficient. Typical numerical values for the river coefficient are quoted in Table 4.5; the river coefficient is quoted for 1 km of river hence the units are m²/d per 1.0 km of river (other lengths of river reach can be selected). Most of the values in Table 4.5 are for rivers in UK, but estimates for other rivers have been made from published information.

Analytical solutions

A number of analytical solutions for the influence of pumped boreholes on river flows have been derived; a review is presented by Hunt (1999). When an infinitely long straight river fully penetrates the aquifer, a Theis image well solution can be used to calculate the loss from the river,

$$\frac{\Delta Q}{Q_w} = \text{erfc}\left(\sqrt{\frac{SL^2}{4Tt}}\right) \tag{4.4}$$

In this equation the river depletion is ΔQ, the pumped discharge is Q_w, with the distance from the well to the river equal to L. The transmissivity, storage coefficient and time since the start of pumping are respectively T, S and t. This conceptual model of a river fully penetrating the aquifer does not represent most field situa-

tions since it supposes that the pumping has no effect on the far side of the river.

Consequently Hunt (1999) developed an alternative analysis in which the stream-bed cross-section has horizontal and vertical dimensions which are small compared to the saturated aquifer thickness. This conceptual model is an improvement over the image well approach. The solution infers that seepage rates from the river into the aquifer are linearly proportional to the change in piezometric head across a semi-pervious layer which represents partial clogging of the stream-bed. One major drawback of Hunt's formulation is that it is not possible to represent the constant loss from a perched river (similar to Condition A' in Figure 4.1). Rushton (1999) shows that for perched rivers, the analytical solution without allowance for disconnection predicts river losses that are too high.

4.3.4 River–aquifer interaction in practice

Estimation of river coefficients

The technique of using river coefficients to describe river–aquifer interaction is illustrated in Figure 4.11b. Whenever possible, river coefficients should be estimated from field measurements of gains or losses in flows along the river or stream, combined with information about differences between river surface levels and groundwater heads. When river flows are primarily groundwater-fed, the increase or decrease in flow along a river can be deduced from flow measurements, either using a measuring structure or from current meter readings based on the velocity-area method (Shaw 1994). When there are substantial runoff components, baseflow separation techniques (Birtles 1978) are necessary.

A methodology for estimating river coefficients is illustrated using data from the River Meden in Nottinghamshire, UK, which flows from Permian strata across the outcrop of the Sherwood Sandstone (Figure 4.12a). Detailed information about the aquifer system is presented in Section 9.2. Flows in the river are strongly influenced by Budby Pumping Station which is 0.25 km from the river and abstracts 14 Ml/d. Flow gaugings on the River Meden were carried out on a number of occasions; a typical set of gauging results is plotted in Figure 4.12b. Flows increase as the river crosses onto the sandstone with a decrease in flows upstream of Budby PS and a further increase downstream of the pumping station. River flows at each location were determined using current meter

Table 4.5 River coefficients; river coefficients are defined as the gain to a river for a 1 km reach for groundwater head 1 m above river surface level

Location	River	Coefficient (m²/d/km)	Aquifer	Bed	Bed width (approx)	Basis of coefficients
Nottingham-shire	Idle	2000	Sherwood Sandstone	silty	15 m	Field measurements of losses plus groundwater model
	Meden	800–2000		alluvium	10 m	Accretion measurements along river plus model (Section 4.3.4)
Shropshire	Tributaries of River Severn	200–2000	Sherwood Sandstone	alluvium often silty	1–20 m	Continuous and spot flow gauging involving baseflow separation plus regional modelling
Cotswolds	Leach	200–23 000	Gt Oolite Limestone	clean	2–15 m	Flow gauging (part of river dries) plus groundwater model
	Coln	2000–20 000		clean	5–10 m	Flow gauging and groundwater model
Charing	Streams	700	Sand and clay	alluvium	1–3 m	Measured losses as river crosses Folkestone Beds
Lincolnshire	Slea	800–10 000	Lincolnshire Limestone	alluv/gravel	2–10 m	Gaugings plus evidence of river drying up
	West Glen	3000–6000		alluvium/clean	3–10 m	High coefficient in upper 2 km with high runoff recharge
	Holywell Bk	2500		alluvium/clean	1–3 m	Gaugings plus model; see Figure 11.11
Berkshire Downs	Kennet	5000	Chalk	alluvium/clean	5–10 m	Accretion measurements and groundwater model
	Winterbourne	500*–4000	Chalk	ditto	1–4 m	Accretion measurements, * for non-perennial reaches
East Kent	Lower Dour	2400–10 000	Chalk	alluvium	5–10 m	Partly from accretion results, also groundwater modelling
	Upper Dour	1200	Chalk	alluvium/clean	1–5 m	Largely non-perennial, accretion results and modelling
Kildare, Eire	Tully	500–1000	lmst gravel	alluvium	2–5 m	Groundwater modelling
Australia	Doyleston Drain	≈7000	sand/gravel	silt/gravel	2.5 m	From Hunt et al. (2001): pumping test with river flow measurements, estimates from analytical solution
Kansas	Arkansas	>5000	alluvium	alluvium	5.5 m	Estimated from measured losses (Sophocleous et al. 1988)
Nigeria	Yobe	15 000–65 000	alluvium	alluvium	20–35 m	Coefficient changes, low river levels to flood (Figure 12.3)

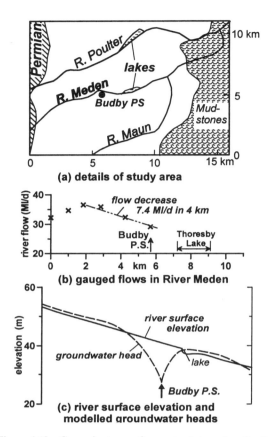

Figure 4.12 Groundwater–surface water interaction for the River Meden

readings; the accuracy of the flows is estimated to be ±5 per cent. Total flows are typically in the range 25–40 Ml/d. From a number of flow gauging records the loss as the river approaches and passes Budby PS is estimated as 5.6 ± 2.4 Ml/d over a length of river of 4 km; this is equivalent to a loss in the range 1.4 ± 0.6 Ml/d per kilometre (Rushton and Tomlinson 1995).

Since this decrease in flow occurs because the river is perched above the water table, most of the loss is due to a unit vertical hydraulic gradient beneath the river. Provided that the elevation of *C* in Figure 4.11b is 1.0 m below river surface level, the river coefficient is numerically equal to the loss of water from the river. When developing a regional groundwater model for this aquifer system, trial values of river coefficient were in the range 0.8–2.0 Ml/d/m per kilometre length of river. The coefficient which gave the best overall simulation of groundwater heads was 1.0 Ml/d/m per

km. When the river coefficient is set to this value in the groundwater model, the modelled groundwater heads and river surface elevations are as plotted in Figure 4.12c.

For a river losing water to an underlying aquifer, accurate values of the river coefficient are necessary but for a river gaining water from the aquifer the inflows are not very sensitive to the magnitude of the river coefficient. In upland streams, flows are frequently more sensitive to stream surface elevations. The Southern Lincolnshire Limestone aquifer system (Section 11.3) is an example of complex river–aquifer interaction with non-perennial stream sections.

River–aquifer coefficients in MODFLOW

The formulation for river–aquifer interaction defined in Figure 4.11b is included in many regional groundwater models, including MODFLOW. In the original version of MODFLOW, river coefficients (or stream-bed conductivity) could be read in directly but with many of the graphical interfaces currently available, the river coefficient is estimated using the diagrams in Figure 4.13. The actual river in Figure 4.13a is idealised as shown in Figure 4.13b. Using the description by Anderson and Woessner (1992) of the RIV package in MODFLOW, a stream-bed conductivity *CRIV* is calculated from the vertical hydraulic conductivity and thickness of the riverbed sediments K_r and M, the width of the riverbed channel W, and the incremental length of river L:

$$CRIV = K_r LW / M$$

$$[\text{units } L^2T^{-1} \text{ for a specified length of river}] \quad (4.5)$$

The flow *from* river *to* aquifer *QRIV* is defined by the equations

$$QRIV = CRIV(HRIV - h) \quad \text{for } h > RBOT \quad (4.6a)$$

$$QRIV = CRIV(HRIV - RBOT) \quad \text{for } h \le RBOT$$
$$(4.6b)$$

where *h* is the groundwater head in the vicinity of the river (the groundwater head in the cell containing the river) and *RBOT* is the bottom of the streambed sediment. This relationship is the same form as the equations in Figure 4.11b but note that *QRIV* is of opposite sign to Q_r.

Difficulties can arise with this approach of estimating the river coefficient (streambed conductivity) solely from the nature and properties of streambed sediments.

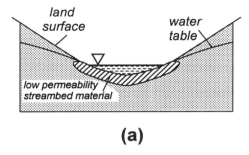

(a)

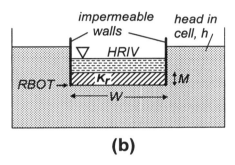

(b)

Figure 4.13 Representation of river–aquifer interaction in MODFLOW: (a) actual geometry of river section, (b) idealisation used in MODFLOW formulation. Reprinted from *Applied Groundwater Modeling*, Anderson and Woessner, copyright (1992) with permission from Elsevier Science

If there is no streambed sediment, how is the conductivity calculated? Furthermore, the study of losses from canals showed that when the regional water table intersects the sides of the canal the loss is *not* proportional to the width of the canal, or to the vertical hydraulic conductivity and thickness of any lining or deposits in the bed of the canal (comment (c) of Section 4.3.3). Instead, the flow between river and aquifer depends primarily on the properties of the aquifer, the dimensions and nature of the riverbed and the nature of the regional groundwater flow. When the river is perched above the water table (Condition A′ in Figure 4.1), the loss is proportional to the width of the river but unsaturated conditions between the river and the saturated water table may also influence the loss, as indicated in Figure 4.8.

In summary, if the river is perched above the regional water table, Eq. (4.5) provides a first estimate of the river coefficient *RC* or *CRIV*. However when the regional water table intersects the sides of the river with either the river feeding the aquifer or the aquifer

feeding the river, Eq. (4.5) should not be used. Whenever possible, the coefficient relating river flows to groundwater head difference should be based on field information.

In the above discussion, it is assumed that the depth of water in the river and hence the area of contact between the river and aquifer does not change. For rivers which flood during the rainy season and almost cease to flow during the dry season this is not an appropriate assumption, since the depth of water in the river and hence the river coefficient changes substantially between these two extremes. A specific example of large changes in the river coefficient based on a Nigerian case study is considered in Section 12.2.2.

River augmentation

River augmentation from boreholes can be used either to support the river flows for ecological reasons (Rushton and Fawthrop 1991) or to maintain flows in rivers which are used for public water supply (Hardcastle 1978). A number of schemes in Britain are described by Downing *et al.* (1981). In designing an augmentation scheme, the aim is to maximise the flow in the river and minimise the flow re-entering the aquifer due to the additional pumping; in practice, some re-circulation does occur. An analysis of idealised conditions by Oakes and Wilkinson (1972) shows that the net gain (which is defined as the increase in river flow divided by the quantity abstracted from boreholes) depends on the transmissivity and storage coefficient of the aquifer, the permeability of the riverbed, the distance from the supply boreholes to the river and the effect of aquifer boundaries.

River augmentation is implemented at locations where river flows in summer are low. Consequently, unless schemes are carefully designed, they may be only partially successful. An example of the problems encountered in developing a river augmentation scheme is provided by the Thames Groundwater Scheme. The basic concept of the scheme is to support flows in the tributaries of the River Thames at times of low flow so that water can be abstracted downstream for public supply. For a pilot project during the development of the scheme, five boreholes were constructed close to the perennial part of the River Lambourn with four boreholes in the non-perennial Winterbourne sub-catchment (Figure 4.14a). Initially pumping took place from the Lambourn group of boreholes; the Winterbourne group were switched on about a month later (Figure 4.14b). River flows were measured at Shaw Gauging Station which is just upstream of the

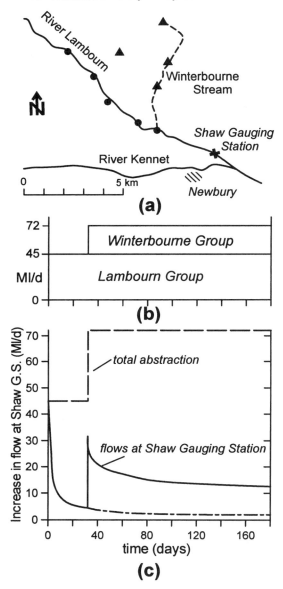

Figure 4.14 Thames groundwater scheme, Lambourn and Winterbourne Pilot Scheme: (a) location of boreholes, (b) quantity of water pumped from well fields, (c) increase in flow deduced from hydrograph at gauging station

confluence with the River Kennet. When the boreholes close to the Lambourn were pumped, the net gain fell significantly, as indicated by the rapid reduction in the additional flows at Shaw (Figure 4.14c). When the Winterbourne boreholes commenced operation, there was a substantial gain in flows at Shaw followed by a steady reduction until the end of the test. One hundred and eighty days after the start of pumping the net gains for the Lambourn and Winterbourne wells were respectively about 10 per cent and 50 per cent.

Experience gained from this pilot scheme led to the positioning of the unconfined boreholes in dry valleys where the transmissivity is sufficient to provide yields of 5–9 Ml/d. Water is then taken by pipeline to below the perennial head of the rivers. This pilot scheme is part of the Thames Groundwater Scheme investigation, which is discussed further in Section 11.2.

Augmentation schemes often fail to achieve their original objectives, as illustrated by the following examples.

* River support has been introduced on the River West Glen (see Section 11.3). The augmentation location is about 1.0 km below Essendine downstream of the Limestone outcrop (Figure 11.6). Once the augmentation commenced, losses in the augmented flows were identified. Three main causes for these losses became apparent from more detailed investigations. The first is that river flows on gravels which provide a route for augmented water to be transferred to the Limestone outcrop. Second, the extensive river gravel deposits transmit water to gravel pits alongside the river from which water evaporates. Third, the river and associated gravel deposits cross minor aquifers; water originating from the river enters these minor aquifers.
* River support has also been implemented in the chalk aquifer of the Candover catchment in Southern England, as described by Headworth *et al.* (1982) and in Section 11.6.2. The natural groundwater head fluctuations of 3–5 m per year are lower than for similar chalk aquifers. This is due to a very permeable zone at the top of the saturated aquifer with specific yield of 0.03–0.05. Below this permeable zone the specific yield takes the more normal value for chalk of about 0.01 (Rushton and Rathod 1981). Once this high permeability zone is dewatered, the yield of the river support boreholes falls off.
* The importance of the augmentation location is also demonstrated by a scheme intended to improve flows through Sleaford in Lincolnshire, UK (Hawker *et al.* 1993). As with the West Glen, there are extensive

river deposits which contain sands and gravels. Not only do the sands and gravels provide a route to the underlying limestone aquifer, but also water flows through the river gravels beneath and by the side of the river channel. When predicting the effectiveness of a river augmentation scheme, account should be taken of the river cross-section together with the cross-sectional area of the permeable river deposits multiplied by their effective porosity. It is through this total effective cross-section that the augmented water will flow.

Further examples of groundwater augmentation of river flows in the UK are presented in Downing *et al.* (1981).

Intermittent flow

The preceding discussion has been concerned with rivers and streams which flow continuously for part or all of their length. In semi-arid areas, where rainfall is often irregular and scarce, flow in the stream or river will only occur for part of the year, while in arid regions river flow occurs during sporadic years. Consequently the principles of river–aquifer interaction as described above need to be modified to represent intermittent flows.

Kruseman (1997), in a detailed review of studies of intermittent flow, provides valuable information about the difficult task of assessing the magnitude of infiltration from intermittent streams and rivers. Although they are dry for most of the year, when heavy rainfall occurs the flows can be very high and carry significant volumes of sediment. The loss from the stream to aquifer can be described by conditions (iii) and (iv) and the relationship **BCD** of Figure 4.11. However, significant changes can occur in the river coefficient which determine the maximum loss from the stream (i.e. the maximum infiltration). During the initial stages when the streambed is dry, infiltration occurs according to an exponential relationship. An equation proposed by Evenari *et al.* (1971) is that

$$In = In_c + (In_0 - In_c)e^{-kt} \tag{4.7}$$

where In_0 is the initial infiltration rate, In_c is the final infiltration rate, k is a constant and t is the time since infiltration commenced.

Features that determine the infiltration rate into riverbed deposits include:

- the lithology of the channel sediments in the source area which influences the hydraulic conductivities,

- infiltration rates increase with streamflow velocity and water temperature,
- infiltration rates decrease with increases in the suspended sediment input,
- infiltration rates reduce as the water table approaches the ground surface,
- stream channel infiltration increases with increasing flow duration.

Further information and references are presented by Kruseman (1997). However, the wide variety of situations in which intermittent flow occurs means that there is no alternative to field investigations when studying a particular situation.

4.3.5 Lakes and wetlands

Lakes and wetlands occur when the groundwater table intersects and lies slightly above the ground surface. When studying groundwater-fed lakes and wetlands it is necessary to consider both the surface water and the groundwater components. Domenico and Schwartz (1997) distinguish between *discharge lakes* which collect groundwater and *flow-through lakes* where water enters on one side and leaves from the other. They quote a number of studies which illustrate the impact of lakes on groundwater systems.

Figure 4.15 contains conceptual diagrams of possible interaction between groundwater and lakes. Figure 4.15a refers to a valley in an aquifer. The water table intersects the sides of the valley and forms a lake. Groundwater inflows occur from both sides into the lake; apart from surface water inflows and outflows the only additional outflow is evaporation.

In Figure 4.15b the regional water table has a gradient from left to right and is located above the bottom of a valley in an aquifer; a lake forms with groundwater-fed inflows from the left and groundwater outflows to the right. In Figure 4.15c the valley is above the natural water table but a surface water inflow causes a lake to form in the bottom of the valley; water seeps out of the bed and sides of the lake resulting in a groundwater mound. Figure 4.15d refers to an effectively impermeable dam constructed on the valley side of a low permeability strata; water seeps through silt deposits, then through an unsaturated zone until it reaches the regional water table.

Kuusisto (1985), in a detailed review of the hydrology of lakes, suggests that the proportion of groundwater flow in the total water balance of lakes varies widely. Groundwater flow to or from large lakes in humid regions can often be ignored but for mountain,

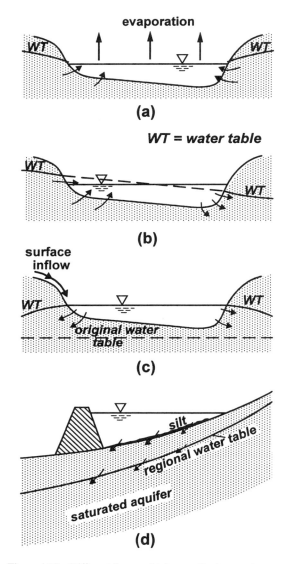

Figure 4.15 Different forms of lake–aquifer interaction; see text for explanations

is considered by Cedergren (1967) among others. Winter (1999) quotes many examples of the use of vertical section models to interpret groundwater interaction with lakes. Other valuable insights about lakes in a regional groundwater setting can be gained from the important paper of Tóth (1963) and the IAHS publication edited by Engelen and Jones (1986).

There are few studies in which the influence of regional and local groundwater conditions on wetland responses have been quantified. Statements are sometimes made that a wetland is directly influenced by abstraction from an underlying aquifer, whereas a careful study of the nature and vertical hydraulic conductivity of the material between the wetland and the aquifer shows that additional abstraction from the aquifer has a minimal effect on the wetland. On the other hand, abstraction from an aquifer is certain to reduce groundwater flows to springs and streams.

Restrepo *et al.* (1998) present a conceptual model for surface water–groundwater interaction in wetlands; a routine is described which can be coupled with the groundwater model, MODFLOW. The wetland simulation module represents flow routing, export and import of water and evapotranspiration from wetlands. Wilsnack *et al.* (2001) consider the groundwater hydrology of a surficial aquifer system in Florida to determine the influence of regulated canals, municipal well-fields and mining activities. They use the MODFLOW wetlands package of Restrepo *et al.* (1998) to provide a reasonable representation of field conditions for a period of 18 months. A study by Winter and Rosenberry (1995) considers wetlands that form on a topographical high.

Devising field experiments to provide insights into groundwater flows in the vicinity of lakes and wetlands requires ingenuity. For instance, Cartwright *et al.* (1979) use piezometer readings to establish upward flow through the base of a lake. The following two case studies are selected to indicate the type of information required for the study of lakes and wetlands.

Lakes in Upper Thames Catchment, the UK

The first example refers to lakes formed by gravel workings in the upper reaches of the River Thames; many of the lakes are used for recreational purposes. The interaction between the aquifer, river and lakes must be specified carefully. Figure 4.16 illustrates the three water balances which must be considered.

Figure 4.16a refers to the *overall water balance* for an area including aquifer, lake and river. The components considered in the overall balance are precipitation,

piedmont and steppe lakes, the groundwater component may be of the same order as surface inflows and outflows. Groundwater outflow from lakes is less likely than groundwater inflow; however, in mountainous lakes on glacio-fluvial formations and lowland bog lakes or fens, outflow often occurs. Significant groundwater outflows are more likely to occur from man-made lakes where a dam is used to artificially raise the water level. Seepage through and around embankment dams

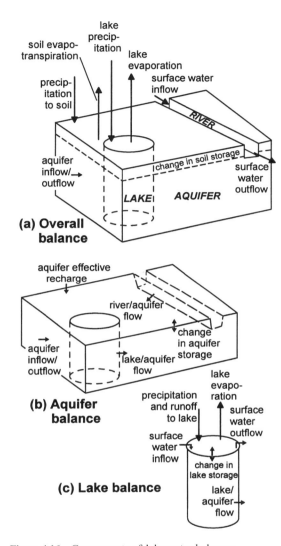

soil evapo-transpiration

lake precip-itation

lake evaporation

precip-itation to soil

surface water inflow

RIVER

aquifer inflow/outflow

change in soil storage

surface water outflow

LAKE   AQUIFER

(a) Overall balance

aquifer effective recharge

river/aquifer flow

change in aquifer storage

aquifer inflow/outflow

lake/aquifer flow

lake evapo-ration

precipitation and runoff to lake

surface water outflow

surface water inflow

(b) Aquifer balance

change in lake storage

lake/aquifer flow

(c) Lake balance

Figure 4.16 Components of lake water balances

evaporation and evapotranspiration, change in soil and aquifer storage, change in river flow and net aquifer inflow/outflow across the boundaries of the study area. Figure 4.16b illustrates the components of the *aquifer water balance*; apart from the aquifer inflow/outflow for the study area, the main input is the effective recharge. Flow also occurs between the lake, the river and the aquifer. A further important component is the change in aquifer storage. Because the water table is not far from the ground surface at certain times of the year, it is possible that evaporation can occur

directly from the saturated aquifer; this possibility is not illustrated in the figure. Figure 4.16c shows the components of the *lake water balance*; inflows and outflows can occur due to surface water flows and interaction with the aquifer. Precipitation to the lake is an important inflow while evaporation from the lake surface is a significant outflow, especially during the summer months.

A major factor influencing lake levels during summer months is evaporation from the lake surfaces. This leads to a considerable fall in lake water levels, which has detrimental effects on fishing and water sports. Figure 4.17 shows the winter and summer lake levels and water table elevations on a typical cross-section of an aquifer in which gravel extraction has led to the formation of many lakes; the gravel lies above low-permeability Oxford Clay. Important features are discussed below.

- There is an overflow between Lakes A and B; in the winter (the unbroken line) the capacity of the overflow is insufficient to take all the water, hence Lake A's winter water level is just above the pipe. The overflow does not operate in summer.
- Oxford Clay has been placed on the downstream side of Lake B to minimise seepage losses. Consequently both winter and summer water tables between Lakes B and C are below Lake B water levels.
- Lakes D and E are on either side of the headwater stream of the River Thames. During the winter the high level in Lake D maintains flows in the River Thames but in summer, water levels in Lakes D and E are below the bed of the River Thames, with the result that any flow in the river seeps into the aquifer and then to the lakes.
- In fact the River Thames is in an artificial channel which was constructed to supply water to a mill. The original channel, which is now called Swill Brook, is to the right of Figure 4.17. During both winter and summer there are flows in Swill Brook.

This discussion illustrates how fieldwork with measurements of lake water levels, groundwater heads, inflows and outflows are all required to develop an understanding of the important mechanisms which govern flows in a system where lakes form a considerable proportion of the area. In developing a regional groundwater model, the lake can be represented as a region with very high hydraulic conductivity (1000 m/d or more) and with a specific yield of 1.0. Features such as overflows are represented as springs.

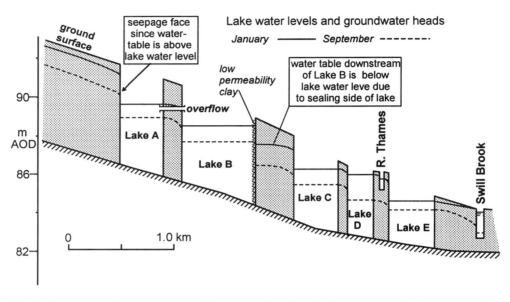

Figure 4.17 Typical cross-section through upper Thames gravel aquifer showing summer and winter water tables and lake levels

Pollardstown Fen

The second example relates to a wetland which is part of a groundwater-fed fen in the Republic of Ireland, namely Pollardstown Fen in County Kildare. The Fen, which consists of peat, marls (containing calcium carbonate) and a grey sticky clay, has been formed in the valley of a limestone gravel aquifer. The fen deposits have a thickness of 12–15 m and are underlain by gravels. Figure 4.18 indicates the main flow processes. Recharge is transmitted through the aquifer leaving either through springs at the margins of the Fen, or by vertically upward flow through the low permeability Fen deposits. The maintenance of conditions both at the margins of the Fen and in the main part of the Fen is crucial for sustaining the important ecological balance.

- At the margins of the Fen, the critical features for the ecologically important species are steady spring flows of calcium-rich groundwater at a constant temperature. Due to the relatively high permeability of the aquifer, approximately uniform conditions are maintained at the Fen margins throughout the year.
- Towards the middle of the Fen, a slow but steady upward flow is required to ensure that the soil remains moist during the summer. Limited field measurements suggest an excess head in the gravel

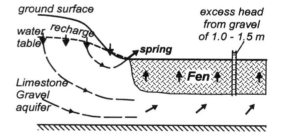

Figure 4.18 Conceptual model for Pollardstown Fen

beneath the Fen of 1.0–1.5 m. From the hydraulic conductivities of the peat, marls and sticky clay the effective vertical hydraulic conductivity is estimated to be between 0.002 and 0.005 m/d. The fact that the soil remains moist but does not become waterlogged indicates that the upward flows just balance summer evapotranspiration.

Since Pollardstown Fen has flowing springs and slow upward flows through the fen deposits, a diverse ecosystem has developed. An important feature of the sustainability is that there is effective man-made drainage within the Fen; water from the Fen is used to support flows in the Grand Canal.

When representing the Pollardstown Fen area in a numerical model, the gravel aquifer is represented as three layers; the lowest layer continues beneath the Fen. In locations where the middle layer coincides with the Fen, low horizontal and vertical hydraulic conductivities represent the Fen deposits. The uppermost layer wedges out at the edge of the Fen. Water can move vertically upward through the Fen deposits due to the higher groundwater head which occurs beneath the Fen in the lowest layer. Water can also leave the uppermost layer through spring mechanisms on the edge of the Fen.

Representation of lakes in MODFLOW

A methodology of including lake–aquifer interaction in a numerical model, using the MODFLOW code, is described by Cheng and Anderson (1993). It is based on a balance in the lake between precipitation to and evaporation from the lake surface, stream inflow to the lake and stream outflow from the lake and groundwater–lake interaction; these components are indicated in Figure 4.16. The following description uses the notation of Cheng and Anderson (1993); some of the symbols have different meanings elsewhere in this book.

Interaction between the lake and aquifer is described by the following equations which are of the same form as river–aquifer interaction (Figure 4.11b). The model nodes may underlie lake nodes or they may be situated horizontally adjacent to the lake. Consider cell L out of a total of n cells interacting with the lake; i, j and k are row, column and layer number of the Lth cell. The lakebed seepage to the Lth cell is

$$q_L = Cond_L \left(HLAKE^{old} - h_{i,j,k} \right) \quad h_{i,j,k} > LBOT_L \tag{4.8a}$$

$$q_L = Cond_L \left(HLAKE^{old} - LBOT_L \right) \quad h_{i,j,k} \leq LBOT_L \tag{4.8b}$$

where $Cond_L$, the lakebed conductivity, is estimated from the vertical hydraulic conductivity divided by the distance from the lake to the cell i, j, k and multiplied by the area through which the flow occurs, $HLAKE^{old}$ is the elevation of the lake surface at the previous time step, $LBOT_L$ is the lakebed elevation.

The transient lake water balance includes precipitation PPT_L, evaporation EV_L, inflows and outflows to the lake Q_{in} and Q_{out}; the out of balance of these components equals the change in lake level multiplied by

the total lake area $TAREA_{LAKE}$. Therefore the new lake elevation $HLAKE^{new}$ can be calculated from

$$
\begin{aligned}
HLAKE^{new} &= HLAKE^{old} \\
&+ \Delta t \left\{ Q_{in} - Q_{out} + \sum_{L=1}^{n} [(PPT_L - EV_L) AREA_L - q_L] \right\} \\
&\div TAREA_{LAKE}
\end{aligned}
\tag{4.9}
$$

Provision is also made in the package to allow for part of the lake to become dry.

4.3.6 Impact of groundwater abstraction on springs, rivers, lakes and wetlands

Abstraction of groundwater is likely to lead to a diminution of groundwater flows from springs and into rivers, lakes and wetlands. However, the reduction or cessation of pumping from an aquifer may not produce immediate beneficial effects for these surface water features.

Springs are sensitive to groundwater exploitation. In central London, there were a number of groundwater-fed artesian springs. However, over-exploitation of the groundwater resources beneath London resulted in the cessation of these artesian flows and a lowering of the deep water table so that it reached more than 50 m below ground level. Now, groundwater pumping has decreased; this is leading to a rise in the deep water table with the resultant risk of flooding of the underground railways and a potential risk of settlement of buildings (CIRIA 1989, Cox 1994, Howland *et al.* 1994). Another example is Jinan, the capital of Shandong Province in China. In this city the actual abstraction is more than 85 per cent of the available resource with the result that the city, once known as the City of Beautiful Springs, can no longer rely on the springs; this is a threat to the tourist industry (Zhang 1998).

Bromsgrove is a small town in the Midlands, UK. Its industrial development depended largely on water mills which relied on springs issuing from the sandstone aquifer (see Section 10.5; Rushton and Salmon 1993). Following extensive groundwater abstraction, the flows in these streams are now very small; currently the abstraction rate is about 90 per cent of the rainfall recharge. Reducing the abstraction from the aquifer would lead to a recovery in the water table elevation with springs starting to flow again. However, rising water tables are likely to lead to undesirable side-effects where urban and industrial development has occurred in locations which historically had water tables close to, or at, the ground surface.

Detailed studies have been carried out into the benefits of reducing abstraction from the Southern Lincolnshire Limestone (see Section 11.3). Reductions in abstraction to 60 per cent of the current values would enhance the winter stream and river flows but would provide only minimal improvements to low summer flows (Rushton 2002). Low summer flows are primarily due to the inability of the aquifer to store water during the winter and release water during the summer.

Another example of the consequences of abstraction on surface water features is provided by the Sava River in Zagreb, Croatia (Berakovic 1998). The Sava River acts as the main source of recharge to the alluvial aquifer in the Zagreb area. Until 1970, river water levels responded to annual fluctuations and longer term climatic changes but there was no clear long-term decline. However, from the early 1970s there has been a reduction in both the average and minimum water levels; by the mid-1990s they had fallen by about 2 m. This is caused partly by increasing abstraction but also by a lowering of the river-bed, primarily due to construction work. During dry summers the river collects drainage water of poor quality. The reduction in groundwater levels and the ingress of poor-quality water are putting at risk the future plans of supplying most of Zagreb's water supply from well fields.

It may take a considerable time for groundwater conditions to recover from over-exploitation. The Lower Mersey Sandstone aquifer (Section 9.4.5) is a typical example. Due to pumping for more than 150 years, at rates higher than the recharge, natural springs ceased to flow. As mining of the groundwater continued, poor-quality water was drawn in from the River Mersey and Mersey Estuary to the south of the study area. A mathematical model was developed to represent the historical exploitation of the aquifer. Subsequently the model was used to explore whether conditions could revert towards the un-exploited state following reductions in abstraction. The simulations showed that, even if groundwater abstraction ceased, it would be several decades before springs would start to flow again (Figure 9.21).

4.4 DRAINS

4.4.1 Introduction

The purpose of drains is to remove excess water and thereby lower the water table. This is an important method for maintaining irrigated land fit for agricultural use. An example of a successful tile drainage scheme is Mardan Saline Control and Reclamation Project (SCARP) in Pakistan (WAPDA 1984). The area is irrigated by canals. Prior to the installation of drains, typical water table elevations varied between 0.3 m and 1.0 m below ground surface. Due to the high water table, the crop yield was poor. A drainage system was designed to take away water equivalent to an average recharge rate of 3 mm/d. Steady state and time-variant methods of analysis were used. Also it was recognised that the effectiveness of the scheme could be restricted by the capacity of the drainage pipes. A typical section and plan of part of the scheme is shown in Figure 4.19; the design is based on the following criteria:

- the desirable depth to water table to provide sufficient aeration of the root zone,
- the depth of the water table to restrict upward capillary flow,
- the depth of drain placement for the least cost per unit area,
- the available depth of the outlet drains, topography and minimum pipe slopes.

There are sufficient deep open drains and topographical gradients for Mardan SCARP to operate without pumps; see the main drain to the left in Figure 4.19a and the topographical contours. In terms of increases in crop yields, it is an outstanding success. However, there are questions as to whether the design of the drainage system is satisfactory. During the dry season or when there is little irrigation, the outlet pipes into the large drainage ditches run partially full. On the other hand, during the rainy season or when there is significant application of irrigation water to the fields, the drains run full under a surcharge. These different responses highlight issues which must be considered when designing perforated drains. As well as the question about the ability of drains to lower the water table, it is essential to consider the flows within the drainage pipes and how they change with time. In addition, conditions in the vicinity of the drain and the drainage envelope should be examined.

4.4.2 Field evidence

There have been a number of field and laboratory studies to assess the effectiveness of drains. Information collected includes the elevation of the water table, conditions in the drains and flows through drains.

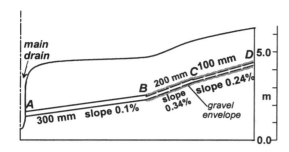

(a) section

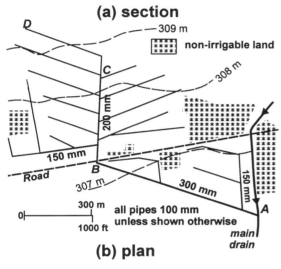

(b) plan

Figure 4.19 Mardan SCARP, an example of a successful gravity drainage project

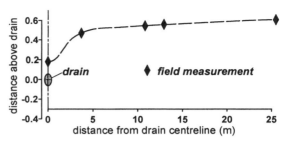

Figure 4.20 Location of water table above drain from field results of Talsma and Haskew (1959)

However, in most field and physical model tests there is a failure to correctly position observation piezometers in the vicinity of the drain. Reliable information about the actual position of the water table can only be obtained if piezometers just penetrate the water table. When the piezometers are positioned correctly, the water table is usually found to be above the drain. The location of the water table *above* the drain is shown clearly in the results of Talsma and Haskew (1959) plotted in Figure 4.20. In a more recent study by Susanto and Skaggs (1992), field evidence is quoted where the water table was 0.27 m immediately above the drain centre-line and midway between the drains it was 0.46 m above the centre-line of the drains. They conclude that, 'these results are contrary to the drainage theory, which normally assumes that the head losses near the drain are relatively low and that the water table intersects the drain.'

Field information about the *hydraulic* conditions *within* drains is limited. In some situations drains run partially full, but there is also evidence of drains running under a surcharge for many months. Detailed monitoring of a drain laid adjacent to a canal showed that, after pumping for more than a year, the hydraulic heads in the drainage lines were always above the top of the drains.

There is also a lack of information about the magnitude of flows from drains. Reference has been made in Section 4.4.1 to the changing flows from the outlet pipes in Mardan SCARP. For pumped drainage schemes, initial pumped discharges need to be substantially higher than the design values to attain a rapid fall in the water table during the early stages of operation of the pumps. In some instances there are two pumps in the pumphouse, one is intended as a standby. Nonetheless, at the start of drainage both pumps are required. All this evidence indicates that, under relatively constant recharge conditions, the rate at which water is withdrawn from the drains decreases with time.

4.4.3 Formulation of drainage problems

In this section a complete formulation of a time-variant drainage problem is presented. It is assumed that the groundwater velocities parallel to the drains are small compared to the velocities on a cross-section perpendicular to the drains, hence the important flow mechanisms are represented by considering a vertical section (Figure 4.21). In the following analysis it is assumed that the soil below the water table is fully saturated, above the water table the pressure is atmospheric. Note that the origin of the co-ordinates is at the base of the aquifer beneath the left-hand drain.

Two-dimensional groundwater flow: the differential equation describing the flow within the saturated zone in the x-z plane is (Eq. (2.50))

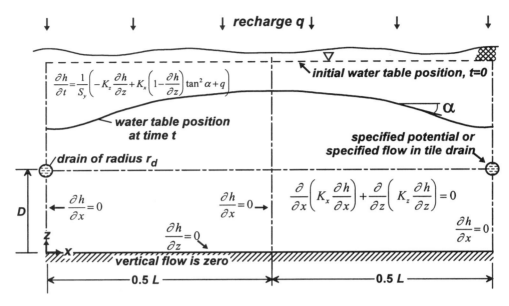

Figure 4.21 Mathematical formulation of tile drainage problem for time-variant conditions. Reprinted from *Journal of Hydrology* **183**, Khan and Rushton, Reappraisal of flow to tile drains, pp. 351–66, copyright (1996) with permission from Elsevier Science

$$\frac{\partial}{\partial x}\left(K_x\frac{\partial h}{\partial x}\right)+\frac{\partial}{\partial z}\left(K_z\frac{\partial h}{\partial z}\right)=S_S\frac{\partial h}{\partial t} \qquad (4.10)$$

The specific storage, S_s is included because confining conditions apply below the water table. However, confined storage effects are small compared to the impact of the specific yield at the water table, hence the term on the right-hand side is set to zero in the remainder of this discussion.

Conditions at the base: at some depth below the drains there is usually a zone of lower hydraulic conductivity; it is assumed that the flow crossing this zone is small enough to be neglected, hence a zero flow condition is set on this boundary with

$$\frac{\partial h}{\partial z}=0 \quad \text{at } z=0 \qquad (4.11)$$

Axes of symmetry: provided that the drains are equally spaced and at the same elevation, also the ground surface is effectively horizontal, vertical axes of symmetry can be defined through the drains and midway between the drains. On these axes of symmetry

$$\frac{\partial h}{\partial x}=0 \quad \text{at } x=0,\ 0.5\,L \text{ and } L \qquad (4.12)$$

Conditions at the drain: the drain is an *internal* boundary. Unless the drain is of a large diameter or the recharge is very small, the water table does not intersect the drain. Two alternate conditions may apply at the drain, each condition can occur in the same drain at different times.

Condition (a): the drain may be partially or just full of water without a surcharge, hence the pressure is atmospheric and the groundwater potential is known. Assuming that the drain is just full, the potential (hydraulic head) around the circumference of the drain equals

$$h=D+r_d \qquad (4.13)$$

where D is the height of the drain centre-line above the base and r_d is the drain radius. These are the conditions on the inside of the drainage pipe. Due to the resistance to flow of groundwater from outside to inside the drainage pipe or the effect of an envelope of material surrounding the pipe, corrections may be needed to the hydraulic conductivity in the vicinity of the pipe, see *maximum water table elevation* below.

Condition (b): an alternative condition at the drain is that the capacity of the pipe to transmit water governs the magnitude of the inflow from the aquifer hence the groundwater potential within the pipe is unknown. The condition is therefore that the flow is limited to Q_{max} (per unit length of drain). When this condition applies there is a surcharge in the pipe hence the groundwater head on the circumference of the pipe

$$h > D + r_d \tag{4.14}$$

When the flow falls below the maximum flow Q_{max} the specified groundwater potential condition applies.

Conditions at the moving water table: due to a vertical recharge q (units L/T) water flows to the water table, water also flows from the water table into the aquifer due to water released from storage as the water table falls. The moving water table conditions are introduced in Section 2.5.2; Figure 2.24b shows how a particle of water on the water table will move from A to A′ during time δt. The *increase* (z is positive upwards) in elevation of the water table δh_{wt} on a vertical line equals

$$\delta h_{wt} = \delta t (v_z + v_x \tan \alpha + q) / S_Y \tag{4.15}$$

where v_x and v_z are the groundwater velocities in the x and z directions and α is the angle of the water table slope below the horizontal.

For steady-state conditions, this equation reduces to

$$v_z + v_x \tan \alpha + q = 0 \tag{4.16a}$$

A second condition, used to identify the location of the water table for steady-state conditions, is that atmospheric pressure occurs at the water table. Therefore, from the definition of groundwater potential, Eq. (2.1), the condition at the water table is

$$h_{wt} = z \tag{4.16b}$$

Initial conditions: realistic initial conditions are important when defining a drainage problem. The most likely condition is that the water table is at, or close to, the ground surface before the drains start to operate. This is represented as a specified constant groundwater head throughout the saturated region.

Maximum water table elevation: if drains are incorrectly spaced, or the capacity of the drains is insufficient, or if there is very high rainfall, the water table may rise to the ground surface. This is represented by defining a maximum elevation of the water table; if at any location the calculated water table position is above

this maximum elevation, it must be reduced to the maximum value.

Conditions in the vicinity of a drain: in recognition of the need to ensure minimum resistance to flow in the vicinity of a drain where the velocities are high, drainage envelopes are often provided. Considerable effort has been expended in studying the hydraulics of water moving through drain perforations and the surrounding envelope, which is usually a more permeable material. There are two physical responses in the vicinity of drains (Willardson 1974, Skaggs 1978) which tend to balance out:

- entry losses as the water finds its way through the perforations into the drain pipe, Panu and Filice (1992) describe analytical solutions to this problem,
- reduced losses due to the higher effective hydraulic conductivities resulting from the drainage envelope which might be gravel, sand, a fabric wrap, geotextiles or other man-made materials (Dierickx 1993, Dierickx *et al.* 1995).

From a number of field tests, Susanto and Skaggs (1992) conclude that the combined effect of the two physical responses can be represented by an effective drain radius and an altered hydraulic conductivity in the vicinity of the drain. For the particular location where they carried out their studies, the hydraulic conductivities in the vicinity of the drains were generally an order of magnitude lower than the remainder of the soil. In the numerical model this effect can be included using Eq. (4.20c) (which is explained in Section 4.4.5) to calculate a modified hydraulic conductivity.

The above discussion concerning the formulation of drainage problems demonstrates that these problems are complex. Numerical methods allow the inclusion of the above conditions but there are several approximate analytical and semi-analytical solutions which are described in the next section. Further information about alternative methods of analysis can be found in Schilfgaarde (1974) and Youngs (1984). A paper by Lovell and Youngs (1984) provides comparisons of the water table elevation midway between drains calculated using a number of alternative approaches.

4.4.4 Approximate methods of analysis for steady-state problems

Using Dupuit theory

The most direct method of analysing the movement of recharge from the water table to tile drains is based on the Dupuit theory (Section 2.3.7). The Dupuit equation can be written as

$$\frac{d}{dx}\left(\frac{d\{h^2\}}{dx}\right) = \frac{-2q}{K} \qquad (4.17)$$

Writing $\{h^2\}$ in brackets signifies that the term is integrated as a single variable.

First, the way in which the Dupuit theory is used to analyse flow into ditches due to recharge is considered, see the left-hand side of Figure 4.22a. The ditches are represented by the boundary conditions,

$h = D$ at $x = 0$ and at $x = L$

Integrating Eq. (4.17) and substituting the boundary conditions leads to an equation for the water table elevation

$$h^2 = D^2 + qx(L - x)/K \qquad (4.18)$$

This is often called the *ellipse equation*. The maximum height of the water table occurs at $x = 0.5L$ with

$$h_{max} = \left(D^2 + qL^2/4K\right)^{1/2}$$

The Dupuit theory assumes that the flow is everywhere horizontal. Therefore at the ditch the water enters over the full depth of the ditch as shown to the left of Figure 4.22a. The ellipse equation is also used for an approximate analysis of flow to tile drains as sketched on the right-hand side of Figure 4.22a. When the horizontal flow arrows inferred by the Dupuit theory are plotted, it is clear that most of the flow arrows are below the drain and hence will not enter the drain. Consequently, the drain is not as efficient at collecting water as indicated by the Dupuit theory. Therefore, when the standard Dupuit theory is applied to drains, it underestimates the maximum height of the water table.

Use of equivalent depth

A number of workers have used a modified ellipse equation to include the effect of convergent flow to small diameter drains. Hooghoudt assumed that the flow in the vicinity of the drains is radial up to a distance of 0.707D from the drain. Elsewhere flows are assumed to be predominantly horizontal so that the Dupuit approximation is applicable. Hooghoudt's analysis led to the concept of an *equivalent depth* D_e as shown in Figure 4.22b. This imaginary depth, when

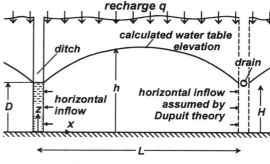

these diagrams represent a vertical section

(a) conventional ellipse

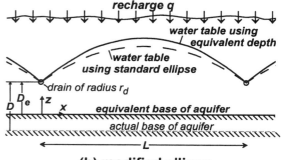

(b) modified ellipse

Figure 4.22 Approaches to the analysis of tile drainage problems: (a) using the Dupuit–Forchheimer equation, (b) modified method using equivalent depths

used with the ellipse equation of the Dupuit approach, leads to a good approximation for the maximum water table elevation midway between the drains. Moody (1966) developed Hooghoudt's approach further, with the equivalent depth expressed as follows:

$$D_e = \frac{D}{\left(1 + \frac{D}{L}\left(2.25\ln\frac{D}{r_d} - c\right)\right)} \qquad 0 < D \le 0.31\,L \quad (4.19a)$$

where the drain radius is r_d and the term c is defined by

$$c = 3.55 - 1.6\frac{D}{L} + 2\left(\frac{D}{L}\right)^2 \qquad (4.19b)$$

$$D_e = \frac{L}{2.55\left(\ln\frac{L}{r_d} - 1.15\right)} \qquad D > 0.31\,L \quad (4.19c)$$

In Figure 4.23 comparisons are made between Figure 4.23a, a numerical solution obtained using the two-dimensional numerical technique of Section 4.4.5 and Figure 4.23b, a solution for the same problem using the equivalent depth technique. For this particular example, $L = 20$ m, the drains of radius $r_d = 0.05$ m are $D = 2.0$ m above the base of the aquifer and $q/K = 0.06$, hence the equivalent depth from Eq. (4.19a) is 1.25 m. In Figure 4.23b, lines of equal groundwater head are drawn as vertical lines up to a distance of $0.707 D$ from the drain. Although these equipotentials are very different from the curved equipotentials deduced from the two-dimensional numerical solution of Figure 4.23a, the maximum water table elevation of 3.51 m is close to the value for the two-dimensional solution of 3.48 m. However, in the vicinity of the drain, the agreement is less satisfactory. According to the Dupuit theory with an equivalent depth, the water table inter-sects the drain at an elevation of 2.0 m above the actual base of the aquifer whereas the numerical solution gives a water table elevation over the drain of 2.38 m. Detailed comparisons between the Moody approach using an equivalent depth and two-dimensional numerical solutions indicate that for smaller values of L/D or larger values of q/K the equivalent depth approach becomes less satisfactory (Khan and Rushton 1996, paper I).

4.4.5 Numerical solutions

Numerical techniques are well suited to the analysis of tile drainage problems. A whole range of practical situations can be represented including steady state or moving water table conditions, the water table reaching the ground surface, drains having specified hydraulic heads and/or maximum flow conditions, the effect of drain perforations, the provision of a drainage envelope and the effect of layers of different hydraulic conductivity within the aquifer. This section considers the use of a finite difference technique to represent some of the more important features of tile drainage problems as identified from the field evidence and precise mathematical formulation of Eqs (4.10) to (4.16). Full details can be found in Khan and Rushton (1996). Alternative numerical solutions are described by Childs (1943), Gureghian and Youngs (1975), Miles and Kitmitto (1989) and Yu and Konyha (1992).

In this section the analysis of steady state problems is considered first, then three important aspects of time-variant problems are examined:

- the calculation for a moving water table is explained using a representative example,
- the inclusion of maximum flow conditions within a drain and/or specified head conditions on the periphery of the drain,
- the prevention of the groundwater table moving above the ground surface.

The influence of all three time-variant conditions is examined in Section 4.4.6 using a representative problem.

(i) Steady state analysis

For the analysis of a steady state problem, a vertical plane of unit thickness in the y direction is considered. The relevant equations are Eq. (4.10) with the specific storage set to zero, together with conditions at the base and on the axes of symmetry as described by Eqs (4.11)

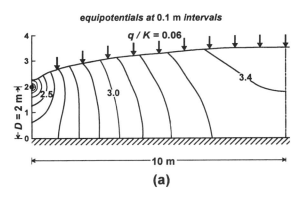

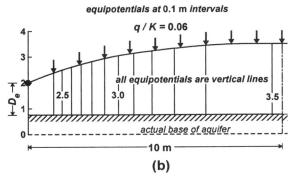

Figure 4.23 Solution for a steady-state drainage problem: (a) two-dimensional numerical solution (b) solution based on equivalent depth according to Moody's formula. Reprinted from *Journal of Hydrology* **183**, Khan and Rushton, Reappraisal of flow to tile drains, pp. 351–66, copyright (1996) with permission from Elsevier Science

and (4.12). A standard finite difference approach is used as described in Section 2.7.4, with the resulting simultaneous equations solved by an iterative technique. The position of the water table is unknown but the two equations, representing zero atmospheric pressure and continuity of flow (Eq. (4.16)), must be satisfied. A computer program which can be used to determine the position of the water table is described by Rushton and Redshaw (1979).

In the vicinity of a tile drain, the flow is predominantly radial. This can be included directly in the finite difference formulation by introducing a *corrected* hydraulic conductivity K_C between the node representing the drain and the surrounding nodes. The derivation is illustrated in Figure 4.24. Consider the horizontal flow Q_x in a vertical plane of unit thickness, between nodal points for a square grid of sides $\Delta x = \Delta z$ due to a groundwater head difference between nodes a and b of Δh

$$Q_x = K\, \Delta z(\Delta h/\Delta x) = K\, \Delta h \qquad (4.20a)$$

Using the Thiem equation (Eq. (2.34)), the total radial flow from a circle of radius r_1, with a groundwater potential h_1, to a radius r_2 at potential h_2 equals

$$Q = 2\pi K\,(h_1 - h_2)/\ln(r_1/r_2)$$

Hence radial flow through one quadrant Q_q (see Figure 4.24b), towards a drain of radius r_d from a radial distance Δx is described by

$$Q_q = 0.25[2\pi K\, \Delta h/\ln(\Delta x/r_d)] \qquad (4.20b)$$

When a modified hydraulic conductivity is introduced for the mesh intervals surrounding the drain

$$K_C = [0.5\,\pi/\ln(\Delta x/r_d)]K \qquad (4.20c)$$

with K in Eq. (4.20a) replaced by K_C, an expression identical to Eq. (4.20b) is obtained.

For a square mesh, this correction is made to the hydraulic conductivities in the x and z directions between the drain and the surrounding nodes as shown in Figure 4.24b. For a rectangular mesh the hydraulic conductivities in the x and z directions need to be modified to reflect the different geometries of the segments.

To illustrate the nature of the correction, consider a square finite difference mesh with $\Delta x = \Delta z = 0.25\,\text{m}$. Three possible drain radii of $r_d = 0.15, 0.10$ and $0.05\,\text{m}$ are considered. The corrected hydraulic conductivities are listed below:

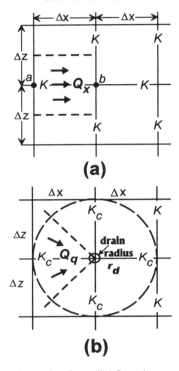

Figure 4.24 is shown with label **vertical plane of unit thickness finite diference mesh shown by horizontal and vertical lines** and parts **(a)** and **(b)**.

Figure 4.24 Correction for radial flow close to a drain: (a) two-dimensional flow in *x-z* plane, (b) radial flow towards drain

$$r_d = 0.15\,\text{m}, \quad K_C = 3.075\,K$$
$$r_d = 0.10\,\text{m}, \quad K_C = 1.714\,K$$
$$r_d = 0.05\,\text{m}, \quad K_C = 0.976\,K$$

When the original hydraulic conductivity is used with $K_C = K$, this is equivalent to a drain radius $r_d = 0.208\,\Delta z$ which for this example equals $0.052\,\text{m}$.

(ii) Time-variant analysis, representation of moving water table

For a time-variant analysis, the movement of the water table is represented using discrete time steps. To illustrate the calculations for the movement of the water table, a specific example is selected; for further details see Khan and Rushton (1996, paper II). In this calculation the vertical hydraulic conductivity is 0.1 m/d with a specific yield of 0.1. The calculation starts from an equilibrium condition with a recharge of 3.0 mm/d, the recharge is then reduced to 1.5 mm/d.

The vertical movement of the water table is calculated from Eq. (2.52):

$$\Delta h = \frac{\Delta t}{S_Y}\left(-K_z\frac{\partial h}{\partial z} + K_x\left(1 - \frac{\partial h}{\partial z}\right)\tan^2\alpha + q\right) \qquad (4.21)$$

Table 4.6 contains calculations for the movement of the water table from time 9.5 days to time 10.0 days for the problem considered in Section 4.4.6. Numerical values refer to two locations, the centre-line between drains and the quarter point midway between the centre-line and a drain.

- The time increment Δt is 0.5 day.
- Water table elevations at a time of 9.5 days are recorded under Step 1.
- Step 2 refers to the calculation of the vertical groundwater head gradient at the water table; a three-point finite difference formula is used with the uppermost node at the water table. Due to the influence of the drains the vertical gradient is larger at the quarter point than the mid-point.
- For Step 3, the slope of the water table at the midpoint is zero due to symmetry but at the quarter point there is a water table slope with $\tan\alpha = 0.0769$ below the horizontal.

- The three components of Eq. (4.21) are calculated in Steps 4, 5 and 6; a positive component is vertically upwards. The correction due to the lateral movement of the particle on the water table is 0.0029 m (Step 5 for the quarter points) compared to the total vertical movement of the particle on the water table of − 0.0163 m.
- The recharge of 1.5 mm/d is equivalent to a vertical upwards component of +0.0075 m (Step 6).
- Changes in water table elevation on the vertical grid lines are calculated in Step 7, the change at the mid-point is greater than at the quarter point.

(iii) Time-variant analysis with maximum flow or specified head at the drain

Conditions in drains change with time. Sometimes the drain runs full under a surcharge, while at other times the hydraulic head in the drain remains relatively constant. Section 4.4.3 describes the *alternative conditions which can hold in a drain*. When including a drain in a numerical solution, the drain must be represented as a specified outflow if the calculated head in the drain is above the top of the drain. However, if the flow is less than the specified value, then the specified head condition applies. In many practical situations, the maximum flow condition applies when the water table is close to the ground surface; specified maximum flows also occur when there is a substantial recharge. However, as the water table falls, conditions in the drain change to a specified head. A methodology

Table 4.6 Calculation of movement of water table between 9.5 and 10 days for the mid point and quarter point between drains; $K_x = K_z = 0.1$ m/d, $S_Y = 0.1$, $q = 1.5$ mm/d, $\Delta t = 0.5$ d

Step	Location	Mid-point	Quarter point
1	water table elevation at 9.5 days (m)	4.9052	4.4390
2	$\partial h/\partial z$	0.0279	0.0327
3	$\tan\alpha$	0.0000	0.0769
4	$-\dfrac{\Delta t}{S_Y}K_z\dfrac{\partial h}{\partial z}$	−0.0140	−0.0163
5	$\dfrac{\Delta t}{S_Y}K_x\left(1 - \dfrac{\partial h}{\partial z}\right)\tan^2\alpha$	0.0000	0.0029
6	$\dfrac{\Delta t}{S_Y}q$	0.0075	0.0075
7	$\Delta h = $ Step 4 + Step 5 + Step 6	−0.0065	−0.0059
8	water table elevation at 10 days (m)	4.8987	4.4331

for calculating the head loss in drainage pipes, which collect water along their length, is described by Yitayew (1998). The example considered in Section 4.4.6 illustrates how conditions in the drain alter with time.

(iv) Time-variant analysis, water table limited by ground surface elevation

When drains are not operating or when significant infiltration occurs, there is a likelihood that part or all of the water table will reach the ground surface. This can be represented directly in the numerical solution by setting a maximum value for the water table elevation. If this elevation is exceeded during calculations of the water table elevation (the calculation procedure of Table 4.6), the water table elevation is reduced to the ground surface elevation.

4.4.6 Representative time-variant analysis

To illustrate the significance of conditions (ii), (iii) and (iv) of Section 4.4.5, consider the representative example of Figure 4.25 in which datum is the impermeable base. Parameter values are as follows:

Drains:
  distance between drains = 50 m
  radius of drains = 0.05 m
  elevation above impermeable base = 2.0 m
  maximum discharge = 0.17 m³/d/m
  minimum head in drain = 2.05 m
Aquifer:
  hydraulic conductivity = 0.1 m/d
  specific yield = 0.1
Initial conditions and recharge:
  initial water table elevation above the base = 5.5 m
  recharge = 0.0, 0.005, 0.002 m/d for successive 40-day periods

Figure 4.25 illustrates certain results from the model simulation.

Figure 4.25a shows the water table elevation on the cross-section at 0, 40 and 80 days.

Figure 4.25b refers to groundwater heads at three locations and how they change with time. The three lines represent the calculated water table elevation midway between the drains, the water table elevation above the drains and the hydraulic heads in the drains. The ground surface, which is the maximum head, is indicated by the long dashes.

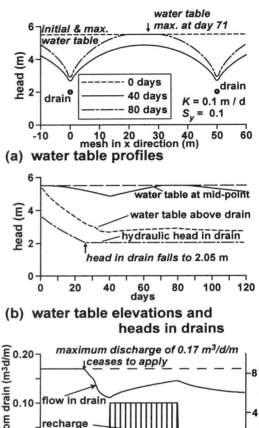

(a) water table profiles

(b) water table elevations and heads in drains

(c) recharge and flow in drains

Figure 4.25 Representative numerical solution illustrating time-variant behaviour including specified maximum flow or specified head in drain and maximum elevation of water table

Figure 4.25c shows how the flows in the drains change with time (the maximum discharge is shown by the broken line); the recharge at different times is also plotted.

Important information, which can be observed in these graphs, is summarised below.

• During the *first forty days*, when there is no recharge, the water table falls from the initial maximum position; the fall above the drains is greater than the fall midway between the drains. The maximum flow condition of $0.17 \, m^3/d/m$ applies until day 26 (Figure 4.25c); thereafter the hydraulic head in the drain remains at 2.05 m (Figure 4.25b) with the flow decreasing, increasing and then decreasing again to $0.12 \, m^3/d/m$ at day 120.

• From *day 40 to day 80* the recharge is 0.005 m/d. After 71 days the water table rises to the maximum value midway between the drains; the maximum water table elevation above the drain is 2.92 m and occurs on day 80.

• From *day 80 to day 120* the recharge falls to 0.002 m/d. The water table midway between the drains falls from a maximum at day 83.5, and the flow in the drains continues to fall during this period.

This example demonstrates that, with the formulation and numerical techniques introduced above, complex time-variant drainage problems can be studied.

4.4.7 Interceptor drains

A review of interceptor drain schemes is used to illustrate the dangers of inadequately formulated conceptual models. Interceptor drains are often used on sloping ground to collect water as it moves downgradient. In a study by Willardson *et al.* (1971) in the USA, interceptor drains were introduced to collect water lost from a canal. Lining the canal was not feasible, consequently interceptor drains were positioned down the hydraulic gradient from the canal; the field situation is illustrated in Figure 4.26a. A numerical model using the resistance network technique was chosen to investigate this problem. Results were expressed as a ratio of the water intercepted by the drains to that lost from the canal. The numerical model is sketched in Figure 4.26b. The analysis was restricted to steady-state conditions with no recharge, an assumption that was valid for that specific environment. One further assumption was that a shallow farm drain in an adjacent field acted as a fully penetrating boundary at a constant groundwater head, thereby limiting the area that needed to be studied. With the limited equipment available in 1971 for solving finite difference equations, this was an appropriate assumption.

Interceptor drains are also being constructed in Pakistan to collect water lost from canals. Figure 4.27 refers to a pumped interceptor drain system constructed in the vicinity of the Chesma Right Bank canal. The elevation of the water table is identified from a number of observation piezometers constructed so that they just penetrate the water table. Field results are plotted for a line of piezometers. The filled squares

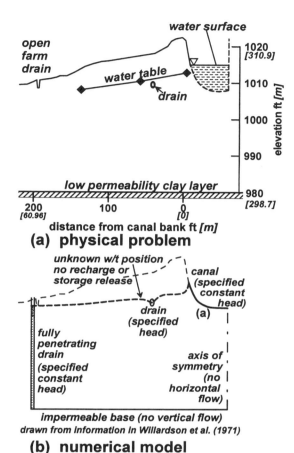

Figure 4.26 Interceptor drain in USA: (a) physical details, (b) representation by numerical model. Reproduced by permission of ASAE from Willardson *et al.* (1971)

joined by the dashed line refer to a cross-section perpendicular to the canal through the pumped sump. These results were recorded one year after the commencement of pumping from the interceptor drain. Of particular importance is the piezometer which penetrates beneath the bed of the canal; it was drilled where a bridge crosses the canal. The readings confirm that the canal is perched above the regional water table. For most of the time, both the main and standby pumps were operating. The following insights can be gained from Figure 4.27.

• Even after pumping for a year, the water table was 1.6 m above the drain; surcharge conditions occurred in the drain hence flows continued to be limited by

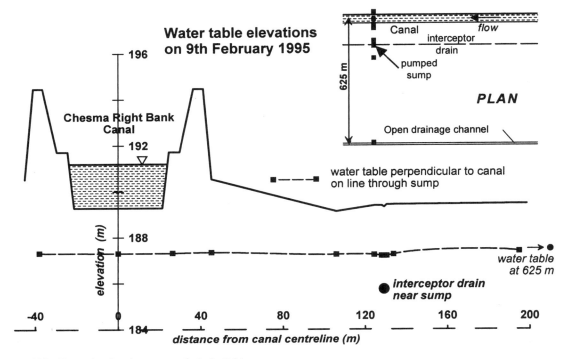

Figure 4.27   Example of an interceptor drain in Pakistan

the maximum flow conditions in the drain or the maximum discharge from the pumps.

- The water table is lower than the bed of the canal, so that the canal is perched. There are two possible causes; one is the presence of silt in the bed of the canal, the second is that the base of the canal is above the regional groundwater table of the surrounding area.
- When the interceptor drain was designed, it was assumed that it would only intercept water lost from the canal whereas in practice it also collects water from the overlying water table on either side of the drain.
- If the drainage system is designed to intercept canal losses and also to collect recharge to the water table, the capacity of the drain (and pump, if required) should be double or treble the average recharge so that there is sufficient spare capacity to lower the water table and to control the water table during periods of high recharge.

Features to be included in analysis

Detailed designs of interceptor drains in Pakistan, which have been carried out by international consul-

tants, are based on numerical model analysis using the same assumptions as Willardson *et al.* (1971). Figure 4.28a illustrates the assumptions inherent in the numerical model analysis at a specific location. An interceptor drain is located 30 m from the canal centre-line and 4.0 m below the canal water level. The lateral boundary, at only 72 m from the canal centre-line, is a specified groundwater head, which is equivalent to a channel which extends to the full depth of the aquifer, the groundwater head in the channel is 6.0 m below the canal water level. In addition, there is a lower impermeable boundary. Without the interceptor drain, water moves from the canal to the lateral boundary. When the interceptor drain is included in the model it collects some of the water which would otherwise move to the fully penetrating channel. Due to the physically unrealistic lateral boundary of the fully penetrating channel, the results of the above analysis fail to represent the actual field situation.

Figure 4.28b has been prepared to show features which *must* be included in a conceptual model of interceptor drains in the vicinity of a canal:

- the possibility of the canal being perched above the regional water table,

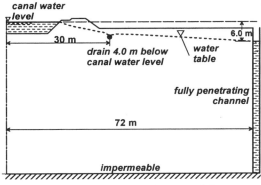

(a) unsatisfactory numerical model

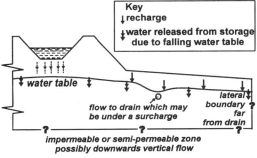

(b) necessary features for conceptual model

Figure 4.28 Models of interceptor drains: (a) unsatisfactory model used for analysis, (b) features that must be considered when preparing conceptual models

- the drain collects both water released from storage at the overlying water table and recharge to the water table,
- a realistic lateral boundary to represent the regional setting; this boundary may be 500 m or more from the canal centre-line,
- the underlying impermeable boundary may not be appropriate in alluvial aquifers if water is pumped from deeper layers within the aquifer system.

Illustrative example

To illustrate the type of analysis that should be carried out to assist in the understanding of the various flow processes associated with interceptor drains, the representative example of Figure 4.29 is introduced. This example has similarities with the Chesma Right Bank canal. There are two interceptor drains, the first 50 m

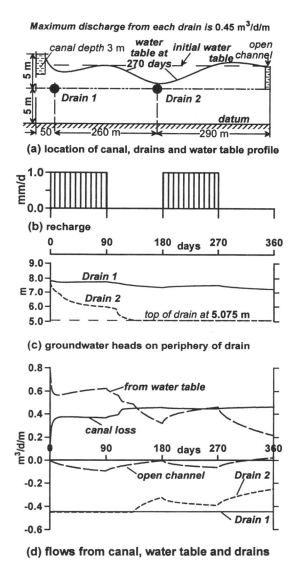

(d) flows from canal, water table and drains

Figure 4.29 Representative example of time-variant interceptor drain behaviour using a numerical model

and the second 310 m from the canal centre-line. The lateral boundary is a partially penetrating channel 5 m deep at 600 m from the canal centre-line. The analysis covers a period of one year (taken as 360 days) with recharge of 0.001 m/d from day 0 to day 90 and between days 180 and 270 but no recharge at other times (Figure 4.29b). The average recharge of 0.0005 m/d over a distance of 600 m is equivalent to 0.30 m³/d/per metre length of canal. In an attempt to provide sufficient pumping capacity, an initial analysis was carried out with the maximum pumping rate for *each* drain equal to $0.3\,m^3/d/m$ (i.e. total pumping capacity is double the average recharge). However, this pumping rate proved to be insufficient to control the water table, hence the maximum discharge from each drain is set at $0.45\,m^3/d/m$. The calculation starts with the water table at an elevation of 8.0 m.

Comments are made below about important features of the results.

- The water table position after 270 days is shown in Figure 4.29a. Drain 2 operates successfully with the result that the water table above this drain is at 5.8 m. However, Drain 1 is unable to lower the water table sufficiently, consequently the maximum water table between Drains 1 and 2 is slightly above the initial water table elevation.
- Figure 4.29c shows the hydraulic head in each drain. For Drain 2 the head is lowered to the top of the drain, 5.075 m, by day 134, thereafter the discharge from the drain is less than the maximum of $0.45\,m^3/d/m$, see Figure 4.29c. For Drain 1, surcharge conditions always occur; the lowest hydraulic head is 7.36 m.
- The upper half of Figure 4.29d shows that the quantity of water lost from the canal (the unbroken line) gradually increases as the drains continue to operate. The quantity of water drawn from the water table decreases, but is strongly influenced by the recharge periods. The lower part of the figure refers to the discharge from Drain 2 which falls below the maximum of $0.45\,m^3/d/m$ but Drain 1 continues to operate at the maximum discharge rate.
- The situation could be improved by increasing the maximum discharge for Drain 1; Drain 2 operates successfully.

4.5 IRRIGATED RICEFIELDS

4.5.1 Introduction

The efficiency of irrigated ricefields in many parts of the world is low; frequently less than 50 per cent of the water supplied is used by the crops (Bos and Nugteren 1982). Although the bed of the ricefield is puddled to reduce seepage losses, field losses are high (Walker and Rushton 1984, 1986, Tuong *et al.* 1994). Field evidence shows that losses are especially high in surface water irrigation schemes with a rotational system of water allocation. In such schemes the ricefield is used to store water for periods of up to 10 days until the next release of water. The concept of using water lost from a rice-

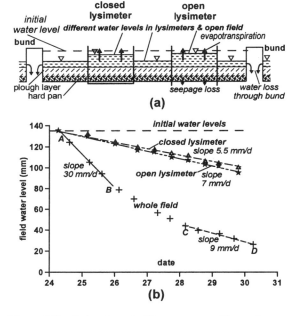

Figure 4.30 Estimation of different causes of losses from a flooded ricefield using lysimeters: (a) cross-section, (b) water levels in lysimeter and field

field as artificial recharge to the underlying aquifer by deliberately over-irrigating has been tested by Tsao (1987).

Reasons for high losses have been identified from careful fieldwork with lysimeters. Figure 4.30 indicates how losses were quantified in the field by monitoring falls in water levels. Figure 4.30a illustrates the arrangements of closed and open lysimeters while Figure 4.30b shows the fall of water level in the lysimeters and in the ricefield with an initial depth of water of 140 mm. The experiments demonstrate that different losses occur from a lysimeter with a closed base, a lysimeter with an open base and from the ricefield. The loss from the lysimeter with closed base is 5.5 mm/d due to evapotranspiration alone. For the lysimeter with the open base, losses of about 7 mm/d occur due to evapotranspiration and seepage through the bed of the ricefield, with the loss through the bed of the ricefield approximately 1.5 mm/d. For the actual ricefield, the initial loss is about 30 mm/d; the only possible cause of this high loss is seepage through the bunds. The flow through the bunds is therefore equivalent to about 23 mm/d. The nature of this flow is illustrated in Figure 4.30a by the flow arrows through the bunds. Significant losses through the bunds can be confirmed by the simple expedient of covering the bunds with plastic sheet. The losses from the field then become similar to those for the lysimeter with the open base.

A further insight from these experiments is that the rate of fall of the water surface in the field decreases as the depth of water in the field becomes smaller, whereas the losses from the closed and open lysimeters are largely independent of the depth of water. The significance of this result is discussed further in Section 4.5.4, *varying depth of water in the field*.

4.5.2 Formulation of flow processes in ricefields

A computational model for losses from flooded ricefields is presented in Figure 4.31; a vertical section perpendicular to the bund is considered. Because the main purpose of the study is to investigate losses through the bund, the plough layer and hard pan are assumed to be effectively impermeable. Flow can occur from the water layer through the bund to the underlying aquifer. At a distance L either side of the centre-line of the bund, the groundwater head in the underlying aquifer is specified. Therefore, flow occurs from the water layer, through the bund and towards the two specified head boundaries.

The mathematical formulation involves the equation for steady state flow in a vertical section through a saturated aquifer (see Section 2.5),

$$\frac{\partial}{\partial x}\left(K_x \frac{\partial h}{\partial x}\right) + \frac{\partial}{\partial z}\left(K_z \frac{\partial h}{\partial z}\right) = 0 \qquad (4.22)$$

together with appropriate boundary conditions. On the top, sides and bottom of the plough layer and hard pan, there is no flow crossing, hence $\partial h/\partial x = 0$ or $\partial h/\partial z = 0$. There are also four specified head boundaries. Two specified heads apply where the water layer is in contact with the bund, the condition of $h = 0.16$ m represents a water depth of 0.16 m. Also the regional groundwater conditions are defined by specified head conditions.

There are also three boundaries where free water surface conditions apply. For a free water surface (or water table) the position of the boundary is unknown but two conditions apply; one condition is that the pressure is atmospheric (hence $h = z$), the second is that no flow crosses the boundary. These free water surface conditions are discussed further in Sections 2.5 and 4.4.3.

Due to the unknown position of the three free surface boundaries, this is a difficult problem to

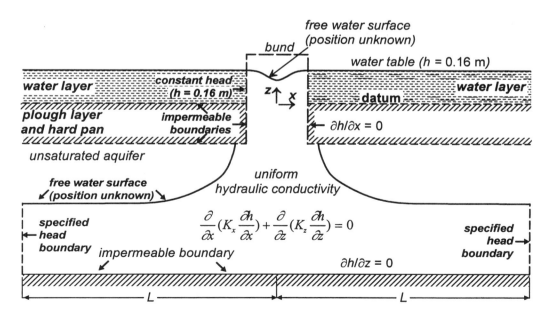

Figure 4.31 Conceptual model and mathematical formulation for losses from a flooded ricefield. Reprinted from *Journal of Hydrology* **71**, Walker and Rushton, Water losses through the bunds of irrigated ricefields interpreted through an analogue model, pp. 59–73, copyright (1984) with permission from Elsevier Science

analyse. The solutions described below were obtained with an electrical analogue resistance network to solve the finite difference equations. An iterative technique is used to converge to the correct locations of the free surfaces (Walker and Rushton 1984).

4.5.3 Representative results

Results for a representative situation are included in Figure 4.32a, which shows the conditions in the bund while Figure 4.22b shows how water flows through the bund to join the regional groundwater flow in the aquifer which, for this example, is from left to right. The height of water in each field is 0.16 m above the puddled clay layer while the groundwater heads in the aquifer at ±10 m from the bund centre-line are 0.44 m and 1.36 m below the top of the plough layer.

Figure 4.32a shows the distribution of horizontal velocities on the vertical sides of the bund; velocities are highest just above the plough layer. Between the bunds, the vertical velocities are almost uniform. The average velocity between the bunds can be estimated from the 0.0 and –0.1 m groundwater head contours which are on average 0.12 m apart. The width of the

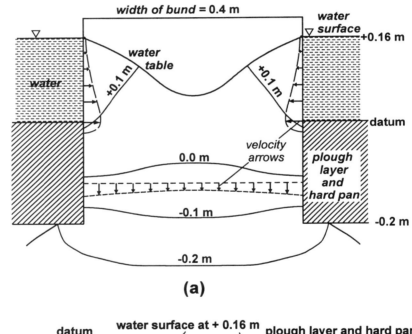

(a)

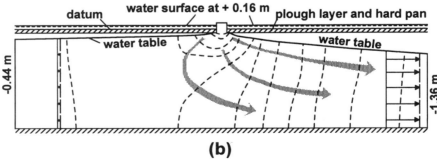

(b)

Figure 4.32 Results for water losses from ricefield deduced from an analogue model: (a) flow through the bund, (b) flow into the aquifer with a regional groundwater flow from left to right. Reprinted from *Journal of Hydrology* **71**, Walker and Rushton, Water losses through the bunds of irrigated ricefields interpreted through an analogue model, pp. 59–73, copyright (1984) with permission from Elsevier Science

bund is 0.4 m, therefore the loss through *one side of the bund* equals

half width of bund × hydraulic conductivity
    × head gradient = 0.2 × 0.8 × 0.1/0.12
    = 0.133 m^3/d/m length of bund,

where the hydraulic conductivity of the bund is taken as 0.8 m/d.

If the rice field has an area of 30 m by 20 m (hence a perimeter of 100 m), the total loss from a single field through the bunds would be 0.133 × perimeter = 0.133 × 100 = 13.3 m^3/d.

Dividing this by the area of the field of 600 m^2 leads to a loss equivalent to 22.2 mm/d. This is an approximate calculation which indicates that the loss due to flow through the bunds as determined from the numerical model is of the same order as the loss measured during field experiments.

Figure 4.32b illustrates the manner in which the water is effectively sucked through the bund to join the underlying regional groundwater flow. Consequently, the bund is an efficient means of drawing water from the ricefield. Many examples of high losses in different parts of the world are quoted in Walker and Rushton (1986) and Tuong *et al.* (1994).

4.5.4 Ricefields with different geometry and water levels

The examples considered above refer to adjacent ricefields at the same elevation with identical water levels. This section considers other configurations.

Varying depth of water in the field

The example in Section 4.5.3 refers to a depth of water in each field of 0.16 m; does the loss depend on the depth of water in the field? Insights can be gained from the lysimeter results in Figure 4.30. Two straight lines can be drawn through the field results for the fall in the ricefield water surface. The first line, AB, which refers to the early stages of the test, indicates that the water level in the field fell about 30 mm in one day when the water depth in the field was between 140 mm and 90 mm above the plough layer. This demonstrates significant losses through the bunds. However, five days after the start of the experiment when the depth of water in the field was between 40 and 30 mm, the rate of fall in the water level reduced to 9 mm in a day; this is only slightly larger than the loss from the open

lysimeter of about 7 mm/d. These results suggest that the loss from the ricefield through the bunds depends critically on the depth of water in the field.

The mathematical model described in Section 4.5.3 was used to explore the influence of the depth of water in the ricefield; more detailed information can be found in Walker and Rushton (1984). The results are summarised in Figure 4.33; the vertical axis refers to the depth of water in the ricefield and the horizontal axis represents the loss through the bund. Results, obtained from the model are shown by the solid triangles. In Figure 4.33 three cases are highlighted and indicated by the letters (a) to (c).

(a) The depth of water in the field is 160 mm. For the parameter values of Section 4.5.3, the total loss through both sides of the bund is 0.27 m^3/d/m. The sketch shows that flow from the field through the bund is continuous across the full width of the bund, with the lowest point of the water table between the bunds about 40 mm below field water level.

(b) For a depth of water in the field of 80 mm the loss is 0.16 m^3/d/m. The diagram for the flows is similar to that of (a) but the depth of the lowest point of the water table in the bund is more than 200 mm below field water level, this is below the base of the plough layer and hard pan.

(c) For a depth of water in the field less than 80 mm, there is no longer a continuous flow of water across the full width of the bund; this is indicated in sketch (c). Consequently the loss through the bund is much smaller.

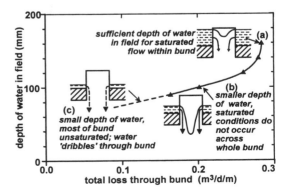

Figure 4.33 Influence of different depths of water in the ricefield

These findings are confirmed by field experience. With a rotational surface water irrigation system, the field is used to store water; the depth of water is likely to be more than 100 mm and the losses are high. However, for groundwater irrigation where water can be supplied every day, the depth of water rarely exceeds 50 mm and the losses are much smaller.

Differing depths of water and terraced ricefields

Beds of adjacent ricefields are often at different elevations with different water depths. When ricefields are at the same elevation but have different depths of water, the field observations of Figure 4.34a show that the water table falls from each of the water surfaces to a minimum within the bund. Numerical model solutions confirm this water table shape.

When the bed of the fields and the water levels are at different elevations (Figure 4.34b), there is still a flow from the flooded field on each side into the bund. It is only with a very large difference in water surface elevations between the fields that some water may flow from the higher to the lower field through the bund, although the major flow continues to be through each bund to the underlying aquifer.

If there is a long series of terraced ricefields, careful consideration must be given to the overall groundwater heads. Figure 4.34c shows fields at the top and at the bottom of a terrace. The upper fields on the left-hand side lose water, as discussed above. However, for the lower fields, the regional groundwater head in the underlying aquifer is likely to be above the water surface in the fields. Consequently, water flows from the aquifer into the lower fields. Field observations, demonstrating this process, are described by Lowe and Rushton (1990). Methods of estimating the magnitude of flows from ricefields into aquifers are described in Chapter 4 of Simmers (1997).

4.6 CONCLUDING REMARKS

This chapter reviews a number of different forms of groundwater–surface water interaction. In each case the conceptual models of the interaction have been reviewed. For example, with horizontal drains the conventional approach of assuming that the water table intersects the drain is shown to be invalid. Insights from one form of groundwater–surface water interaction can be used for other situations. Losses of water from lined canals do not simply depend on the hydraulic conductivity of the lining; the properties and flow processes in the underlying aquifer also influence the losses. This leads to questions about the validity of the procedure often followed in MODFLOW of calculating the river coefficient (river conductance) directly from the dimensions and hydraulic conductivity of the bed deposits.

Groundwater flows to or from rivers or canals differ depending on whether they are *connected to* or *disconnected from* the aquifer. When there is direct connection between the river or canal and the aquifer, there are several properties which determine the magnitude of the flow including the dimensions and hydraulic conductivity of the bed of the river or canal, the properties and dimensions of the underlying aquifer and the nature of the outflow from the aquifer. Higher flows through underlying low permeability strata are particularly significant. When the river or canal is perched well above the water table, disconnected conditions occur. The nature of any lining or bed deposits then becomes the most important property. Unsaturated conditions can occur between the canal or river and the aquifer water table.

When using models to estimate losses from a surface water body to an aquifer it is essential to ensure that the lateral boundaries are a substantial distance from the surface water body. In earlier studies, when computing power was limited, lateral and underlying boundaries were often located too close to the surface water body. With the greatly increased computing

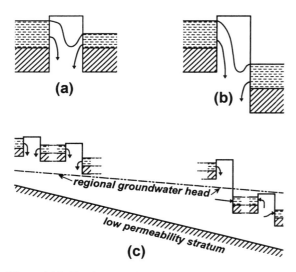

Figure 4.34 Further practical examples: (a) different field water levels, (b) fields on either side of the bund at different elevations, (c) a series of terraced fields

power available today this is not an acceptable procedure; it can lead to serious errors in predictions, as indicated by the study of interceptor drains.

Determining the magnitude of coefficients (or conductances) to describe groundwater–surface water interaction is difficult. Every effort must be made to obtain field information about the gains or losses in the surface water flows. There are many practical difficulties because the groundwater components are often a small proportion of the total surface flows. Nevertheless, reasonable approximations to the groundwater components can be made with careful fieldwork, as demonstrated in a number of the case studies described in the following chapters.

PART II: RADIAL FLOW

The analysis of radial flow towards pumped boreholes is often restricted to the estimation of aquifer parameters such as the transmissivity, storage coefficient and leakage coefficient. However, by studying flow towards a pumped borehole, an understanding can be gained of the flow processes within the aquifer system and close to the borehole. There are two stages in the interpretation of information gained when pumping an aquifer; first, developing a conceptual model of the aquifer flow processes; second, selecting suitable methods of analysis so that field measurements can be interpreted to provide quantitative information. Due to the large range of possible flow processes which can occur in the aquifer and at the borehole, interpretation of the field information requires a substantial number of analytical and numerical techniques.

In the four chapters in Part II, alternative methods of studying flow towards pumped boreholes are reviewed. Various idealisations are introduced so that analytical or numerical solutions can be used to represent the actual field situation. The earliest approach was to assume steady-state radial flow in a confined aquifer. Time-variant radial flow can be studied using the Theis analysis for confined aquifers with the associated curve-matching techniques; many field problems have been studied using this approach. More complex analytical solutions have also been developed. In addition, numerical techniques of representing radial flow to boreholes and wells are available. For each method of analysis, it is essential to consider how the inherent idealisations and simplifications limit the application of each technique to specific field problems.

The contents of the four chapters can be summarised as follows. Chapter 5 is concerned with radial flow without vertical flow components; analytical and numerical solutions are described for the pumping and recovery phases of pumping tests. Further issues include leaky aquifer response, delayed yield, interfering boreholes, overflowing wells and the effect of boundaries or changes in aquifer parameters.

Chapter 6 considers large diameter wells, which are commonly used in developing countries. The key to understanding large-diameter wells is the effect of well storage; methods of analysing pumping tests in large-diameter wells are presented. The reliability of large-diameter wells when used to irrigate a crop over a growing season is also examined.

Chapter 7 focuses on situations where vertical flow components are important; analytical and numerical techniques of analysis are presented and applied to case studies. Extensive use is made of a relatively straightforward two-zone numerical model which has proved to be invaluable in understanding and quantifying flows for many aquifer systems. The chapter concludes by demonstrating that aquitard storage is significant in many practical situations.

Chapter 8 uses the approaches discussed in Chapter 7 to gain further insights and understanding of techniques such as step pumping tests and packer tests; the discussions are illustrated by case studies. The interpretation of water level data from pumping boreholes is explored. The reliable yield of a number of aquifer systems is examined, especially when the hydraulic conductivity varies with saturated depth. Artificial recharge using injection wells and the hydraulics of horizontal wells are also considered.

Some of the more surprising findings in the study of radial flow are highlighted below, with relevant section numbers given in brackets.

- Observation boreholes in the vicinity of pumped boreholes and open over a significant depth of the aquifer, cause such a disturbance to aquifer flows that measured drawdowns are frequently unreliable (5.1.6).
- Deviations from type curves at later times, which are usually explained as no-flow or recharge boundaries, may be due to changes in hydraulic conductivity or storage coefficient with radius; alternatively, the

Table II.1   Key topics and section numbers where information can be found

Topic	Sections where topic is discussed
Field data, pumping and recovery	5.1, 5.15.3, 6.4, 7.4.6, 8.4.1, 8.5.3, 8.6.2, 8.7.2
Readings from individual piezometers	5.1, 7.4.6, 7.5.4, 8.5.3, 8.6.2, 8.8.3
T or S varying within aquifer	5.3, 5.4.4, 5.11, 5.13, 5.16, 7.3.3, 8.2.5, 8.7, 8.7.3
Changing saturated thickness	5.12, 6.6.1, 8.5.2, 8.7.2, 8.8.3
Changing between confined/unconfined	5.7, 5.16, 7.2.4, 8.2.5
Storage in aquitard	7.5.2, 7.5.3, 7.5.4
Boundary effects	5.10, 6.6.1, 6.6.2, 7.3.3
Impact of well storage	5.1.4, 6.2, 6.3, 6.6, 7.4.6
Well losses	5.5.4, 7.4.6, 8.2, 8.5.3, 8.5.4, 8.6.4
Seepage face	5.1.4, 6.3.3, 8.2.5, 8.5.4

deviations may be caused by interfering boreholes (5.11, 5.15).

• Increasing drawdowns can continue in an observation borehole after the pump has been switched off; this occurs in large-diameter wells (6.4) and weathered-fractured aquifers (7.4.6).

• Tubewells exert water table control over greater distances when there are extensive lateral zones of low hydraulic conductivity (7.3.3).

• The classical assumption of leaky aquifer analysis, that a vertical gradient is set up instantaneously from the top to bottom of an aquitard, is not valid in many practical situations; when aquitard storage is included it is easier to draw water from the bottom of aquitard at the start of pumping (7.5.3).

• Step-drawdown pumping tests are more appropriate for unconfined and leaky aquifers than for confined aquifers (8.2).

Specific topics are often discussed in several sections in Part II. Some of the more important topics are listed in Table II.1 above, together with the section numbers where information can be found. In each chapter, summary tables are provided of the contents.

5

Radial Flow to Pumped Boreholes – Fundamental Issues

Studies of radial flow to pumped boreholes under steady and unsteady conditions were among the earliest groundwater problems to be analysed. Steady-state and time-variant flow in confined aquifers were considered by Thiem (1906) and Theis (1935); analytical solutions were derived which can be used to estimate aquifer parameters from field pumping tests. Subsequently, a wide range of alternative aquifer and borehole conditions were considered, leading to a substantial number of analytical and numerical solutions. Table 5.1 lists the situations reviewed in the first part of this chapter; radial flow problems are classified in Section 5.2. However, the first section of this chapter refers to a specific pumping test for which extensive field information is available. This example is presented at the start of the discussion on radial flow to illustrate the complex flow processes which occur in practical radial flow situations.

5.1 AQUIFER RESPONSE DUE TO PUMPING FROM A BOREHOLE

5.1.1 Introduction

In real-life situations, the flow through aquifers to pumped boreholes is often far more complex than the conditions implicit in analytical solutions. This is illustrated by considering a specific pumping test in an unconfined sandstone aquifer for which measurements were made in individual piezometers. This detailed field study illustrates the complicated flow processes in an unconfined aquifer, including the different responses of individual piezometers at varying depths. A careful examination of the field response also allows the development of conceptual models. It is then possible to

consider how the idealisations and simplifications, inherent in many of the conventional methods of analysing radial flow to pumped boreholes, limit their applicability to actual field problems. This introductory problem also illustrates how an open observation borehole in the vicinity of a pumped borehole can disturb the flow processes within the aquifer so that the measurements in the observation borehole do not represent the actual conditions in the aquifer.

5.1.2 Details of field study

Important insights into flow processes in the vicinity of a pumped borehole were gained from a detailed study of pumping from an unconfined aquifer at Kenyon Junction in the Lower Mersey Permo-Triassic Sandstone aquifer of north-west England. The critical question is the magnitude of the specific yield of this heavily exploited aquifer. Rather than following the conventional approach of providing a number of observation boreholes at increasing radial distances from the pumped borehole, individual piezometers were constructed at varying depths within a single borehole. Drawdown measurements were obtained for the water table and at increasing depths within the aquifer with the deepest piezometer some 10 m below the penetration of the pumped borehole. Results were obtained during both the pumping and recovery stages. Detailed information about the tests can be found in Rushton and Howard (1982) with further information in Walthall and Ingram (1984).

The pumped borehole of 600 mm diameter is 86 m deep with a rest water level 32 m below ground level. For the test, the pumping rate was 1619 m³/d for one day, and recovery was monitored for a further day.

Groundwater Hydrology: Conceptual and Computational Models. K.R. Rushton
© 2003 John Wiley & Sons Ltd   ISBN: 0-470-85004-3

Table 5.1   Conditions discussed in Sections 5.3 to 5.8

Condition	Analytical (A) and/ or Numerical (N)	Section
Steady radial flow in confined aquifers	A	5.3
Unsteady radial flow in confined aquifers	A	5.4
Unsteady radial flow in confined aquifers	N	5.5
Recovery phase for unsteady radial flow	A	5.6
Pumping and recovery phases, confined/ unconfined aquifer	A & N	5.7
Leaky aquifer without storage in aquitard	A & N	5.8

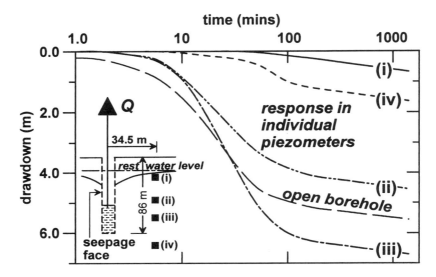

Figure 5.1   Drawdown curves for observation piezometers at Kenyon Junction (Rushton and Howard 1982). Reprinted from *Ground Water* with permission of the National Ground Water Association. Copyright 1982

The piezometers were all located within a 200-mm diameter borehole 100 m deep, positioned at 34.5 m from the pumped borehole. A total of six standpipes were installed in this borehole; the location of the four piezometers considered in this discussion are shown in Figure 5.1. Each standpipe piezometer consisted of a 19 mm plastic Cassagrande piezometer with a high air entry plastic filter. The tips of the piezometers were surrounded by filter sand; bentonite pellets were used to provide seals between the filter zones. The water levels in the standpipe piezometers under non-pumping conditions showed a difference of 1.0 m over a vertical distance of 62 m. Similar vertical hydraulic gradients have been noted under non-pumping conditions by Brassington (1992) and by Price and Williams (1993). Groundwater heads were lower in the deeper piezom-

eters, reflecting the generally downward flow due to the long-term pumping from the aquifer.

5.1.3 Measurements in pumped borehole and observation piezometers

Information about this test including drawdowns for four of the piezometers and for the pumped borehole is presented in Figures 5.1 and 5.2. The four piezometers for which results are plotted are located as follows:

piezometer (i): the top section of the borehole monitoring the actual water table,
piezometer (ii): positioned 20 m below rest water level,
piezometer (iii): positioned 43 m below rest water level,

piezometer (iv): positioned 64 m below rest water level which is 10 m below the penetration of the pumped borehole.

Note that time is plotted to a logarithmic scale with the drawdowns plotted to an arithmetic scale; this approach is similar to the Cooper–Jacob plot (see Section 5.4.3).

For the first stage of this analysis, each of the drawdown curves is examined in turn.

- Considering first the *pumped borehole*, the unbroken line in Figure 5.2 indicates that the water level falls rapidly with a drawdown of 6 m after one minute and about 12 m after three minutes; by ten minutes the fall is more than 20 m. For times greater than 100 minutes the drawdown increases only slowly; at 100 minutes the drawdown is 33.2 m while at 1440 minutes (1 day) it is 34.2 m.
- *Piezometer (i)* is close to the water table; Figure 5.1 indicates that up to 30 minutes the water table is unaffected at 34.5 m from the pumped borehole. Thereafter there is a steady decline which approaches a straight line on a *log-arithmetic* plot. At 1440 mins the drawdown is less than 0.7 m.
- Next, consider *piezometer (iii)* which is at the same depth as the water column in the pumped borehole; this piezometer undergoes the largest drawdowns. Unlike the pumped borehole, there is not an immediate response; drawdowns commence after two minutes. Between three and thirty minutes the drawdowns increase rapidly but with a decreasing trend between 30 and 200 minutes. From 200 to 1440

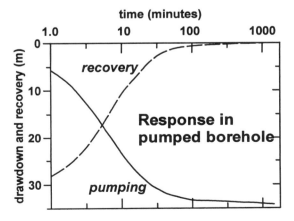

Figure 5.2 Response in pumped borehole at Kenyon Junction

minutes the line representing the drawdowns is parallel to the response for the water table piezometer.
- *Piezometer (ii)* has an elevation between piezometers (i) and (iii), but its response is similar to that of the deeper piezometer (iii). Drawdowns are smaller than piezometer (iii) because it is at the same elevation as the seepage face of the pumped borehole.
- *Piezometer (iv)* is more than 10 m below the base of the pumped borehole; drawdowns in this deeper piezometer are smaller than in piezometers (ii) and (iii). The initial response is also slower than for piezometers (ii) and (iii). This probably results from low-permeability layers of fine-grained silty sandstone between the base of the pumped borehole and piezometer (iv) (Walthall and Ingram 1984). Note that for all four piezometers the slope of the line from 200 to 1440 minutes is similar.

5.1.4 Conceptual models of flows for different times

From information about drawdowns in the pumped borehole and the four piezometers, it is possible to derive conceptual models of the flow towards the pumped borehole. Three diagrams are prepared for conditions after three minutes, 300 minutes and 30 minutes from the start of pumping. The selection of early then late and finally intermediate times (in the literature these are often referred to as *segments* of the time-drawdown curves) is to assist in the development of conceptual understanding of the flow processes.

1. *Response after three minutes of pumping*: conditions at this time are illustrated in the schematic cross-section of Figure 5.3a. None of the piezometers at 34.5 m from the pumped borehole show any response. During this initial stage of the test, water is taken from the easiest source, namely well storage. This can be verified by the following calculation. The abstraction rate of 1619 m^3/d is equivalent to 3.37 m^3 in three minutes. Dividing by the cross-sectional area of the borehole of 0.283 m^2 (radius 0.3 m) leads to a drawdown in the pumped borehole of 11.9 m if all the water is taken from well storage. This is close to the actual drawdown in the pumped borehole of 12.2 m. The removal of water from the borehole with no other changes in the aquifer is illustrated in Figure 5.3a.
2. *Response after 300 minutes of pumping*: it is helpful to consider next the conditions at 300 minutes. Figure 5.3c illustrates the drawdowns in the pumped borehole and in the four observation piezometers.

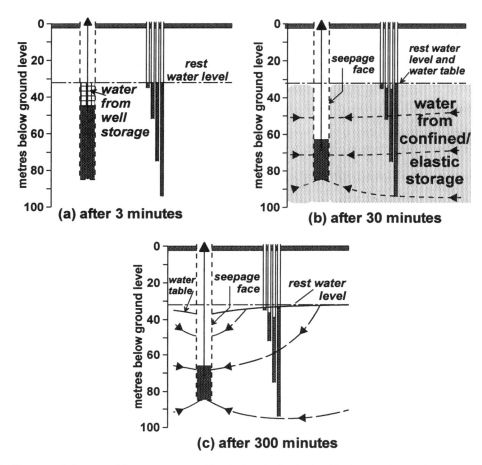

Figure 5.3 Conceptual diagram of flows to a pumped borehole: (a) after 3 min, (b) after 30 min, (c) after 300 min

The drawdown in piezometer (i) of 0.4 m indicates a fall in the water table which results in a release of water from unconfined storage. At the well face the fall in water table is probably about 2 m below rest water level. The drawdown in the pumped borehole of more than 32 m is far larger; consequently there is a substantial seepage face above the pumped water level.

Since the drawdown in the pumped borehole is much larger than the drawdowns in the observation piezometers, there are horizontal components of flow towards the pumped borehole both into the water column in the borehole and across the seepage face. In addition, vertical components of flow can be identified from the differing groundwater heads in the four piezometers at 34.5 m from the pumped

borehole. Since groundwater flow components are in the direction of decreasing groundwater heads (increasing drawdowns), vertical flow components occur from the water table, piezometer (i), down to piezometer (ii) and further down to piezometer (iii). However there is an upward component of flow from piezometer (iv) to piezometer (iii). From this information and from rules of flow net construction (Rushton and Redshaw 1979) approximate flow lines are constructed as shown by the broken lines in Figure 5.3c.

A further important finding is that, after 200 minutes, all the piezometers show a similar rate of fall, thereby indicating that an approximately steady flow system has been set up in the aquifer with the increase in drawdown primarily due to the lowering of the water table as water is released from *unconfined storage*.

3. *Response after 30 minutes of pumping*: this is illustrated in Figure 5.3b. At this intermediate time the water table has hardly fallen; see piezometer (i). To identify the relative importance of factors which influence the flow processes at this time, consider the volume of water that has been pumped from the borehole during 30 minutes; this equals 30 × 1619/1440 = 33.7 m³. This is only partially due to well storage since the drawdown in the pumped borehole of about 31 m is equivalent to 8.8 m³. In addition, substantial falls occur in groundwater heads in piezometers (ii), (iii) and (iv); see Figure 5.1. Due to the resultant decrease in confining pressure, water is released from *confined/elastic storage*. To obtain a very approximate estimate of the volume of water released from confined storage, assume that there is an average fall in groundwater head of 2.0 m out to a radial distance of 100 m (the fall is actually greater closer to the borehole and less at 100 m). With a confined storage coefficient of 0.0003 (this is the value from the numerical model analysis of Rushton and Howard 1982), the fall in groundwater head multiplied by the plan area and the confined storage coefficient is equivalent to 18.8 m³. Hence well storage plus the estimated release from confined storage totals 27.6 m³; this is of a similar magnitude to the 33.7 m³ taken from the pumped borehole. Therefore a major factor at 30 minutes is the water released from confined storage. This mechanism is illustrated in the conceptual drawing of Figure 5.3b.

To summarise, in the early stages, well storage dominates with a seepage face forming on the face of the pumped borehole due to the large fall in pumped water level. By 30 minutes, release from confined/elastic storage is the most important source of water. After 300 minutes, water released from unconfined storage as the water table falls becomes the major source of water.

5.1.5 Recovery phase

Information about the recovery phase can be found for the pumped borehole in Figure 5.2 and for the observation piezometers in Figure 5.4. The broken line in Figure 5.2 shows the rapid recovery in water level in the pumped borehole. After three minutes the water level in the pumped borehole rises by 11 m; this is only slightly less than the fall in the pumped water level of 12.2 m during the first three minutes of pumping. This

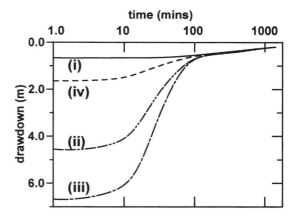

Figure 5.4 Recovery curves for observation piezometers at Kenyon Junction (Rushton and Howard 1982). Reprinted from *Ground Water* with permission of the National Ground Water Association. Copyright 1982

demonstrates that during the early stages of recovery, water continues to enter the borehole to refill well storage at almost the same rate as when the pump was operating. After 10 minutes the recovery is 77 per cent complete.

Figure 5.4 shows the first day of recovery in the observation piezometers. The recovery in piezometer (i), which represents conditions at the water table, is slow with no measurable change until after 30 minutes; this is similar to the slow initial response during the pumping stage when well storage and confined/elastic storage play an important role. From 30 to 1440 minutes there is a slow recovery, but recovery is incomplete at the end of one day. For piezometers (ii) and (iii), which are below the water table but at depths less than the penetration of the pumped borehole, recovery is slow to start but occurs rapidly between 10 and 100 minutes. Soon after 100 minutes the curves for the water table piezometer and the deeper piezometers become close; drawdowns in the pumped borehole are also of similar magnitude. For piezometer (iv), which is deeper than the bottom of the pumped borehole, recovery is slow but after 150 minutes the results for this deep observation piezometer are indistinguishable from the other piezometers.

Conceptual understanding of conditions in the aquifer system during recovery can be gained from a careful examination of Figures 5.2 and 5.4. At the moment the pump is switched off, water continues to enter the borehole at the same rate since it is the drawdown in a pumped borehole which causes the inflow of

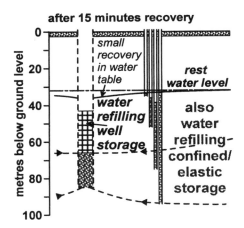

Figure 5.5 Conceptual model of aquifer flows fifteen minutes into recovery

water, not the operation of the pump. This is confirmed by the small changes in groundwater heads during the first few minutes of recovery. With significant flows of water towards the borehole, the recovery in the borehole is rapid. This is the reason for the rapid recovery observed in most operational boreholes.

Figure 5.5 is a conceptual diagram of conditions after 15 minutes of recovery. The volume of water required to refill well storage is shown in the diagram as 'water refilling well storage'. At fifteen minutes, recovery is also occurring in observation piezometers (ii), (iii) and (iv), indicating that groundwater enters from more distant parts of the aquifer to refill the confined/elastic storage. After refilling well storage and confined/elastic storage, water from some distance from the pumped borehole begins to refill the unconfined storage. Consequently, it is only after 30 minutes that there is any recovery of the water table. This recovery is slow due to the substantial volumes of water required to replenish the unconfined storage.

5.1.6 Results for open borehole instead of individual piezometers

Prior to the installation of the individual piezometers the original open observation borehole was used to monitor drawdowns during pumping test. An open borehole is supposed to average or integrate the groundwater heads. Consideration of the different shapes of drawdown curves for the four piezometers in Figure 5.1 indicates that it is not possible to define an

average response. The field results for the open borehole are shown by the broken line in Figure 5.1. Up to 30 minutes the drawdowns in the open observation borehole are greater than any of the individual piezometers. From 100 days onwards, the response is midway between piezometers (ii) and (iii), but it is not an average *of all four piezometers*.

To understand why the open observation borehole provides unpredictable results, the impact of this open borehole on the aquifer flow processes must be examined. In terms of aquifer response, the open observation borehole acts as a major vertical fissure extending from the water table to 10 m below the penetration of the pumped borehole. Consequently it provides a direct vertical hydraulic connection within the aquifer system. A conceptual diagram of the open borehole disturbing flows within the aquifer system is presented in Figure 5.6a. The figure represents conditions at about six minutes after the start of pumping. At this time the pumped borehole begins to draw water from confined/elastic storage (Figure 5.3b). At the location of the open observation borehole, the effective storage coefficient is 1.0 since the borehole is full of water and the water surface is open to the atmosphere. Since the open borehole is a column of water penetrating through the whole aquifer, water is drawn vertically through this open hole. In effect the open observation borehole acts as a secondary pumping well as indicated by the flow arrows at the top of the observation borehole in Figure 5.6a.

A basic principle of experimental physics is that the procedure of making a measurement must not disturb the process that is being observed. The open observation borehole clearly violates this principle since it causes serious disturbances to the flow patterns in the aquifer. An alternative way of explaining the impact of an open observation borehole is to consider the equipotentials (lines of equal groundwater head) as water flows to the pumped borehole. Flow lines towards a pumped borehole are shown in Figure 5.3c; the equipotential lines are typically at right angles to the flow lines, as indicated by the broken lines in Figure 5.6b. However, the open observation borehole is, itself, a vertical line of equal groundwater head (i.e. a vertical equipotential), hence it prevents the formation of curved equipotentials. Consequently, the lines of equal groundwater head, and also the flow lines, are severely distorted in the vicinity of the open observation borehole.

The distortion of the flow pattern due to the presence of open observation boreholes is a frequent cause of unreliable field data from pumping tests. There is no

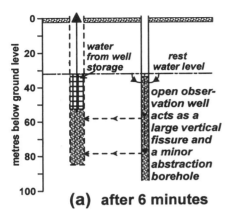

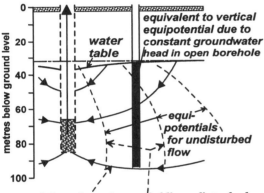

Figure 5.6 Disturbance in groundwater flow caused by the presence of an open observation borehole: (a) flow processes at six minutes, (b) impact on equipotential and flowlines at 300 minutes

method of correcting for the flow disturbance caused by an open borehole. Whenever possible, two separate boreholes should be provided (or two piezometers within a single borehole), one just penetrating below the water table and the other penetrating to the full depth of the pumped borehole but with only the lowest two metres open to the aquifer. If only one observation borehole can be provided, it should be designed either to monitor the water table or to monitor the deep aquifer response.

5.1.7 Analysis of pumping test results

Various methods of analysis (these methods are introduced in subsequent sections) have been used for the pumping tests at Kenyon Junction. In addition to the data described above, field results are available for step tests, constant rate tests and the recovery phase for the pumped borehole alone or for the pumped borehole plus the open observation borehole. When analytical techniques were used, the estimated transmissivity varied from 36 to 916 m^2/d and the storage coefficient from 0.00022 to 0.088 (Walthall and Ingram 1984). With such a wide range of results, it is not acceptable to quote average values since the failure to obtain consistent results is a clear indication that the methods of analysis are not suitable for the actual field situation. This assertion must be emphasised most strongly. If resource assessments are based on aquifer parameters which are averages of values obtained using a range of analytical techniques for pumping test analysis, none of which is consistent with the actual field conditions, these resource assessments are likely to be unreliable.

Why are the estimated parameter values so varied and therefore unreliable? One major cause is the failure of the open observation borehole to monitor actual conditions within the aquifer; instead it causes a disturbance to the aquifer flows. Second, the methods of analysis are frequently inconsistent with the actual aquifer response. Major features, which must be included in any method of pumping test analysis, include the following.

- Recognition that there are two storage coefficients, confined storage which applies over the full depth of the aquifer and the specific yield which operates at the water table.
- Inclusion of well storage: during the early stages of pumping, the quantity of water taken from the aquifer is reduced due to well storage; during the early stages of the recovery phase, water flows from the aquifer into the borehole to replenish well storage.
- Vertical components of flow, which occur as water is drawn from the water table downward into the aquifer and laterally to the pumped borehole, are often significant.
- Conditions at the borehole such as the presence of a seepage face, varying hydraulic conductivities with depth, or a non-uniform distribution of flows into the borehole all have an impact on the aquifer response.

Some of the more advanced analytical methods allow for several of the above features, but none are sufficiently flexible to take account of all of these features.

An alternative approach is to use numerical models. A paper by Rushton and Howard (1982) refers to the use of a two-zone numerical model to represent the pumping the test at Kenyon Junction. In a subsequent paper, Rathod and Rushton (1991) provide full details of an improved version of the two-zone model. In the numerical model, the following properties, identified from the conceptual diagrams of Figures 5.3a–c, are included:

• confined storage and specific yield,
• horizontal and vertical components of flow,
• well storage,
• a greater proportion of the water entering the pumped borehole from the lower part of the aquifer,
• continuous analysis of pumping followed by recovery phase,
• delayed yield operating during the pumping phase.

Further details of the two-zone model and its application in a variety of situations can be found in Section 7.4; parameter values for the Kenyon Junction test are quoted in Table 7.5.

5.1.8 Summary

The Kenyon Junction test is selected as the introductory field pumping test since it highlights the actual flow processes which can occur in what appears to be a relatively straightforward field situation. The test highlights the significance of well storage, confined/elastic storage and unconfined storage and the manner in which these storages are significant at different times during the pumping and recovery phases. The test also demonstrates the importance of horizontal and vertical components of flow and the development of a seepage face in the borehole. The need for individual observation piezometers and the unreliability of open observation boreholes are also emphasised.

5.2 CLASSIFICATION OF RADIAL FLOW PROBLEMS

Radial flow through an aquifer to boreholes or wells is strongly influenced by the contrast between the large areal extent of a typical aquifer compared with the small cross-sectional area of a borehole. Consequently, groundwater velocities in the vicinity of a borehole

are likely to be high (as a first approximation they are inversely proportional to the radial distance from the borehole), hence conditions in the vicinity of the borehole are critical to the development of an understanding of the impact of pumped boreholes. In the following discussion the term *borehole* will normally be used, although the terms *tubewell* and *well* will also be used especially when the structure has been constructed by hand.

Steady state flow in a confined aquifer towards boreholes can be analysed using the Thiem formula (Eq. (2.37)). However, the important contribution by Theis (1935) provided a practical method of analysing time-variant groundwater flow to a pumped borehole. Following this fundamental paper, there have been a large number of studies of time-variant radial flow to boreholes with a variety of different aquifer conditions (see, for example, Kruseman and de Ridder 1990, Hall 1996, Batu 1998). The focus of most of these studies has been the analysis of data obtained from pumping tests to estimate the aquifer parameters. Fewer investigators have considered the actual flow conditions in the vicinity of individual boreholes or the long-term yield of the borehole-aquifer system. Although brief reference is made in Part II (sections 5.4, 5.6, 5.8, 5.10 and 5.17) to the classical approach of pumping test analysis, the main emphasis is the study of practical issues related to borehole or well performance, flow mechanisms within complex aquifer systems and long-term yields.

An important issue is to ascertain what idealisations can be made to simplify the analysis yet still retain the important features of the flow processes. Examples of a *few* of the commonly adopted idealisations are considered below and sketched in Figure 5.7.

(a) *Confined aquifer*: with impermeable strata above and below, radial flow (indicated by the flow arrows) occurs towards the *fully penetrating borehole*; analysis is in terms of the radial co-ordinate *r* measured from the axis of the borehole.

(b) *Traditional leaky aquifer*: the main aquifer is underlain by an impermeable stratum, overlain by a low permeability layer (*aquitard*) which in turn is overlain by another aquifer. Provided that the borehole fully penetrates the main aquifer, it is possible to describe the flow processes in the main aquifer in terms of the radial co-ordinate *r* with the vertical flow through the overlying low permeability stratum dependent on the drawdown in the main aquifer.

(c) *Unconfined aquifer*: with a fully penetrating borehole and a small water table drawdown; flow is

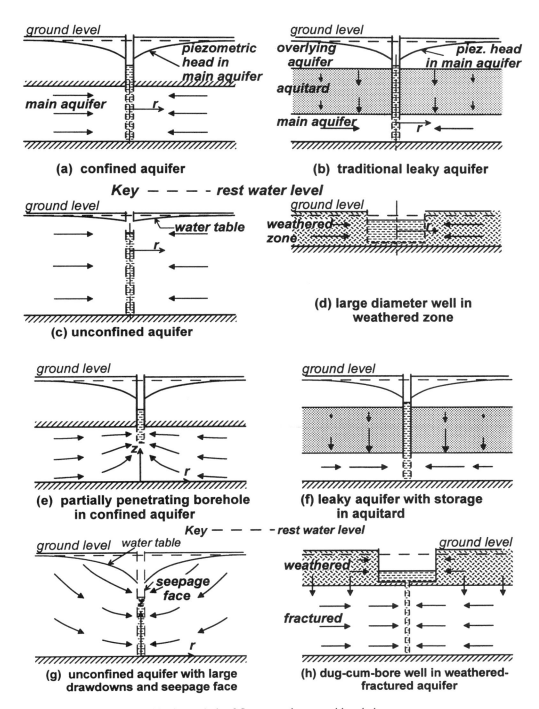

Figure 5.7   Idealisations introduced in the analysis of flow towards pumped boreholes

assumed to be radial, the saturated thickness of the aquifer is assumed to remain constant although a correction can be made for a decrease in saturated depth in the vicinity of the borehole.

(d) *Large diameter well in weathered zone of an aquifer*: radial flow is assumed to occur towards the large diameter well.

(e) *Confined aquifer* with flow towards a *partially penetrating borehole*: in the vicinity of the partially penetrating borehole, there are upward components of flow; consequently the analysis should be in terms of the radial co-ordinate r and the vertical co-ordinate z.

(f) *Leaky aquifer with allowance for storage in the aquitard*: the traditional assumption of a constant vertical flow through the full depth of the aquitard (the upper and lower arrows in (b) are of the same magnitude) is not correct; instead there are high vertical velocities at the bottom of the aquitard with lower velocities towards the top of the aquitard (hence the smaller flow arrows toward the top of the aquitard). Special techniques are required for this situation.

(g) *Unconfined aquifer with fully penetrating borehole*: there is a sizeable fall in the water table with a seepage face above the borehole water level. In recognition that water is released from storage at the water table the flow lines are curved, consequently the analysis is in terms of the co-ordinates r and z.

(h) *Weathered-fractured aquifer with a dug-cum-bore well*: in the weathered zone the flow is predominantly horizontal towards the dug well; also a vertical flow occurs from the weathered to the fractured zone. In the fractured zone the flows are mainly horizontal towards the bore well (although in practice the fractures may not all be horizontal).

In examples (a) to (d), the flow in the main aquifer can be described adequately as horizontal flow, consequently it is acceptable to work in terms of the radial co-ordinate r. These and other similar problems are discussed in Chapters 5 and 6. Flow processes for examples (e) to (h) are more complex; they are discussed in Chapter 7 with several practical examples introduced in Chapter 8.

Before any attempt is made to analyse data associated with a particular pumping test, it is essential to prepare sketches similar to those in Figure 5.7 to identify the important flow processes within the aquifer system.

When using results from a pumping test to determine aquifer parameters, field values of drawdown in the pumped borehole and observation boreholes (if available) are plotted on *log-arithmetic* or *log-log* graph paper. Often the field data are compared with a number of 'type curves' (a line or series of lines derived from analytical solutions plotted on graph paper to the same scales) to decide which type curve most nearly matches the field response. This approach may lead to unreliable parameter values when a type curve is selected which does not represent the actual flow conditions within the aquifer system. Consequently, it is necessary to consider carefully the hydrogeology of the area to identify the likely flow processes; the shape of the curve of the field drawdowns is just one item to be considered when selecting the method of analysis.

Pumping tests are usually specially devised experiments in which the aquifer is pumped for a specified time; the pump is then stopped and the aquifer is allowed to recover. The aim is to have the aquifer at rest before the test starts with no other boreholes operating within the zone of influence during the test or recovery phases. Alternatively, if there are boreholes operating in the zone of influence, they should pump at a constant rate before the test, during the test and during the period when recovery readings are taken.

At many pumping stations, due to the continuous use of the boreholes for public supply, it is not possible to switch off the pumps for a sufficient time to achieve *at-rest* initial conditions. In such circumstances an alternative procedure can be adopted in which borehole drawdowns are monitored during pumping. The pump is then switched off, with monitoring of drawdowns continuing during the recovery stage and when the pump is switched back on. Figure 5.8a is a representative example of water levels in an observation borehole; these results are analysed in Section 5.5 using a numerical model.

Sections 5.3 to 5.8 describe the study of pumping from confined and leaky aquifers with uniform parameters. Brief reference is made to the basic theories and the analysis of pumping tests using conventional techniques; for detailed information about pumping test analysis the reader is referred to Kruseman and de Ridder (1990), Hall (1996) and Batu (1998). The use of numerical methods to study flow towards a pumped borehole in a confined aquifer is introduced; the differing approaches to the analysis of pumping tests using conventional and numerical methods is highlighted. Further examples of issues which arise when pumping from single-layer aquifers are considered in Section 5.9 to 5.17.

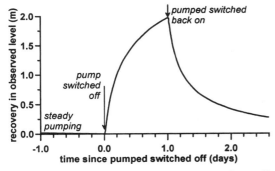

(a) field values of water levels in observation well

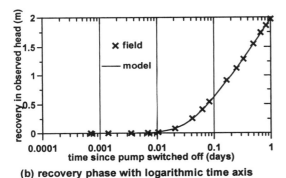

(b) recovery phase with logarithmic time axis

Figure 5.8 Pumping test in a borehole which initially pumps at a constant rate, is switched off for a time and then pumping recommences

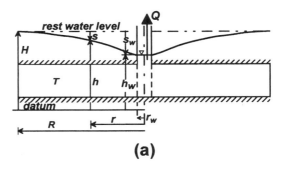

(a)

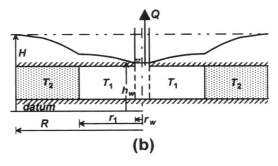

(b)

Figure 5.9 Steady radial flow to a borehole: (a) in uniform confined aquifer, (b) with changed transmissivity at larger radial distances

$$s_w = \frac{Q}{2\pi T}\ln\!\left(\frac{R}{r_w}\right) \qquad (5.2)$$

A more general equation for groundwater heads h_1 and h_2 at radial distances of r_1 and r_2 from the pumped borehole is

$$h_2 = h_1 - \frac{Q}{2\pi T}\ln\!\left(\frac{r_1}{r_2}\right) \qquad (5.3)$$

or $\quad s_1 = s_2 + \dfrac{Q}{2\pi T}\ln\!\left(\dfrac{r_2}{r_1}\right) \qquad (5.4)$

Groundwater head (or drawdown) curves around pumped boreholes are usually sketched showing a gentle curve from a distant boundary to the borehole (Figure 5.9a is a good example) but these sketches fail to emphasise the true variation of groundwater head with radius. The following simplified example demonstrates how changes in groundwater head with radius are concentrated close to the pumped borehole.

5.3 STEADY RADIAL FLOW IN CONFINED AQUIFERS

The simplest form of radial flow relates to a confined aquifer with constant transmissivity T in which steady-state conditions apply (Figure 5.9a). If there is a constant discharge Q from a well of radius r_w and the groundwater head at a distant boundary radius R is H, the groundwater head at the well equals

$$h_w = H - \frac{Q}{2\pi T}\ln\!\left(\frac{R}{r_w}\right) \qquad (5.1)$$

This equation, usually referred to as the Thiem equation, can be derived by integrating the steady state form of Eq. (2.61). Alternatively, the change in groundwater head can be expressed as the drawdown in the well s_w below the undisturbed groundwater head:

Consider steady state flow in a confined aquifer with a transmissivity of $500\,m^2/d$ towards an abstraction borehole pumping at $1000\,\pi\,m^3/d$; this value of pumping rate is selected so that $Q/2\pi T = 1.0$. If the groundwater head at a radial distance of $10\,km$ is $50\,m$ above datum, the groundwater heads can be calculated from Eq. (5.4) at appropriate locations between the outer boundary and the pumped borehole of radius $0.1\,m$. For the total distance from the outer boundary of $10\,000\,m$ to the borehole at $0.1\,m$, the fall in groundwater head equals $\ln(10\,000/0.1)$ or $11.513\,m$; for each *tenfold* decrease in radius the fall in groundwater head equals $\ln(10.0)$ or $2.303\,m$. The groundwater head distribution is summarised in Table 5.2. Note that the change in groundwater head between $10\,km$ and $1.0\,km$ is exactly the same as the change in head between $1.0\,m$ and $0.1\,m$. It is this steep fall in groundwater head of $2.303\,m$ in a horizontal distance of only $0.9\,m$ which means that aquifer conditions close to the borehole are of critical importance in understanding the borehole response.

Another equation can be derived from Eq. (5.3) which allows an estimate to be made of the effect of changed transmissivities in the vicinity of a pumped borehole in confined aquifers. If the transmissivity from the well radius r_w to an intermediate radius r_1 is T_1 and the transmissivity from r_1 to the outer radius R is T_2 (see Figure 5.9b), the pumped discharge can be calculated from the difference between the maximum head H and the head in the pumped borehole h_w,

$$Q = \frac{2\pi T_1 T_2 (H - h_w)}{T_1 \ln(R/r_1) + T_2 \ln(r_1/r_w)} \tag{5.5}$$

In Figure 5.9b the transmissivity close to the borehole T_1 is greater than T_2, hence the slope of the groundwater head curve is flatter for $r < r_1$.

The opposite situation is represented in the following calculation. Taking the same numerical parameter

Table 5.2 Variation in groundwater head from outer boundary at $10\,km$

Radius (m)	Groundwater head (m)	Drawdown (m)
10 000	50.000	0.000
1 000	47.697	2.303
100	45.395	4.605
10.0	43.092	6.908
1.0	40.790	9.210
0.1	38.487	11.513

values as used for Table 5.2 but with groundwater heads in the pumped borehole and at the outer boundary of 38.487 and $50.0\,m$ respectively and with the transmissivity T_1 between 1.0 and $0.1\,m$ *reduced* from $500\,m^2/d$ to $100\,m^2/d$ (this could simulate clogging of the aquifer during the drilling of the borehole), the pumping rate is reduced to $1745.3\,m^3/d$ compared to a discharge of $3141.6\,m^3/d$ for a constant transmissivity. Consequently, deterioration of the aquifer in the vicinity of a pumped borehole can have a considerable effect on the borehole yield.

5.4 ANALYTICAL SOLUTIONS FOR UNSTEADY RADIAL FLOW IN CONFINED AQUIFERS

5.4.1 Analytical solutions

This section briefly reviews analytical solutions for radial flow in confined aquifers; detailed information about the derivation of the equations and alternative methods of matching field to analytical solutions can be found in Todd (1980), Kruseman and de Ridder (1990), Walton (1987) and Hall (1996).

The differential equation describing radial flow in an aquifer of constant transmissivity is

$$\frac{\partial}{\partial r}\left(T\frac{\partial s}{\partial r}\right) + \frac{1}{r}T\frac{\partial s}{\partial r} = S\frac{\partial s}{\partial t} + q \tag{5.6}$$

Two further physical parameters are included in addition to those for steady flow analysis of Section 5.3, the confined storage coefficient S and the time since pumping started t.

The well-known solution for Eq. (5.6) due to Theis (1935) for unsteady flow in a uniform confined aquifer (hence $q = 0$) is the foundation for many studies of groundwater flow to pumped boreholes. The drawdown at any radial distance r from the pumped borehole

$$s = \frac{Q}{4\pi T}W(u) \tag{5.7}$$

where the well function $W(u) = \int_u^\infty \frac{e^{-x}}{x}dx$ and $u = \frac{r^2 S}{4Tt}$

In deriving Eq. (5.7) a number of important assumptions are made. All too often these assumptions are not carefully considered before deciding whether the Theis analysis is appropriate for the problem under consideration. The assumptions inherent in the Theis analysis are as follows:

- the aquifer is confined so that no water enters from above or below the main aquifer,
- the aquifer is homogeneous, isotropic and of uniform thickness,
- the aquifer is of infinite areal extent and the pumped borehole has an infinitely small radius,
- the piezometric surface is horizontal at the start of the test,
- the borehole fully penetrates the aquifer and is pumped at a constant rate,
- there are no well losses in the vicinity of the borehole,
- observation boreholes are open over the full depth of the aquifer,
- groundwater flow can be described by Darcy's Law; the groundwater density and viscosity remain constant.

When using the Theis equation to analyse a pumping test, drawdowns are measured in an observation borehole for the duration of the test. Consequently there will be at least twenty field measurements, yet there are only two unknown parameters, the transmissivity T and the storage coefficient S. Since the number of data points is far greater than the two unknowns, some form

of *fitting* method has to be introduced to determine the parameter values. There are two commonly used fitting methods for analysing the pumping phase of a confined aquifer test.

5.4.2 Theis method using log-log graphs

In the first method field data are plotted on graph paper which uses a logarithmic scale for both time and drawdown. This field data can then be matched against the theoretical curve drawn using identical graph paper. Figure 5.10 indicates how the two curves are superimposed. To obtain a good match, the graph of the field data is moved horizontally and vertically over the theoretical curve until the best fit is achieved; these two *degrees of freedom* are equivalent to the two unknowns of transmissivity and storage coefficient. Examples of the use of the Theis curve-matching method can be found in many books including Kruseman and de Ridder (1990). Note that when traditional type curves are plotted, drawdowns increase upwards. However, elsewhere in this book, field data of drawdowns are plotted as positive downwards, i.e. in the same direction as the actual change in water level.

In many situations it is not possible to obtain an

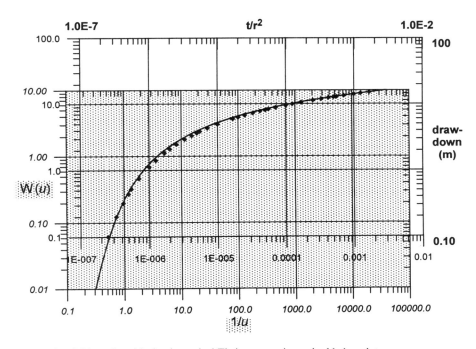

Figure 5.10 Comparing field results with the theoretical Theis curve using a double log plot

exact match between all the field results and the Theis curve; compromises have to be made. Perhaps at small times or towards the end of the test the field readings do not match the theoretical curve, hence the field readings for these times are ignored. However, the field data for later times may provide valuable insights into boundary conditions. Similarly early time data may not fit the type curve because of the influence of well storage.

One major disadvantages of this curve matching technique is that logarithmic scales are used for both drawdown and time. Although it is appropriate to use a logarithmic scale for time, the logarithmic scale for drawdowns means that the smaller initial drawdowns are plotted to a larger scale with the result that they are given undue emphasis.

5.4.3 Cooper–Jacob technique

In an alternative approach due to Cooper and Jacob (1946), an infinite series expansion of the infinite integral for $W(u)$ is used:

$$W(u) = -0.5772 - \ln(u) + u - \frac{u^2}{2\,2!} + \frac{u^3}{3\,3!} - \frac{u^4}{4\,4!} + \ldots.$$

(5.8)

In the Cooper–Jacob method, all but the first three terms of the series are ignored, which means that the method only applies to data for which u ($= r^2 S/4Tt$) is less than 0.01. Field data are plotted with drawdown on an arithmetic scale and time on a logarithmic scale (Figure 5.11). A straight line is plotted through the data points on this *log-arithmetic* plot. The two unknowns of transmissivity and storage coefficient are calculated from the slope of the line Δs (where Δs is the increase in drawdown for a tenfold increase in time) and the intercept of the line with the time axis, t_0.

The equations used to estimate the transmissivity and storage coefficient are:

$$T = \frac{2.30Q}{4\pi\Delta s} \qquad S = \frac{2.25Tt_0}{r^2}$$

(5.9)

Once values of the transmissivity and storage coefficient have been estimated, the time corresponding to $u = 0.01$ is calculated from

$$t_{0.01} = \frac{r^2 S}{0.04T}$$

(5.10)

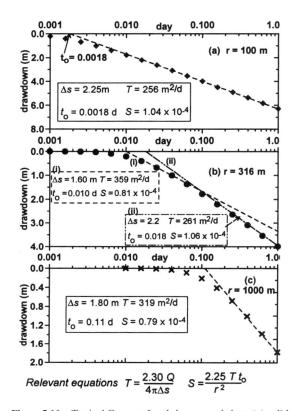

Figure 5.11 Typical Cooper–Jacob *log-normal* plots: (a) valid analysis with single straight line, (b) invalid analysis where two straight lines are plotted, (c) invalid result where the condition $u \leq 0.01$ is violated

All data points at smaller times should be ignored and a new straight line drawn through the remaining points. Frequently this second step is ignored, thereby invalidating the results.

Figure 5.11 illustrates how this procedure should be followed; the data points plotted in the figure are obtained from a radial flow numerical model (see Section 5.5) representing pumping from a confined aquifer. Input parameter values for the numerical model are:

$T = 250\,\mathrm{m^2/d}$, $S = 0.0001$ and $Q = 1000\,\pi\mathrm{m^3/d}$; the duration of pumping is 1.0 day.

The three plots refer to observation piezometers at 100 m, 316 m and 1000 m. There is no difficulty in plotting a straight line through the points for $r = 100\,\mathrm{m}$ (Figure 5.11a). Using the slope of the line and the intercept of the straight line with the time axis, the estimated values of transmissivity and storage coefficient are $256\,\text{m}^2/\text{d}$ and 0.000104. For (b), at a radial distance of $316\,\text{m}$, two straight lines can be drawn; for the earlier times (i) the transmissivity and storage coefficient are $359\,\text{m}^2/\text{d}$ and 0.000081 while for the later times (ii) the values are $261\,\text{m}^2/\text{d}$ and 0.000106. For curve (c) at $r = 1000\,\text{m}$ it is only possible to draw a straight line through the last four points; the calculated transmissivity and storage coefficient are $319\,\text{m}^2/\text{d}$ and 0.000079 respectively.

To confirm whether it is appropriate to use the Cooper–Jacob method, the critical time when $u = 0.01$ varies according to the radius:

$$r = 100\text{m}, \quad t_{0.01} = 0.1 \text{ day},$$
$$r = 361\text{m}, \quad t_{0.01} = 1.0 \text{ day},$$
$$r = 1000\text{m}, \quad t_{0.01} = 10.0 \text{ day}.$$

If lines are drawn through field readings for times earlier than those quoted above, unreliable results will probably be obtained. Consequently for $r = 100\,\text{m}$, where all values after 0.1 day are acceptable, a suitable straight line can be drawn and reliable results are obtained. For $r = 316\,\text{m}$, two possible lines have been drawn; because line (ii) is based primarily on the last four values which are for times close to the critical time of 1.0 day, acceptable values of transmissivity and storage coefficient are obtained. However, line (i), which appears to be a reasonable straight line through some of the data points, is not acceptable. For $r = 1000\,\text{m}$ the Cooper–Jacob method should not be used. This issue of inadmissible values for the Cooper–Jacob method is stressed because, despite clear statements in the literature, the method is often used when data points are at too short a time to be valid.

5.4.4 Transmissivity and storage coefficient varying with radius

For certain pumping tests, two or more estimates of transmissivity are derived from alternative methods of analysis based on the same set of data. Also, when there are measurements in a number of observation boreholes, alternative values of the transmissivity and storage coefficient are quoted for different radial distances from the pumped borehole. Multiple values of T and S cannot be correct since the basis of the Theis theory is that the transmissivity and the storage coefficient are constant from the pumped borehole to the observation borehole and on to infinity. If a single value of transmissivity cannot be obtained, this means

that one of the assumptions of the Theis theory is being violated (it may be an effect other than changes in transmissivity). The only reliable conclusion when the field data do not lie on a straight line for the *logarithmic* plot or fail to match part of the theoretical curve with a *log-log* plot is that *the Theis theory does not fully represent actual field conditions*. Values of transmissivity or storage coefficient deduced from such an analysis may be seriously in error; furthermore insights about other features of aquifer response (e.g. boundary effects) may be missed.

In practice, aquifer parameters are not uniformly distributed. A number of authors have investigated pumping test responses when aquifer properties vary; their work can provide insights into the reasons why different parameter values are deduced during pumping test analysis. Barker and Herbert (1982) considered an idealised aquifer in which the aquifer parameters within a radial distance R from the pumped borehole are transmissivity T_1 and storage coefficient S_1 with corresponding values at radial distances greater than R equalling T_2 and S_2 (see Figure 5.12). Analytical expressions for the drawdowns, s_1 and s_2, in the inner and outer regions at large times are

$$\frac{4\pi T_2}{Q}s_1 = \ln\frac{2.25T_2 t}{S_2 R^2} + \frac{2T_2}{T_1}\ln\frac{R}{r} \qquad (5.11)$$

and

$$\frac{4\pi T_2}{Q}s_2 = \ln\frac{2.25T_2 t}{S_2 R^2} \qquad (5.12)$$

The equation for s_1 indicates that at large times the aquifer approaches a steady-state condition within the radius R with drawdowns depending on T_1 and T_2 whereas the equation for s_2 indicates that the drawdowns correspond to an aquifer having aquifer

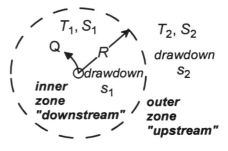

Figure 5.12 Example of pumping from inner region with aquifer properties T_1 and S_1, and outer region with properties T_2 and S_2

properties T_2 and S_2. Hence the Cooper–Jacob plot for s_2 at long times can be used to estimate the aquifer parameters beyond the radius R.

Further work on this topic is presented in Butler (1988); in a more recent study, Jiao and Zheng (1997) have considered the sensitivity of observation borehole responses to the aquifer properties in the region between the pumped borehole and the observation borehole (the *downstream* or inner zone) and the properties at and beyond the observation borehole (the *upstream* or outer zone). Transmissivity estimates are sensitive to the parameters of the outer zone whereas the storage coefficient is more difficult to estimate and depends mainly on the storage coefficient in the outer zone but is also influenced by the inner zone storage coefficient. The issue of changes in transmissivity and storage coefficient with radial distance from the pumped borehole is considered further in Section 5.11.

5.5 NUMERICAL SOLUTION FOR UNSTEADY RADIAL FLOW

As an alternative to analytical techniques, numerical solutions based on a discrete space, discrete time approximation can be used. The differential equation for time-variant radial flow in Section 2.4.4 can be written as

$$\frac{\partial}{\partial r}\left(mK_r \frac{\partial s}{\partial r}\right) + \frac{m}{r} K_r \frac{\partial s}{\partial r} = S \frac{\partial s}{\partial t} + q \qquad (5.13)$$

In many practical situations, the radial hydraulic conductivity K_r and/or the storage coefficient S vary with radius. Furthermore, the borehole has a finite radius with well storage effects; in addition, most aquifers are of finite extent. During pumping, conditions may change between confined and unconfined at some locations. Consequently, assumptions of the Theis analysis, such as the borehole having an infinitely small radius or the aquifer extending to infinity, rarely apply. When a numerical model is used, in which both the radial and time co-ordinates are divided into discrete steps, a finite well radius, an outer boundary and other features can be represented.

5.5.1 Theoretical basis of the numerical model

In developing the numerical model it is helpful to use an alternative variable in the radial direction; the variable a is defined so that

$$a = \ln(r) \qquad (5.14)$$

Consequently Eq. (5.13) can be rewritten as

$$\frac{\partial}{\partial a}\left(mK_r \frac{\partial s}{\partial a}\right) = Sr^2 \frac{\partial s}{\partial t} + qr^2 \qquad (5.15)$$

This equation can be expressed in finite difference form with the radial co-ordinate divided into a mesh which increases logarithmically so that the increment Δa is constant; the time also increases logarithmically. Figure 5.13 illustrates the discrete mesh in both space and time. The differential equation can be written in backward difference finite difference form

$$\frac{1}{\Delta a}\left\{mK_{r,n-1} \frac{s_{n-1} - s_n}{\Delta a} - mK_{r,n} \frac{s_n - s_{n+1}}{\Delta a}\right\}_{t+\Delta t}$$
$$= \frac{S_n r_n^2}{\Delta t}\{s_{n,t+\Delta t} - s_{n,t}\} + qr_n^2 \qquad (5.16)$$

where $K_{r,n-1}$ refers to the hydraulic conductivity *between* nodes $n-1$ and n, $K_{r,n}$ refers to the hydraulic conductivity *between* nodes n and $n+1$ with the storage coefficient S_n referring to node n.

In the finite difference equation there are three unknown drawdowns at the new time $t + \Delta t$ with one known drawdown at the previous time t. This leads to a set of simultaneous equations which can be solved using a Gaussian elimination technique. In unconfined

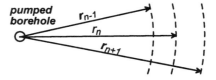

(a) radial increments

(b) radial and time increments

Figure 5.13   Basis of the radial flow numerical model

aquifers, the saturated depth m is a function of the unknown drawdown s, therefore the simultaneous equations are solved iteratively four times with modified values of the saturated depth based on the most recent estimates of the drawdowns.

It is convenient to introduce the concept of equivalent hydraulic resistances which are defined as the inverse of the coefficients in Eq. (5.16).

equivalent radial hydraulic resistances $H_{n-1} = \Delta a^2/(K_{r,n-1}m)$
    and $H_n = \Delta a^2/(K_{r,n}m)$
equivalent time-storage coefficient resistance $T_n = \Delta t/(S_n r_n^2)$

Allowance can be made for well storage by modifying the equivalent time-storage coefficient resistance so that the storage coefficient within the well is unity (Rushton and Redshaw 1979),

$$T_1 = 2\Delta t \, \Delta a/r_w^2$$

Rathod and Rushton (1984) provide a computer program in BASIC for the radial flow numerical model.

5.5.2 Comparison of analytical and numerical solutions

In the Theis analytical solution the well radius is infinitely small while the aquifer extends to infinity. However, in the numerical solution, a well radius is defined with the storage coefficient within the well set to unity; the radial distance to the outer boundary is also defined. To demonstrate how the numerical model can be used to simulate pumping from a confined aquifer, consider the following example: $T = 250 \, \text{m}^2/\text{d}$, $S = 0.0001$, $Q = 1000 \, \pi \, \text{m}^3/\text{d}$ and the radial distance to the observation well, $r = 100 \, \text{m}$; hence $Q/4\pi T = 1.0$ and

$u = r^2 S/4Tt = 1/(1000t)$. In Table 5.3, comparisons are made between the Theis solution and three numerical solutions,

- the first numerical solution has a small borehole of radius of 0.001 m, six logarithmic mesh intervals for a tenfold increase in radius and ten logarithmic time steps for a tenfold increase in time; the outer boundary is at 100 000 m,
- the second is identical to the first but with a borehole radius of 0.1 m,
- the third is identical to the first but with twenty logarithmic intervals for tenfold increases in both radius and time.

The results of Table 5.3 demonstrate that a numerical model, with six logarithmic intervals for a tenfold increase in radius and ten logarithmic time steps for a tenfold increase in time, provides satisfactory agreement with the analytical solution; differences from the Theis analytical solution increase from 0.01 to 0.05 m.

5.5.3 Example of the use of the numerical model

The usefulness of the numerical model will be illustrated by the example of Figure 5.8 which refers to a supply borehole which is switched off for one day with measurements taken of the drawdowns both during the recovery and when the pump is switched back on. In the observation borehole at 200 m from the pumped borehole, the fall in the water level for the day immediately before the pump is switched off is only 0.02 m; the observation well water level when the pump is switched off is taken as datum. During the one day when the pump is not operating, the recovery in the observation borehole is 1.97 m (Figure 5.8a). Once the pump is switched back on, water levels start to fall

Table 5.3 Comparison of analytical and numerical solutions

Time (day)	Drawdown (m) at 100 m			
	Theis	Numerical $r_{well} = 0.001$ m	Numerical $r_{well} = 0.1$ m	Numerical with smaller mesh and time intervals
0.001	0.22	0.21	0.18	0.22
0.010	1.82	1.78	1.77	1.80
0.100	4.04	3.99	3.99	4.01
1.000	6.33	6.28	6.28	6.31
10.00	8.63	8.59	8.59	8.61
100.0	10.94	10.89	10.89	10.91

again but after one day of pumping, the level in the observation well is still more than 0.4 m above the level when the pump is switched off.

The numerical model in Section 5.5 is used to reproduce this aquifer response. It is not known for how long the pump had been operating before it was switched off. In the numerical model the pump is run for 10 days since an increase to 20 days made no difference. Therefore the pumping scheme in the model is:

from −10.0 to 0.0 days, Q = 1500 m³/d
from   0.0 to 1.0 days, Q = 0.0 m³/d
from   1.0 to 2.6 days, Q = 1500 m³/d

Because Figure 5.8a uses arithmetic scales, the rapid responses just after the pump is switched off and just after the pump is switched back on are not clearly displayed. However, when a logarithmic timescale is used for the recovery (Figure 5.8b), the rapid response is shown clearly. After a number of trial simulations, the following aquifer parameters were deduced: transmissivity = 180 m²/d, confined storage coefficient 0.0005.

As an alternative to the use of the numerical model, it is possible to superimpose three sets of calculated values from the Theis analytical solution, the first set of values relating to a pumping rate of +1500 m³/d from the starting time (corresponding to day −10.0) and continuing for 12.6 days, the second of −1500 m³/d from a time of 10.0 to 12.6 days and the third of +1500 m³/d from a time of 11.0 to 12.6 days.

5.5.4 Representation of well losses

Well losses occur when the drawdowns in a well are higher than those which would occur due to horizontal laminar flow towards a fully penetrating borehole. A detailed discussion on well losses can be found in Section 8.2; the purpose of this discussion is to consider how well losses can be included in radial flow models. Examples of the occurrence of well losses include:

- changes in hydraulic conductivity close to the well (this is discussed in Section 5.3 where a development of the Thiem formula, Eq. (5.5), is used to investigate the influence of lower hydraulic conductivities close to a well),
- the occurrence of a seepage face; see, for example, the Kenyon Junction test in Section 5.1,
- clogging of an injection borehole as described in Section 8.6.4,
- partial penetration of a borehole, see Section 7.3.2.

Well losses can be represented as additional drawdown in the vicinity of the pumped borehole. This additional drawdown is represented in the numerical model by modifying the equivalent horizontal hydraulic resistance between the node representing the borehole (node 2) and the first nodal point within the aquifer (node 3). The well loss factor is a multiplier for the horizontal hydraulic resistance. If the well loss factor equals 1.0, the aquifer is unaffected. A well loss factor of 10 indicates severe clogging or a large seepage face, whereas a factor of less than 1.0 indicates that some form of well development has occurred which increases the hydraulic conductivity close to the well. The well loss factor is usually determined by a trial and error procedure during model refinement. Well loss factors greater than 10.0 suggest severe deterioration of the aquifer, gravel pack or well screen.

5.6 ANALYSIS OF THE RECOVERY PHASE FOR UNSTEADY RADIAL FLOW IN CONFINED AQUIFERS

Valuable information can be gained from measurements in the pumped borehole and any observation boreholes during the recovery period after the pump is switched off. The recovery phase can be analysed either using analytical or numerical techniques.

The Theis analytical equation for drawdown, Eq. (5.7), assumes that once pumping starts, it continues to an infinite time. Consequently, to represent the cessation of pumping when the abstraction rate equals zero, an imaginary borehole is introduced at the same location as the pumped borehole with a pumping rate equal to but of opposite sign to the original pumping rate, Figure 5.14. This imaginary injection commences at time $t' = 0$ when the pump stops; see the lower part of Figure 5.14b. Drawdowns due to the real and imaginary pumping, when combined, are equal to the drawdowns which occur when a pump operates for a limited period:

$$s = \frac{Q}{4\pi T}W(u) - \frac{Q}{4\pi T}W(u') \qquad (5.17)$$

where $u = \dfrac{r^2 S}{4Tt}$ and $u' = \dfrac{r^2 S}{4Tt'}$

Using a truncated form of the series expression for drawdowns (Eq. (5.8)), in a similar manner to the Cooper–Jacob approach with the restriction that $u' \leq 0.01$, Eq. (5.17) can be rewritten as

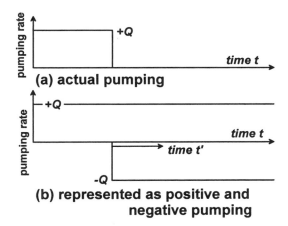

Figure 5.14 Representing recovery as a negative pumping which commences as the pump is switched off

$$s = \frac{2.30Q}{4\pi T} \log \frac{t}{t'} \qquad (5.18)$$

Therefore a *log-arithmetic* graph plotting log t/t' against the drawdowns during recovery should be a straight line with a slope

$$\Delta s = \frac{2.30Q}{4\pi T} \qquad (5.19)$$

As an alternative, the recovery phase can be analysed using a numerical model in which the analysis is continuous from the start of the pumping phase to the end of the recovery phase.

5.7 COMPARISON OF ANALYTICAL AND NUMERICAL TECHNIQUES OF PUMPING TEST ANALYSIS

This section explores the basic difference in approach between the analytical and numerical methods of interpreting data from pumping tests. In 'curve-fitting techniques' the field results are plotted and matched against the results of analytical solutions. There are several alternative techniques for estimating aquifer parameters from the pumping phase (*log-log* plots or *log-arithmetic* plots) and further techniques for the recovery phase (see Sections 5.4 and 5.6).

In the numerical method a single numerical model is used to obtain a best fit between the field and modelled results for *both* the pumping *and* the recovery phases

using the same model parameters. A trial and error technique is adopted to obtain a best fit; single values of the hydraulic conductivity and storage coefficients are used for the whole simulation of both the pumping and recovery phases.

An example is introduced to *explore* the alternative approaches of the analytical and numerical solutions. The purpose of this example is not to determine which method is *better* but to illustrate the differences in approaches (in fact, there is a bias toward the numerical method since the data used for the analysis are generated using a numerical model). The representative example refers to a confined aquifer of uniform thickness pumped at a rate of 1500 m³/d for two days with additional data collected for a further two days during recovery. An observation borehole at a radial distance of 43 m from the pumped borehole is used to record the aquifer response. Drawdowns for the pumping phase are plotted to a *log-arithmetic* scale in Figure 5.15a while the results for the recovery phase are plotted as for a Theis recovery analysis in Figure 5.15b with a timescale of $\log_{10}(t/t')$. Note that the recovery is incomplete after 2 days with a residual drawdown of 0.67 m.

The Cooper–Jacob method can be used to analyse the pumping phase; the broken line in Figure 5.15a shows an acceptable fit through the data points. With $\Delta s = 1.45$ m (the change in drawdown for a tenfold increase in time), the calculated transmissivity is $T = 189$ m²/d (equivalent to a hydraulic conductivity of 1.26 m/d for a saturated thickness of 150 m). The intercept with the horizontal axis corresponds to $t_o = 0.003$ day; this leads to a confined storage coefficient of $S = 0.00076$. Using these values of T and S, the time equivalent to $u = 0.01$ (see Eq. (5.10)) is 0.17 day, hence the straight line in Figure 5.15a is consistent with this minimum time. Even though an apparently satisfactory match can be obtained for the pumping phase, there is no possibility of fitting a straight line through the recovery plot of Figure 5.15b. The inability to draw a straight line through the recovery data indicates that there is an important feature which has not yet been identified.

For the numerical analysis a different form of graphical presentation is used with the drawdown and then the recovery plotted to *log-arithmetic* scales with the recovery immediately to the right of the pumping phase (Figure 5.16). When the numerical model is used to attempt to reproduce the data for *both* the pumping and recovery phases, no match is possible. The best fit based on the pumping phase only is shown by the broken line which is obtained with a transmissivity of

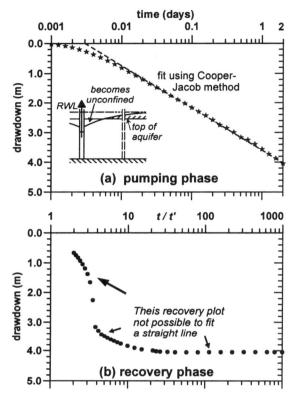

Figure 5.15 Results for pumping test in a confined aquifer: (a) drawdowns during pumping phase shown by ★ with broken line representing best fit for Cooper–Jacob analysis, (b) recovery phase with drawdowns shown by ●

192 m²/d (hydraulic conductivity of 1.28 m/d) and a storage coefficient of 0.00075. Although the fit with the pumping phase is satisfactory, the fit for the recovery phase is unacceptable.

The inability of the numerical model to match the recovery data occurs because there is important information which has not yet been included in the analysis. Although at the start of the test the aquifer is confined, the confining head is only 3 m above the top of the aquifer; see the inset of Figure 5.15a. Consequently, unconfined conditions develop around the pumped borehole with a substantial change in storage coefficient of more than two orders of magnitude. The development of unconfined conditions in the vicinity of the pumped borehole during the pumping phase has a strong influence on the recovery phase; recovery is far from complete two days after the cessation of pumping due to the time taken to refill the unconfined storage. The numerical model simulation with the changing conditions between confined and unconfined is indicated in Figure 5.16 by the unbroken line. The relevant parameters are

aquifer hydraulic conductivity	= 1.0 m/d
aquifer thickness	= 150 m
confined storage coefficient	= 0.0005
specific yield	= 0.05
borehole radius	= 0.2 m
distance to no flow boundary	= 20 km
initial groundwater head above top of aquifer	= 3.0 m

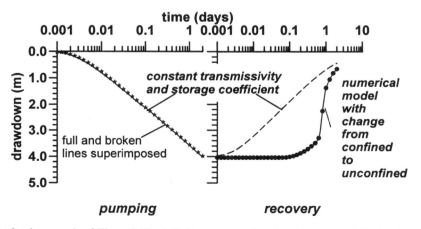

Figure 5.16 Data for the example of Figure 5.15a plotted to log-normal scales with recovery following immediately after the pumping phase. Broken line shows best fit using the numerical model (assuming that the aquifer is always confined); continuous line represents numerical model with changes between confined and unconfined conditions

The value of the transmissivity from the Cooper–Jacob analysis of $189\,m^2/d$ is of the same order as the correct value of $150\,m^2/d$; the confined storage coefficient of 0.00076 is not very different from the correct value of 0.0005. However, relying solely on the pumping phase would mean that the crucially important feature of conditions changing between confined and unconfined is not identified. Using a numerical model with a single set of parameters to match both the pumping and recovery phases is a more reliable approach since it provides both values of the aquifer parameters of transmissivity and storage coefficient and also additional information about the aquifer conditions.

5.8 ANALYSIS OF LEAKY AQUIFERS WITHOUT STORAGE IN THE AQUITARD

Confined aquifer conditions with *no flow* entering the aquifer from above or below rarely occur in practice. Leaky aquifer conditions with some water entering through an overlying or underlying layer occur more frequently. This section will consider the classical leaky aquifer situation; more complex conditions such as the inclusion of storage in the aquitard or falling water tables in the overlying aquifer are considered in Sections 7.5, 8.5.2 and 10.6.1.

The assumptions of classical leaky aquifer theory are illustrated in Figure 5.17. Of particular importance is the initial condition that the groundwater head in the main aquifer is identical to the water table in the overlying aquifer. Furthermore, any drawdown in the overlying unpumped aquifer is assumed to be sufficiently small so that the water table in the overlying aquifer remains at a constant elevation. When pumping from the main aquifer commences, drawdowns occur, this sets up groundwater gradients across the aquitard between the main aquifer and the unchanging water table in the overlying aquifer. The vertical velocity in the aquitard is calculated from Darcy's Law as

$$v = K' \frac{s}{m'} \tag{5.20}$$

where K' is the vertical hydraulic conductivity and m' the thickness of the aquitard, s is the drawdown in the main aquifer. It is assumed that this vertical velocity is set up instantaneously throughout the full thickness of the aquitard. Therefore the recharge from the aquitard into the main aquifer is

$$q = K' \frac{s}{m'} \tag{5.21}$$

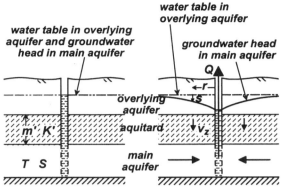

(a) initial conditions **(b) conditions during pumping**

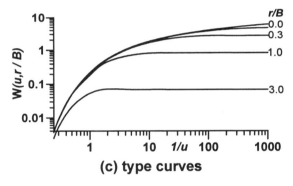

(c) type curves

Figure 5.17 Classical leaky aquifer theory: (a) initial conditions in the aquifer system, (b) conditions during pumping, (c) leaky aquifer type curves

This recharge is substituted into the differential equation for time-variant radial flow towards a borehole (Eq. (5.6)).

Solutions for classical leaky-aquifer pumping tests were derived by Jacob (1946); a convenient group of terms is

$$B = (Tm'/K')^{1/2} \tag{5.22}$$

where T is the transmissivity of the main aquifer.

The analytical solution is written as

$$s = \frac{Q}{4\pi T} W\left(u, \frac{r}{B}\right) \tag{5.23}$$

where the term $W(u, r/B)$ is a function containing an infinite integral (Kruseman and de Ridder 1990). The relevant type curves for classical leaky aquifer theory for a *log-log* scale are plotted in Figure 5.17c.

When analysing a pumping test in a leaky aquifer there are three parameters to be determined: the transmissivity, the storage coefficient and the leakage factor *B*. Consequently there are three *adjustments* when matching the curves, vertical and horizontal movement plus selecting the particular shape of curve which best matches the field results. Curve matching is far easier if the pumping test continues for a sufficiently long time for drawdowns to stabilise. When the effect of aquifer storage becoming negligibly small, most of the water is supplied by leakage through the aquitard.

Including leakage in the radial flow numerical model is straightforward. The expression for vertical flow through the aquitard into the main aquifer replaces *q* in the finite difference form of the radial flow equation (where $K_v \equiv K'$)

$$\frac{1}{\Delta a}\left\{ mK_{r,n-1}\frac{s_{n-1}-s_n}{\Delta a} - mK_{r,n}\frac{s_n-s_{n+1}}{\Delta a}\right\}_{t+\Delta t}$$
$$= \frac{S_n r_n^2}{\Delta t}\{s_{n,t+\Delta t} - s_{n,t}\} + K_{v,n}\left\{\frac{s_{n,t+\Delta t}}{m'}\right\}r_n^2 \qquad (5.24)$$

Note that in the final term representing the vertical leakage, it is preferable to specify the drawdown at time $t + \Delta t$; this is more likely to lead to a stable solution than if the drawdown is specified at time *t*.

An example of a long term pumping test in a leaky aquifer is presented in Figure 5.18; the Sherwood sandstone aquifer is overlain by the Mercia Mudstone formation. Note that in Figure 5.18, the drawdowns are presented on a *double arithmetic plot*; this form of plot is chosen to highlight the difference during the recovery phase between the non-leaky Theis analysis (shown by the broken line) and the faster recovery for a leaky aquifer. The match for the pumping and the recovery phase using the numerical model, as described above, is shown in Figure 5.18. In the analysis, $T = 49.1\,\mathrm{m^2/d}$, $S = 0.00052$, $r/B = 0.022$; for an aquitard thickness of 210 m, the effective vertical hydraulic conductivity is 0.00010 m/d. Note that the duration of the test needs to be at least 450 days to obtain a reasonable estimate of the vertical hydraulic conductivity. The information from this test was crucial to the development of a realistic groundwater model of the Nottinghamshire Sherwood Sandstone aquifer; for further information see Section 9.2 and Rushton *et al.* (1995).

5.9 FURTHER CONSIDERATION OF SINGLE-LAYER AQUIFERS

In most practical situations, field conditions are different from the assumptions of the classical confined or

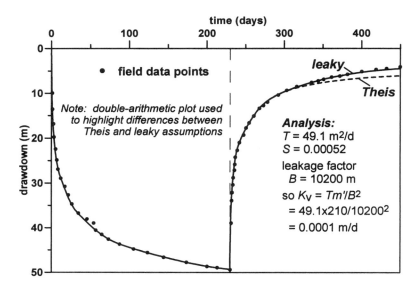

Figure 5.18 Example of analysis of pumping and recovery phases for a leaky aquifer test. Information about test, observation borehole is 225 m from pumped borehole which is 430 m deep, saturated thickness of Mercia Mudstone is 210 m, abstraction rate 4.6 Ml/d for 230 days

Table 5.4 Conditions discussed in Sections 5.10 to 5.17

Condition	Analytical (A) and/or Numerical (N)	Section
Aquifer with restricted dimensions (boundary effects)	A & N	5.10
Change in transmissivity or storage coefficient with radius	N	5.11
Changing saturated depth	N	5.12
Changing hydraulic conductivity with depth	N	5.12
Varying abstraction rates	A & N	5.13
Overflowing boreholes	A & N	5.14
Interfering boreholes	A & N	5.15
Conditions changing between confined and unconfined	A & N	5.16
Delayed yield	A & N	5.17

leaky aquifer theory described in the preceding sections. A number of different circumstances which occur in practice are listed in Table 5.4. For certain problems it is possible to represent these conditions using analytical solutions, but at other times it is necessary to use numerical solutions as indicated in the second column of Table 5.4.

5.10 AQUIFER WITH RESTRICTED DIMENSIONS (BOUNDARY EFFECTS)

The Theis theory assumes that the aquifer is of infinite extent, but most real aquifers are of limited extent. In the early stages of analysis, it is advisable to check how far the effect of pumping will spread; this is especially important with confined aquifers. As a representative example, consider an aquifer with transmissivity of $500 \, \text{m}^2/\text{d}$, confined storage coefficient of 0.0001 and a borehole pumping at 3141.6 $(1000\pi) \, \text{m}^3/\text{d}$; the drawdowns at 5 km from the pumped borehole are considered. After one day

$u = r^2 S/4Tt = 1.25$, $W(u) \approx 0.16$ (from tables e.g. Kruseman and de Ridder 1990), hence $s = (Q/4\pi T)$ $W(u) = 0.08$ m or 8 cm.

This calculation shows that the impact of pumping reaches 5 km within one day. Unless the aquifer extends well beyond 5 km with unchanged parameters, the Theis equation should not be used for times greater than about 0.5 day.

For both analytical and numerical methods of solution, image well approaches can be used for aquifers which are of limited extent due to the presence of boundaries. Full descriptions of image well techniques are given in Bouwer (1978), Todd (1980) and Kruseman and de Ridder (1990). There are many practical examples of boundaries across which no flow occurs such as the presence of a fault limiting the extent of the aquifer. This form of boundary can be represented by an imaginary *abstraction* well positioned at an equal distance beyond the boundary. For a recharge boundary where the groundwater head is effectively constant, an imaginary *recharge* well is positioned at an equal distance on the other side of the boundary. A word of caution is necessary about the use of recharge boundaries because the image well approach represents a boundary which can provide an unlimited source of water over the full depth of the aquifer, a situation which rarely occurs in practice.

For more than one boundary, multiple image wells are required. An example of image wells for three perpendicular boundaries is shown in Figure 5.19. Two of the boundaries, *AB* and *CD*, are impermeable no-flow boundaries while *BC* is a recharge boundary (a deep river or canal). At location P there is a pumped borehole with an abstraction rate $+Q$; the location of the borehole is defined by the dimensions x' and y'.

To represent these three boundaries, a series of image wells are required as described below.

- The recharge boundary *BC* is represented by a fictitious recharge well R at a distance y' beyond the recharge boundary, the discharge from the fictitious recharge well is $-Q$.
- The no-flow boundary *CD* is represented by an image well P_1 at a distance of x' to the right of *CD* with a discharge of $+Q$.

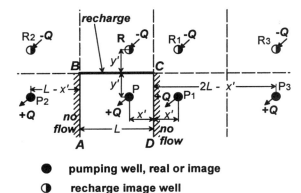

● **pumping well, real or image**

◑ **recharge image well**

Figure 5.19 Example of image well analysis when a pumped borehole is bounded by two impermeable boundaries and one recharge boundary

- The no-flow boundary AB is represented by image well P_2 at a distance of $L - x'$ to the left of AB with a discharge of $+Q$.
- It is also necessary to provide an image well for P_2 about the boundary CD, this is the image well P_3 at a distance of $2L - x'$ to the right of CD with a discharge of $+Q$.
- Further image wells, P_4, P_5, P_6, etc. are required along this horizontal line, the number of image wells depends on the maximum time period.
- For each of the image wells P_1, P_2, P_3, etc., image recharge wells R_1, R_2, R_3, etc. are required at a distance y' above the extended line passing through the recharge boundary BC.

For four boundaries a two-dimensional array of image wells is required; theoretically there should be an infinite number of image wells. In practice, the number of image wells can be restricted. Nevertheless, with a large number of image wells, some discharging and others recharging, there is a combination of many positive and negative terms. Unless great care is taken in the selection of the number of image wells there is a risk of numerical errors due to the addition of many positive and negative terms (Chan 1976).

When pumping occurs from a well field in which the impact of regional groundwater flow is small (extensive alluvial aquifers are one example), each well or borehole has an area of its own from which it can draw water. This situation can be represented by no-flow boundaries between this well and neighbouring wells. As an alternative to image wells, the effect of sur-

rounding wells can be represented directly using a radial flow numerical model with a circular outer no-flow boundary. An equivalent radius is selected so that the circular area is equal to the area associated with the individual well. Examples when this approach proved to be helpful include large diameter wells (Section 6.6).

Pumping test drawdowns often exhibit a flattening of the time-drawdown curve on a *log-arithmetic* plot; this is usually considered to indicate a recharge boundary. An increasing slope of the drawdown curve is attributed to an impermeable boundary. In practice, these changes in the slope of the drawdown curve may occur due to changes in the transmissivity or storage coefficient; this is considered in the following section. Other reasons for the changes in slope include the response in unconfined aquifers where the overlying water table acts as an upper boundary; see the field results of Figure 5.1.

5.11 CHANGE IN TRANSMISSIVITY OR STORAGE COEFFICIENT WITH RADIUS

In very extensive aquifers it may be realistic to assume that the transmissivity and storage coefficient are effectively constant from the pumped borehole to beyond the distance for which the pumping has any effect. However, in many practical situations, the transmissivity and storage coefficient do have different magnitudes within the influence of the pumped borehole. The radial flow numerical model is used to explore the effect on drawdowns of changes in transmissivity or storage coefficient at some distance from the pumped borehole. In the following examples, sudden changes are made to the transmissivity and/or the storage coefficient in a similar manner to Figure 5.9b. Initially the discussion will focus on the changed shapes of the time-drawdown curves; subsequently the results of specific examples are presented in tabular form.

To demonstrate the impact of changes in the transmissivity or storage coefficient at some distance from the pumped borehole, a series of *log-arithmetic* drawdown curves for an observation borehole are presented in Figure 5.20. The diagrams show the drawdowns during both the pumping and recovery phases. The unbroken lines refer to constant parameters with the broken lines indicating drawdowns due to changed parameter values beyond a specified distance from the pumping borehole. In Figure 5.20a the decrease in transmissivity leads to increased drawdowns at larger times compared to the straight line for the constant transmissivity aquifer; this response is usually associated with a barrier boundary. An increased transmis-

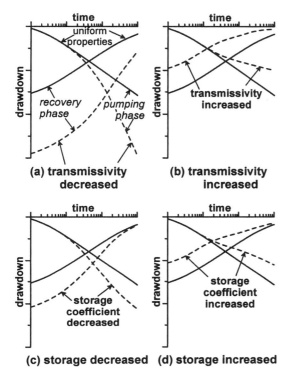

Figure 5.20 Effect of changed aquifer parameters at a specified distance from the pumped borehole; full line constant parameters, broken line changed parameters beyond specified radius

sivity (Figure 5.20b) results in the drawdowns tending to more constant values; this is usually attributed to a recharge boundary. However, the most striking finding is the significantly different responses at the end of the recovery phase. The reduced transmissivity leads to a residual drawdown three times the uniform transmissivity case, whereas the increase in transmissivity has a residual drawdown only one third of the uniform transmissivity. When the storage coefficient is decreased (Figure 5.20c), the response is similar to decreased transmissivity during the pumping phase (Figure 5.20a), but the response is distinctly different during the recovery phase, with the residual drawdown at the end of the recovery phase almost identical to the uniform parameter situation. There are similarities between increased storage and transmissivity during the pumping phase (Figures 5.20d and b), but different responses during recovery.

The effects of changed transmissivities and/or

storage coefficients are explored in more detail by considering specific examples. For each example the aquifer is confined. The uniform parameters are transmissivity $250\,m^2/d$, confined storage coefficient 0.0001 with an abstraction rate of $1000\pi\,m^3/d$ for 1.0 day and recovery for a further 1.0 day. For the non-constant parameter examples, the parameters change at 300 m. Results are quoted in Table 5.5 for observation boreholes at 100 and 1000 m at times of 0.01, 0.1 and 1.0 day during the pumping phase and at similar times during the recovery phase. For example, in No. 1 the transmissivities and storage coefficients remain at the same constant values to a radial distance in excess of 100 km. For Group B, in examples 2 to 5 the transmissivity is reduced or increased beyond 300 m (the multiplying factors are 0.1, 0.3, 3.0 and 10.0); for ease of comparison the values for constant transmissivity and storage coefficient are repeated in the middle of group B. With Group C, in examples 6 to 9, the storage coefficient is decreased or increased beyond 300 m (again using factors of 0.1, 0.3, 3.0 and 10.0). There are two additional examples, Group D (Nos 10 and 11), in which both the transmissivity and storage coefficient change beyond 300 m; in the first example the multiplying factors for transmissivity and storage are both 10.0, for the second example the factors are both 0.1. Much can be learnt from a careful study of the results; a number of insights are discussed below.

- Reductions in transmissivities beyond a specified radius cause a rapid increase in drawdowns (Nos 2 and 3) while increased transmissivities lead to a levelling off in the drawdowns during pumping (Nos 4 and 5).
- Decreases in the storage coefficient have a roughly similar effect during the pumping phase to reductions in transmissivities although the effect of changed transmissivities is more marked (compare runs 2–3 and 6–7); similar conclusions hold for increases in transmissivity and storage coefficient (Nos 4–5 and 8–9).
- The changed transmissivities have a major effect on the recoveries; the residual drawdowns at the end of the recovery for one day (Nos 2–5) vary at 100 m from 6.18 m for a tenfold reduction in transmissivity to 0.07 m for a tenfold increase in transmissivity.
- Changes in storage coefficient (Nos 6–9) have little effect on the final recovery; residual drawdowns at 100 m only vary between 0.77 and 0.56 m. This feature is of great importance in identifying whether changed transmissivities or changed storage coefficients are the dominant effect.

Table 5.5 Effect of changed transmissivities or storage coefficients beyond 300 m. Abstraction rate $1000\,\pi\,\text{m}^3/\text{d}$, observation wells at 100 m and 1000 m

No.	Transmissivity m²/d <300 m	>300 m	Storage <300 m	>300 m	Drawdown (m) at 100 m — time / pumping 0.01	0.10	1.00	recovery 1.01	1.10	2.00	Drawdown (m) at 1000 m — pumping 0.01	0.10	1.00	recovery 1.01	1.10	2.00
1	*250*	*250*	*0.0001*	*0.0001*	*1.78*	*4.01*	*6.30*	*4.53*	*2.39*	*0.71*	*0.00*	*0.22*	*1.80*	*1.81*	*1.67*	*0.65*
Group B changes in transmissivity																
2	250	25	0.0001	0.0001	1.81	6.70	21.09	19.37	15.22	6.18	0.00	0.00	2.49	2.52	2.85	3.33
3	250	75	0.0001	0.0001	1.80	5.33	11.79	10.02	6.78	2.27	0.00	0.06	2.77	2.80	2.95	1.81
1	*250*	*250*	*0.0001*	*0.0001*	*1.78*	*4.01*	*6.30*	*4.53*	*2.39*	*0.71*	*0.00*	*0.22*	*1.80*	*1.81*	*1.67*	*0.65*
4	250	750	0.0001	0.0001	1.77	3.18	4.00	2.23	0.85	0.24	0.00	0.25	0.94	0.94	0.72	0.23
5	250	2500	0.0001	0.0001	1.76	2.68	2.93	1.17	0.26	0.07	0.00	0.17	0.40	0.40	0.24	0.07
Group C changes in storage coefficients																
6	250	250	0.0001	0.00001	1.82	5.28	8.47	6.66	3.30	0.77	0.00	1.30	3.91	3.90	2.72	0.75
7	250	250	0.0001	0.00003	1.80	4.67	7.42	5.63	2.86	0.75	0.00	0.70	2.87	2.88	2.28	0.72
1	*250*	*250*	*0.0001*	*0.0001*	*1.78*	*4.01*	*6.30*	*4.53*	*2.39*	*0.71*	*0.00*	*0.22*	*1.80*	*1.81*	*1.67*	*0.65*
8	250	250	0.0001	0.0003	1.77	3.48	5.34	3.59	1.96	0.65	0.00	0.03	0.95	0.96	0.99	0.55
9	250	250	0.0001	0.0010	1.75	3.02	4.41	2.67	1.47	0.56	0.00	0.00	0.30	0.31	0.34	0.35
Group D combination of changes in transmissivities and storage coefficient																
10	250	2500	0.0001	0.001	1.74	2.50	2.71	0.97	0.22	0.07	0.00	0.03	0.18	0.18	0.16	0.06
11	250	25	0.0001	0.00001	1.76	4.03	10.29	8.58	6.67	3.23	0.00	0.00	0.01	0.01	0.01	0.09

• When both the transmissivity and storage coefficient are increased tenfold (No. 10), the drawdowns are generally slightly less than when the only change is an increase in the transmissivity (No. 5). On the other hand, when the transmissivity and storage coefficient are both reduced tenfold (No. 11), the drawdowns are smaller than when the transmissivity is decreased (No. 2) but larger than when the storage is decreased (No. 9).

In practical situations it is unlikely that the changes in transmissivity or storage coefficient will have radial symmetry. Nevertheless, the above approach provides an indication of the nature of changes in transmissivity and/or storage coefficient. This allows estimates to be made of the likely impact on long-term resource availability of any changes in aquifer parameters.

5.12 CHANGING SATURATED DEPTH AND CHANGING PERMEABILITIES IN UNCONFINED AQUIFERS

In many unconfined aquifers changes occur in the saturated depth both due to seasonal changes in recharge and due to pumping. These changes in saturated depth can have a dramatic impact on the yield of an aquifer. Pumping causes a withdrawal of water from unconfined storage which results in a lowering of the water table leading to a decrease in the saturated depth in the vicinity of the pumped borehole. Consequently, part of the aquifer, which could initially transmit water to the borehole, is no longer available since it has been dewatered. This is especially serious when any high conductivity zones in the upper part of the aquifer are dewatered. Typical examples of the severe drop in yield due to a reduction in the saturated depth were observed in a Miliolite Limestone aquifer (Rushton and Raghava Rao 1988) and a shallow alluvial aquifer (Powell *et al.* 1983) in India. In the UK, the Chalk of the Berkshire Downs (Rushton and Chan 1976) and the Yazor gravels (Rushton and Booth 1976) are further examples. Most of these examples are discussed further in Chapter 8.

Figure 5.21 is a representative example of the impact of decreases in saturated depth on the yield of an aquifer. The aquifer can be idealised as having three zones (or layers) of different hydraulic conductivity with a productive zone towards the middle of the aquifer:

upper zone; thickness 12.0 m, $K = 50$ m/d, $S_Y = 0.03$
middle zone; thickness 1.0 m, $K = 200$ m/d, $S_Y = 0.09$
lower zone; thickness 10.0 m, $K = 10$ m/d, $S_Y = 0.01$

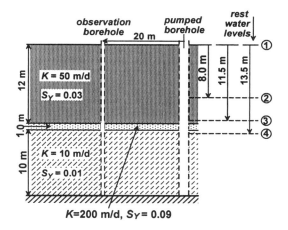

Figure 5.21 Representative example of layered aquifer system

Further information is presented in Table 5.6. The transmissivity depends on the saturated depth. When only the lower zone is saturated the transmissivity is 100 m²/d, when the middle and lower zones are both saturated the transmissivity increases to 300 m²/d, when the whole aquifer is saturated the transmissivity becomes 900 m²/d. This form of varying transmissivity can be included directly in the radial flow numerical model. At each node the transmissivity is calculated depending on the current drawdown; the well radius is 0.2 m with an outer boundary of zero drawdown at 10 km.

Four alternative rest water levels are considered in Table 5.6; they highlight the effect of different initial saturated depths. With the rest water level at the top of the aquifer system, Example 1, or at 8.0 m from the top of the aquifer, Example 2, the pumping regime of 2000 m³/d for 0.25 day followed by recovery for 0.75 day and a further pumping phase of 2000 m³/d for 0.25 day can be maintained without excessive drawdowns. However, for initial rest water levels of 11.5 m and 13.5 m below the top of the aquifer system, Examples 3 and 4, the abstraction rates are reduced to 700 m³/d and 360 m³/d respectively to prevent excessive drawdowns. Drawdowns are quoted in Table 5.6 for the pumped borehole and an observation borehole at 20 m at times of 0.25 and 1.25 days. In addition the discharge per unit drawdown is calculated for 0.25 day.

For Example 1, with the full saturated depth of the aquifer system available, the drawdowns are small with a discharge per unit drawdown in the pumped borehole

Table 5.6   Yield of zoned aquifer with different initial water levels; all the elevations or drawdowns are expressed as distances below the rest water level for that example. Pumping from 0.00 to 0.25 day and 1.00 to 1.25 day

Example	1	2	3	4
Rest water level m	0.00	8.00	11.50	13.50
Initial saturated depth m	23.00	15.00	11.50	9.50
Initial transmissivity m/d	900	500	325	95
Pumping rate m^3/d	**2000**	**2000**	**700**	**360**
Pumped drawdown at 0.25 d (m)	2.76	10.28	6.50	5.69
Discharge/drawdown ($m^2/d/m$)	725	195	108	63
Pumped drawdown at 1.25 d (m)	2.81	11.51	6.68	5.86
Drawdown at 20 m at 0.25 d (m)	0.67	1.04	0.45	0.80
Drawdown at 20 m at 1.25 d (m)	0.71	1.12	0.48	0.87

of 725 m³/d/m. In Example 2, with the rest water level 8.0 m below maximum, the pumping rate of 2000 m³/d can just be maintained; there is a significant reduction in discharge per unit drawdown to 195 m³/d/m. When the rest water level is 0.5 m above the high permeability layer, Example 3, this highly permeable zone is quickly dewatered close to the borehole and hence the pumping rate is set at 700 m³/d and the discharge per unit pumped drawdown is 108 m³/d/m. A further reduction in pumping rate is required when the initial water level is below the highly permeable zone, Example 4. The achievable pumping rate is 360 m³/d (less than one-fifth of the pumping rate when the aquifer is initially fully saturated), the discharge per unit pumped drawdown is 63 m³/d/m.

These results confirm the experience of many farmers that, following a good rainy season when water tables are close to ground surface, high well yields can be achieved and maintained for several months. However, as the water table falls, so that it approaches or becomes below a higher permeability zone, the sustainable yield falls off rapidly.

5.13 VARYING ABSTRACTION RATES

Variations in the abstraction rate occur in many pumping tests. Increases or decreases in abstraction can occur due to the pump characteristic, adjustments to the control valves, variation in output of the pump motors and breaks in the power supply. If airlifting is used for a pumping test, a substantial decrease in discharge with drawdown is certain to occur.

Analytical solutions are available for variable borehole abstraction rates. For example, Abu-Zied and Scott (1963) consider an exponential decay function to describe the change in abstraction while Hantush (1964) developed solutions for various abstraction rates including exponential and hyperbolic decays. Lai *et al.* (1973) derive expressions for the drawdown in a pumped borehole where there is a change in pumping rate or when the pumping rate is linearly proportional to time up to a specified time when it becomes constant. Sharma *et al.* (1985) consider the case of a linear decay $Q(t) = Q_0(1 - \alpha t)$ where α is a discharge parameter; they describe a modified form of the *log-log* curve matching technique.

In practice, changes in abstraction rate rarely conform to the above mathematical expressions. Rather than trying to derive a mathematical expression to fit a particular situation it is advantageous to adopt a Kernel function technique (Section 6.3) or use the radial flow numerical model with the abstraction rate varying to reflect rates measured during the test (for example, the case study in Section 7.4.6).

Account must be taken of varying abstraction rates when carrying out pumping test analysis. Figure 5.22 shows the difference in drawdown when the discharge rate falls in proportion to the pumped well drawdown; the pumping rate over 3.25 days falls by just over 20 per cent. The manner in which the drawdown curve tends to the horizontal would normally be interpreted as being due to a recharge boundary rather than reducing abstraction. Even if the abstraction rate decreases by 10 per cent, a distinctive change in slope occurs compared to the straight-line plot due to constant abstraction.

5.14 OVERFLOWING ARTESIAN BOREHOLES

When an aquifer is under artesian pressure, boreholes will flow naturally without the need for pumping. Before the borehole is allowed to flow there is a

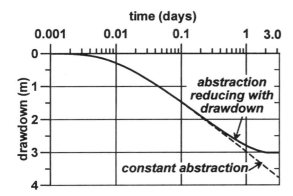

Figure 5.22 Observation well drawdowns resulting from a reduction in abstraction rate which is proportional to the pumped drawdown; at 3.25 days abstraction rate is 0.79 times starting value

piezometric head above ground level. On opening the outlet valve, the head in the borehole drops and water flows out.

In the approach developed by Jacob and Lohman (1952), the head in the borehole is assumed to fall instantaneously. They derive a method of analysis for a sudden and constant drawdown in the overflowing borehole and propose a straight-line technique of estimating the transmissivity similar to the Cooper–Jacob analysis for constant discharge. They develop their analysis using a non-dimensional parameter

$$\alpha = tT / Sr_w^2 \tag{5.25}$$

where r_w is the radius of the overflowing borehole. The discharge Q, due to an instantaneous drawdown in the borehole s_w, can be calculated from the equation

$$Q = 2\pi Ts_w G(\alpha) \tag{5.26}$$

Values of $G(\alpha)$ for $10^{-4} \le \alpha \le 10^{12}$ are quoted by Jacob and Lohman (1952). They studied several boreholes in a low transmissivity aquifer; the estimated transmissivities compared favourably with values deduced from recovery analysis.

One important feature of this analytical approach is the assumption of an instantaneous fall in the head in the artesian borehole which corresponds to an instantaneous infinite discharge. In practice the discharge from the borehole is limited due to the restrictions of the borehole diameter and any valve-gear. In the field problems studied by Jacob and Lohman, this assump-

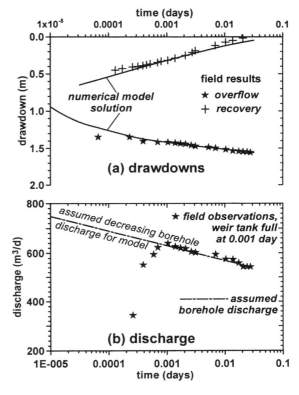

Figure 5.23 Numerical model analysis of a field overflow test showing the match between field and modelled drawdowns and the assumed discharge from the overflowing borehole (Rushton and Rathod 1980). Reprinted from *Ground Water* with permission of the National Ground Water Association. Copyright 1980

tion did not lead to serious errors since the borehole yields were low, ranging from 1–30 gallons/min (5.5–165 m³/d); achieving a virtually instantaneous fall in head was not difficult. However, for tests in higher-yielding aquifers, the time taken to open the valve and the restrictions in flow due to the pipe work means that an instantaneous fall in head will not occur.

Rushton and Rathod (1980) describe a modified overflow test carried out in a limestone aquifer in Eastern England. To overcome uncertainties due to the time taken to open the valve and restriction on the flow caused by the pipe work, readings were taken of *both* the head in the borehole and the flows from the borehole. The overflow and recovery phases were both monitored; the results are included in Figure 5.23. Since the weir tank, which was used for measuring the flows, did not fill until 0.001 day, the calculated flows before this time are not used. The chain-dotted line drawn in Figure 5.23b represents the overflowing borehole discharge for the whole test; during the recovery phase the discharge is zero.

In a radial flow numerical model simulation of this test, the discharge taken from the node representing the 0.1 m diameter borehole corresponds to the chain-dotted line of Figure 5.23b. The match in Figure 5.23a between the field drawdowns (discrete symbols) and numerical model (continuous lines) is adequate; note that the overflowing phase and the recovery phase are represented by a single run of the numerical model. The poor match for the first field reading occurs because of significant oscillations in the water column in the borehole as soon as the valve was opened. The transmissivity is estimated to be $450 \pm 130 \, \mathrm{m^2/d}$ and the storage coefficient 0.0002 ± 0.0001.

5.15 INTERFERING BOREHOLES

Ideally, pumping tests should be carried out in aquifers from which no other boreholes are pumping. However, in many practical situations there are other pumped boreholes in the vicinity of the test site; these are termed *interfering boreholes*. This discussion on the impact of interfering boreholes on pumping test interpretation is in three parts. Initially the effect of an interfering borehole on observed pumping test data is explored. Second, expressions are derived to assess the magnitude of the impact; in confined aquifers an interfering borehole 1.0 km from the test site can still have a sizeable effect. Finally, a case study is considered in which deviations from the Cooper–Jacob straight line were thought to be due to changes in aquifer parameter values; subsequently the effect of other pumped boreholes was identified as the cause of the deviations.

5.15.1 Introduction to interference

Pumping test analyses are usually based on the assumption that the aquifer is in a state of equilibrium before the test starts. Frequently there are other pumping boreholes which cause a disturbance in the aquifer; the effects of these interfering boreholes are often ignored because they are considered to be at too great a distance to have a major influence. Even if an interfering borehole is relatively close, it is assumed that its effect can be ignored provided that the interfering borehole discharge is maintained at a steady rate during the test.

The effect of interfering boreholes is illustrated by a typical example, Figure 5.24a, in which a test borehole **T** is at a radial distance r_T from an observation borehole **Obs**. In addition there is a second abstraction borehole **I** at a distance of r_I from the observation borehole. Since the distance of interfering borehole **I** to the observation borehole is ten times the distance from the test borehole to the observation borehole, the effect of the interfering borehole may be small. The interfering borehole starts pumping at time t_o before the start of the test in borehole **T**; the actual test consists of a pumping and recovery phase. During the whole of the test the interfering borehole pumps at a constant rate; see Figure 5.24b. The symbols used to describe the test and the parameter values on which the plots of Figure 5.24 are based are as follows.

Borehole	Pumping rate	Starting time		Distance to observation borehole
I	$Q_I \, 1000\pi \, \mathrm{m^3/d}$	0	0.0 day	$r_I \, 100 \, \mathrm{m}$
T	$+Q_T \, 500\pi \, \mathrm{m^3/d}$	t_o	1.0 day	$r_T \, 10 \, \mathrm{m}$
T	$-Q_T \, -500\pi \, \mathrm{m^3/d}$	t_r	2.0 day	$r_T \, 10 \, \mathrm{m}$

The transmissivity and confined storage coefficient of the aquifer are $250 \, \mathrm{m^2/d}$ and 0.0001 respectively. Note that in the above table, pumping in the test borehole, which starts at time t_0 and finishes at time t_r, is represented as a pumping rate of $+Q_T$ from a time t_0 to infinity and a second pumping rate of $-Q_T$ from time t_r to infinity; this is similar to the Theis recovery analysis of Section 5.6 and Figure 5.14. Individual drawdowns are calculated from tables of the well function for each of the above three pumping conditions, the *total* drawdowns, obtained by superposition of the individual drawdowns, are plotted as an unbroken line in Figure 5.24c.

However, these are not the results that will be observed during the test. Measurements of the drawdown commence when the test borehole starts pumping, therefore zero drawdown (the observer's datum) coincides with the broken horizontal line of Figure 5.24c. The information actually obtained by the observer is shown in Figure 5.24d. By plotting the 'observed drawdowns' of Figure 5.24d on a *logarithmic* scale (Figure 5.25), straight lines are not obtained. In fact, the results are of a similar shape to Figures 5.20a or c, suggesting that there is a decrease in transmissivity and/or storage coefficient at some distance from the test borehole.

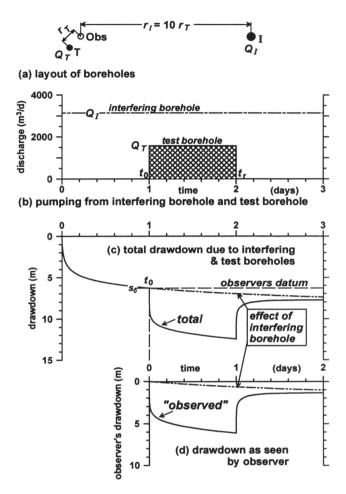

Figure 5.24 Example of the influence of an interfering borehole on pumping test response: (a) layout of boreholes, observation **Obs**, test **T** and interfering **I**, (b) pumping rates for interfering borehole (operates for 3 days) and test borehole (pumps between day 1 and 2), (c) drawdowns with origin of the graph at the start of pumping from interfering borehole with broken line indicating the effect of interfering borehole, (d) what is observed when the datum for drawdowns is taken as the start of the pumping in the test borehole (broken line shows effect of interfering borehole) (Rushton 1985). Reprinted from *Ground Water* with permission of the National Ground Water Association. Copyright 1985

5.15.2 Theoretical analysis of interfering boreholes

The following theoretical analysis refers to a similar situation to that above, pumping rates and timing for the Interfering and Test boreholes are shown in Figure 5.26. The analysis is based on the superimposition of the effects of the interfering and test boreholes using the Theis theory (see also Rushton 1985). When calculating the drawdowns, apart for very small times, it is acceptable to take the first three terms of the expansion

for $W(u)$ (Eq. (5.8)). The terms can be rearranged to express the drawdown due to pumping from a borehole as:

$$s = \frac{Q}{4\pi T}\left[\ln\frac{2.246Tt}{r^2 S} + \frac{r^2 S}{4Tt}\right] \qquad (5.27)$$

For times between the start of the test pumping t_0 and the start of recovery t_r, the total drawdown s_t in the

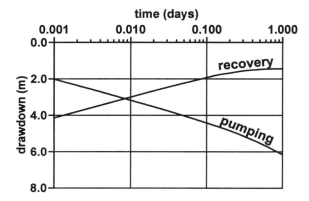

Figure 5.25 Observer's record of pumping test (diagram (d) of Figure 5.24) presented as a log-arithmetic plot

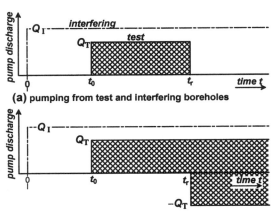

(a) pumping from test and interfering boreholes

(b) method of representing test and interfering boreholes

Figure 5.26 Theoretical analysis of interfering boreholes, details of pumping and recovery

observation borehole due to the interfering borehole Q_I and the test borehole Q_T is

$$s_t = \frac{Q_I}{4\pi T}\left[\ln\frac{2.246Tt}{r_I^2 S} + \frac{r_I^2 S}{4Tt}\right]$$
$$+ \frac{Q_T}{4\pi T}\left[\ln\frac{2.246T(t-t_0)}{r_T^2 S} + \frac{r_T^2 S}{4T(t-t_0)}\right] \quad (5.28)$$

where the discharges and radial distances are defined in Figures 5.24 and 5.26. However, the observer will take as datum the drawdown s_0 at time t_0, when the test starts; s_0 is due to the discharge Q_I from the interfering borehole,

$$s_0 = \frac{Q_I}{4\pi T}\left[\ln\frac{2.246Tt_0}{r_I^2 S} + \frac{r_I^2 S}{4Tt_0}\right] \quad (5.29)$$

Subtracting s_0 from the total drawdown s_t gives, with some rearrangement, the test drawdown s_T as measured by the observer,

$$s_T = \frac{Q_I}{4\pi T}\left[\ln\frac{t}{t_0} + \frac{r_I^2 S}{4T}\left(\frac{t_0-t}{t_0 t}\right)\right]$$
$$+ \frac{Q_T}{4\pi T}\left[\ln\frac{2.246T(t-t_0)}{r_T^2 S} + \frac{r^2 S}{4T(t-t_0)}\right] \quad (5.30)$$

The second term in Eq. (5.30) is the expression for the drawdowns due to the test borehole alone with no interfering borehole, hence the error in drawdown s_e due to the interfering borehole is

$$s_e = \frac{Q_I}{4\pi T}\left[\ln\frac{t}{t_0} + \frac{r_I^2 S}{4T}\left(\frac{t_0-t}{t_0 t}\right)\right] \quad (5.31)$$

To appreciate the significance of this error, consider an aquifer with transmissivity and storage coefficient of 250 m²/d and 0.0001 respectively and a pumping rate from the interfering borehole of 2000π m³/d. At time $t = 2t_0$ (assuming the duration of the test is the same as the time for which the interfering borehole pumps before the test) it is possible to calculate the error in the drawdown due to the interfering borehole at different radial distances from the observation borehole. The results are listed in the second column of Table 5.7.

The second column of Table 5.7 shows that, for a confined aquifer with a storage coefficient of 0.0001, the interfering borehole has effectively the same impact whether the distance to the interfering borehole is 10 m or 1000 m; even at a distance of 3000 m there is still a substantial effect. Results for two further values of the storage coefficient are contained in columns three and four of Table 5.7; for an unconfined aquifer with a storage coefficient of 0.01, interfering boreholes at 100 m or beyond have no effect.

5.15.3 Practical example of the significance of interfering boreholes

A series of pumping tests were carried out at Ashton Keynes; water was abstracted from the confined Great Oolite limestone aquifer. This discussion will concen-

Table 5.7 Dependence of error in drawdown on storage coefficient and distance to interfering borehole

Radial distance (m) to interfering borehole	Error in drawdown (m)		
	$S = 0.0001$	$S = 0.001$	$S = 0.01$
10	1.386	1.386	1.336
30	1.386	1.381	0.936
100	1.386	1.336	–
300	1.381	0.936	–
1000	1.336	–	–
3000	0.936	–	–

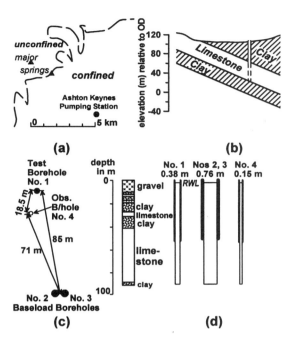

Figure 5.27 Details of Ashton Keynes test: (a) location of interface in limestone aquifer between confined and unconfined regions, (b) cross-section through aquifer system, (c) layout of boreholes, (d) strata and casing details for individual boreholes (Gonzalez and Rushton 1981). Reprinted from *Ground Water* with permission of the National Ground Water Association. Copyright 1981

trate on a three-day test. Additional details, including information about a step test and a long duration test, can be found in Gonzalez and Rushton (1981).

Information about the limestone aquifer system and the boreholes is presented in Figure 5.27a, which shows the location of Ashton Keynes pumping station relative to the outcrop. Figure 5.27b is a sketch of the limestone and the clays underlying and overlying the limestone. The site layout is shown in Figure 5.27c; note that there are two baseload boreholes as well as the pumping and observation boreholes. Diagrams of construction and strata logs of the four boreholes are displayed in Figure 5.27d.

For the three-day test, readings of drawdown were obtained in the test borehole and the observation borehole; during the test the baseload borehole pumped at a constant rate. Readings were also taken for the first day of recovery. The results are plotted in a *logarithmic* form in Figure 5.28. For both the pumping and recovery phases, the field readings approach a straight line after 0.01 day (14.4 mins). Using the conventional Cooper–Jacob analysis for the pumping phase, the calculated values for the aquifer parameters are:

$$T = 1060 \text{ m}^2/\text{d}, \ S = 0.006$$

However, at later times the data for both pumping and recovery phases show deviations from straight lines; the deviations are similar to those for Figure 5.20c which are due to a reduced storage coefficient at some distance from the test borehole. Attempts were made to reproduce these curves by introducing a change in storage coefficient with radius. By selecting a storage coefficient of 0.005 up to a radius of 400 m but decreased to 0.0005 for greater distances, a good match to the pumping and recovery phases can be obtained.

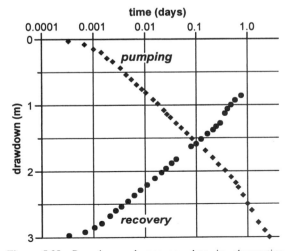

Figure 5.28 Pumping and recovery data in observation borehole for three-day pumping test at Ashton Keynes

However, there is no physical reason for changes in storage coefficient.

In the analysis thus far, no account has been taken of the baseload boreholes shown in Figure 5.27c. The pumping test report states that the baseload boreholes were pumped at a constant rate throughout the test and therefore their influence can be ignored. However, a careful examination of the pumping records at Ashton Keynes Pumping Station provided the following evidence.

- Normally the baseload boreholes (which were used alternately) are not pumped continuously for 24 hours each day; 18 to 20 hours is a more normal pattern.
- Water is pumped to a service reservoir, which contains sufficient water for about 24 hours' supply.
- There was no other location to which water from the baseload boreholes could be directed.
- Consequently until a few days before the start of the test there was little pumping so that the service reservoir was almost empty. Pumping of the baseline boreholes started about 2.5 days before the start of the test.
- Throughout the test the baseload boreholes pumped at a constant rate; the baseload pumping rate fell slightly at the start of recovery due to the limited storage in the service reservoir. After 0.3 day into the recovery the rate reduced significantly; this is the cause of the change in slope after 0.3 day into the recovery.

Table 5.8 lists the pumping rates from the baseload and test boreholes.

In Figure 5.29, the unbroken line represents an analysis using the radial flow numerical model of Section 5.5 with the pumping rates of Table 5.8. The drawdowns in the observation borehole equal the sum of drawdowns due to the test and baseline boreholes. Zero drawdown is taken as the water level in the observation borehole as the test starts. Note in particular the drawdowns which would have occurred due to the test borehole alone (shown by the broken line and labelled baseload not included). The difference between the unbroken and broken lines is a consequence of the influence of the baseline boreholes. Therefore the influence of baseload boreholes can be ignored only if they pump at a constant rate for a long time before the start of the test, with continued steady pumping after the completion of the test. Since pumping at the constant rate at the baseload boreholes only started about 2.5 days before the start of the test, they caused increasing

Table 5.8 Abstraction rates and timings for test and baseload boreholes

Time from start of test (days)	Abstraction rate (m³/d)	
	Test borehole	Baseload borehole
? to −2.5	0	0
−2.5 to 0.0	0	9220
0.0 to 0.625	4265	9220
0.625 to 3.0	4070	9220
3.0 to 3.3	0	9114
3.3 to 4.0	0	8500

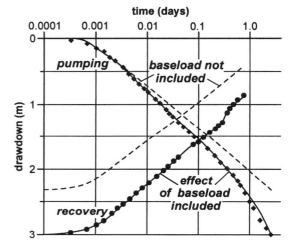

Figure 5.29 Numerical model results (broken and unbroken lines) which reproduce pumping and recovery phases in the observation borehole for the Ashton Keynes test (Gonzalez and Rushton 1981). Reprinted from *Ground Water* with permission of the National Ground Water Association. Copyright 1981

drawdowns in the observation borehole throughout the test.

5.16 CONDITIONS CHANGING BETWEEN CONFINED AND UNCONFINED

Conditions changing between confined and unconfined have an impact on the response of a pumped borehole; this is demonstrated by the example in Section 5.7. A closed-form analytical solution for pumping from a borehole when the 'aquifer undergoes

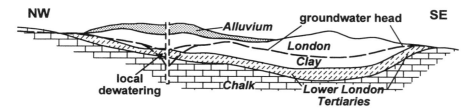

Figure 5.30 Section through the Chalk of the Braintree area, UK. *Journal of Hydrology* **62**, Reprinted from Rushton and Senarath, A mathematical model study of an aquifer with significant dewatering, pp. 143–58, copyright (1983) with permission from Elsevier Science

conversion from artesian to water table conditions' was developed by Moench and Prickett (1972), who demonstrated the differences between the Theis analysis for constant storage coefficient and the effect of conversion to water table conditions in the vicinity of a pumped borehole.

The importance of changes from confined to unconfined conditions became apparent during a study of a chalk aquifer in the catchment of the River Pant which is overlain by the Lower London Tertiaries which in turn are overlain by the London Clay, Figure 5.30 and Rushton and Senarath (1983). Due to the London Clay cover, which allows very little water to pass through, the resources of the chalk aquifer were expected to be severely restricted. Prior to 1970 a moderate yield was obtained from boreholes in the Chalk, primarily due to water entering the Chalk in the north-west. Abstraction increased by about 70 per cent in the 1970s; the existing boreholes provided the water but with increases in pumped drawdowns of more than 20 m. Early in the 1970s the water table fell below the base of the London Clay confining layer into the Lower London Tertiaries. The situation towards the end of the 1970s in the vicinity of a representative borehole is sketched in Figure 5.30.

A significant decline in pumped water levels is often an indication that the aquifer is being over-exploited. Therefore a detailed field study was carried out supported by regional groundwater modelling. The Lower London Tertiaries are not considered to be an aquifer in the sense of transmitting water. Nevertheless, the specific yield of the Lower London Tertiaries is about 0.05, provided that drainage occurs slowly. Therefore the mechanism which supported the increased yields was identified as the slow dewatering of the Lower London Tertiaries into the Chalk, with water flowing through the Chalk to the pumped boreholes.

Incorporation of this mechanism in a regional groundwater model was essential. It was decided to

represent the increasing dewatered areas around boreholes as modified storage coefficients. The manner in which this was achieved is illustrated in Figure 5.31. The *first part* of the analysis involves the development of expressions for the equivalent storage coefficients due to a change between confined and unconfined conditions. Consider the differential equation for steady state regional groundwater flow, omitting the time-dependent terms from Eq. (2.54b),

$$\frac{\partial}{\partial x}\left(T_x \frac{\partial h}{\partial x}\right) + \frac{\partial}{\partial y}\left(T_y \frac{\partial h}{\partial y}\right) = -q$$

This equation can be written in finite difference form for the square mesh of Figure 5.31a with uniform transmissivity T, using the approach of Section 2.7. If $\Delta x = \Delta y$,

$$h_1 + h_2 + h_3 + h_4 - 4h_0 = Q/T \qquad (5.32)$$

where Q, the total discharge from node 0, equals $-q\Delta x \Delta y$. Taking the average groundwater head at the four nodes surrounding the central node 0,

$$h_{av} = 0.25(h_1 + h_2 + h_3 + h_4)$$

thus $\quad h_{av} - h_0 = 0.25 Q/T \qquad (5.33)$

Considering next the steady-state radial flow to a pumped borehole in an aquifer with constant transmissivity, from Eq. (5.4)

$$h_{av} - h_0 = (Q/2\pi T)\ln(r_{av}/r_0)$$

but the radial distance to nodes h_1, h_2 etc. is Δx, therefore

$$h_{av} - h_0 = (Q/2\pi T)\ln(\Delta x/r_0) \qquad (5.34)$$

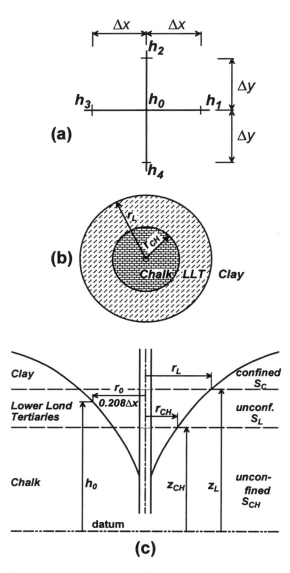

From Eqs (5.33) and (5.34), following rearrangement

$$r_0 = \exp(-0.5\pi)\Delta x = 0.208\Delta x \qquad (5.35)$$

Therefore the groundwater head at node 0 represents the conditions in a borehole of radius 0.208 times the mesh spacing. This situation is illustrated in Figure 5.31c.

The groundwater head at the face of the borehole h_w will be lower than the computed value h_0 and can be calculated from the standard radial flow equation,

$$h_w = h_0 - (Q/2\pi T)\ln(0.208\Delta x/r_W) \qquad (5.36)$$

The *second part* of the analysis relates to the determination of the radial distances at which the water table crosses the interface between different stratum. Figures 5.31b and c show areas around a pumped borehole where unconfined conditions occur in the Chalk (the inner circle of radius r_{CH}), where unconfined conditions occur in the Lower London Tertiaries (the area from r_{CH} to r_L) and the area beyond r_L where confined conditions occur due to the London Clay.

To calculate the radius at which the drawdown curve at a particular time intersects the interface between the Lower London Tertiaries and the overlying Clay at an elevation of z_L (Figure 5.31c)

$$r_L = r_0 \exp[(2\pi T/Q)(z_L - h_0)]$$
$$= 0.208\Delta x \exp[(2\pi T/Q)(z_L - h_0)] \qquad (5.37)$$

A comparable equation applies for the radius of the intercept between the drawdown curve and the interface between the Chalk and the Lower London Tertiaries. However, because r_{CH} is less than r_0 the relevant equation is:

$$r_{CH} = 0.208\Delta x\{\exp[(2\pi T/Q)(h_0 - z_{CH})]\}^{-1} \qquad (5.38)$$

Multiplying the areas in Figure 5.31b by the appropriate storage coefficients (specific yield for Chalk S_{CH}, specific yield for Lower London Tertiaries S_L and confined storage coefficient of London Clay S_C) and dividing by the area represented by a nodal point, the equivalent storage coefficient S_{eq} can be estimated as

$$S_{eq} = \frac{1}{\Delta x^2}[\pi r_{CH}^2 S_{CH} + \pi(r_L^2 - r_{CH}^2)S_L$$
$$+ (\Delta x^2 - \pi r_L^2)S_C] \qquad (5.39)$$

The development of this representation of changing conditions between confined and unconfined storage

Figure 5.31 Representation of dewatering of various strata leading to modified equivalent storage coefficient: (a) finite difference mesh, (b) plan showing areas associated with different storage coefficients, (c) section through borehole indicating how the radii associated with different storage coefficients are calculated. *Journal of Hydrology* **62**, Reprinted from Rushton and Senarath, A mathematical model study of an aquifer with significant dewatering, pp. 369–87, copyright (1983) with permission from Elsevier Science

coefficients in the vicinity of a pumped borehole proved to be the key to understanding the unexpectedly high yields of boreholes in the confined chalk aquifer in Figure 5.30 (Rushton and Senarath 1983). Soon after this study was carried out, there was a substantial reduction in the quantity of water pumped from the chalk aquifer. Recovery in groundwater heads was more rapid than expected; the rapid recovery was due to a reduction in the specific yield of the Lower London Tertiaries to about 0.01. This reduction from 0.05 during dewatering to 0.01 during recovery could occur due to consolidation of the formation or air-entrapment.

5.17 DELAYED YIELD

5.17.1 Background

'Delayed yield' is written in inverted commas because it is a term which describes a number of different phenomena. It was introduced by Boulton (1963) to describe the delayed release of water as the water table falls (also called slow gravity drainage). The concept of delayed yield was introduced to explain the difference between responses predicted by the Theis theory and the form of time-drawdown curve observed in the field for unconfined aquifers. The complex response of unconfined aquifers to pumping has already been introduced through the field data for the Kenyon Junction test (see Section 5.1 and especially Figure 5.1). Figure 5.32 contains two typical time-drawdown curves for unconfined aquifers plotted to a *log-log* scale. The unbroken line represents the situation where the specific yield is three orders of magnitude greater than the confined storage coefficient; the broken line relates to

a specific yield about ten times the confined storage coefficient. These curves have a sigmoid shape.

Boulton (1963) had two main reasons for introducing delayed yield. The first was that an explanation was required for the sigmoid shape of time-drawdown curves. The second reason was based on geotechnical observations that water drains slowly from a soil sample when a reduction in the water table elevation occurs. However, if the water table lies within a very permeable gravel or sand, delayed drainage is not observed. Boulton's study led to analytical solutions which have now been included in computer packages. A flexible computer package has been prepared by Hall (1996); his discussion on the substantially different parameter values deduced by different workers using the same field data is instructive.

In the published discussion on Boulton's (1963) paper, contributors suggested that vertical flows within the aquifer are an alternative cause of the apparent delayed yield; this issue was subsequently considered by Neuman (1972). During the 1970s there was an extensive debate about the causes of the 'delayed yield' effect; see, for example, Neuman (1979). Bouwer (1979) reviewed existing physical reasons for delayed yield and proposed that soil water hysteresis is a possible cause, although elastic response of the aquifer and restricted air movement above the water table may also contribute to the phenomenon.

To summarise, there are several possible reasons for the 'delayed yield' effect:

- the phenomena of elastic storage and specific yield both apply in unconfined aquifers,
- in an unconfined aquifer there are flow components in both radial and vertical directions,
- there are therefore different responses at different depths within the aquifer,
- there is often a delay in the release of water from storage as the water table falls,
- open observation boreholes may provide unreliable information (see Figure 5.1).

Methods of analysing pumping tests in unconfined aquifers using analytical solutions and type curves can be found in Kruseman and de Ridder (1990) and Hall (1996). The discussion below is concerned with the physical processes of delayed yield as first introduced by Boulton (1963) and the inclusion of delayed yield in radial flow numerical models. Time-variant flow in unconfined aquifers in the *r-z* plane with both radial and vertical flow components is considered in Section 7.2.

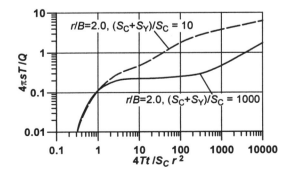

Figure 5.32 Typical delayed yield curves

5.17.2 Inclusion of delayed yield in radial flow numerical model

The following explanation of the inclusion of delayed yield in a numerical model highlights the physical processes which are assumed to occur. According to Boulton (1963) a change in drawdown, δs during a time period from t to $t + \delta t$ leads to an instantaneous release of water $S_C \delta s$ (where S_C is the confined storage coefficient) plus a delayed drainage which at a later time t' equals

$$\delta s \alpha S_Y \exp[-\alpha(t'-t)] \tag{5.40}$$

where S_Y is the specific yield and α is the reciprocal of the delay index (units d^{-1}).

When this process is represented in a numerical model, the time is divided into discrete time steps; see Figure 5.33. The drawdown increases by Δs_n during the nth time step as the time increases from $t_n - \Delta t_n$ to t_n. This increase in drawdown leads to three components of water available for flow through the aquifer.

1. A quantity of water is released due to the confined storage coefficient; this equals

$$S_C \frac{\Delta s_n}{\Delta t_n}$$

2. Inflow occurs at the water table resulting from delayed yield of previous drawdown increments.

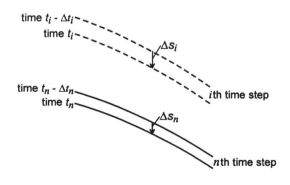

Figure 5.33 Discrete time steps in method for including delayed yield in numerical models

Due to a drawdown Δs_i during the ith time step ($i < n$) a contribution reaches the water table during the nth time step,

$$\alpha \Delta s_i S_Y \exp[-\alpha(t_n - t_i)]$$

The total inflow during the nth time step from all the previous drawdown increments can be determined from the summation,

$$\alpha S_Y \sum_{i=1}^{n-1} \Delta s_i \exp[-\alpha(t_n - t_i)] \tag{5.41}$$

3. During the current time step there is a contribution at the water table due to the specific yield which is factored according to the delayed yield concept. From Eq. (5.40), the contribution during the current time step, where t_s is the time since the start of the current time step, equals

$$\Delta Q = \int_0^{\Delta t_n} ds \alpha S_Y \exp[-\alpha(\Delta t_n - t_s)] \tag{5.42}$$

Assuming a linear change in drawdown during the time step,

$$ds = (\Delta s_n / \Delta t_n) dt_s$$

then Eq. (5.42) becomes

$$\Delta Q = (\Delta s_n / \Delta t_n)\alpha S_Y \int_0^{\Delta t_n} \exp[-\alpha(\Delta t_n - t_s)] dt_s$$
$$= (\Delta s_n / \Delta t_n) S_Y [1 - \exp(-\alpha \Delta t_n)] \tag{5.43}$$

Therefore the contribution during the current time step can be written in terms of an equivalent storage coefficient which equals $S_Y [1 - \exp(-\alpha \Delta t_n)]$.

Assembling the components from (1), (2) and (3) above,

$$\{S_C + S_Y[1 - \exp(-\alpha \Delta t_n)]\} \frac{\Delta s_n}{\Delta t_n}$$
$$+ \alpha S_Y \sum_{i=1}^{n-1} \Delta s_i \exp[-\alpha(t_n - t_i)] + q \tag{5.44}$$

By considering the first term and then the second and third terms together, the effect of delayed yield can be expressed as an effective storage coefficient S' where

$$S' = S_C + S_Y[1 - \exp(-\alpha \Delta t_n)] \tag{5.45}$$

plus an effective recharge q' where

$$q' = \alpha \, S_Y \sum_{i=1}^{n-1} \Delta s_i \exp[-\alpha(t_n - t_i)] + q \qquad (5.46)$$

Using an approach introduced by Ehlig and Halepaska (1976), Eq. (5.46) can be written in a more convenient form for evaluation,

$$q' = \alpha S_Y \left\{ \sum_{i=1}^{n-2} \Delta s_i \exp[-\alpha(t_{n-1} - t_i)] \right.$$
$$\left. + \frac{\Delta s_{n-1}}{\alpha \Delta t_{n-1}} [1 - \exp(-\alpha \Delta t_{n-1})] \right\} \exp(-\alpha \Delta t_n) + q \qquad (5.47)$$

When a single layer radial flow model is used, delayed yield can be represented with the right-hand side of Eq. (5.16) written in terms of the equivalent storage coefficient and equivalent recharge,

$$\frac{S'_n r_n^2}{\Delta t} \{ s_{n,t+\Delta t} - s_{n,t} \} + q' r_n^2 \qquad (5.48)$$

5.17.3 Comparison of analytical and numerical solutions for a field example with delayed yield

As an example of the inclusion of delayed yield in pumping test analysis, the problem introduced by Boulton (1963) is reanalysed. The aquifer, which has a saturated depth of about 27 m, was pumped at a rate of 5880 m³/d for 50 hours. Drawdowns were measured in two observation wells positioned at 22 m and 168 m from the pumped borehole. The field results are indicated by circles in Figure 5.34. Using his type curves, Boulton carried out separate analyses for the data from the two observation wells. The resultant parameter values are quoted in the first and second rows of numbers in Table 5.9.

The difficulty with Boulton's values is that, apart from the hydraulic conductivity K, all the other parameters take different values at $r = 22$ m compared to $r = 168$ m. Such differences are unlikely, for instance changes in the specific yield from 8.8 per cent to 1.98 per cent are not physically plausible. This is confirmed by a numerical model simulation in which variations in confined storage coefficient and specific yield with radius are introduced so that the actual values at 22 m and 168 m coincide with the values in the first two lines of Table 5.9. The resultant drawdowns showed such differences from field values that they are not plotted in Figure 5.34.

The next step was to use parameter values through-

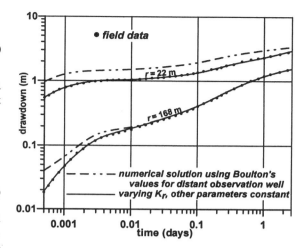

Figure 5.34 Re-analysis of unconfined pumping test from Boulton (1963) using radial flow model, parameter values recorded in Table 5.8

Table 5.9 Aquifer properties for the field example in Boulton (1963)

	K (m/d)	S_C	S_Y	α (d⁻¹)
Boulton $r = 22$ m	129	0.0027	0.088	0.077
Boulton $r = 168$ m	129	0.00063	0.0198	0.057
Numerical $r = 22$ m	*250*	*0.0006*	*0.016*	*0.05*
Numerical $r = 168$ m	*130*	*0.0006*	*0.016*	*0.05*

out the whole aquifer based on Boulton's calculated values for the distant observation well at $r = 168$ m. The resultant drawdowns for each of the observation wells are indicated by the chain-dotted lines in Figure 5.34. The match at the distant observation well is satisfactory apart from the earlier times, whereas drawdowns at the closer observation well are too large. There is field evidence which supports the likelihood of higher hydraulic conductivities closer to the pumped borehole; consequently a new set of parameters, as indicated by the final two lines in Table 5.9, were deduced. The match at each observation well, which is shown by the unbroken lines, is good. Therefore the key to achieving a match at each observation well with consistent parameter values is an increasing hydraulic conductivity towards the pumped borehole but with all the other parameters remaining constant with radius.

5.18 CONCLUDING REMARKS

This chapter considers certain of the fundamental issues which arise due to radial flow towards pumped boreholes. In the initial field example of Section 5.1, data from a series of piezometers on a vertical section are used to develop an understanding of the complexity of the flow processes in the aquifer. Conceptual models are developed which demonstrate the importance of identifying the three different sources of stored water that are released to balance the water withdrawn by the pump.

The remainder of the chapter focuses on situations where vertical flows do not need to be represented explicitly. There are many different physical situations due to the nature of any overlying strata, the presence of boundaries, changes in the saturated depth or variations in the rate at which water is withdrawn from the borehole. Impermeable overlying strata result in confined conditions whereas overlying aquitards result in leaky aquifer conditions. In unconfined aquifers, delayed drainage at the water table can influence the aquifer response. The presence of boundaries to the aquifer, or changes in aquifer properties with radius, modify the response in observation wells and may restrict the long-term yield of the aquifer system. Interfering boreholes within the zone of influence of pumped boreholes have a similar impact. A further feature which can influence the radial flow is the decrease in saturated depth especially close to the pumped borehole. This is particularly important when zones or layers of higher hydraulic conductivity are dewatered. Additional issues addressed here include variations in the abstraction rate due to changes in pumped discharge or when the aquifer is under an artesian pressure and the discharge from the borehole decreases as the artesian head falls. The importance of the recovery phase when the pumping ceases is also highlighted. Tables 5.1 and 5.4 summarise these alternative conditions.

For some of the aquifer conditions, analytical solutions are available; these analytical solutions are summarised in the text with special emphasis on the approximations inherent in the formulation. Derivations and the practical applications of analytical methods are described thoroughly in Kruseman and de Ridder (1990), Hall (1996) and Batu (1998). As an alternative, more flexible approach for the analysis of field problems, a numerical model for radial flow is presented. The numerical model can be used for all situations for which analytical solutions are available and also for many other field situations which are not consistent with the assumptions of the analytical methods. Tables 5.1 and 5.4 indicate whether analytical and/or numerical computational models are appropriate. For analytical approaches, curve-fitting techniques are generally used; automatic or visual curve-fitting techniques using computer software are becoming more widely available. For comparisons between field data and numerical model results for both the pumping and recovery phases, sensitivity analyses are generally adopted to obtain parameter values which provide the closest agreement for all observation locations and all phases of the test.

The fundamental issues involved in radial flow to pumped boreholes are presented in Chapter 5. In Chapter 6, large diameter wells are considered. Due to the dominant effect of well storage, alternative methods of analysis are required. Greater complexities in the flow mechanisms within the aquifer system also occur when vertical flows are represented explicitly; four alternative formulations are introduced in Chapter 7 with applications to practical situations described in Chapter 8.

6

Large Diameter Wells

6.1 INTRODUCTION

Large diameter dug wells are widely used, especially in developing countries; often they have a diameter of several metres. Before the availability of powered pumps, dug wells were used for centuries without deepening since lifting water more than a few metres was not practicable. However, once powered pumps became available, water could be pumped out from greater depths; consequently well water levels (and usually the aquifer water table) fell, requiring deepening of the wells.

Large diameter wells were drawn to the attention of the writer due to apparently anomalous results from an observation well about 15 m from a large diameter well. After pumping ceased, the water level in the pumped well began recovery but the water levels in the observation well continued to decline for several hours. This indicated that the large diameter well continued to draw water from the aquifer even though the pump was not operating. Subsequent work has led to a greater understanding of the advantages of large diameter wells.

A further important issue is the spacing between large diameter wells. When one farmer digs a well and obtains a plentiful supply of water, a neighbouring farmer often digs a similar well, possibly as close as 20 m from the original well. This will usually lead to a reduction in yield from the original well. The need for sufficient spacing between wells is illustrated by a case study of *agrowells* in Sri Lanka (Section 6.6.2).

Both analytical and numerical methods can be used to study the response of large diameter wells. As an alternative to closed form analytical solutions, an analytical technique using Kernel functions has proved to be valuable. With the Kernel function approach it is possible to include the effect of the seepage face which

Table 6.1 Large diameter well topics considered in Chapter 6

Condition	Analytical (A) and/or Numerical (N)	Section
Description of flow processes for large diameter wells	–	6.2
Three analytical methods for large diameter wells	A	6.3
Analysis of pumping tests using observation well data	N	6.4
Varying abstraction rates in large diameter wells	N	6.5
Three examples of agricultural use of large diameter wells	N	6.6

occurs between the water table and the well water level. Numerical methods are also appropriate when interpreting field results for large diameter wells. They allow the representation of the decreasing pumping rates with increasing drawdowns which often occur with farmers' pumps. Numerical methods are also of great value when analysing the operation of a large diameter well for an entire growing season. Table 6.1 contains a summary of topics considered in Chapter 6.

6.2 DESCRIPTION OF FLOW PROCESSES FOR LARGE DIAMETER WELLS

To understand the operation of large diameter wells it is necessary to distinguish between three quantities:

pumping rate: the rate at which water is pumped from the well, Q_P,

Groundwater Hydrology: Conceptual and Computational Models. K.R. Rushton
© 2003 John Wiley & Sons Ltd ISBN: 0-470-85004-3

well storage: the rate at which water is taken from storage within the well, Q_S,

aquifer flow: the rate at which water flows from the aquifer into the well, Q_a.

The discharge from the pump is supplied by water from well storage *plus* water from the aquifer:

$$Q_P = Q_S + Q_a$$

This equation also holds when the pump is not operating, then $Q_P = 0$ so that $Q_S = -Q_a$.

Well storage has a dominant impact on the response of large diameter wells. Figure 6.1 illustrates the various stages in a daily cycle. The quantity of water pumped from the well, the quantity of water taken from well storage and the quantity of water drawn from the aquifer, are plotted in Figure 6.2. Details of the well, aquifer parameters and pumping rate are given in Figure 6.2; note that the pumping rate is 500 m³/d for five hours (0.208 day). Important features of the six stages in Figures 6.1 and 6.2 are discussed below.

(i) At the start of the cycle, there is no pumping from the aquifer (time = 0.0 day); the water level in the well is at the same elevation as the water table with no residual drawdown.

(ii) Soon after pumping commences (0.03 day), most of the water discharged by the pump is taken from well storage with small flows from the aquifer. The

relative magnitudes of the flows can be identified in Figure 6.2.

(iii) Midway through the pumping phase (0.1 day) substantial quantities of water are still taken from well storage, but an increasing quantity of water is taken from the aquifer.

(iv) Immediately after the pump has stopped (0.208 day), due to differences between the groundwater head in the aquifer and the water level in the well, water continues to flow from the aquifer into the well. This water is used to refill well storage, as indicated by the negative signs of water from well storage for times greater than 0.208 day.

(v) Part way through recovery (0.3 day), significant inflows continue from the aquifer to the well; this leads to a rise in the well water-level as well storage is refilled.

(vi) Towards the end of recovery (0.9 day), there is a small inflow from the aquifer to the well due to the much smaller difference between the aquifer water table and the well water level. Recovery is not totally complete, since at the end of the day there is a small residual drawdown.

If a challenge is presented to devise a pumping methodology so that most of the water is drawn from an aquifer when the pump in the well is not operating, the likely response is that this is not feasible. Yet this is precisely how a large diameter well operates. As indicated

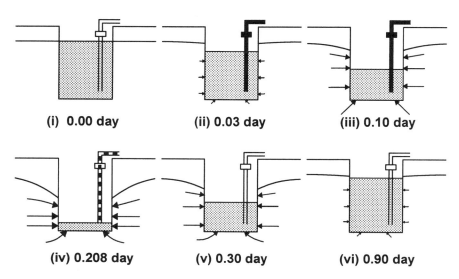

(i) 0.00 day   **(ii) 0.03 day**   **(iii) 0.10 day**

(iv) 0.208 day   **(v) 0.30 day**   **(vi) 0.90 day**

Figure 6.1   Stages in the daily cycle of operation of a large diameter well

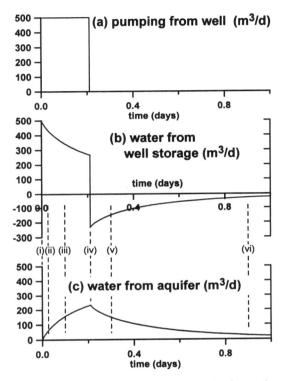

well diameter = 7.0 m
initial sat. depth = 8.0 m
K = 7.5 m/d, S = 0.025
pumping rate = 500 m³/d
for 5 hrs (0.208 d)

(a) pumping from well (m³/d)

(b) water from well storage (m³/d)

(c) water from aquifer (m³/d)

Figure 6.2 Flows associated with the daily cycle of operation of a large diameter well, vertical lines (i) to (vi) correspond to the diagrams in Figure 6.1

by Figure 6.2c, the quantity of water drawn from the aquifer while the pump is operating (the area under the curve to the left of 0.208 day) is only one-third of the total quantity drawn from the aquifer.

6.3 ANALYTICAL SOLUTIONS FOR LARGE DIAMETER WELLS

6.3.1 Conventional analyses

Analytical solutions for drawdowns in or around a large diameter well have been developed by Papadopulos and Cooper (1967) and Papadopulos (1967). The analysis allows for the possibility of different diameters

for the large diameter well and for the bore into the aquifer. Consequently dug-cum-bore wells can be analysed; a dug-cum-bore well is sketched in Figure 6.8. Numerical values of the drawdown for large diameter wells derived from the equations of Papadopulos and Cooper (1967) can be printed from a program prepared by Hall (1996).

Due to the dominant influence of well storage during the initial stages of pumping from a large diameter well, it is only possible to use the data for the later stages of the pumping phase to estimate aquifer parameters. This is demonstrated in Figure 6.3 where the results of the Papadopulos and Cooper analysis are presented on a double logarithmic plot but with the x-axis equal to $4Tt/r_w^2$ rather than the conventional approach of using $1/u = 4Tt/r_w^2 S$. The choice of this alternative x-axis demonstrates the small differences in the early stages of pumping between curves for storage coefficients ranging from 0.00001 to 0.1.

6.3.2 Drawdown ratio method

Due to the difficulty in identifying the appropriate storage coefficient, an alternative approach has been devised to examine deviations from the initial linear time-drawdown response due to the dominant effect of well storage. This can be achieved by examining ratios of drawdowns. Considering a typical example, during the early stages of pumping the well drawdown at 0.01 day divided by the drawdown at 0.004 day equals 2.5 if all the water is taken from well storage. At times of 0.1 day and 0.04 day the ratio of drawdowns is less than 2.5. This occurs because at 0.1 day more water is taken from the aquifer and less from well storage resulting in a reduced rate of fall in well water level. The drawdown in well water level at 1.0 day is greater than that at 0.4 day, but if most of the abstracted water is taken from the aquifer, the ratio of drawdowns at these two times may be less than 1.5. If all the water is taken from the aquifer without any increase in drawdown, the drawdown ratio would fall to 1.0.

These ratios of well drawdowns are written as $s_t/s_{0.4t}$; when plotted against $4Tt/r_w^2$ they form a series of curves as shown in Figure 6.4. A comparison with the double logarithmic plot of Figure 6.3 (note that in Figure 6.3, the timescale $4Tt/r_w^2$ extends from 0.01 to 10 000 whereas in Figure 6.4 the range is from 0.01 to 100) shows that the use of the well drawdown ratio $s_t/s_{0.4t}$ leads to curves of a more distinctive shape with the differences between the curves becoming apparent for a non-dimensional time of $4Tt/r_w^2 = 0.01$ compared to $4Tt/r_w^2 = 0.3$ for the double logarithmic plot.

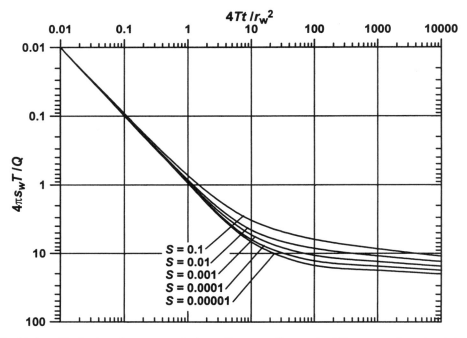

Figure 6.3   Double logarithmic plot of type curves for large diameter well; note the different non-dimensional time axis

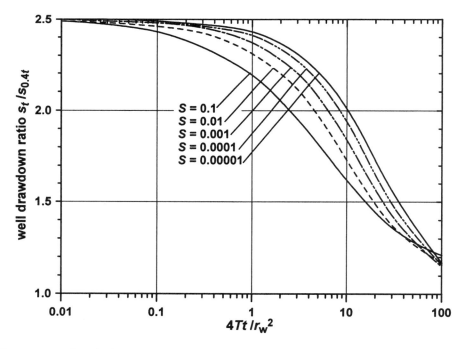

Figure 6.4   Type curves replotted using well-drawdown ratio (Rushton and Singh 1983). Reprinted from *Ground Water* with permission of the National Ground Water Association. Copyright 1983

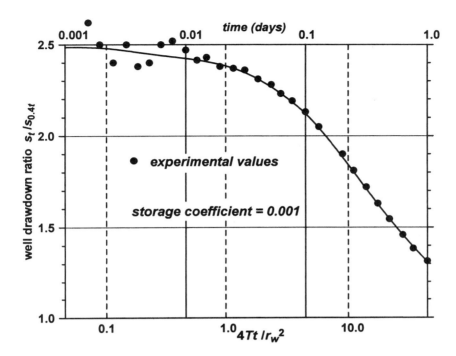

Figure 6.5 Practical example of use of well-drawdown ratio (Rushton and Singh 1983). Reprinted from *Ground Water* with permission of the National Ground Water Association. Copyright 1983

An example of the application of the method is presented in Figure 6.5 (Rushton and Singh 1983); the solid circles represent values of the ratio $s_t/s_{0.4t}$. For times of 0.01 day (14.4 mins) or less, the well water level drawdowns are small so that accurate values of the well drawdown ratio $s_t/s_{0.4t}$ cannot be obtained, hence the scatter in the discrete values. At longer times the results show a smooth change. Attempts were made to match the data against the curves for different storage coefficients of Figure 6.4; the best match occurs with $S = 0.001$ as shown in Figure 6.5. When matching the curves, only horizontal movement is necessary; this is more straightforward than matching with a double logarithmic plot where horizontal and vertical movement is required. For a non-dimensional time of $4Tt/r_w^2 = 1.0$, the corresponding actual time is 0.021 day; this allows a direct calculation of the transmissivity for the well of radius 3.0 m,

$$T = \frac{r_w^2}{4t} = \frac{3.0 \times 3.0}{4.0 \times 0.021} = 102\,\mathrm{m^2/d}$$

6.3.3 Alternative methods including Kernel function techniques

Mace (1999) reviews a large number of analytical and numerical methods for estimating the hydraulic conductivity for large diameter wells. In particular he illustrates how slug test interpretation methods can be used to analyse the recovery phase. An alternative analytical approach, using Kernel functions, has been developed by Patel and Mishra (1983); the technique is extended to include the recovery phase by Mishra and Chachadi (1985). The basic approach of the discrete Kernel function method is that time is divided into a number of discrete time steps. As shown in Figure 6.6, the rate at which water is drawn from the aquifer varies for each time step. This can be represented with the abstraction commencing at the start of the step but with an equal but opposite abstraction starting at the end of the step; Figure 6.7b. This technique is similar to the Theis recovery method (Section 5.6) where the recovery phase is represented by a negative abstraction which commences when the abstraction ceases.

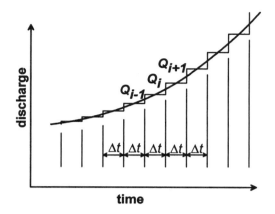

Figure 6.6 Representation of changing discharge rate as a series of steps for Kernel function approach (Rushton and Singh 1987). Reprinted from *Ground Water* with permission of the National Ground Water Association. Copyright 1987

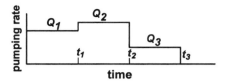

(a) required stepped abstraction rate

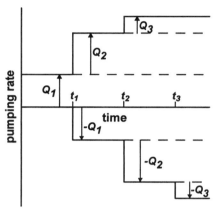

(b) representing stepped abstractions using Theis well function with +ve and -ve abstraction rates

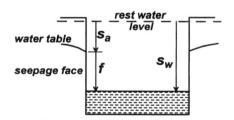

(c) water level drawdown, aquifer drawdown and seepage face

Figure 6.7 Representing stepped discharge rates as positive and negative discharges and inclusion of seepage face (Rushton and Singh 1987). Reprinted from *Ground Water* with permission of the National Ground Water Association. Copyright 1987

Figure 6.7a shows three abstraction increments of Q_1, Q_2 and Q_3 covering time increments 0 to t_1, t_1 to t_2 and t_2 to t_3. As shown in Figure 6.7b this is represented as three positive and three negative abstraction rates which continue to infinity. Consequently the drawdown at radius r is determined from the following equation,

$$4\pi T_s = Q_1\langle W[r^2 S/4Tt] - W[r^2 S/4T(t - t_1)]\rangle$$
$$+ Q_2\langle W[r^2 S/4T(t - t_1)] - W[r^2 S/4T(t - t_2)]\rangle$$
$$+ Q_2\langle W[r^2 S/4T(t - t_2)] - W[r^2 S/4T(t - t_3)]\rangle$$
$$+ \cdots \tag{6.1}$$

in which $W[\ ]$ is the well function of Eq. (5.7) and $\langle\ \rangle$ signifies that the term is only included when the expression in square brackets [ ] is positive.

The next step is a *well water balance* in which the proportion of water drawn from the aquifer can be estimated. During a typical discrete time step, the well drawdown increases by Δs_w. This increase in well drawdown releases water from well storage and also draws water from the aquifer; these two flow components must equal the quantity of water pumped from the well during the time step (this will be zero during the recovery phase). The quantity of water supplied by well storage can be calculated directly from the cross-sectional area of the well and the drawdown increment. Hence for the kth time step, the effective discharge taken from well storage $Q_{s,k}$ can be deduced from the equation

$$Q_{s,k} = \Delta s_w \pi r_w^2 / \Delta t \tag{6.2}$$

where Δs_w is the drawdown increment in the well during the kth time step of Δt.

Consequently, the drawdown in the well at the end of the kth time step is given by the equation

$$s_w(r_w, k\Delta t) = s_w(r_w, (k-1)\Delta t) + Q_{s,k}\Delta t / \pi r_w^2 \tag{6.3}$$

The value of $s_w(r_w, (k-1)\Delta t)$ is known from the previous time step; $s_w(r_w, k\Delta t)$ can be calculated from a more general form of Eq. (6.1) for the drawdown in the aquifer at the well face.

$$s_a(r_w, k\Delta t) = \left\{ \sum_{i=1}^{k-1} Q_{a,i}(W[k-i+1] - W[k-i]) \right.$$
$$\left. + Q_{a,k}W[r_w^2 S/4T\Delta t] \right\} \Big/ 4\pi T \qquad (6.4)$$

In this equation $W[k-i+1]$ is a shorthand form of $W[r_w^2 S/4T(k-i+1)\Delta t]$.

Provided that there is no seepage face,

$$s_w(r_w, k\Delta t) = s_a(r_w, k\Delta t) \qquad (6.5)$$

A second equation relates the pump discharge to the rate at which water is taken from well storage plus the withdrawal from the aquifer:

$$Q_P = Q_S + Q_a \qquad (6.6)$$

When Eqs (6.3) and (6.4) are substituted in Eq. (6.5) they form, with Eq. (6.6), two simultaneous equations in Q_S and Q_a. A program in BASIC, which solves these equations, calculates drawdowns in the pumped well and at other radial distances from the pumped well, is included in Rushton and Singh (1987).

In a further developments of the Kernel function method by Rushton and Singh (1987) a seepage face is incorporated in the analysis. The drawdown of the well water level equals the aquifer drawdown at the well face plus the seepage face; Figure 6.7c. Therefore, Eq. (6.5) is modified to become

$$s_w(r_w, k\Delta t) = s_a(r_w, k\Delta t) + f(r_w, k\Delta t) \qquad (6.7)$$

where $f(r_w, k\Delta t)$ is the height of the seepage face. In their analysis Rushton and Singh (1987) defined the seepage face height as a function of the discharge rate; the appropriate coefficients are estimated from field measurements. Comparisons of simulations of a suitably monitored field test, with and without an allowance for the seepage face, showed that analyses which ignore the seepage face are likely to underestimate the transmissivity by about 25 per cent while the storage coefficient is only one-fifth of the correct value.

With all these methods of analysis it is assumed that the saturated depth remains constant. However, since many large diameter wells are in unconfined aquifers

with low transmissivity and specific yield, a decrease in saturated depth is likely to occur. Numerical models should be used when the decrease in saturated depth is significant.

6.4 NUMERICAL ANALYSIS OF LARGE DIAMETER WELL TESTS USING OBSERVATION WELL DATA

Radial flow numerical models can be used to analyse a wide variety of pumping tests in large diameter wells. The basic numerical model for radial flow towards a well is described in Section 5.5. The large radius of the well with the appropriate well storage is directly included in the standard numerical model together with variations in the saturated depth. The seepage face can be represented as a modified hydraulic conductivity at the well face (Section 5.5.4). In this discussion, the numerical model is used to analyse pumping tests where pumping and recovery data are available in both the pumped well and in an observation well.

The methodology is introduced with reference to a test in a dug-cum-bore well; details can be found in Figure 6.8. For the dug-cum-bore well, the minimum radius in the numerical model is that of the bore; an artificial well storage coefficient is introduced to represent the well storage of the overlying dug well. The effective storage is equal to the ratio of the square of the dug well radius to the square of the bore well radius; in this example the ratio is 494.

A careful examination of the field data shows that drawdowns in the large diameter well are small in the early stages of pumping. If well storage effects were not significant, the *log-arithmetic* plot of Figure 6.8a would be a straight line. Of even greater interest is the recovery curve for the observation well. When the pump is switched off, drawdowns continue to increase reaching a maximum at *about three hours after* the pumping stopped. This occurs because water continues to be drawn into the well to refill well storage even though pumping from the well ceases.

Initial attempts were made to match the field results for the pumping phase to the Papadopulos and Cooper (1967) type curves. It was difficult to be sure of the best fit to the type curves; the transmissivity was estimated to be in the range 25–100 m^2/d, no reliable estimate of the aquifer storage coefficient could be made.

When the radial flow numerical model was used, after a number of trials, the match shown by the unbroken lines of Figure 6.8 was obtained. In the early stages of attempting to match field and modelled results, the recovery phase of the observation well was

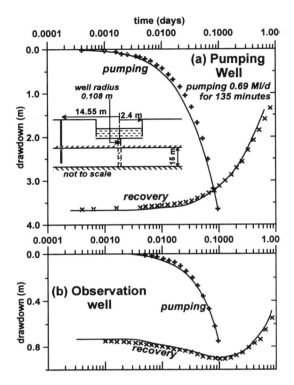

Figure 6.8 Details of a test in a dug-cum-bore well, field values indicated by discrete symbols, numerical model results by continuous lines (Rushton and Holt 1981). Reprinted from *Ground Water* with permission of the National Ground Water Association. Copyright 1981

used as the main test of the adequacy of the numerical model. Note that in Figure 6.8 the match for all four curves is obtained with a single set of aquifer parameters. From a sensitivity analysis (Rushton and Holt 1981), the range of parameters which give an adequate match for pumping and recovery phases in the large diameter well and in the observation well are

$$\text{transmissivity} = 26 \pm 2\,\text{m}^2/\text{d};$$
$$\text{storage coefficient} = 0.0008 \pm 0.0002$$

The transmissivity of $26\,\text{m}^2/\text{d}$ is consistent with the low permeability of certain hard rock aquifers and the abstraction rate of $691\,\text{m}^3/\text{d}$ for 135 minutes (equivalent to almost $65\,\text{m}^3$) is a good yield from such a low transmissivity aquifer.

When no observation well data are available, the pumped well data for both the pumping and recovery

phases should be used. In this alternative approach, the pumped well drawdowns can also be used to estimate the quantity of water withdrawn from well storage during a specified time. When the water withdrawn from well storage is subtracted from the pumping rate (which is zero during recovery) an estimate can be made of the varying rate at which water is taken from the aquifer. With these two items of information, the changing water level in the large diameter well and the rate at which water is withdrawn from the aquifer, comparisons can be made with *two* outputs from the numerical model. A field example is described by de Silva and Rushton (1996).

6.5 ANALYSIS WITH VARYING ABSTRACTION RATES

The issue of varying abstraction rates from pumped boreholes is considered in Section 5.13; for large diameter wells, when a farmer's pump is used, the impact of the varying abstraction rate becomes even more significant. Consideration of the basic theory of centrifugal pumps highlights why there is a decreasing pump discharge with increasing drawdown.

For a centrifugal pump, assuming that there are no losses, the discharge Q can be calculated from

$$Q = C_{\text{pump}}\sqrt{(u^2 - 2gH)} \tag{6.8}$$

where C_{pump} is a constant related to the design of the pump

u is the velocity at the tip of the vane and

H is the total lift, i.e. the vertical distance between inlet water level and outlet.

Figure 6.9 illustrates how the pump characteristic of Eq. (6.8) is applied to a large diameter well. The graph with axes as broken lines and values and axis titles in italics represents the characteristic curve. The coefficients are arranged so that when there is no lift of water, the discharge is $1000\,\text{m}^3/\text{d}$ but when the lift is $H = 10.0\,\text{m}$, all the energy is used in holding up the column of water so that there is no discharge.

The two small diagrams (a) and (b) of Figure 6.9 show two extreme conditions in a large diameter well. In Figure 6.9a, the pump has just started with the well water level identical to the rest water level (*RWL*), the lift is $2.0\,\text{m}$ and the discharge equals $894\,\text{m}^3/\text{d}$. A new axis is drawn on the right, indicating that the zero drawdown condition corresponds to a lift of $2.0\,\text{m}$; the solid line superimposed on the broken line of the pump characteristic curve represents the pumped discharge-

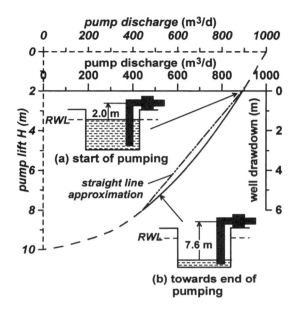

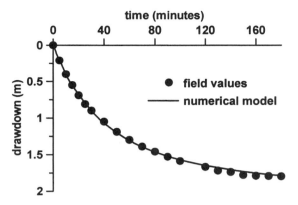

Figure 6.10 Analysis of a field test with decreasing abstraction rates

Figure 6.9 Diagram showing how the pump characteristic of a centrifugal pump is converted to a well drawdown-discharge curve

drawdown curve for the well. Diagram (b) represents conditions towards the end of the pumping phase when the water level in the well approaches the bottom of the pump suction pipe. The drawdown is 5.6 m, the lift $H = 7.6$ m, hence there is a reduced discharge of 489 m³/d. This analysis ignores pump and pipe losses. In practical situations the discharge with a lift of 7.6 m will be lower due to losses.

Field readings (Rushton and Singh 1983) of discharge for different well drawdowns support the form of curve shown in Figure 6.9. Reliable data are difficult to obtain, therefore an approximation of a straight line relationship can be made over the range of operation of the pump in the large diameter well.

This form of relationship was used to analyse an actual test in a weathered Deccan Trap aquifer in India. The initial abstraction rate of 1360 m³/d fell to 737 m³/d for a drawdown of 1.6 m. This falling abstraction rate can be represented by the linear relationship,

$$Q = 1360(1 - 0.286 s_w) \, \text{m}^3/\text{d} \qquad (6.9)$$

where s_w is the drawdown in the well.

Figure 6.10 shows the match between field data and a radial flow numerical model simulation using the above linear decrease in abstraction rate. The well

diameter is 5.0 m; the match was obtained with a transmissivity of 230 m²/d and a storage coefficient of 0.0025. The agreement between field data and model results is encouraging.

Rather than attempting to identify the reduction in abstraction with pumped drawdown, it is possible to develop a field technique of ensuring that the quantity of water taken from the well remains constant even though the pump discharge decreases. This can be achieved by returning some of the pumped water back into the well and ensuring, by means of a measuring device such as a weir tank, that the actual quantity of water leaving the well remains constant. Further details can be found in Athavale *et al.* (1983). This approach can also be used to reduce the effective withdrawal rate from the well, thereby ensuring that the test lasts for a longer period. The increased number of readings over a longer time period makes the analysis of the test more reliable.

6.6 USE OF LARGE DIAMETER WELLS FOR AGRICULTURE

The information and techniques described in the preceding sections of this chapter can lead to estimates of the transmissivity, storage coefficient, well loss and the nature of the seepage face associated with large diameter wells. With this information, the long-term operation of a well and aquifer can be examined when the well discharge is used to irrigate crops.

A representative example is introduced below to illustrate the methodology of analysing the response of a well during a complete growing season. The findings are used to explain successes and failures of Agrowells in Sri Lanka. A second example considers the reduction in yield during the growing season of large diameter wells in a Miliolite limestone aquifer.

6.6.1 Representative problem

Details of the representative problem are as follows.

- During and immediately following the rainy season, a crop of rice is grown which is partly rain-fed but supported by surface water irrigation from reservoirs.
- When the rainy season crop is harvested, there is sufficient water stored in the shallow aquifer (largely due to the water losses through the bunds of the ricefields, as explained in Section 4.5) for a mainly or totally groundwater-irrigated crop. In this analysis it will be assumed that there is no rainfall during the growth of this second crop.
- All the farmers would like to grow an irrigated crop, but experience shows that there is insufficient water; instead it is necessary to limit the number of wells.
- For the purpose of this study, the following information is available. Properties of the aquifer are that the horizontal hydraulic conductivity is 6.0 m/d with an initial saturated depth of 8.0 m, the specific yield is 0.025. A seepage face will form which depends on the flow from the aquifer into the well; the seepage face height in metres is defined as $0.0067Q_a$. The maximum permissible pumped drawdown is 7.2 m, which is equivalent to 0.8 m of water remaining in the bottom of the well for pump suction.
- The questions are to decide on the radius of the large diameter well, the spacing between wells and the pumping rate over a growing season of 100 days, assuming that the same volume of water is required each day.

A trial and error method is used with the findings summarised in Table 6.2. The first stage is to decide upon a well spacing. Figure 6.11 shows an idealised arrangement of large diameter wells; in practice, the distribution of the wells may be less regularly spaced yet the basic approach can still be used. If the spacing between wells is d, the area associated with an individual well is shown shaded; the square array of chain-dotted lines indicates the location of no-flow boundaries when the same discharge occurs from each well. The cross-section in Figure 6.12 illustrates the interference between wells. Rather than using a technique such as image wells to represent these boundaries, the analysis is carried out using the radial flow model (Section 5.5) with an equivalent outer no-flow boundary at radius R (see Figure 6.11) such that the circular area is equal to the shaded area associated with a single well. The effective outer radius for the radial flow model,

$$R = d/\sqrt{\pi} \tag{6.10}$$

Three stages are followed in an exploration of the number, spacing and discharge of the large diameter wells.

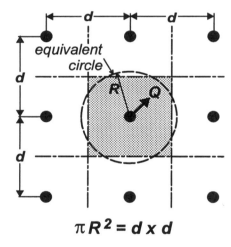

Figure 6.11 Idealised well spacing showing the equivalent radius R

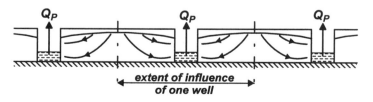

Figure 6.12 Cross-section through wells showing interference between wells

Table 6.2 Summary of alternative simulations for operating large diameter wells during a 100-day growing season; *d* is the regular well spacing, *R* is the equivalent outer radius in the numerical model, $s_{w,f}$ is the drawdown in the well at the end of the first pumping phase

Run No.	r_w m	Q_P m³/hr	time hour	d m	R m	$s_{w,f}$ m	Fail day	Comment
A(i)	1.5	20.0	2.5	100	56.4	4.3	11	Large initial d/down, increase r_w
A(ii)	3.0	20.0	2.5	100	56.4	1.4	27	All d/downs large, so reduce Q_P
A(iii)	3.0	10.0	2.5	100	56.4	0.7	63	Resource too small, larger spacing
B(i)	3.0	20.0	5.0	300	169.3	2.6	56	Large initial d/down, increase r_w
B(ii)	6.0	20.0	5.0	300	169.3	0.7	86	Well w/l fast fall, smaller spacing
C(i)	3.0	12.5	5.0	250	141.1	1.6	95	Just fails, small increase in r_w
C(ii)	3.5	12.5	5.0	250	141.1	1.3	100	OK, final well d/down 6.96 m

A Each farmer has a well: typical areas of small farms are one hectare (100 m by 100 m), therefore if each farmer has a well the equivalent outer radius R = 56.4 m. Assuming that the irrigation requirement is 5 mm/day, the quantity required to irrigate the whole plot is $0.005 \times 10\,000 = 50.0$ m³ each day; this discharge can be achieved by pumping at 20 m³/hr (a typical rating for a suction pump) for 2.5 hours. This pumping rate is used for *Run A(i)*, the well radius is chosen to be 1.5 m. Using the numerical model described in Section 5.5, with no recharge and the representation of a variable saturated depth, the drawdown in the well at the end of the *first* pumping phase of 2.5 hours is $s_{w,f} = 4.3$ m. This drawdown, of more than 50 per cent of the initial depth of water column in the well, is a cause for concern. By day 11 the drawdown reaches 7.2 m (90 per cent of the original water column); this is taken as failure of the well. The results for this and subsequent simulations are summarised in Table 6.2.

For the second trial, *Run A(ii)*, the well radius is doubled to 3 m in an attempt to reduce the drawdowns in the pumped well; this leads to a drawdown at the end of the first phase of 1.4 m. However, the recovery between each pumping phases is far from complete, with the result that the maximum permissible drawdown is reached during day 27. To ascertain whether a sustainable pumping regime can be devised with the wells positioned on a square grid of sides 100 m, the pumping rate from the well is halved to 10 m³/hr, *Run A(iii)*. The drawdown of the well water level at the end of the first phase of 0.7 m is promising, but failure still occurs at 63 days. This suggests that the interference between wells leads to a shortage of water at longer times, hence a wider spacing between wells is required.

B Increase the spacing between the wells: for *Runs*

B(i) and *B(ii)*, the spacing between the wells is increased to 300 m, the equivalent outer impermeable radial boundary is at 169.3 m from the pumped well. In *Run B(i)* the well radius remains at 3.0 m, and the pumping rate is set back to the original value of 20.0 m³/hr but the duration is increased to 5.0 hours. As indicated in Table 6.2, the drawdown at the end of the first day is 2.6 m and failure occurs during day 56. Since the drawdown at the end of the first pumping phase is high at 2.6 m, this is corrected by increasing the well radius to 6.0 m in *Run B(ii)*, leaving the other parameters unchanged. The drawdown at the end of the first phase is reduced to 0.7 m and failure occurs during day 86, indicating that the simulation is approaching a satisfactory selection of well spacing, well radius and pumping regime. Therefore for this example, a more detailed study is made of the changing drawdowns.

Time-drawdown plots for the water level in the pumped well and for the drawdown at the boundary between adjacent wells, are presented in Figure 6.13a. The solid circles show that the pumped well drawdowns increase more rapidly after 30 days; in fact the daily increase in drawdown at 85 days is double that at 35 days. The reason for this increased rate of drawdown is apparent from Figure 6.13b which represents a radial section through the aquifer and well from the outer no-flow boundary at 169.3 m to the well face at a radius of 6.0 m and within the well. A logarithmic scale is used in the radial direction. At the end of the second pumping phase, i.e. at 1.208 day, the drawdowns are small, as shown by the broken line; at the well face there is a seepage face. However at the end of the pumping phase for day 85, large drawdowns have occurred, leading to a substantial reduction in the saturated depth. This decrease in saturated depth is also illustrated in Figure 6.12.

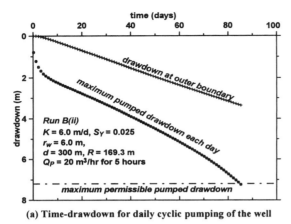

(a) Time-drawdown for daily cyclic pumping of the well

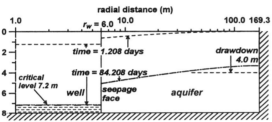

(b) Radial cross-section showing well water level and the water table elevation towards the beginning (broken line) and towards end (dash-dotted line) of irrigation period

Figure 6.13 Study of long-term response of large diameter well for run No. B (ii)

Conflicting responses occur when attempting to withdraw substantial quantities of water from a shallow unconfined aquifer. Large drawdowns are required to release water from storage at the water table. On the other hand, a reasonable saturated thickness is required, especially close to the pumped well, to move water through the aquifer into the well. With the spacing of 100 m, the interference between the wells was too great, whereas with the 300 m spacing it is difficult to maintain an adequate saturated thickness close to the well. This suggests that a well spacing between 100 m and 300 m is advisable; for the next trial a spacing of 250 m is adopted.

C Aim to abstract 50 per cent of the stored water: thus far there has been no clear basis for the selection of the pumping rate. However, from Figure 6.13b, it does appear that a possible target is to pump out 50 per cent of the water stored in the aquifer; the appropriate drawdown target at the outer boundary is indicated by the 4.0 m arrow on the right hand boundary. This is not

an easy target to achieve. Note must be taken of the distorting effect of using a logarithmic radial scale; the area beyond 100 m accounts for 65 per cent of the area from which release of water from storage is necessary. For *Run C(i)*, the well spacing is reduced to 250 m and the abstraction rate is calculated as 50 per cent of the total stored water distributed over 100 days. Hence the daily discharge equals

$$0.5 \times \text{volume of aquifer} \times \text{specific yield} \div 100 \text{ days}$$
$$= 0.5 \times 8 \times 250 \times 250 \times 0.025 / 100 = 62.5 \text{m}^3$$

Therefore, the pumping rate is set at 12.5 m³/hr for five hours. The drawdown at the end of the first phase is 1.6 m and failure occurs during day 95. Only a small change is needed to achieve the 100 days of pumping. For *Run C(ii)* with the well radius increased to 3.5 m, the pumped drawdown at 100 days of 6.96 m is less than the maximum permissible value of 7.2 m.

There is no justification in further refinement of these simulations; the arrangement of *Run C(ii)* is one feasible solution. An arrangement with a well spacing of 200 m and an abstraction planned to remove 50 per cent of the stored water would be equally valid. But what determines the quantity of water that can be abstracted safely by the farmers? A simple calculation indicates how much water is available. If the average depth of water table lowering is 4.0 m in 100 days, this is equivalent to 0.04 m per day. With a specific yield of 0.025, the quantity available is 0.04 × 0.025 = 0.001 m/d or 1.0 mm/d. For a crop water requirement of 5.0 mm/d, this means that only 20 per cent of the area can be irrigated. If a higher cropping intensity is attempted, failure is certain to occur.

6.6.2 Agrowells in Sri Lanka

The representative example described in Section 6.6.1 has direct relevance to the successful development of large diameter wells for dry-season irrigation in Sri Lanka (see front cover). During the dry season, water supplies from irrigation reservoirs are unreliable and do not provide sufficient flexibility for the farmer to grow a second crop. However, under the auspices of the Agricultural Development Authority, large diameter wells have been constructed and operated effectively. The key to the successful operation can be gained from the following information.

- The wells are constructed in the weathered zone; the depth to the underlying hard rock is usually about 7.5 m.

- None of the wells are constructed on high ground.
- During the rainy season the fields are used for a flooded rice crop, this leads to significant recovery of groundwater levels due to recharge from the ricefields.
- Before construction commences, a trial pit at least 7.5 m deep is excavated; the column of water in the pit should be at least 5 m.
- During construction, the sides of the excavation stand unsupported; the well is lined with brickwork, often the mortar between the bricks is deliberately missing.
- The wells penetrate to the underlying hard rock and often extended a short distance into the hard rock.
- Diameters of the wells are in the range 5–7.5 m.
- Water from the wells is used carefully; the pumps have a yield of 440 m³/d (18.3 m³/hr), and operation depends on the crop water requirements. Typically the pump operates for three to six hours in a day. Towards the end of the irrigation season the farmer might find that there is still water in his well; this allows him to grow a late season cash crop covering a small area.
- Finance was only provided for a limited number of farmers to construct wells; typically one in five farmers were chosen to construct an agrowell. Consequently most farmers, who had a plot of about 1 ha, had neighbours without a large diameter well. Therefore the effective well spacing was about 200 m with only about 20 per cent of the total area irrigated during the dry season; the remaining plots were barren (see front cover).

The situation in Sri Lanka is similar to *Run C(ii)*; therefore the system worked satisfactorily, at least for the farmers who had wells. Some of the farmers without wells managed to obtain loans to construct their own wells; this led to a reduced spacing between wells. They aimed to grow crops over the whole of their plots but, as indicated by *Runs A(ii)* and *(iii)*, interference between the wells meant that the wells ran dry before the end of the growing season. This case study reinforces the argument that the first step in planning irrigation using large diameter wells is to estimate the total volume of water available (50 per cent of the stored water is a realistic target for a shallow aquifer). This defines the area that can be irrigated. With shallow aquifers which are refilled each year, it is unlikely that the cropping area can exceed 30 per cent. The yield of the wells is not a basis for planning the irrigation; the well yields often exceed the long-term available resource.

6.6.3 Case study of a Miliolite limestone aquifer

A Miliolite limestone aquifer in Western India has supported domestic and limited agricultural demands for many centuries; pumping rates from large diameter wells of 3000 m³/d can be maintained under high groundwater levels. However, there is a serious problem with maintaining even the smallest yield during dry summers; this has serious consequences when the well is used to irrigate a summer crop (Rushton and Raghava Rao 1988).

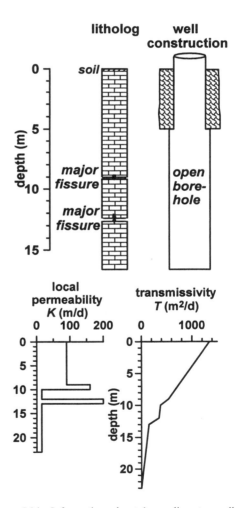

Figure 6.14 Information about large diameter wells in a Miliolite Limestone aquifer in western India. Reproduced by permission of IAHS Press from Rushton and Raghava Rao (1988)

To investigate the aquifer properties, a borehole of 0.45 m diameter was drilled into the Miliolite limestone aquifer. When a pumping test was carried out under high water levels, the pumping rate of 5200 m³/d led to a well drawdown of only 0.13 m after 4 hours pumping. Information gained during the drilling includes the identification of two major fissures at 9 and 13 m below ground level; Figure 6.14. A diagram was prepared showing the approximate distribution of the local permeability and hence the transmissivity variation with saturated depth.

When the water table is above the upper fissure, the transmissivity is greater than 500 m²/d. With a transmissivity of this magnitude there is no difficulty in drawing water into the large diameter well. However, when the water table is below the lower of the two fissures, the transmissivity is 150 m²/d or less. With this reduced transmissivity, the yield of the well falls off rapidly. Drilling the well deeper is of no advantage since the well would extend into underlying low permeability beds.

6.7 CONCLUDING REMARKS

The practical advantages of using large diameter wells are demonstrated in this chapter. Due to their characteristic of drawing water from the aquifer into the well even when the pump has ceased operating, large diameter wells are especially useful for the supply of irrigation water. Furthermore, they operate successfully in low transmissivity aquifers.

Interpretation of pumping tests in large diameter wells is difficult due to the dominant effect of well storage. A curve-fitting technique based on an alternative way of plotting the field results provides more reliable values of the aquifer parameters. The application of Kernel function techniques for both the pumping and recovery phases is also described. The flexibility of the radial flow numerical model in representing actual aquifer behaviour is demonstrated by analysing data for the pumping and recovery phases in both observation and pumping wells. When farmers' pumps are used for a pumping test, the assumption of a constant pumped discharge is not valid. Decreasing pumping rates with falling pumping levels are included directly in the numerical model. This permits the analysis of a test in which a substantial fall occurs in the pump discharge.

Once the pumping and recovery phases of a pumping test have been represented successfully by a numerical model, the model can then be used to simulate pumping for longer time periods. A specific example considers the selection of well diameter, pumping rate and spacing between wells in a shallow aquifer which is used to irrigate a crop for a period of 100 days. If the well diameter is too small, the pumping rate is too high or the wells are positioned too close to each other, the well is pumped dry before the end of the irrigation period. The target of withdrawing half of the water stored in the aquifer is an ambitious target which can be achieved by the judicious choice of well diameter, well discharge and well spacing. However, this does not allow farmers to grow crops over the whole of their land; instead, the cropping area must be restricted with small pumped drawdowns during the early weeks of irrigation.

The investigation described above explains the initial success of a large diameter well development scheme in Sri Lanka in which only 20 per cent of the farmers were provided with finance to construct and operate wells. The numerical model also indicated the risk of failure of the scheme if more wells are constructed. The fact that the aquifer can only support an irrigated cropping intensity of 20 per cent, even though the wells can abstract more water, must be recognised and respected by all those planning schemes and providing finance.

7

Radial Flow where Vertical Components of Flow are Significant

7.1 INTRODUCTION

The wide use of the Theis equation has encouraged the view that for pumped wells or boreholes, horizontal flow is the only important component. However, in practical situations vertical flows are significant both for single aquifer systems and for multi-layered aquifer systems. From Figure 7.1 it is clear that vertical flow components are important in:

- Figure 7.1e: a confined aquifer with a partially penetrating borehole,
- Figure 7.1f: a leaky aquifer with storage in the aquitard which means that higher vertical flows occur towards the bottom of the aquitard compared to the flows towards the top of the aquitard,
- Figure 7.1g: an unconfined aquifer in which the flows originate from the water table moving horizontally and vertically through the aquifer; a well face where the flow varies with depth and a seepage face where velocities are reduced, and
- Figure 7.1h: a layered aquifer with a large diameter well in the upper weathered zone and a borewell in the underlying fractured zone.

These examples are typical of practical situations where the correct representation of vertical flow components in the aquifer system is vital.

When vertical flows are significant, conceptual models must be developed which include all the important flow components. Once identification of the essential flow processes has been completed, suitable analytical or numerical methods of solution can be selected. A further aspect of conceptual model development is an assessment of whether the study involves the aquifer response for a relatively short time period while a pumping test is conducted, or a longer time period when water is drawn from the more distant parts of the aquifer system.

All groundwater problems are three-dimensional in space and are time-dependent. However, in many situations it is appropriate to limit the number of dimensions included in the analysis. In Chapters 5 and 6, problems are idealised as involving $[r, t]$ (radial and time) co-ordinates. In this chapter, four alternative approximations are considered.

1. Analyses using the full $[r, z, t]$ (radial, vertical, time) co-ordinates; reference is made to both analytical and numerical methods of solution.
2. There are occasions when it is acceptable (and pragmatic) to use time-instant solutions in terms of $[r, z]$ co-ordinates; the time-instant approach can be used when there is an approximate equilibrium between inflows and outflows so that changes in drawdown with time are small.
3. A third approach to radial-vertical flow problems involves the representation of horizontal flows in individual zones using two or more $[r, t]$ solutions with vertical flows interconnecting these zones; this approach is described as a two-zone analysis $[r, t, v_z]$.
4. The final idealisation is concerned with leaky aquifers in which allowance is made for storage in the aquitard; this involves two interconnected solutions, one of radial flow in the main aquifer and the second of vertical flows in the aquitard. This approach is denoted by $[r, t : z, t]$.

Groundwater Hydrology: Conceptual and Computational Models.   K.R. Rushton
© 2003 John Wiley & Sons Ltd   ISBN: 0-470-85004-3

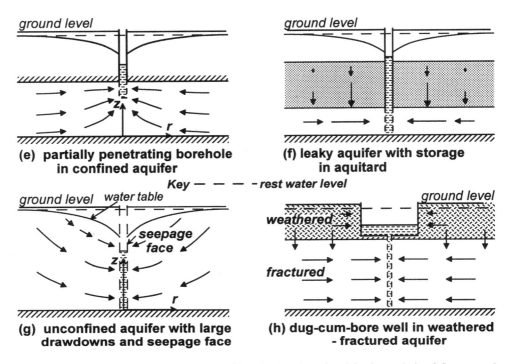

Figure 7.1   Repeat of part of Figure 5.7, a diagram of idealisations introduced in the analysis of flow towards pumped boreholes

Table 7.1   Alternative formulations of radial flow problems in which vertical flow components are significant, together with examples of the use of these techniques

Section	Formulation	Coords	Issues illustrated by case studies
7.2	radial-vertical time-variant	$[r, z, t]$	7.2.4: borehole with solid casing, time-drawdown response at different locations 7.2.4: unconfined aquifer with many low conductivity lenses compared to individual lenses combined as continuous layers
7.3	radial-vertical time-instant	$[r, z]$	7.3.2: reduction in discharge for a specified drawdown due to partial penetration of the borehole 7.3.3: impact of zones of lower and higher hydraulic conductivity on effectiveness of water table control using tubewells
7.4	two-zone model	$[r, t, v_z]$	7.4.6: contribution to borehole discharge from well storage and from upper and lower aquifer zones 7.4.6: pumping from the fractured zone of a weathered-fractured aquifer; quantities of water drawn from weathered zone
7.5	leaky with aquitard storage	$[r, t: z, t]$	7.5.3: influence of aquitard storage on an aquifer system in Riyadh; increase and subsequent decrease in flows from aquitard with time 7.5.4: slow response of aquitard piezometer due to pumping from underlying aquifer 7.5.4: ignoring aquitard storage results in too high an estimate of vertical hydraulic conductivity of aquitard

Chapter 7 considers the formulation plus analytical and numerical solutions for these four alternative approaches to radial-vertical flow problems. The contents of Chapter 7 are summarised in Table 7.1; the table also lists case studies which are selected to illustrate the alternative formulations.

7.2 RADIAL-VERTICAL TIME-VARIANT FLOW [r, z, t]

Sections 7.2.1 to 7.2.3 provide information about the mathematical formulation, analytical solutions and numerical methods for time-variant flow in the vertical plane; practical examples are presented in Section 7.2.4.

7.2.1 Mathematical formulation

Important features of the analysis of flow in [r, z, t] co-ordinates are illustrated in Figure 7.2 which refers to a partially penetrating borehole in an unconfined aquifer. The vertical axis z has its origin at the impermeable base of the aquifer. Rather than working in terms of the groundwater head h, the changing conditions are defined in terms of the drawdown s below the original or rest water level which is at an elevation of H_O above the base of the aquifer. The drawdown s varies with radius, depth and time. At a radius r the height of the water table (or free surface) above the base of the aquifer is H, the equivalent *water table* drawdown is s_{wt}. The manner in which drawdowns are a function of both r and z is demonstrated by the observation piezometer to the right of the borehole in Figure 7.2. The groundwater head where the piezometer is open to the aquifer (the bottom of the piezometer) is below the water table, therefore the drawdowns s is greater than the water table drawdown at the piezometer location.

An axisymmetric formulation is used; consequently the following differential equation represents continuity and Darcy's Law in terms of the drawdown s and the co-ordinates r, z, t:

$$\frac{K_r}{r}\frac{\partial s}{\partial r} + \frac{\partial}{\partial r}\left(K_r\frac{\partial s}{\partial r}\right) + \frac{\partial}{\partial z}\left(K_z\frac{\partial s}{\partial z}\right) = S_s\frac{\partial s}{\partial t} \tag{7.1}$$

Note that the specific storage S_S and the radial and vertical hydraulic conductivities K_r and K_z can be functions of r and z.

The following boundary conditions apply.

1. At the base of the aquifer, $z = 0$, there is no flow in the vertical direction hence

$$\text{at } z = 0, (\partial s/\partial z) = 0 \tag{7.2}$$

2. For analytical solutions, the outer boundary condition is that of zero drawdown at a considerable radial distance from the pumped borehole,

$$\text{as } r \to \infty, s = 0 \tag{7.3a}$$

When a numerical technique is used, there will be a finite outer boundary. The usual approach is to ensure that the outer boundary at radius R is at a sufficient distance from the pumped borehole to have a negligible effect, hence

$$\text{at } r = R, \ s = 0. \tag{7.3b}$$

3. On the water table two conditions must be satisfied, the physical condition that the pressure is atmospheric and a kinematic boundary condition related to the motion of a point which lies on the water table; these conditions are introduced in Section 2.5.2. On the water table the condition that the groundwater head is atmospheric, i.e. the pressure $p = 0$, means that $H = z$, where z is the vertical height above datum. Since the current analysis is in terms of drawdowns, the drawdown of the water table s_{wt} below the rest water level is

$$s_{wt} = H_O - H = H_O - z \tag{7.4a}$$

The dynamic condition for the vertical movement δH of the water table during a time step δt is derived in Chapter 2 as

$$\delta H = -\frac{\delta t}{S_Y}\left(K_z\frac{\partial h}{\partial z} - K_r\frac{\partial h}{\partial r}\frac{\partial H}{\partial r} - q\right) \tag{2.53}$$

In terms of drawdowns, this becomes

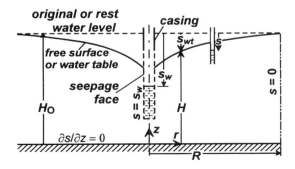

Figure 7.2 Unconfined flow in the r–z plane

$$\delta s_{wt} = -\frac{\delta t}{S_Y}\left(K_z\frac{\partial s}{\partial z} + K_r\frac{\partial s}{\partial r}\frac{\partial s_{wt}}{\partial r} + q\right) \qquad (7.4b)$$

4. Four alternative conditions can hold on the face of the borehole or on the axis beneath the borehole:

   (a) When there is solid casing there is no horizontal flow, therefore

$$\partial s/\partial r = 0 \qquad (7.5a)$$

   (b) On the seepage face, which is between the water level in the borehole and the intersection of the water table with the borehole (or the bottom of the solid casing), the pressure is atmospheric, thus

$$s = H_O - z \qquad (7.5b)$$

   (c) Within the water column in the borehole, the groundwater head and hence the drawdown is constant and equal to s_w, therefore

$$s = s_w \qquad (7.5c)$$

   (d) Finally, on the axis beneath the borehole there is no radial flow, hence

$$\partial s/\partial r = 0 \qquad (7.5d)$$

5. Initial conditions are usually that at zero time, $t = 0$, the drawdowns are zero throughout the domain, i.e. $s = 0$.

7.2.2 Analytical solutions

Detailed studies have been carried out by a number of investigators into pumping from unconfined aquifers. In particular, explanations were required of the different segments of observation borehole time-drawdown curves (e.g. Figure 5.1 of the Kenyon Junction test). Proposed reasons for the shape of these segments include the difference between elastic storage and the release of water as the water table falls (the specific yield), the delay in the release of water as the water table falls (delayed yield) and vertical flow components which result in differences in drawdown with depth (Section 5.15).

In certain of the analytical solutions the aquifer is represented in terms of the radial co-ordinate r using the vertically integrated (or depth-averaged) equation. Boulton (1963) developed an analytical solution for early and late time aquifer response; he also incorpo-

rated delayed yield (see Section 5.17) to represent the time taken for water to be released from storage at the water table. The inclusion of delayed yield helped to explain the shape of the time-drawdown curves. In the 1970s there was an extensive debate about the flow processes in unconfined aquifers focusing on the relevance of delayed yield and the significance of vertical components of flow within the aquifer (see for example Gambolati (1976) and Neuman (1979)). It is the author's view that both delayed yield and vertical flow components occur in most situations. This discussion will focus on the contributions of Boulton and Neuman.

In an early analytical solution, which includes both radial and vertical flow components, Boulton (1954) assumed that the specific storage is negligible compared to specific yield. His analysis is based on a fixed domain solution in which the saturated depth remains constant. The borehole is of infinitely small radius acting as a line sink with the aquifer of infinite areal extent. The water table is assumed to be horizontal with no recharge, hence Eq. (7.4b) simplifies to

$$S_y\frac{\partial s_{wt}}{\partial t} = -K_z\frac{\partial s}{\partial z} \qquad (7.6)$$

This approach is developed further by Neuman (1972) to include the specific storage. In subsequent papers the analysis is extended to include partial penetration (Neuman 1974) and anisotropic conditions (Neuman 1975). The mathematical formulation is summarised below.

The differential equation in terms of the drawdown $s(r, z, t)$ (see Figure 7.3) for constant K_r and K_z (homogeneous-anisotropic) becomes

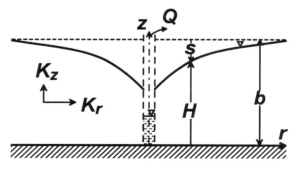

Figure 7.3 Definition diagram for Neuman's analysis of an unconfined aquifer

$$K_r \frac{\partial^2 s}{\partial r^2} + \frac{K_r}{r} \frac{\partial s}{\partial r} + K_z \frac{\partial^2 s}{\partial z^2} = S_s \frac{\partial s}{\partial t} \qquad (7.7a)$$

with the initial condition of zero drawdown throughout the aquifer system therefore,

$$s(r,z,0) = 0 \quad \text{hence} \quad H(r,0) = b \qquad (7.7b)$$

Note that Neuman uses the symbol b for the original saturated thickness.

Boundary conditions are that

$$s(\infty,z,t) = 0 \qquad (7.7c)$$

$$\partial s(r,0,t)/\partial z = 0 \qquad (7.7d)$$

and on the water table

$$K_r \frac{\partial s}{\partial r} n_r + K_z \frac{\partial s}{\partial z} n_z = \left(S_y \frac{\partial H}{\partial t} - q \right) n_z$$

at (r,H,t) \qquad (7.7e)

where n_r and n_z are the components of the unit outer normal in r and z directions

and $\quad H(r,t) = b - s(r,H,t)$ \qquad (7.7f)

On the face of the well which has an infinitely small radius,

$$\lim_{r \to 0} \int_0^H r \frac{\partial s}{\partial r} ds = -\frac{Q}{2\pi K_r} \qquad (7.7g)$$

So that this set of non-linear equations can be solved, they are linearised by assuming that the drawdown s remains much smaller than H; consequently the boundary condition for the water table, Eq. (7.7e), is shifted to the horizontal plane $z = b$. Therefore this becomes a fixed domain problem, similar to that of Boulton (1954) but with the specific storage included. By applying Laplace and Hankel transformations, an approximate solution can be obtained. The drawdown $s(r, z, t)$ (for details of the equations see Neuman 1972) can be expressed in terms of the discharge Q the transmissivity T and the following non-dimensional parameters,

$$\sigma = S_S b / S_Y \qquad z_D = z/b$$
$$b_D = b/r \qquad K_D = K_z / K_r$$
$$t_S = Tt / S_S b r^2 \qquad S_D = 4\pi Ts/Q$$

By integration over the depth of the aquifer, an expression is derived for the average drawdown in an observation borehole perforated over the full depth of the aquifer. Tabulated values and graphical plots of

Neuman's solution are presented by Kruseman and de Ridder (1990); values can also be deduced from a computer program by Hall (1996).

A number of similar analytical solutions have been developed for the study of pumping tests in unconfined aquifers. Do the assumptions introduced in these analytical solutions mean that they are suitable for use with field problems? Assumptions in the solution of Neuman (1972) include:

- many of the assumptions are similar to the Theis analysis, such as an aquifer of infinite radial extent and constant thickness which lies above an impermeable stratum with the initial water table horizontal,
- all the drawdowns including the drawdown in the pumped borehole are small compared to the saturated aquifer thickness,
- the borehole is of infinitely small radius and fully penetrates the aquifer; no allowance is made for the presence of a seepage face or well losses,
- water is released instantaneously as the water table falls, no account is taken of the delayed drainage of water. Subsequently Moench (1995) has extended Neuman's solution to allow for the gradual release of water from storage at the water table (Moench's paper also provides a helpful commentary concerning the physical validity of Boulton's expression for delayed yield),
- the analysis allows for both horizontal and vertical components of flow.

Provided that these assumptions are appropriate, Neuman's analytical solution is suitable for the determination of aquifer parameters. However, actual conditions in many practical problems are not consistent with these assumptions. Brown (1997) made comparisons between numerical solutions and the analytical solution of Neuman (1972); he shows that inaccuracies can occur in the calculation of vertical flows when the water table drawdown is not negligible compared to the aquifer thickness. Despite these limitations, the analytical fixed domain solutions are useful in developing an understanding of the response of unconfined aquifers.

A valuable comparison between parameter values deduced from the analytical solutions of Neuman (1974) and Moench (1995) is presented by Chen and Ayers (1998). They analyse data from a test in the Platte River Valley near Grand Island, Nebraska using an inverse solution technique for parameter estimation. Data from fourteen observation piezometers are analysed. When delayed yield is included in the analysis (i.e. using the Moench technique) the radial hydraulic conductivity is lower by about 20 per cent, the vertical hydraulic conductivity is typically a factor of two higher, the confined storage coefficients are slightly lower and the specific yield is about 10 per cent higher.

7.2.3 Numerical methods

Time-variant solutions to represent groundwater flow in the r-z plane can be obtained using numerical techniques. Important contributions have been made by Neuman and Witherspoon (1971) using the Finite Element method, Lennon *et al.* (1979) using the Boundary Integral Element method and Luther and Haitjema (1999), using an Analytical Element solution. For each of these analyses the specific storage is ignored; also there is a restricted distance to the outer boundary where a zero drawdown condition is enforced. In practice, the distance to the outer boundary should be at least 1000 times the well radius.

Particular care is required in representing the moving water table, especially in the vicinity of the pumped borehole if the water table intersects a seepage face. The following discussion is concerned with finite difference solutions. Equations (7.4a–7.4b) are the governing equations for a moving water table, but how are they enforced in a numerical model? One approach is that Eq. (7.4b) is used to calculate the position of the water table at the end of the time step; the new water table is represented as specified drawdowns on the vertical grid lines. An alternative approach proposed by Rushton and Redshaw (1979) and developed by Brown (1997) is to use Eq. (7.4b) to determine the inflow at each node on the water table.

A detailed explanation of the methodology of representing a moving water table in the x-z plane as a succession of specified head boundaries is presented in Section 4.4.5(ii); a numerical calculation is included to illustrate the procedure. A similar approach is used in the r-z plane; steps in the calculation are summarised below.

1. The elevation of the water table on each vertical line of nodes is defined and represented as a specified drawdown where $s_{wt} = H_O - z$; for the initial calculation $s_{wt} = 0$, but for subsequent time steps the water table is a curved boundary.
2. Obtain a solution for the finite difference form of Eqs (7.1) to (7.3) and (7.5) (see Sections 2.7 and 5.5.1); the water table is represented as a curved boundary with specified drawdowns. Adjustments are required to the individual mesh intervals which

intersect the water table and the associated finite difference equations.
3. Determine values of the vertical gradient at water table nodes, then, using Eq. (7.46), calculate the vertical velocity at each water table node from the finite difference form of

$$v_{\bar{z},wt} = \frac{1}{S_Y}\left(K_z \frac{\partial s}{\partial \bar{z}} - K_r \frac{\partial s}{\partial r} \frac{\partial s_{wt}}{\partial r}\right) \tag{7.8}$$

where $\bar{z}$ is measured vertically downwards.
4. From the vertical velocity and water table drawdown at time t, calculate the drawdown at time $t + \Delta t$

$$s_{wt,t+\Delta t} = s_{wt,t} + v_{\bar{z},wt}\Delta t$$

The above steps are repeated until the final time is reached.

Two difficulties can occur during the calculations. When the water table is close to a mesh point, certain coefficients in the finite difference equations become large so that there is a risk of errors occurring when solving the simultaneous equations and predicting the drawdown of the water table for the next time step. Second, care is required in defining the location of the intersection of the water table and seepage face.

Rather than representing the water table as a specified drawdown, the water table conditions can be defined as *specified nodal inflows* with the mesh adjusted to represent the water table location. Equation (7.8) defines the vertical velocity at the water table, and the inflow at each node is obtained from the velocity multiplied by the nodal area, $2\pi r_0^2 \Delta a$. The individual steps in the simulation are similar to those listed above but for steps (1) and (4) specified flow conditions are enforced. This approach of enforcing inflows at the water table nodes has proved to be less liable to instabilities than the representation of the water table as specified drawdowns.

7.2.4 Examples of numerical solutions in [*r*, *z*, *t*]

The following examples, based on the finite difference technique, demonstrate how the inclusion of both horizontal and vertical flows leads to an understanding of the impact of partial penetration of a borehole and the influence of layering on the yield of an aquifer.

Case study: borehole with solid casing, time-drawdown response at different locations

Consider conditions in a confined aquifer containing a borehole which has blank casing in the upper part of

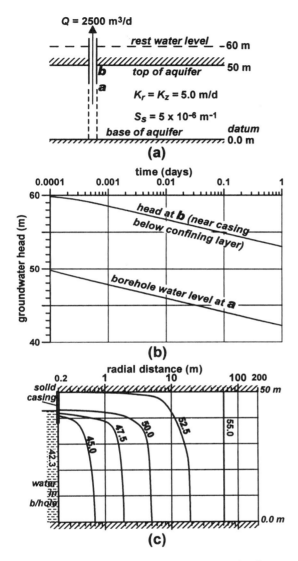

Figure 7.4 Numerical analysis in [*r*, *z*, *t*], example of a partially penetrating well: (a) details of the problem, (b) variations in groundwater head with time, (c) groundwater head contours after one day

the borehole with an open hole beneath. Details of a representative problem are given in Figure 7.4a. The aquifer thickness is 50 m; at the start of the test there is a confining head of 10.0 m. With datum taken as the bottom of the confined aquifer the initial groundwater head everywhere in the aquifer is 60.0 m. The aquifer is homogeneous and isotropic with a hydraulic conductivity of 5.0 m/d and a specific storage of $0.000005 \, \text{m}^{-1}$

(hence the confined storage coefficient is 0.00025). The aquifer extends to 1000 m in the radial direction where the groundwater head remains at 60 m.

For this example the blank casing of inner diameter 0.175 m extends 10 m below the upper confining layer, hence the open hole extends for 40 m. Figure 7.4b shows changes of conditions with time at two strategic locations when the pumping rate is $2500 \, \text{m}^3/\text{d}$.

Location (a) in Figure 7.4a refers to the water level in the borehole; there is a rapid fall in the water level in the aquifer immediately adjacent to the open section of the borehole. By 0.0001 day (0.144 min) the pumped water level has fallen just over 10 m to 49.8 m. The pumped water level continues to fall steadily so that by 1.0 day it is at 42.3 m. Provided that the pumped water level remains above 40 m, a seepage face will not develop.

Location (b) is adjacent to the blank casing immediately below the impermeable overlying layer. At a time of 0.0001 day, the fall in groundwater head is small, but by 1.0 day the groundwater head has fallen almost seven metres to 53.07 m. If this groundwater head falls below 50.0 m the water pressure would become less than atmospheric and some form of dewatering would occur.

To help in visualising conditions due to the presence of blank casing, the groundwater head distribution at 1.0 day is plotted in Figure 7.4c. Note that a logarithmic scale is used in the radial direction but in the vertical direction the scale is arithmetic. Equipotentials are steeply curved in the vicinity of the end of the blank casing. For the time period covered by this simulation, a seepage face does not form, nor do unconfined conditions occur beneath the confining layer. After about 45 days, unconfined conditions start to develop at the top of the aquifer.

Case study: comparison of responses of unconfined aquifers (a) containing many low conductivity lenses and (b) combining the individual lenses as continuous layers

Many aquifers contain several lenses of low hydraulic conductivity within a generally high conductivity material; how does the aquifer response differ if the individual lenses are represented as a series of continuous low conductivity layers? Figure 7.5a refers to a pumped borehole in an unconfined aquifer of hydraulic conductivity 1.0 m/d containing many clay lenses, each 5 m thick, with horizontal and vertical hydraulic conductivities of 0.001 m/d. The radial co-ordinate is plotted to a logarithmic scale. In Figure 7.5b all the clay lenses

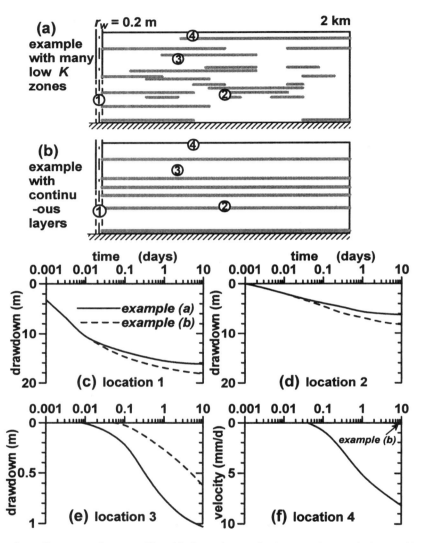

Figure 7.5 Comparison of responses for an aquifer with discontinuous clay lenses and an equivalent problem with continuous clay lenses: (a) and (b) details of problems, (c) to (f) time-drawdown plots at four different locations

are combined as six continuous layers each 5 m thick with the same total lateral extent as the discontinuous lenses. The aquifer is 100 m thick and extends to 2000 m with a no-flow outer boundary. The borehole of radius 0.2 m fully penetrates the aquifer with solid casing in the upper 50 m. For each example the initial condition is zero drawdown everywhere.

In Figures 7.5c–f, comparisons are made between drawdowns or vertical velocities at four locations as indicated on Figure 7.5a and b. The following insights can be gained from a comparison of the results.

Figure 7.5c refers to the drawdown within the pumped borehole at location 1: the differences in time-drawdown curves are not large even though the simulation continues for ten days. For Figure 7.5b with the continuous layers of lower hydraulic conductivity, a slightly larger pumped drawdown is required to draw water to the borehole.

Figure 7.5d refers to drawdowns at location 2, which is 20 m from the pumped borehole and at a depth of 70 m: the drawdown curves are of similar shape to location 1 although the drawdowns are smaller. This indicates that there is a good connection via horizontal high hydraulic conductivity layers between the pumped borehole and location 2.

Figure 7.5e refers to drawdowns at location 3 close to the solid casing at 3.5 m from the axis of the borehole and at a depth of 30 m: the relatively rapid response for Figure 7.5a indicates that there is a good connection via sand layers between locations 1 and 3. The response with the continuous layers of Figure 7.5b is slower and more muted.

Figure 7.5f refers to the vertical Darcy velocity on the water table at location 4 which is 6.3 m from the borehole axis: for Figure 7.5a there is no response until 0.05 day (more than 1 hour), later the vertical velocity and hence the rate of fall in the water table increases steadily. For Figure 7.5b with the continuous layers there is only a very small response after 10 days, the response is almost too small to be plotted.

This example demonstrates that combining individual zones as a series of continuous layers results in similar responses when there is direct connection through higher conductivity zones to the pumped borehole. However, when a continuous low-conductivity layer isolates part of the aquifer from the pumped borehole, the responses are significantly different.

7.3 RADIAL-VERTICAL TIME-INSTANT $[r, z]$

7.3.1 Principles of the approach

In a *time-instant* radial-vertical approach, a pseudo-steady state analysis is used to represent part of the time-variant response; a detailed discussion of the basis of the time-instant approach is presented in Section 12.4. For many problems of flow to a borehole, recharge approximately balances the abstraction so that pseudo-steady state conditions apply. An example is provided by a waterlogged alluvial aquifer in Pakistan where the changes in water table elevations are small compared to pumped drawdowns of tens of metres. A further example is the third stage of the pumping test with multiple piezometers at Kenyon Junction (see Figure 5.1 and 5.3c); the fall in water table after 1 day is 0.7 m compared to the pumped borehole drawdown of 34 m and deep piezometer drawdowns approaching 7 m. Furthermore, all the drawdowns within the aquifer change at a similar rate. Consequently, certain practical situations are consistent with

the assumption of pseudo-steady state conditions with little movement of the water table.

With pseudo-steady state conditions the confined storage contribution to flow is small, hence the specific storage term of Eq. (7.1) can be neglected. Therefore the appropriate differential equation is

$$\frac{K_r}{r}\frac{\partial s}{\partial r} + \frac{\partial}{\partial r}\left(K_r\frac{\partial s}{\partial r}\right) + \frac{\partial}{\partial z}\left(K_z\frac{\partial s}{\partial z}\right) = 0 \tag{7.9}$$

where K_r and K_z can be functions of r and z. Numerical solutions for specific problems can be obtained using finite difference or finite element techniques.

Two case studies are selected here to illustrate the use of the *time-instant* approach when radial and vertical flow components are important.

7.3.2 Case study: reduction of discharge due to partial penetration of borehole

This first problem is concerned with the effect of partial penetration of an abstraction borehole in an isotropic confined aquifer. For the representative example of Figure 7.6a, information about the geometry and aquifer parameters is recorded on the figure. The lateral boundary condition is zero drawdown at a radial distance of 1000 m from the pumped borehole. The effect of borehole penetration on yield is explored by representing, in a numerical model, different borehole penetrations, each having the same pumped drawdown of $s_w = 7$ m. The discharge required to generate this drawdown is used as a measure of the effectiveness of the borehole in collecting water.

The numerical model solves the equation

$$K_r\frac{\partial^2 s}{\partial r^2} + \frac{K_r}{r}\frac{\partial s}{\partial r} + K_z\frac{\partial^2 s}{\partial z^2} = 0 \tag{7.10}$$

using techniques similar to those outlined in Sections 2.7 and 5.5.1 with the following boundary conditions:

* on the outer boundary at a radial distance R, the drawdown is zero, therefore

  at $r = R$, $s = 0.0$

* since this is a confined aquifer, on the upper and lower boundaries there is no vertical flow, hence

  at $z = 0$ and $z = b$, $\partial s / \partial z = 0$

* beneath the borehole, the radial flow is zero:

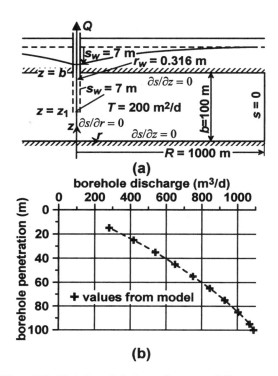

(a)

(b)

Figure 7.6 Variation of discharge from a partially penetrating borehole: (a) details of the problem, (b) change of discharge with penetration

for $0 \leq z \leq z_1,$ $\partial s/\partial r = 0$

• on the open section of the borehole, the drawdown is specified as

for $z_1 \leq z \leq b,$ $s_w = 7.0$ m;

The discharge at the borehole can be estimated from the integral of the radial approach velocities obtained from the numerical model,

$$Q = 2\pi r_w \int_{z=z_1}^{z=b} K_r \frac{\partial s}{\partial r} dz \qquad (7.11)$$

When the borehole is fully penetrating the flow is horizontal, hence the discharge can be calculated from the Thiem formula (Eq. (5.1)). Using parameter values from Figure 7.6a,

$$Q = 2\pi s_w T / \ln(R/r_w) = 1091 \ \text{m}^3/\text{d}$$

This value is also obtained from the [r, z] numerical model. When the model is used to represent partial penetration but with the pumped drawdown retained at 7.0 m, the discharge decreases as the penetration decreases; the results are shown in Figure 7.6b. If the borehole penetrates 80 per cent of the aquifer depth, the discharge is reduced to 89 per cent of the maximum value while a penetration of 40 per cent leads to a discharge of 55 per cent of the maximum. Results for penetrations less than 17.5 m are not included due to insufficient mesh intervals to represent adequately the convergent flows to small open sections of the borehole. The graph of Figure 7.6b provides valuable information about the limited effectiveness of partially penetrating boreholes.

7.3.3 Case study: effectiveness of water table control using tubewells

This second study, pumping from an aquifer with zones of different hydraulic conductivity, was prompted by the need to understand the limited success of water table control in waterlogged areas of Pakistan. The purpose of the SCARP (Saline Control and Reclamation Project) tubewells (boreholes) in Pakistan is to control the water table in areas which have become waterlogged due to canal irrigation. One crucial question is the selection of the spacing between tubewells to achieve effective water table control.

The conceptual model used to understand more about the SCARP tubewells is presented in Figure 7.7. The borehole penetrates to an elevation of z_1 (measured from the base of the aquifer); in the upper part of the borehole a solid casing extends downwards to an elevation z_2, with a slotted screen between z_1 and z_2. Another important feature is that the water table (the upper boundary of the aquifer) is represented as having zero drawdown. This is an appropriate assumption since the water losses from the canals and return flow from irrigation roughly balance the abstraction thereby ensuring that the water table remains at an approximately constant elevation.

The mathematical formulation of the problem (adapted from Eqs (7.1)–(7.5)) is as follows.

1. The differential equation describing flow in the aquifer system

$$\frac{K_r}{r}\frac{\partial s}{\partial r} + \frac{\partial}{\partial r}\left(K_r \frac{\partial s}{\partial r}\right) + \frac{\partial}{\partial z}\left(K_z \frac{\partial s}{\partial z}\right) = 0 \qquad (7.9)$$

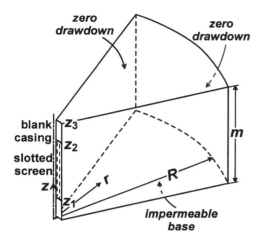

Figure 7.7 Conceptual model used to investigate SCARP tubewells. Reproduced by permission of IAHS Press from Cookey *et al.* (1987)

is solved using a finite difference numerical technique.

2. At a distant boundary $r = R$, the drawdown is zero, hence $s = 0$.
3. At the base of the aquifer, $z = 0$, there is no vertical flow hence $\partial s/\partial z = 0$.
4. Beneath the penetration of the borehole and where the borehole has solid casing, there is no horizontal flow, therefore for $0 \le z \le z_1$ and $z_2 \le z \le z_3$, $\partial s/\partial r = 0$.
5. Where slotted casing is present the drawdown s_w is constant in the borehole. The borehole drawdown is unknown but a specified discharge Q is withdrawn.
6. On the upper boundary, $z = z_3$, the drawdown of the water table is zero, i.e. $s = 0$.

In this study the following parameter values apply:

depth of aquifer	$= 100\,\mathrm{m}$
radius of borehole	$= 0.1\,\mathrm{m}$
length of blank casing	$= 17.5\,\mathrm{m}$
penetration of the borehole	$= 77.5\,\mathrm{m}$
outer radius of aquifer	$= 10\,\mathrm{km}$
hydraulic conductivity of the aquifer	$= K\,\mathrm{m/d}$
pumping rate from borehole	$= 1000K\,\mathrm{m^3/d}$

From a numerical model solution, drawdowns at all nodal points are calculated; from these drawdowns, flows can be calculated. Selected results are presented in Figure 7.8 for the problem as defined above. In a

second problem (Figure 7.9) layers of higher and lower hydraulic conductivity are included. These plots provide the essential information to assess the critical factors which determine the effectiveness of the borehole in water table control.

For each problem four diagrams have been prepared: (a) shows the physical arrangement, (b) plots the approach velocities to the well face, (c) shows the vertical velocities on the upper boundary (the water table), and (d) indicates the quantities of water entering the aquifer system from the upper zero drawdown boundary; the lines of the bar chart represent inflows for each nodal point of the numerical model (there are four nodal points for a tenfold increase in radius).

Problem A: Figure 7.8 presents results for Problem A, in which the aquifer is homogeneous and isotropic with the hydraulic conductivity everywhere equal to K. Points to note from Figure 7.8 include the following.

- The approach velocities to the borehole, diagram (b), are greatest at the top and bottom of the slotted section of the casing.
- The vertical velocities at the water table, diagram (c), are approximately constant to a radial distance of 10 m, but then reduce, becoming very small at radial distances greater than 100 m.
- These vertical velocities do not give a clear picture of the quantities of water moving from the water table into the aquifer; that is why diagram (d) is prepared, showing the total flow per node on a logarithmic scale. When the vertical velocities of diagram (c) are multiplied by the areas associated with the logarithmic mesh spacing (the area associated with the node at 100 m is one hundred times the area associated with the node at 10 m), significant flows can be identified up to the node at 178 m.

Problem B: The results for the homogeneous isotropic aquifer of Problem A are not consistent with the observed impact of the SCARP tubewells. Figure 7.8d indicates that most of the water is drawn from the water table within a radial distance of 178 m with only a small contribution at 316 m and beyond. However, the experience in Pakistan is that a tubewell penetrating about 80 m achieves water table control for distances up to 800 m and beyond. Drillers logs indicate layers or lenses of lower and higher hydraulic conductivity within these alluvial aquifers.

To explore the impact of layers of different hydraulic conductivity, *Problem B* is devised with two low permeability layers 1.0 m thick having vertical hydraulic

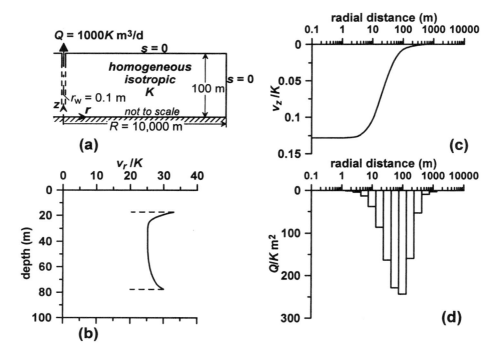

Figure 7.8 Results for Problem A, the impact on the aquifer of a tubewell pumping in a homogeneous isotropic aquifer: (a) details of the problem, (b) approach velocities at the tubewell, (c) vertical velocities from the water table, (d) vertical flows from the water table at nodal points. Reproduced by permission of IAHS Press from Cookey *et al.* (1987)

conductivities of $0.002\,K$; these layers are at 12.5 and 57.5 m below the top of the aquifer. In addition there is a zone of increased hydraulic conductivity between 32.5 and 37.5 m below the top of the aquifer with a hydraulic conductivity of $10\,K$. The results of this simulation are presented in Figure 7.9.

Important aspects of the results include the following.

- A high approach velocity occurs through the zone of high hydraulic conductivity, shown in diagram (b); in the high conductivity zone the approach velocity is $v_r/K = 145$ compared to less than 20 elsewhere along the slotted casing (the significance of high approach velocities on well losses is considered in Section 8.2.5). The more uniform approach velocities for the homogeneous aquifer are also shown on the diagram.
- Diagram (c) indicates a significant change in the vertical velocities from the water table compared to the velocities for the homogeneous isotropic aquifer. Without the low permeability zone the maximum v_z/K equalled 0.128; the presence of the low perme-

ability layer reduces the maximum value to 0.0071. Since the vertical velocity for *Problem A* is eighteen times the value for the layered aquifer, it cannot be plotted on diagram (c).

- The combination of low and high conductivity zones also leads to more water being drawn from the water table at greater distances from the pumped borehole (Figure 7.9d). Between radial distances of 237 and 580 m the flow from the water table totals more than half of the pumped well discharge. This distribution should be compared with the broken line which refers to the homogeneous aquifer.
- The overall conclusion is that a spacing between dewatering wells of 1600 m (1 mile) is realistic for long-term water table control in aquifers which contain layers of lower and higher hydraulic conductivity since the effect of each borehole can extend to at least 1000 m.

The results for several alternative aquifer non-heterogenities are described by Cookey *et al.* (1987). The assumption that a layered aquifer can be repre-

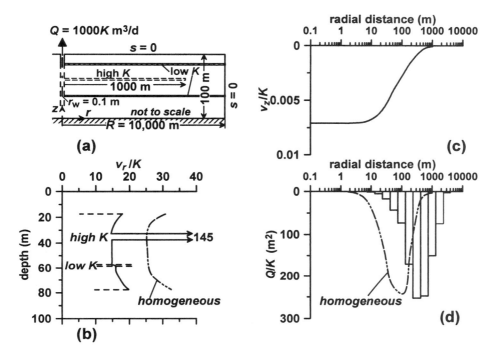

Figure 7.9   Results for Problem B, the impact on the aquifer of a tubewell pumping in an aquifer with layers of higher and lower hydraulic conductivity: (a) details of the problem, (b) approach velocities at the tubewell, (c) vertical velocities from the water table, (d) vertical flows from the water table at nodal points. Reproduced by permission of IAHS Press from Cookey *et al.* (1987)

sented as being anisotropic is also explored. Drawdowns and flow patterns for an anisotropic aquifer are very different to the layered situation, indicating that an anisotropic formulation should not be used to study layered aquifer systems.

7.4 TWO-ZONE APPROXIMATION $[r, t, v_z]$

7.4.1 Introduction

The importance of vertical flow components associated with pumped boreholes is highlighted in the introduction to this chapter. Time-variant and time-instant techniques have been considered. However, there are many practical situations where there is insufficient information about parameter values to justify a full $[r, z, t]$ solution. Also, when there are significant changes in groundwater head with time, the time-instant approach is not suitable. Moreover, the primary purpose of many studies is to understand and identify the flow processes with little need for accurate aquifer parameter estimation. Therefore, an alternative simpler

approach, such as the *two-zone radial flow model*, is often of great value in *exploring* moderately complex flow processes towards pumped boreholes (Rathod and Rushton 1991).

This discussion on the two-zone model opens with an examination of typical field situations. The important flow processes for each situation are identified from conceptual models; this is followed by descriptions of the representation of each of the field situations using the same basic two-zone model. Details of the equivalent numerical model are presented and the corresponding flow equations derived; a computer program in FORTRAN is included in the Appendix. Typical examples of the use of the two-zone model are described; parameter values for two-zone models for a number of different problems are presented.

7.4.2 Examples of formulation using the two-zone model

In the first example, Figure 7.10(a), the aquifer system consists of an upper unconfined aquifer, a lower

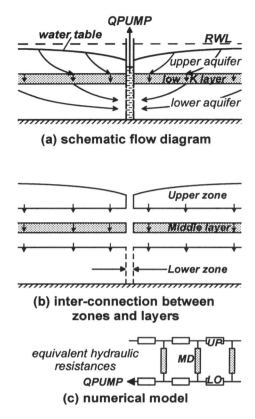

(a) schematic flow diagram

(b) inter-connection between zones and layers

(c) numerical model

Figure 7.10 Conceptual and numerical formulation as a two-zone model for an aquifer with a layer of low hydraulic conductivity with water abstracted from below the low conductivity layer

confined aquifer and a layer of low hydraulic conductivity between the upper and lower aquifers. A borehole, with solid casing through the upper aquifer and the intermediate low conductivity layer, draws water from the lower aquifer. Typical flowlines from the water table, through the upper aquifer, vertically through the low conductivity layer and into the lower aquifer are sketched in the diagram. In Figure 7.10b the three parts of the aquifer system are drawn separately; the flows *between* the individual zones and layers are also indicated. The upper aquifer is represented by the *Upper* aquifer *zone*, the resistance to horizontal flow in the Upper zone varies according to the saturated thickness of the Upper zone. Low hydraulic conductivity layers between the upper and lower aquifers are represented by the *Middle* low permeability *layer*; vertical flows

occur through this low permeability layer. Water is transferred from the Middle low permeability layer into the *Lower* aquifer *zone*; water in the Lower zone is drawn to the pumped borehole. Figure 7.10c shows how the two-zone model represents the flows in the zones and layers as defined in Figure 7.10a and b. The Upper zone *UP* is represented by equivalent horizontal hydraulic resistances, for the Middle zone *MD* vertical hydraulic resistances are used to represent the vertical flow from the Upper zone to the Lower zone *LO*. The Lower zone consists of equivalent hydraulic resistances which transfer water to the pumped borehole where the discharge is *QPUMP*.

In the second example, Figure 7.11a, there is an *Overlying* low permeability leaky layer above the *Upper* aquifer. Between the *Upper* and *Lower* aquifer zones there is a *Middle* low permeability layer. This is a common situation in alluvial aquifers and in other multiple aquifer systems. When the aquifer system is redrawn as different units (Figure 7.11b), there are now four parts, the *Overlying* layer, OL, the *Upper* and *Lower* zones, and the *Middle* layer. For an alluvial aquifer, the more permeable material determines the horizontal hydraulic conductivities of the Upper and Lower zones while the clays and silts define the vertical hydraulic conductivities of the Overlying and Middle layers. Usually, a complex aquifer can be idealised into these two zones and two layers. The representation of groundwater heads and flows at a specific location will not be precise, but the general response of the aquifer can be explored using this approach. The numerical model for this example, Figure 7.11c, is identical to the previous example apart from the pumped borehole drawing water from both the Upper and Lower zones and vertical flows occurring through the Overlying layer.

7.4.3 Details of the two-zone model

The next task is to define the conceptual features of the two-zone model to represent each of the characteristics of the typical field problems. Figure 7.12 is a diagram of the idealised aquifer system including an Overlying layer and the rest water level *RWL* which normally does not change with time. Flow processes associated with the four zones or layers are described below.

1. The Overlying leaky low permeability layer *OL*: vertical flow through this layer depends on the classical leaky aquifer theory; Section 5.8. Flow through this layer is proportional to the head difference between the rest water level in the aquifer

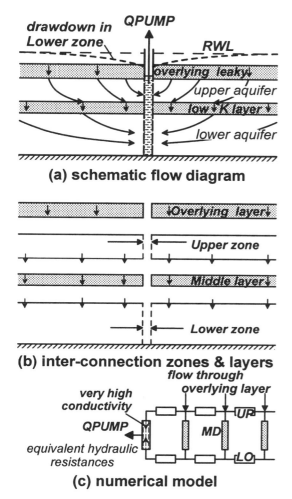

(a) schematic flow diagram

(b) inter-connection zones & layers

(c) numerical model

Figure 7.11 Conceptual and numerical formulation as a two-zone model for an aquifer with a layer of low hydraulic conductivity and an overlying low permeability layer with water abstracted from both aquifer zones

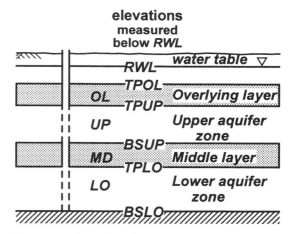

Figure 7.12 Idealisation of an aquifer into a system of zones and layers (Rathod and Rushton 1991). Reprinted from *Ground Water* with permission of the National Ground Water Association. Copyright 1991

is open in the Upper zone, water will be abstracted via the borehole.

3. The Middle layer *MD*: water passes vertically through this layer between the Upper and Lower aquifer zones with the flow depending on the difference in groundwater heads between the two zones. The main impact of the Middle layer is that a significant groundwater head difference between the Upper and Lower zones may occur due to the movement of water through the Middle layer.

4. The Lower aquifer zone *LO*: water is transferred from the Upper aquifer zone through the Middle layer, water is also released due to confined storage properties. Water can also be withdrawn from this zone by the pumped borehole.

Elevations of the interfaces of the zones and layers are all defined as vertical distances *below* rest water level. The depth below *RWL* of top (*TP*) and base (*BS*) of the Upper (*UP*) zone are written as *TPUP* and *BSUP*, as indicated in Figure 7.12. For the lower aquifer the elevations are *TPLO* and *BSLO*. For a complete specification of the interfaces it is necessary to define the top of the Overlying layer, *TPOL*.

The two-zone model is based on a discrete space–discrete time approximation similar to that used for the single layer radial flow aquifer model of Section 5.5. In effect a radial flow model is used for the Upper aquifer zone and a second radial flow model for the

above the Overlying layer and the groundwater head in the Upper aquifer zone.

2. The Upper aquifer zone *UP*: this zone may receive water from the Overlying layer, or if it is unconfined as in Example 1, from water released from storage as the water table falls (delayed yield can be included). Rainfall recharge is another possible source of water. Water is also taken into or released from confined storage. In addition, water is lost from the Upper zone due to downwards flows through the Middle layer. Finally, if the borehole

Lower aquifer zone. These two models are interconnected at each nodal point through the Middle layer. As with the single layer radial flow model of Section 5.5, each layer is terminated at an outer boundary either as a zero flux or a zero drawdown condition. Zero flux conditions often apply when there is interference with neighbouring boreholes or the aquifer is of limited extent. A zero flux boundary may have little effect during a pumping test but will be important when considering long-term pumping from the aquifer. Different boundary conditions for the Upper and Lower zones can occur; for example, there may be contact between a surface water body and the Upper zone whereas the Lower zone may extend to far greater radial distances.

Conditions at the face of the borehole depend on whether the borehole can take water from the Upper zone alone, the Lower zone alone or both zones; the borehole is likely to have solid casing through low conductivity layers. If water is taken from both zones, this is simulated by introducing a high vertical conductivity at the face of the borehole between the two zones as illustrated in Figure 7.11c. Well storage effects are included automatically by extending the mesh into the borehole with a storage coefficient of 1.0 as explained in Section 5.5. Well losses are represented as changed horizontal hydraulic conductivities immediately outside the face of the borehole; for further information see the final sub-section of 5.5, and Section 8.3.

Table 7.2 lists all the dimensions and aquifer parameters required for the two-zone numerical model. If there is clear field evidence of changing parameter values with radius, they should be included. Often there are uncertainties about the actual magnitudes of the parameters, yet from the physical properties of the different strata and published information about similar problems, reasonable first estimates of parameter values can be made. The importance of the individual parameters can be explored using sensitivity analyses while attempting to match the model and field drawdowns during the pumping and recovery phases.

7.4.4 Discrete space–discrete time equations for the two-zone model

In Section 5.5, the equations for the single zone numerical model are derived using finite difference approximations. However, for the two-zone model with vertical flows through the overlying and middle layers, the equations will be derived using a lumping argument so that it is possible to appreciate the physical significance of the terms. Identical equations are obtained using a more formal finite difference approach. The derivation of the equations is carried out with reference to Figure 7.13.

The following steps are required to derive the flow balances for the Lower zone and the Upper zone of the two-zone model. Steps (a)–(e) refer to the flow balance for the Lower zone; the additional factors for the Upper zone are introduced in step (f). The flow balance for the Lower zone involves

- radial flows from node $n-1$ to n and from $n+1$ to n,
- vertical flow through the Middle layer due to the differences in drawdown between the Upper and Lower zones, also
- water released from storage due to compressibility effects in the Lower zone.

The unknown variables are the drawdowns below the initial rest water levels.

(a) *Radial flow between nodal points*

Using the Thiem equation, Eq. (5.1), the flow in the Lower zone QLO_{n+1} from node $n+1$ to node n can be written using Figure 7.13a as

$$QLO_{n+1} = \frac{-2\pi K_r m[DLO_{n+1} - DLO_n]}{\ln(r_{n+1}/r_n)} \qquad (7.12)$$

Table 7.2 Parameter names for two-zone model

Zone or layer	Thickness	Hydraulic conductivity		Storage coefficient	
		Radial	Vertical	Confined	Unconfined
Overlying	*TPUP – TPOL*	–	*PERMOL*	–	–
Upper	*BSUP – TPUP**	*PERMRUP*	*PERMVUP*	*SCONUP*	–*
Middle	*TPLO – BSUP*	–	*PERMVMD*	–	–
Lower	*BSLO – TPLO*	*PERMRLO*	*PERMVLO*	*SCONLO*	–

* If there is no Overlying layer, the thickness is *BSUP – DUP*() and unconfined storage is *SUNCUP.*

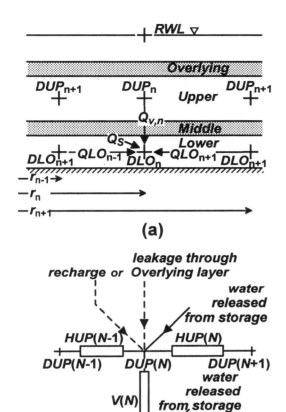

Figure 7.13 Derivation of flow balance equations for two-zone model: (a) flow balance for node *n* of Lower zone, (b) arrangement of equivalent hydraulic resistances and other sources of flow (Rathod and Rushton 1991). Reprinted from *Ground Water* with permission of the National Ground Water Association. Copyright 1991

where the following parameters refer to the Lower zone,

K_r is the radial hydraulic conductivity,
m is the saturated thickness of the zone,
DLO_{n+1} and DLO_n are the drawdowns and
r_{n+1} and r_n are the radial distances for nodes *n*+1 and *n* respectively.

As with the single zone radial flow model (Section 5.5), the mesh in the radial direction increases

logarithmically with the alternative radial variable *a* defined as

$$a = \ln(r)$$

hence $\Delta a = \Delta r/r = \ln(r_{n+1}/r_n)$

Therefore Eq. (7.12) can be rewritten as

$$QLO_{n+1} = \frac{-2\pi[DLO_{n+1} - DLO_n]}{\Delta a/(K_r m)} \qquad (7.13)$$

In a similar manner, radial flow from node *n* − 1 to node *n* equals

$$QLO_{n-1} = \frac{-2\pi[DLO_{n-1} - DLO_n]}{\Delta a/(K_r m)} \qquad (7.14)$$

(b) *Vertical flow through the Middle layer*

In calculating the vertical flow through the middle layer, account must be taken of the area *A* represented by nodal point *n*

$$A = 2\pi r_n \Delta r = 2\pi r_n^2 \Delta a$$

The vertical flow through the Middle layer from the Upper zone can be calculated from Darcy's Law as

$$Q_{v,n} = \frac{-AK_V[DUP_n - DLO_n]}{DZ}$$
$$= \frac{-2\pi\Delta a[DUP_n - DLO_n]}{DZ/(K_V r_n^2)} \qquad (7.15)$$

where K_V is the effective vertical hydraulic conductivity of the Middle layer and *DZ* is the vertical thickness of the Middle layer.

(c) *Contribution from confined storage*

Water released from confined storage S_C depends on the change in drawdown; the drawdown in the Lower zone at node *n* at the *Previous* time step is written as $PDLO_n$:

$$Q_S = \frac{-AS_C[PDLO_n - DLO_n]}{\Delta t}$$
$$= \frac{-2\pi\Delta a[PLDO_n - DLO_n]}{\Delta t/(S_C r_n^2)} \qquad (7.16)$$

(d) *Flow balance for node in Lower Zone*

To satisfy continuity, the four flow components, Eqs (7.13) to (7.16), must sum to zero; when each of the terms is divided by $2\pi\Delta a$

$$\frac{[DLO_{n+1} - DLO_n]}{\Delta a^2 / (K_r m)} + \frac{[DLO_{n-1} - DLO_n]}{\Delta a^2 / (K_r m)}$$

$$+ \frac{[DUP_n - DLO_n]}{DZ / (K_V r_n^2)} + \frac{[PDLO_n - DLO_n]}{\Delta t / (S_C r_n^2)} = 0 \quad (7.17)$$

(e) *Equations in terms of hydraulic resistances*

It is convenient to introduce the concept of equivalent hydraulic resistances (with n altered to N) so that the denominators of the above equations can be written as

radial hydraulic resistance $HLO( ) = \Delta a^2 / (K_r m)$

as shown in Figure 7.13b. $HLO(N-1)$ is between nodes $N-1$ and N while $HLO(N)$ *is* between N and $N+1$.

vertical hydraulic resistance $V( ) = DZ / (K_V r_n^2)$

time-storage coefficient $TSLO( ) = \Delta t / (S_C r_n^2)$

Hence Eq. (7.17) can be written in a form suitable for incorporation in a computer program as

$$\frac{[DLO(N+1) - DLO(N)]}{HLO(N)}$$

$$+ \frac{[DLO(N-1) - DLO(N)]}{HLO(N-1)}$$

$$+ \frac{[DUP(N) - DLO(N)]}{V(N)}$$

$$= \frac{-[PDLO(N) - DLO(N)]}{TSLO(N)} \quad (7.18)$$

(f) *Flow balance for node in Upper Zone*

Derivation of the equation for node N in the Upper zone is similar to the Lower zone but there may also be a contribution due to the release of water at the water table (specific yield) and recharge or flow through the Overlying layer. The equation for the Upper zone is

$$\frac{[DUP(N+1) - DUP(N)]}{HUP(N)}$$

$$+ \frac{[DUP(N-1) - DUP(N)]}{HUP(N-1)}$$

$$+ \frac{[DLO(N) - DUP(N)]}{V(N)}$$

$$= \frac{-[PDUP(N) - DUP(N)]}{TSUP(N)}$$

$$+ \frac{[DUP(N) - RWL]}{VOL(N)} - RECH * R(N) * R(N) \quad (7.19)$$

In the above equation

- the time/storage coefficient $TSUP(N)$ includes the specific yield if the node is unconfined,
- delayed yield effects can be included using the methodology described in Section 5.17,
- $VOL(N)$ is the vertical hydraulic resistance of the Overlying layer (this term will only be included if an Overlying layer is present),
- if there is no Overlying layer, recharge $RECH$ may occur directly to the Upper zone,
- abstraction is represented as a negative recharge at node 1.

7.4.5 Solution of simultaneous equations

Equations (7.18) and (7.19) can be written for each node in the Lower zone and each node in the Upper zone. When boundary conditions at the borehole and at the outer boundary are included, this leads to a set of simultaneous equations which can be solved directly using a technique such as Gaussian elimination. Due to the different orders of magnitude of many of the coefficients, it is advisable to work in double precision.

If unconfined conditions apply in the Upper zone, the radial hydraulic resistances depend on the current saturated depth. However, the saturated depth is a function of the unknown drawdown. Consequently, the simultaneous equations are solved four times for each time step with current values of drawdown at each node used to update the saturated depths.

Calculations start with a very small time step; an initial time step of 10^{-7} day ensures that the initial drawdowns on the face of the borehole are small. A logarithmic increase in time step is used with ten time steps for a tenfold increase in time. A mesh spacing with six logarithmic mesh increments for a tenfold increase in radius is usually adequate. For large diameter wells, twelve mesh intervals for a tenfold increase in radius are recommended.

A flowchart of the program is shown in Figure 7.14; note that the program can be used for a number of pumping phases with the time step returning to the initial value for each phase.

7.4.6 Examples of the use of the two-zone model

Case study 1: contribution to borehole discharge from well storage and from upper and lower aquifer zones

Details of this example including parameter values can be found in Table 7.3 and Figure 7.15a. There is a low permeability layer between the two aquifer zones, also

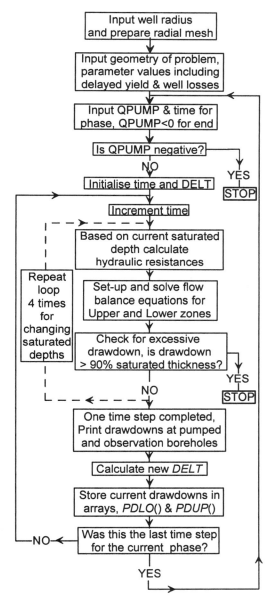

Figure 7.14 Flow chart for two-zone model (Rathod and Rushton 1991). Reprinted from *Ground Water* with permission of the National Ground Water Association. Copyright 1991

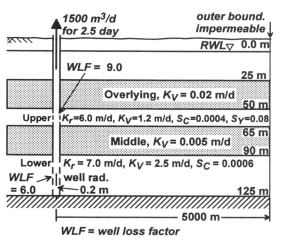

(a) details of example

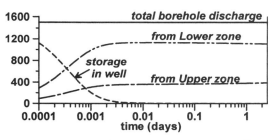

(b) component flows to borehole (m³/d)

Figure 7.15 Typical example of the use of the two-zone model: (a) geometry and aquifer properties, (b) flows into the borehole from the Upper and Lower zones

the Upper aquifer zone is overlain by another low permeability layer. Pumping at a rate of 1500 m³/d from a production borehole of 0.2 m radius continues for 2.5 days. Well loss factors for the Upper and Lower zones are set at 9.0 and 6.0; this means that the radial

hydraulic resistances for the mesh intervals adjacent to the borehole are multiplied by these factors (this is equivalent to dividing the radial hydraulic conductivities by these factors). A zero flow boundary is enforced at 5000 m, thereby limiting the area from which the borehole can draw water.

Variations of drawdowns with time are obtained from the numerical model. Using the equations in Section 7.4.5, flows within the aquifer zones and between the aquifer zones can be estimated. In this discussion, attention will be focused on the component flows into the borehole.

Figure 7.15b shows how the total borehole discharge of 1500 m³/d is made up of three components which change with time. Up to 0.0004 day (0.6 min) the greatest contribution is from water stored within the

Table 7.3 Parameter values for first case study (units m or m/d)

Zone or layer	Thickness	Hydraulic conductivity		Storage coefficient	
		Radial	Vertical	Confined	Unconfined
Overlying	50.0 − 25.0 = 25.0	–	0.02	–	–
Upper	65.0 − 50.0 = 15.0	6.0	1.2	0.0004	0.08
Middle	90.0 − 65.0 = 25.0	–	0.005	–	–
Lower	125.0 − 90.0 = 35.0	7.0	2.5	0.0006	–

borehole. Well storage becomes less important with increasing time; by 0.007 day (10 mins) well storage becomes negligible. Due to the higher transmissivity, flow from the Lower zone is greater than from the Upper zone. The ratio of transmissivities between Lower and Upper zones is 2.72 compared to a ratio of the flows into the borehole of 2.92 when equilibrium conditions are reached.

Case study 2: pumping from the fractured zone of a weathered-fractured aquifer, quantities of water drawn from the weathered zone are of special interest

The second example, Figure 7.16, refers to a weather-fractured aquifer in central India. Full details of the test and analysis can be found in Rushton and Weller (1985) and Rathod and Rushton (1991). This discussion will concentrate on the use of the two-zone model to interpret a pumping test. In the pumping test, water was withdrawn from the fractured zone beneath a weathered zone with a saturated depth of about 10 m. A number of deep observation piezometers and shallow observation wells were used to monitor the test; this discussion will focus on the drawdowns in the pumped borehole and in one shallow observation well.

Figure 7.16a records the drawdowns during pumping for 7.5 hours and recovery for a further 3.5 hours; the field readings are indicated by discrete symbols. After an initial rapid fall, the pumped drawdown quickly approaches equilibrium. This steady drawdown occurs because an air-lifting technique is used to withdraw water. With air-lifting, the discharge decreases as the water is lifted greater vertical distances due to increased pumped drawdowns. The final discharge rate was 288 m³/d but the rate was higher in the early stages. In the absence of precise information, a relationship between the pumped drawdown and the pumped discharge is introduced,

$$Q_{PUMP} = 375\left(1.0 - 0.017 DLO_{well}^2\right) \text{m}^3/\text{d}$$

With this relationship the discharge rate falls from 375 m³/d with zero drawdown to 288 m³/d for a pumped drawdown of $DLO_{well} = 3.7$ m. This is a provisional relationship but it does allow an assessment of the effect of a variable pumping rate.

Parameter values for the two-zone model are listed in Table 7.4; the numerical model drawdowns are plotted as continuous lines in Figure 7.16a. Before examining field and modelled results, it must be appreciated that this is a highly variable hard rock aquifer with water abstracted by air-lifting; drawdowns measurements were made using unsophisticated techniques. Therefore close agreement between field and modelled responses should not be expected. Nevertheless, the agreement is encouraging.

- For the pumped borehole during the pumping phase, there are certain differences between the field and modelled results; this occurs primarily because of the approximate nature of the assumed equation for the reduction in abstraction. Nevertheless, the inclusion of the reduced abstraction rate is essential to approximately represent the pumped water levels.
- For the recovery phase in the pumped borehole there is a good match; during the later stages the match is not as satisfactory but there is an inconsistency in the field results with the drawdowns in the pumped borehole less than those in the shallow observation well.
- With the shallow observation well the drawdowns continue to increase after pumping ceases; this is reproduced by the model.

The main purpose of this analysis is to determine whether pumping from the fractured zone has any impact on the weathered zone. When the test was conducted the accepted understanding was that the fractured zone is independent of the weathered zone and that a 'new' source of water is tapped by drilling into the underlying fractured aquifer. To test this concept, the numerical model results are used to calculate the

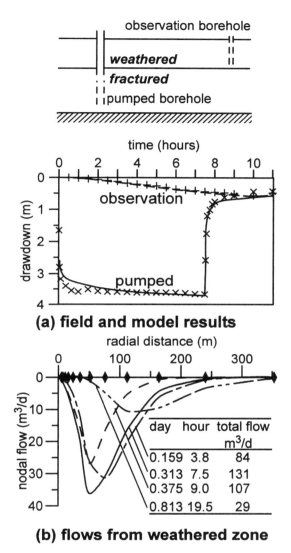

(a) field and model results

(b) flows from weathered zone

Figure 7.16 Details of pumping test in weathered-fractured aquifer: (a) geometry of pumping test, field drawdowns discrete symbols, modelled drawdowns continuous lines, (b) vertical flows between weathered and fractured zones (diamonds indicate the nodes where flows are calculated)

- At 3.8 hours, roughly halfway through the pumping phase, the vertical nodal flows (indicated by the dashed line) have a maximum value at about 50 m from the pumped borehole; the sum of the vertical flows from the weather zone totals 84 m³/d compared to the final abstraction rate from the fractured zone of 288 m³/d.
- By the end of the pumping phase at 7.5 hours (indicated by the unbroken line) the total flow from the weathered zone of 131 m³/d is 45 per cent of the abstraction rate from the fractured zone borehole.
- At 9.0 hours, 1.5 hours after *pumping ceased*, there are still sizeable flows from the weathered to the fractured zone (see chain dotted line) resulting in a continuing fall in the water table in the weathered zone; the total flow of 107 m³/d is 37 per cent of the abstraction rate while the pump was operating.
- Even at 19.5 hours, twelve hours after pumping ceased, there is a small but significant flow of 29 m³/d from the weathered zone to the fractured zone.

These findings demonstrate convincingly that most of the water abstracted from the fractured zone comes from the weathered zone. During the pumping phase, some water is also supplied by the confined storage coefficient of the fractured zone. When pumping ceases, water is drawn from the weathered zone to refill that storage. The significance of this test in terms of long-term resources is discussed in Section 8.5.

Other examples of the application of the two-zone model

The two-zone model has been applied in a wide variety of situations. Parameter values used in certain of these studies are listed in Tables 7.5 to 7.7. Table 7.5 refers to the parameters used in the numerical model simulation of the Kenyon Junction pumping test which is described in Section 5.1; further information can be found in Rushton and Howard (1982) and Rathod and Rushton (1991). Table 7.6 and Section 8.5.4 refer to an aquifer system in Yemen where a sandstone aquifer is overlain by alluvium. Finally, Table 7.7 contains input data when the two-zone model is used to represent the operation of an injection well in an alluvial aquifer in India over a period of 250 days; see Section 8.6.4.

magnitude and timing of flows from the weathered to the fractured zone. Figure 7.16b plots vertical flows from the weathered to the fractured zone during both the pumping phase and the subsequent recovery. The flows are calculated at nodal points of the numerical model (the locations of the nodal points are indicated by the diamond symbols). Flow distributions with radius are quoted at four times.

Table 7.4   Parameter values for the weathered-fractured aquifer (units m or m/d)

Zone or layer	Thickness	Hydraulic conductivity		Storage coefficient	
		Radial	Vertical	Confined	Unconfined
Overlying	0.0 − 0.0 = 0.0	–	0.0	–	–
Upper	13.0 − 0.0 = 13.0	0.50	0.15	0.0001	0.01
Middle	13.0 − 13.0 = 0.0	–	0.3*	–	–
Lower	37.0 − 13.0 = 24.0	4.0	0.6	0.004	–

* Due to zero thickness of the Middle layer, this value has no effect.

Table 7.5   Parameter values for Kenyon Junction sandstone aquifer test (units m or m/d)

Zone or layer	Thickness	Hydraulic conductivity		Storage coefficient	
		Radial	Vertical	Confined	Unconfined
Overlying	0.0 − 0.0 = 0.0	–	0.0	–	–
Upper	45.0 − 0.0 = 45.0*	1.20	0.2	0.0003	0.08†
Middle	45.0 − 45.0 = 0.0	–	0.0	–	–
Lower	90.0 − 45.0 = 45.0	4.0	0.2	0.004	–

* Saturated depth decreases as water table drawdowns occur.
† Delay index $9.0\,\text{d}^{-1}$ (for more information see Rathod and Rushton 1991).

Table 7.6   Parameter values for pumping test in Yemen; see Section 8.5.4 (units m or m/d)

Zone or layer	Thickness	Hydraulic conductivity		Storage coefficient	
		Radial	Vertical	Confined	Unconfined
Overlying	0.0 − 0.0 = 0.0	–	0.0	–	–
Upper	29.0 − 0.0 = 29.0*	2.00	0.20	0.0015	0.10
Middle	32.0 − 29.0 = 3.0	–	0.02	–	–
Lower	94.0 − 32.0 = 62.0	1.50	0.15	0.0005	–

Notes: *Saturated thickness decreases as water table falls. The Upper zone is alluvium, the Middle layer is conglomerate and the Lower zone is sandstone. For further information see Section 8.5.4, Sutton (1985) and Grout and Rushton (1990).

Table 7.7   Parameter values for injection well site in India (see Section 8.6) (units m or m/d)

Zone or layer	Thickness	Hydraulic conductivity		Storage coefficient	
		Radial	Vertical	Confined	Unconfined
Overlying	45.0 − 25.0 = 20.0	–	0.00004	–	–
Upper	85.0 − 45.0 = 40.0	1.25	0.12	0.001	–
Middle	110 − 85.0 = 25.0	–	0.0002	–	–
Lower	145 − 111 = 35.0	3.70	0.37	0.0045	–

Note: this is an alluvial aquifer with a comparatively high clay content.

7.5 INCLUSION OF STORAGE IN AQUITARDS [*r, t: z, t*]

7.5.1 Introduction

Classical leaky aquifer theory (Section 5.8) assumes that as soon as a drawdown occurs in the main aquifer, a uniform vertical hydraulic gradient is set up instantaneously across the full thickness of the aquitard. This vertical flow is assumed to occur from the overlying aquifer, through the aquitard and into the underlying aquifer. The instantaneous vertical velocity throughout the full depth of the aquitard is given by

$$v_z = K' \frac{s}{m'} \tag{7.20}$$

where *s* is the drawdown in the main aquifer relative to the constant water level above the overlying aquifer; the other symbols are defined in Figure 7.17.

The left-hand side of the figure illustrates the classical assumption of constant vertical velocities from the overlying aquifer, throughout the aquitard, then from the aquitard into the main aquifer. However, during the initial stages of pumping, due to the specific storage in the aquitard S_S', the actual velocities are greatest at the bottom of the aquitard and decrease towards the top of the aquitard. Vertical velocities for the early stages of pumping are illustrated on the right-hand side of Figure 7.17. However, at longer times, when an equilibrium condition is approached, the vertical velocities

become effectively constant throughout the depth of the aquitard.

7.5.2 Analytical solutions

Analytical studies of the impact of storage in an aquitard on pumping test analysis have been carried out by Hantush (1960) and Neuman and Witherspoon (1969a,b). In Neuman and Witherspoon (1972) a specific pumping test is considered.

The analysis due to Hantush (1960) considers two alternative leaky aquifer systems. In one the aquitard is overlain by an unpumped aquifer, in the second the aquitard is overlain by an impermeable bed. It is assumed that flow in the aquitard is vertical. Analytical solutions are obtained for early times and late times. An examination of the type curves shows that, when account is taken of aquitard storage, the early-time drawdown in the main aquifer is significantly smaller than when no account is taken of aquitard storage. Drawdowns during the early stages of pumping when aquitard storage is included may be as little as one-tenth of the values when aquitard storage is ignored.

In the studies of Neuman and Witherspoon (1969a,b), allowance is made for storage in the aquitard and a fall in the water table in the overlying aquifer. Expressions are presented which permit the calculation of drawdowns at different depths within the aquitard. A helpful and thorough discussion on the applicability of theories of flows through leaky aquifers is provided by Neuman and Witherspoon (1969b); the consequences of ignoring aquitard storage are considered. There are many important insights in their paper; a number of similar ideas are developed in the case studies of Sections 7.5.3 and 7.5.4.

Using the graphical results from Neuman and Witherspoon (1969b), sets of aquifer and aquitard parameters are selected to illustrate when aquitard storage can be ignored and when it must be considered.

Aquitard storage can be ignored:
distance to observation borehole 40 m; aquifer $T = 400\,\text{m}^2/\text{d}$, $S_C = 0.00016$; aquitard $K' = 0.01\,\text{m/d}$, thickness = 20 m, $S_S' = 4 \times 10^{-6}\,\text{m}^{-1}$,

Aquitard storage must be included:
distance to observation borehole 400 m; aquifer $T = 200\,\text{m}^2/\text{d}$, $S_C = 0.0001$; aquitard $K' = 0.04\,\text{m/d}$, thickness = 40 m, $S_S' = 8 \times 10^{-6}\,\text{m}^{-1}$

Many alternative combinations of aquifer and aquitard parameters could be chosen.

In a further paper, Neuman and Witherspoon (1972)

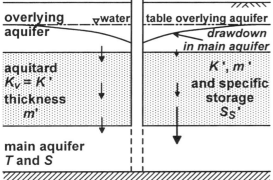

Figure 7.17 Leaky aquifer response using conventional theory or allowing for aquitard storage; figure shows aquifer and aquitard properties and differences in flow processes with and without aquitard storage

consider the practicalities of using analytical solutions to estimate, from pumping tests, aquifer parameters and especially the properties of the aquitard. Of particular importance is their statement that 'a series of observation boreholes in more than one layer is a prerequisite for any reliable evaluation of aquitard characteristics'. They introduce a technique based on the ratio of drawdowns in the aquitard and in the main aquifer at the same radial distance from the pumped borehole. Drawdowns for small values of time are used; this eliminates the impact of some of the unknown non-dimensional parameters which become significant only at later times. A field example from the USA and the practical details of the analysis of the field data are presented.

7.5.3 Case study: influence of aquitard storage on an aquifer system; increase and subsequent decrease in flows from the aquitard

The following discussion on the use of a numerical model to study the effect of storage in an aquitard is based on the PhD thesis of Al-Othman (1991). As part of a pumping test in Riyadh, Saudi Arabia, an observation borehole was constructed in an overlying aquitard. The observed drawdowns in the aquitard were far smaller than expected based on the assumption of an instantaneous uniform vertical gradient throughout the full depth of the aquitard. This finding led to the development of a numerical model which represents the vertical movement of water through the aquitard as well as radial flow through the main aquifer to the pumped borehole. In this analysis it is assumed that only vertical flow occurs in the aquitard. Neuman and Witherspoon (1969c), using an analysis of one aquitard between two aquifers, suggest that the assumption of predominantly vertical flow in the aquitard is acceptable provided that the hydraulic conductivity of the aquifers is more than two orders of magnitude greater than that of the aquitard; in the current study the ratio is 400 : 1. The flow to the pumped borehole in the main aquifer is simulated using a time-variant radial flow numerical model with a logarithmatic mesh spacing in the radial direction (see Section 5.5). The aquitard is represented as a series of vertical columns in which horizontal flow between the columns is not allowed. The vertical velocity in an individual column is represented in terms of drawdowns in the aquitard by the equation

$$K'\frac{\partial^2 s}{\partial z^2} = S_S'\frac{\partial s}{\partial t} \qquad (7.21)$$

Boundary conditions are that at the upper boundary of the aquitard, the drawdown is zero (maintained by the water table in the overlying aquifer) while for the lower boundary, the drawdown equals the corresponding drawdown of the radial flow model. Initial conditions are that the drawdowns are zero everywhere. The flow from the bottom of the aquitard into the main aquifer is calculated from the groundwater head gradient at the bottom of each aquitard 'column'; this flow acts as recharge to the main aquifer.

A specific problem, designed to represent conditions in Riyadh, was analysed using the numerical model; details of the parameter values are as follows:

Aquitard, $m' = 100\,\text{m}$, $K' = 0.05\,\text{m/d}$, $S_s' = 8 \times 10^{-6}\,\text{m}^{-1}$
Aquifer, $m = 25\,\text{m}$, $K = 20.0\,\text{m/d}$, $S_s = 2 \times 10^{-6}\,\text{m}^{-1}$

Twenty mesh intervals in the vertical direction are used to simulate each 'column' of aquitard; the borehole has a radius of 0.1 m.

Figure 7.18 contains important results from this simulation. Figure 7.18b shows how the drawdowns in the aquitard change with time at a radial distance of 21.5 m from the pumped borehole; results are quoted at the base of the aquitard and at 10, 20, 40, 60 and 80 m above the base of the aquitard. The drawdown response at the base of the aquitard is rapid, but considerable delays occur even at 10 m above the base of the aquifer. At 80 m above the base (20 m below the top) of the aquifer, no drawdown occurs until 0.1 day. These delays occur because of storage within the aquitard. In the classical approach without storage in the aquitard, all the drawdown curves are scaled-down versions of the drawdown at the base of the aquitard.

Vertical velocities at the base of the aquitard are plotted in Figure 7.18c; individual plots refer to radial distances of 0.1, 0.316, 2.15, 14.7 and 100 m from the axis of the pumped borehole. These results are remarkable because the velocities from the aquitard in the early stages of pumping are substantially higher than the final values when equilibrium is reached.

Figure 7.19 illustrates the causes of the higher initial flows when aquitard storage is included. Conditions during early times are represented to the left of the borehole and for late times to the right of the borehole. A piezometer in the aquitard penetrating 90 per cent of the aquitard thickness is used to illustrate conditions towards the bottom of the aquitard. During the early stages of pumping, the effect of drawdowns in the main aquifer does not spread far upwards into the aquitard so that little change occurs in the aquitard observation piezometer even though it penetrates almost to the

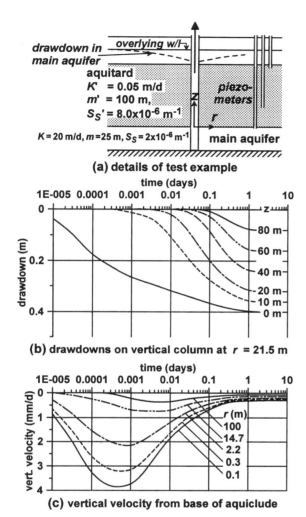

(a) details of test example

(b) drawdowns on vertical column at r = 21.5 m

(c) vertical velocity from base of aquiclude

Figure 7.18 Results from a numerical model representation of a leaky aquifer with storage in the aquitard

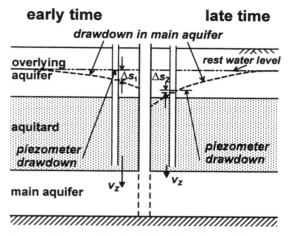

piezometer penetrates 90% of aquitard thickness
early time drawdowns:
  substantial drawdown in main aquifer
  small drawdown in piezometer penetrating 90% aquitard
  large Δs_1, hence large v_z at base of aquitard
late time drawdowns:
  substantial drawdowm in main aquifer
  piezometer drawdown 90% of main aquifer drawdown
  small Δs_2, hence small v_z at base of aquitard

Figure 7.19 Leaky aquifer with storage in aquitard, diagram showing conditions towards the bottom of the aquitard for early and late times

bottom of the aquitard. Since the groundwater head in the observation piezometer is close to the rest water level, the vertical gradient in the lower part of the aquitard is proportional to the difference Δs_1 between the head in the main aquifer and that in the piezometer. For late times, which are illustrated to the right of the borehole, the aquitard storage effects have largely dissipated, so that the vertical head gradient throughout the aquitard is almost uniform. Consequently, the drawdown in the piezometer which nearly penetrates to the base of the aquitard is only slightly less than the drawdown in the main aquifer. The difference in head

between the main aquifer and the deep piezometer is Δs_2. This difference is far smaller than Δs_1 which occurs during early times. Therefore the vertical gradients and hence the vertical flows at the base of the aquitard for radii less than 200 m are higher at early times, but reduce at later times.

Examination of Figure 7.18b, which refers to a radial distance from the pumped borehole of 21.5 m, provides confirmation of these higher initial vertical velocities. The vertical velocity towards the bottom of the aquitard at $z = 5$ m is proportional to the difference between the full line which is the drawdown at the base of the aquifer and the broken line which represents the drawdown at $z = 10$ m. The maximum difference between these drawdowns is 0.25 m compared with a difference of 0.04 m at later times. Since it is easier to collect water from the bottom of the aquitard when the analysis includes aquitard storage, the drawdowns in the main aquifer are smaller for early times than those when classical leaky aquifer assumptions are used.

7.5.4 Influence on aquitard storage of pumping from sandstone aquifers

The following discussion relates to the influence of storage in aquitards on the resources of two sandstone aquifers in the UK. In these case studies the aquitards are of different material; in the first study, the aquitard is a consolidated formation which contains a number of continuous, more permeable layers, whereas in the second case study the aquitard is unconsolidated, mainly clays, but with some sand lenses which may be quite extensive but are not laterally continuous.

Case study 1: slow response of aquitard piezometers due to pumping from underlying aquifer

This case study concerns a 90 m thick Triassic Sandstone aquifer overlain by an aquitard consisting of about 60 m of Mercia Mudstone. The Mercia Mudstone contains low permeability mudstones and siltstones interbedded with layers of sandstone having higher permeabilities. A site at Fradley in the English Midlands was used to examine how much water can move through the aquitard of the Mercia Mudstone to enter the sandstone aquifer. Further information about conditions at Fradley can be found in Boak and Rushton (1993).

Fradley is an operational pumping station. Piezometers were constructed to monitor the water table and the groundwater heads in the middle of the aquitard; records were also kept of drawdowns in the underlying aquifer. Due to variations in actual abstraction rates and the variable duration of pumping, identification of responses in the aquitard is not straightforward. Nevertheless, there is clear evidence of a delayed response for the piezometer in the middle of the Mercia Mudstone; it takes many hours before drawdowns in the observation piezometer in the aquitard responded to changes in pumping from the main aquifer.

From data collection over a period of twenty-one months, information has been extracted to allow analysis of the aquifer and aquitard responses. The two-zone numerical model of Section 7.4 is used to deduce parameter values, listed in Figure 7.20a, which provide an approximate representation of the aquifer behaviour. Figure 7.20b is concerned with the response in a piezometer towards the middle of the aquitard due to an increase in abstraction of 4 Ml/d. The field data cover a period of 60 days; even after 60 days the aquitard piezometer had not reached an equilibrium condition. The numerical model representation of these conditions is shown by the broken line in Figure

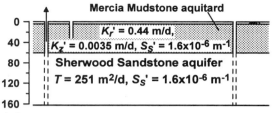

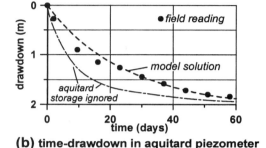

Figure 7.20   Aquifer system at Fradley: (a) physical problem, (b) drawdowns in aquitard piezometer

7.20b; the representation of the second and third data points is not very accurate, but this is likely to be due to variable pumping prior to the period of data analysis. If aquitard storage is ignored, the approximate response is indicated by the chain-dotted line of Figure 7.20b.

Case study 2: Ignoring aquitard storage leads to too high an estimate of the aquitard vertical hydraulic conductivity

In the Fylde aquifer system in north-west England, the Sherwood Sandstone aquifer is overlain by drift consisting mainly of silts and clays with some sand lenses; a representative schematic section, where the drift is typically 60 m thick, is shown in Figure 10.30. The Fylde aquifer is part of the Lancashire conjunctive use scheme. Whenever possible surface water supplies are used but during drought periods very large quantities of water are pumped from the Fylde aquifer. Further information about the regional setting of the Fylde aquifer can be found in Section 10.6.4.

Most of the inflow to the Sherwood Sandstone aquifer occurs by vertically downwards leakage through the drift. This scenario was explored using a multi-layered groundwater model with a classical leaky aquifer response in the drift (there was no allowance for the release of water stored in the drift); the model study covered twenty-five years of historical records. The model successfully reproduced the first two periods of heavy pumping which occurred in drought periods. During the model refinement, parameter values describing the vertical hydraulic conductivity of the drift were deduced. However, the model failed to reproduce the aquifer response for further periods of heavy abstraction.

Failure to represent storage in the aquitard, and therefore to represent the high flows from the bottom of the aquitard during early stages of pumping as illustrated by Figure 7.18c, resulted in a model which introduced too high a value of the vertical hydraulic conductivity for the drift. Furthermore, extensive sand lenses within the drift were not represented in the original model. These sand lenses provided a reliable source of water within the aquitard during the early pumping periods but ceased to provide water once they were dewatered. Once these additional mechanisms were recognised and incorporated in the numerical model, improved simulations of the aquifer response were achieved.

7.6 CONCLUDING REMARKS

Four alternative formulations are considered for radial flow problems in which vertical flow components are significant. These alternative formulations are selected to represent the dominant flow mechanisms which apply in different practical problems. Most of the approaches involve idealisations such as the use of a time-instant (pseudo-steady state) approximation or the representation of a complex aquifer system by two zones with low permeability layers between or overlying the more permeable zones. It is realistic and pragmatic to introduce these idealisations, due to limited information about the nature of most aquifers and the limited number of observation piezometers.

The study of time-variant radial-vertical flow includes a discussion of the two phenomena of delayed drainage at the water table and vertical flow components within the saturated aquifer; they cause similar responses in observation piezometers. Both of these phenomena occur in many practical situations. Numerical solutions for time-variant radial-vertical flow are used for a study of time-drawdown responses in the vicinity of a borehole with solid casing through the upper part of the aquifer. A further investigation relates to an aquifer containing discontinuous layers of lower hydraulic conductivity.

The time-instant approach, with analysis in terms of radial and vertical co-ordinates, is appropriate when investigating situations when there is a balance between discharge from and recharge to the aquifer system, or for the stage of a pumping test in an unconfined aquifer when water is drawn primarily from the water table with only small changes in drawdown with time. Case studies based on numerical models include the reduction in discharge due to partial penetration of a borehole and water table control using tubewells in layered aquifer systems.

When the time-variant nature of the aquifer response is significant, a two-zone approximation can be used to best advantage. Many pumped aquifer systems can be idealised as two main aquifer zones with low permeability layers between or overlying the permeable zones. Detailed descriptions of the derivation of the equations for this approximation are presented; a computer program for the two-zone approach is included in the Appendix. Two case studies are considered in detail. The first study illustrates the complex movement of water between the aquifer zones and the significance of well storage, confined storage and unconfined storage. The second example refers to pumping from a weathered-fractured aquifer system.

The final section in Chapter 7 relates to storage in aquitards. Conventional leaky aquifer theory ignores aquitard storage and assumes that a uniform vertical hydraulic gradient is set up instantaneously through the full depth of the aquitard. Frequently this is an invalid assumption. If aquitard storage is ignored and traditional leaky aquifer theory is used for pumping test analysis, both the transmissivity of the aquifer and the vertical hydraulic conductivity of the aquitard may be over-estimated by a factor of up to five. A methodology is introduced to represent aquitard storage in a radial flow numerical model. Using this methodology successful investigations are carried out into the aquifer responses and the aquifer resources for three aquifer systems in which aquitard storage has a substantial effect.

Certain of the techniques of analysis developed in this chapter are used in Chapter 8 to investigate practical issues related to the operation of boreholes and the yield of aquifer systems.

8

Practical Issues of Interpreting and Assessing Resources

8.1 INTRODUCTION

Pumping tests are generally used to estimate values of the aquifer parameters of transmissivity and storage coefficient. However, a greater understanding of actual conditions in the aquifer systems can often be gained from a critical examination of all the field data. The purpose of this chapter is to demonstrate that many important insights can be obtained from a careful examination of both specially designed pumping tests and long-term records of water levels at pumping stations. The discussion is based on concepts and methodologies developed in Chapters 5 to 7; the contents of this chapter are summarised in Table 8.1.

The first two topics refer to the techniques of step-drawdown pumping tests and packer testing. The true value of step-drawdown pumping tests is gained if, rather than restricting the analysis to the determination of the formation loss and well loss constants, the step tests are used to explore the flow processes in the vicinity of pumped boreholes. Packer tests provide detailed insights into the variation of aquifer properties with depth. Radial-vertical time-instant flow modelling greatly assists in the interpretation of these two field techniques.

The next topic is the interpretation of water level measurements in production boreholes under pumped and non-pumped conditions; information can be gained about aquifer conditions in the vicinity of pumping stations. The long-term reliable yield of an aquifer system can be assessed from pumping tests. Case studies of heavily exploited aquifers are reviewed; reasons for the over-estimation of aquifers resources are discussed. A pilot scheme of artificial recharge in a heavily exploited aquifer using an injection well is also studied; field data from an injection test demonstrate the serious impact of clogging of the injection well. Further examples, when yields from aquifer systems are less than initially anticipated, are provided by chalk and limestone aquifers in which transmissivities vary significantly between high and low water table elevations. Methods of estimating the aquifer parameters and impacts on long-term yields are examined. Finally there is an introduction to horizontal wells; a case study is described of a horizontal well in a shallow coastal aquifer.

8.2 STEP DRAWDOWN TESTS AND WELL LOSSES

8.2.1 Introduction

The efficiency and the realistic yield of a borehole can be assessed using a step-drawdown test in which the discharge rate Q is increased in a series of steps each lasting one to two hours; the drawdown in the well s_w at the end of each step is recorded. Results for a typical step test in an unconfined aquifer are included in Figure 8.1. If there are no well losses, the drawdown increases linearly with discharge rate, but in most practical situations the drawdowns for higher discharge rates are greater than predicted by a linear relationship. Jacob (1947) proposed the non-linear relationship

$$s_w = BQ + CQ^2 \qquad (8.1)$$

where B is the formation loss and C is the well loss coefficient. Although several papers consider methods of analysing step-tests, the actual causes of the well losses are not usually examined. Clarke (1977) suggests that well losses are usually composed of turbulent losses around the well screen and frictional losses in the

Groundwater Hydrology: Conceptual and Computational Models. K.R. Rushton
© 2003 John Wiley & Sons Ltd ISBN: 0-470-85004-3

Table 8.1   Issues and case studies considered in Chapter 8

Issue	Method of analysis	Case studies
8.2 step-drawdown tests	radial-vertical time-instant [r, z]	four examples of field tests in alluvial, chalk and sand-gravel aquifers
8.3 packer testing	radial-vertical time-instant [r, z]	comparison of packer test results with core testing and pumping test slug tests
8.4 production borehole records	–	identifying minimum water table elevation during drought years
8.5 reliable yield	two-layer [r, t, v_z]	weathered-fractured aquifer; tubewells into fractured zone are too effective at collecting water; 8.5.2
		alluvial aquifer with low K in uppermost layer; 8.5.3
		alluvium overlying sandstone; presence of seepage face limits borehole yield; 8.5.4
8.6 injection wells	two layer [r, t, v_z]	pilot artificial recharge scheme in alluvial aquifer, serious consequences of well clogging; 8.6.2–8.6.4
		artificial recharge by injection boreholes, London; 8.6.5
8.7 variable K with depth	two-layer [r, t, v_z]	chalk aquifer of Berkshire Downs, reduced yields with lower water tables
8.8 horizontal wells	analytical and numerical with hydraulic conditions in well	horizontal well in shallow coastal aquifer; determination of inflows along horizontal well

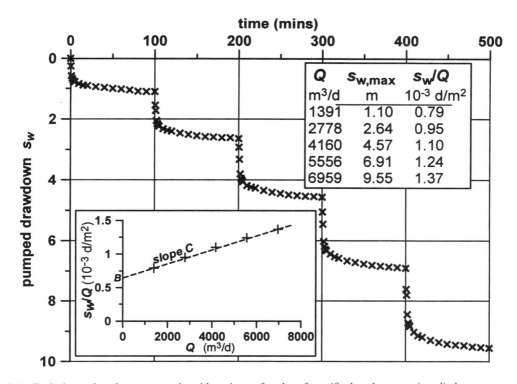

Figure 8.1   Typical step-drawdown test results with an inset of a plot of specific drawdowns against discharge

casing screen. However, the location and operation of most pumped boreholes means that there are several physical phenomena which can influence the draw-down-discharge relationship of a step test. In addition to turbulent losses in the aquifer and around the well screen, the effects of a seepage face, dewatering of fissures in the aquifer and the possibility of conditions changing between confined and unconfined can also influence the response. These features will be explored using a numerical model; in addition field examples are presented showing how these features affect step-drawdown tests.

8.2.2 Confined, leaky or unconfined conditions?

Are step tests suitable for confined, leaky or unconfined aquifers? For classical laminar flow to a fully penetrating well in a *confined* aquifer, the drawdown in the well can be written using the Cooper–Jacob approximation (Section 5.4.1) as

$$s_w = \frac{2.30Q}{4\pi T}\log\left(\frac{2.25Tt}{r_w^2 S}\right) \qquad (8.2)$$

Provided that each step lasts for the same time (i.e. t is constant), the above equation can be simplified

$$s_w = BQ$$

In step tests, pumping usually continues until the draw-down approaches a steady state. However, Eq. (8.2) shows that the pumped drawdown in a confined aquifer increases with time. Yet, in most practical situations, approximately steady pumped drawdowns are achieved. The reason for this can be appreciated by considering the response of leaky and unconfined aquifers.

For a *leaky* aquifer, an equilibrium is reached when storage effects are no longer significant and the pumped drawdown can be estimated from the expression

$$s_w = \frac{Q}{2\pi T}K_0\left(\frac{r_w}{L}\right) \qquad (8.3)$$

where $K_0()$ is a Bessel function and the coefficient $L = (Tm'/K')^{1/2}$ (see Section 5.7 and Eq. (5.22); note that in that Section 7.7 the symbol B is used rather than L as in Eq. (8.3)). Therefore the pumped drawdown is proportional to the discharge, $s_w = BQ$, for times greater than about two hours since equilibrium is approached more rapidly at the pumped borehole than in more distant parts of the aquifer system.

For *unconfined* aquifers there is no equivalent equilibrium equation, but the results of the Kenyon Junction test, especially Figures 5.1 and 5.3c, demonstrate that an equilibrium is approached after about 100 minutes. Consequently, in an ideal situation there is a linear relationship between discharge and pumped drawdown, hence $s_w = BQ$.

The conclusion of this discussion is that equilibrium is reached in step-drawdown tests in leaky and unconfined aquifers but not in confined aquifers. In practice few aquifers meet the strict conditions of a confined aquifer; most aquifers are leaky or unconfined. Consequently the following study of well losses is based on the flow patterns for leaky and unconfined aquifers.

8.2.3 Estimating the coefficients *B* and *C*

It is helpful to take the equation

$$s_w = BQ + CQ^2 \qquad (8.1)$$

as a basis for the study of step-drawdown tests. The estimation of the coefficients B and C is illustrated in Figure 8.1. The box in the top right-hand corner of the figure shows the five discharge rates, the pumped drawdowns at the end of each step of 100 minutes and, in the third column, the ratio s_w/Q. From the numerical values it is clear that the *specific drawdown* s_w/Q is not constant but increases with pumping rate. It is helpful to work in terms of the specific drawdown since, when s_w/Q is plotted against Q, the constants B and C can be estimated directly. Dividing Eq. (8.1) by Q

$$s_w/Q = B + CQ \qquad (8.4)$$

Therefore, as indicated by the graph in the bottom left-hand corner of Figure 8.1, $B = 0.65 \times 10^{-3}$ d/m^2; this value is the intercept with the y axis. The slope of the curve gives the constant $C = 0.103 \times 10^{-6}$ d^2/m^5.

This particular test refers to a sandstone aquifer in which the borehole penetrates 150 m with solid casing for the top 10 m, and the remainder of the borehole is uncased. In such a situation, there are unlikely to be substantial turbulent well losses since there is no gravel pack and no slotted casing. However, one important piece of information is that at the start of the test, the rest water level was 9.1 m below ground level, or 0.9 m above the bottom of the solid casing. This means that during the first step, the pumped water level fell below the solid casing and a small seepage face formed; the length of the seepage face increased during the subsequent steps. Consequently, one of the cases to be studied later refers to the effect of a seepage face on drawdown-discharge relationships. Incidentally, the presence of a seepage face can be identified in the field by listening to water cascading down the borehole when pumping ceases.

8.2.4 Exploring well losses

The radial-vertical time-instant approach of Section 7.3 is used to study the drawdown-discharge relationships for a number of physical situations that occur in unconfined aquifers. Conditions on boundaries are as specified as in Figure 7.7, which is redrawn here as Figure 8.2. Since the duration of pumping in step tests is relatively short, it is realistic to assume zero drawdown at the water table. Conditions on the well face are important:

- in the upper part of the well bore there is blank casing; no flow can occur through the casing,
- if the pumped water level is below the bottom of the blank casing, a seepage face forms between the well water level and the base of the blank casing,
- a third condition of constant drawdown occurs along the column of water in the borehole where there is direct contact with the aquifer or gravel pack.

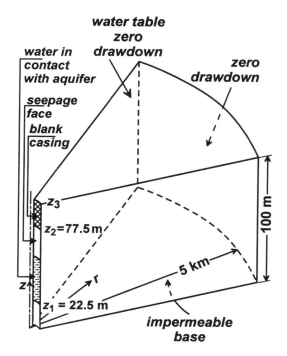

water in contact with aquifer

water table zero drawdown

zero drawdown

seepage face
blank casing

z_3

$z_2 = 77.5\,m$

z

5 km

100 m

$z_1 = 22.5\,m$

r

impermeable base

Figure 8.2 Details of boundary conditions for model used to explore well losses

For most of the simulations, the following parameter values apply:

depth of aquifer = 100 m   hydraulic conductivity
                                              $K = 3.0\,m/d$
diameter of borehole   maximum radius
    = 0.2 m   = 5.0 km
borehole penetration   solid casing to 22.5 m
    = 77.5 m

A series of numerical model solutions are obtained with increasing pumping rates; in each case the resultant pumped drawdown is recorded. The results are plotted as specific drawdown (s_w/Q) against discharge Q. Any deviation from a straight *horizontal* line is an indication of non-linear losses which are conventionally attributed to turbulent flow.

8.2.5 Causes of well loss

A series of possible causes of well losses are investigated; the findings are presented as graphs of specific drawdown against discharge; more detailed information can be found in Rushton and Rathod (1988).

(a) Seepage face in homogeneous aquifer

A sequence of simulations, with pumped drawdowns gradually increasing, provides information about the discharge required to achieve these drawdowns. Provided that the pumped water level is within the solid casing, the discharge increases linearly with the pumped drawdown with the ratio $s_w/Q = 5.7 \times 10^{-3}\,d/m^2$; see Figure 8.3. However, when the pumped drawdown is below the bottom of the solid casing (the situation illustrated in Figure 8.2), a seepage face forms so that less water is drawn from the aquifer. For all the steps in which the drawdown in the pumped well is greater than 22.5 m, seepage faces form and s_w/Q is greater than $5.7 \times 10^{-3}\,d/m^2$. For the final step, with a pumped drawdown of 65 m, $s_w/Q = 7.85 \times 10^{-3}\,d/m^2$.

The type of relationship identified in Figure 8.3 is often ascribed to turbulent losses with a straight inclined line drawn through individual field values (see Figure 8.8d). In many tests there are only three or four field values; results for smaller discharges may not be available. Consequently the available field readings may appear to lie on a sloping straight line. However, for this example there are no turbulent losses; the non-linear behaviour is due solely to the development of the seepage face.

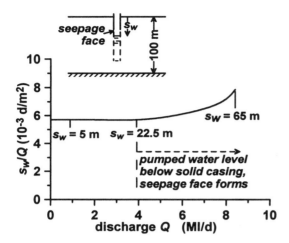

Figure 8.3 Specific drawdown-discharge curve due to the presence of a seepage face. Reproduced by permission of Geological Society, London, from Rushton and Rathod (1988)

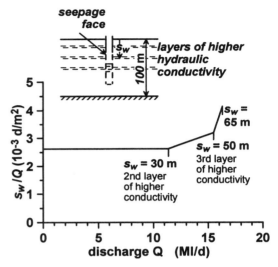

Figure 8.4 Specific drawdown-discharge curve due to the presence of a seepage face and layers of higher hydraulic conductivity. Reproduced by permission of Geological Society, London, from Rushton and Rathod (1988)

(b) Seepage face with dewatered zones of higher hydraulic conductivity

The second example is designed to explore the impact of horizontal layers of higher hydraulic conductivity which are dewatered as the pumped water level declines. This is a phenomenon frequently observed in limestone and chalk aquifers.

Three layers with higher hydraulic conductivity are included in the model at depths of 10, 30 and 50 m below the top of the aquifer; each layer is 1.0 m thick with a hydraulic conductivity fifty times the standard. The same procedure is followed as in (a) above; the results are plotted in Figure 8.4. While the pumped water level is within the casing, the specific drawdown remains at $s_w/Q = 2.62 \times 10^{-3}$ d/m^2.

Significant changes occur as the pumped water level falls below the middle high conductivity layer at a depth of 30 m. Once this high conductivity layer is above the water level in the well, water seeps out rather than being drawn out under a higher lateral hydraulic gradient. As the pumped water level falls further, the contribution from the middle high conductivity layer becomes less significant and the specific drawdown increases further. When the pumped water level falls below 50 m, the lower hydraulic conductivity layer ceases to act efficiently. As shown in Figure 8.4, the specific drawdown increases rapidly with increasing discharge rates.

(c) Non-Darcy flow and deterioration of hydraulic conditions close to borehole

In the third example, the conditions required to reproduce the well loss equation

$$s_w/Q = B + CQ$$

are explored. This is achieved by introducing a zone of 'well loss' surrounding the slotted portion of the borehole, as shown in Figure 8.5. This zone of well loss could correspond to a gravel pack, or to the restriction as water passes through the slots in the borehole casing, or disturbance of the formation during the drilling of the borehole, etc. Two different phenomena will be explored, non-Darcy flow and reduction in the hydraulic conductivity of the aquifer in the vicinity of the borehole.

(i) Non-Darcy flow can occur due to high velocities in the approach to the borehole (Todd 1980). As a first approximation it is assumed that these losses are proportional to the square of the velocity. It is helpful to use the concept of hydraulic resistance to understand how these turbulent flows can be simulated in the numerical model. If the hydraulic resistance remains constant, the head loss doubles with a doubling of the flow; this is the normal Darcy response. If, however, the hydraulic resistance increases linearly with the flow,

the head loss is proportional to the square of the flow. In the standard model, the hydraulic resistance for the mesh interval adjacent to the borehole R_h depends on the geometry and the hydraulic conductivity. The turbulent effect can be represented by changing the hydraulic resistance with the discharge according to a formula of the form

$$R_h' = R_h(1.0 + 0.0025Q)$$

The significance of this assumption in terms of modified hydraulic conductivities is shown in Table 8.2.

When these modified hydraulic conductivities are included in the time-instant radial flow model, the relationship between specific drawdown and discharge is as shown by the unbroken line in Figure 8.5. The slope of this line represents turbulent well losses; the intercept

with the y-axis represents the formation losses. The value for a discharge rate of zero is obtained by extrapolation back to the y-axis; the specific drawdown corresponding to zero discharge is $s_w/Q = 5.7 \times 10^{-3}$ d/m^2. This is identical to the value in Figure 8.3 since no seepage face develops. For all the results plotted in Figure 8.5 the pumped drawdowns are less than the depth of the solid casing of 22.5 m.

(ii) Reduction in the hydraulic conductivity close to the borehole often occurs due to some form of deterioration in the aquifer, gravel pack or slotted casing. This may only change the laminar flow components and not modify the turbulent losses. This situation is represented by modifying, in the numerical model, the horizontal hydraulic conductivities for the four logarithmic mesh intervals between the well face and a radial distance of 1.0 m. The modified hydraulic conductivities are 1.0, 1.8, 2.16 and 2.55 m/d with the mesh intervals beyond 1.0 m remaining at 3.0 m/d. When these changes are combined with the turbulent effects, the resultant specific drawdown-discharge relationship is shown by the broken line in Figure 8.5. The coefficient B, which is the specific drawdown at zero discharge, increases from 5.7 to 7.6 $\times 10^{-3}$ d/m^2, while the slope C retains the same value.

The results of Figure 8.5 demonstrate that:

- the traditional $BQ + CQ^2$ relationship can be represented by a zone of 'well loss' extending over the full depth of the slotted section. Part of the coefficient B relates to formation (or aquifer) losses, but part may be due to clogging in the vicinity of the pumped borehole. The turbulent loss is proportional to the square of the velocity distributed over the depth of the slotted section.

- clogging in the vicinity of the borehole will lead to an increase in the coefficient B; therefore it is incorrect to state that B is due to the formation loss alone.

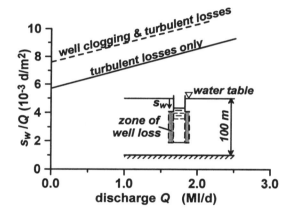

Figure 8.5 Example of a zone of 'well loss' fully surrounding the screened section of a borehole. Reproduced by permission of Geological Society, London, from Rushton and Rathod (1988)

Table 8.2 Inclusion of the effect of non-Darcy flow by modifying horizontal hydraulic conductivity adjacent to the borehole from 0.1 m to 0.178 m with $R_h = 0.3333$

Discharge (m³/d)	Standard R_h (m/d)⁻¹	Additional hydr. res. as fraction of R_h.	Total hydr. res. R_h' (m/d)⁻¹	Equivalent hydr. cond. (m/d)
0	0.3333	0.0	0.3333	3.00
400	0.3333	1.0	0.6667	1.50
800	0.3333	2.0	1.0000	1.00
1200	0.3333	3.0	1.3333	0.75
1600	0.3333	4.0	1.6667	0.60
2000	0.3333	5.0	2.0000	0.50

• well penetration can influence coefficient B. If the location and/or length of the slotted section is changed, this leads to a different value of B.

Driscoll (1986) makes important observations about the differing contributions of laminar and turbulent losses. He introduces a parameter L_p which is the percentage of the total head loss attributed to laminar flow

$$L_p = 100 \times \frac{BQ}{BQ + CQ^2} \qquad (8.5)$$

Driscoll explains that if the assumption by Jacob is correct (that BQ is equal to the aquifer loss and CQ^2 is the well loss), L_p would equal the well efficiency. But Driscoll argues that in testing hundreds of wells, he has found that L_p does not represent the well efficiency, citing turbulent losses occurring in the undisturbed formation around the well. Therefore the terms BQ and CQ^2 may each represent both aquifer losses and well losses. This finding is supported by the modelling results of Figure 8.5.

(d) Turbulent flow in a single fissure

Examples in the above sub-section showed that the traditional step test relationship only holds if the turbulent losses are spread relatively uniformly over the full depth of the slotted section. Consequently, the effect of turbulence in a single fissure will be examined. The location of the fissure is shown in Figure 8.6; the fissure is represented as having a thickness of 5 m, extending for 31 m from the well face with a hydraulic conductivity under laminar flow conditions of 30 m/d, which is ten times the remainder of the aquifer. In practice, the fissure will be much narrower; however, when it is represented in a numerical model, its effect is spread over a vertical mesh interval which, for this model, is 5 m.

Turbulent losses, which are proportional to the square of the velocity, can be represented in a similar manner to the above example by increasing the horizontal hydraulic resistance as the flow increases. The velocities in the fissure decrease with increasing distances from the borehole. Turbulent losses are represented by modified horizontal hydraulic resistances R'_{hf} in the fissure.

$$R'_{hf} = R_{hf}(1.0 + 0.1375 v_f) \qquad (8.6)$$

where R_{hf} is the horizontal hydraulic resistance under laminar flow conditions and v_f is the Darcy velocity in the fissure.

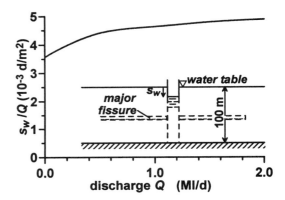

Figure 8.6 Effect of turbulent conditions in a major fissure. Reproduced by permission of Geological Society, London, from Rushton and Rathod (1988)

Simulations using the numerical model are carried out with different discharge rates. Adjustments are made to the hydraulic resistances along the fissure according to the velocities within the fissure; this requires an iterative procedure. The diagram of specific drawdown against discharge is presented in Figure 8.6. For low discharge rates the specific drawdown of 3.57×10^{-3} d/m^2 is smaller than that for the homogeneous aquifer of 5.7×10^{-3} d/m^2. However, as the discharge rate increases to 2000 m^3/d, the fissure velocity in the vicinity of the borehole increases, leading to a substantial increase in the horizontal hydraulic resistance in the fissure as defined in Eq. (8.6). Due to the rapid increase in hydraulic resistance, the fissure becomes ineffective at transmitting water to the borehole; the specific drawdown tends to 4.9×10^{-3} d/m^2.

(e) Conditions changing between confined and unconfined

The final example is not normally considered as an occurrence of well loss, yet it does fall within the classification of a non-constant specific drawdown even though the specific drawdown *decreases* with increasing discharge rates. When an aquifer is initially confined, water is drawn laterally through the aquifer, but if the groundwater head falls sufficiently for unconfined conditions to occur close to the borehole, it becomes easier to collect water with substantial vertical flows from the overlying water table. This situation is illustrated by Figure 8.7; around the borehole the groundwater head falls below the confining layer so that unconfined conditions occur and the specific yield becomes operative in that region.

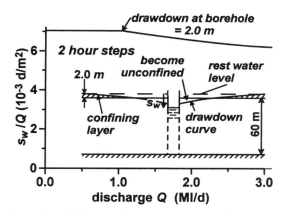

Figure 8.7 Fall in specific drawdown due to changes from confined to unconfined in the vicinity of a pumped borehole. Reproduced by permission of Geological Society, London, from Rushton and Rathod (1988)

Using a time-variant numerical solution $[r, z, t]$, a series of two-hour step tests are simulated in an aquifer which is 60 m deep with initial conditions that the groundwater head is 2.0 m above the confining layer. With a discharge rate of 100 m³/d, the pumped drawdown after two hours is 0.704 m. For a pumping rate of 1000 m³/d (1.0 Ml/d), the pumped drawdown is 7.04 m with the aquifer drawdown just outside the casing immediately below the confining layer equal to 1.91 m (this is less than the confining head of 2.0 m). When the discharge rate is 1047 m³/d, the drawdown at this location increases to 2.0 m so that unconfined conditions begin to develop. For higher discharge rates, unconfined conditions spread further from the pumped borehole. With a discharge rate of 3500 m³/d, the unconfined region extends to 20 m from the borehole centre-line. Since it is now easier for the borehole to collect water from the unconfined area surrounding the borehole, the pumped drawdown is 21.7 m compared to a drawdown of 24.6 m which would occur if the whole aquifer remained confined. Therefore the specific drawdown-discharge curve, Figure 8.7, has declining values for pumping rates greater than 1047 m³/d.

8.2.6 Field examples of well losses

In field situations, there may be a number of reasons for the specific drawdown to change with increasing discharge rates. The four field examples in Figure 8.8 are selected to illustrate how the physical conditions influence the shapes of the specific drawdown-discharge curve.

(a) *Tubewell in an alluvial aquifer*: this aquifer contains lenses of sand and sandy clay. The tubewell extends to 63 m below ground level with solid casing to 28 m and five slotted sections adjacent to the more permeable layers; there is a gravel pack around the tubewell. A straight inclined line similar to Figure 8.5 is drawn through the four specific drawdowns plotted in Figure 8.8a. Significant turbulent losses occur which are probably due to the high velocities close to the borehole in the more sandy layers.

(b) *Injection well in an alluvial aquifer*: this injection well test is described in Section 8.6; the borehole taps two aquifer units which contain sand and some clay lenses. Three step tests were performed; two sets of results are plotted in Figure 8.8b. Result A refers to a step test before the injection test (*IT*); the three specific discharge values lie on a straight inclined line. After the 250-day injection test, when substantial clogging had occurred, a second step test was conducted with specific drawdowns approaching 46×10^{-3} d/m². These results are not plotted in Figure 8.8b because this would require a ninefold increase in the vertical scale. However, the line through the field values is parallel to line A. After cleaning and redevelopment, a third step test was carried out; results are plotted on Figure 8.8b as line C, which has the same slope but is lower than line A. This example demonstrates that turbulent losses remain relatively constant; they are due mainly to high velocities through the gravel pack and slotted casing. However, the coefficient B changes significantly between the different tests, and this is due to the reduction and subsequent increase in the hydraulic conductivities in the vicinity of the well.

(c) *Borehole in a chalk aquifer*: extensive tests were carried out in an unconfined chalk aquifer in the Berkshire Downs, England; further information can be found in Section 8.7. The results for conventional two-hour steps are included in Figure 8.8c; the slope of the line plotted through the three field readings is shallow, indicating that increasing abstraction rates lead to a roughly proportionate increase in pumped drawdowns. Results are also available for a series of tests lasting 14 days. The slope of the line through these field results is far steeper; the primary reason for this is seepage face development during the longer tests with the consequent dewatering of fissures (see Figure 8.4).

(d) *Shallow unconfined gravel aquifer*: the Yazor gravels are shallow with a saturated thickness of only 5 m (Rushton and Booth 1976). Nevertheless, yields of more than 2 Ml/d can be achieved. Figure 8.8d

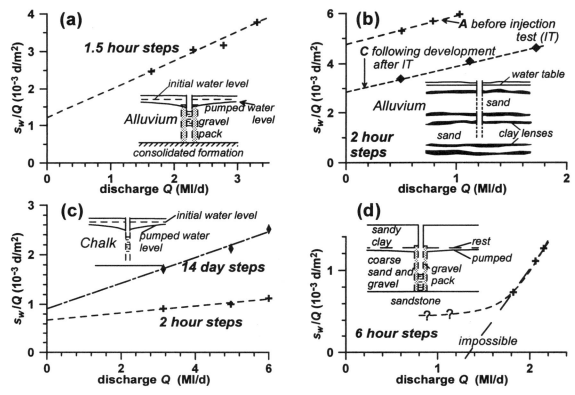

Figure 8.8 Four field examples of step-drawdown tests. Reproduced by permission of Geological Society, London, from Rushton and Rathod (1988)

contains three results from five-hour steps. The values appear to lie on a straight line. A straight line is constructed through the field readings; this line is labelled *impossible* since it crosses the *x*-axis and infers a negative specific drawdown, and hence a negative pumped drawdown for smaller discharge rates! Field observations show that a large seepage face formed; therefore the results for Figure 8.4 can be used as a guide to the likely shape of the curve. The curve plotted in Figure 8.8d is only tentative, but it does indicate that the borehole cannot operate efficiently at discharge rates greater than 1.8 Ml/d.

8.3 PACKER TESTING TO IDENTIFY VARIATIONS IN HYDRAULIC CONDUCTIVITY WITH DEPTH

Packer tests are introduced in Chapter 2; they can be used to explore local aquifer responses including the

variation of hydraulic conductivity with depth. In geotechnical investigations, as a borehole is drilled, sections of the borehole are isolated, often using a single packer. From pumped (or injection) drawdown characteristics, the properties of the isolated section can be identified. Further information can be found in the US Bureau of Reclamation Earth Manual and British Standard 5930. A variety of methods of analysis are used. Some tests are time-variant when a slug of water is displaced and the changes of head with time are measured (see Section 8.3.6); in other methods readings are taken when an equilibrium condition is reached. Factors published by Hvorslev (1951) are used for different geometries of the casing and end conditions.

8.3.1 Conducting packer tests

Extensive investigations in a sandstone aquifer in north-west England indicated that a reliable technique for estimating the hydraulic conductivity of a packered section involves pumping water from between two packers set about 3 m apart so that the drawdown, relative to at-rest conditions, is in the range of 1–3 m. Pumping rates are adjusted so that the drawdown for the first step is approximately 1.0 m; this pumping rate is maintained until a steady drawdown is achieved. The pumping rate is then increased in two steps and decreased in two steps to provide discharge-drawdown relationships under both increasing and decreasing pumping rates. This procedure provides confirmation that the packers are seated correctly and there are no other causes of leakage or non-linear behaviour. Discharge tests are preferred to injection tests to prevent clogging of the borehole and any difficulties which can arise with the quality of injected water (Brassington and Walthall 1985).

Figure 8.9a is a schematic diagram showing the isolation of part of the borehole by two packers; water is pumped from between the packers and the resultant groundwater head h_p compared to the undisturbed head h_o. Pumping water from between the packers also results in changes in hydraulic head in the section of the borehole above the upper packer h_u and below the lower packer h_l. Further information about the practi-

calities of packer testing can be found in Price *et al.* (1982), Walthall and Ingram (1984) and Brassington and Walthall (1985).

8.3.2 Interpretation of packer tests using analytical expressions

A formula was introduced by Hvorslev (1951) to analyse the results obtained from packer tests with corrections due to alternative conditions at the open end of the borehole. For the arrangement of Figure 8.10a, with a section of a borehole of length L between the packers,

$$Q = \frac{2\pi L K_r H}{\ln\left[\dfrac{mL}{2r_w} + \sqrt{1 + \left(\dfrac{mL}{2r_w}\right)^2}\right]} \tag{8.7}$$

where Q is the pumping rate from between the packers, H is the decrease in groundwater head compared to the unpumped condition, r_w is the borehole radius, L is the vertical distance between the packers, K_r is the radial hydraulic conductivity and m is a ratio representing the anisotropy, $m = \sqrt{K_r / K_z}$.

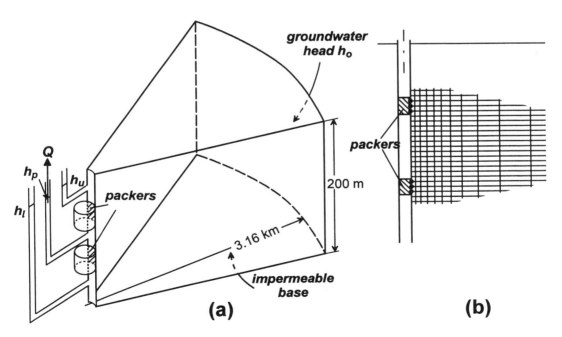

Figure 8.9 Schematic diagram of packers in a borehole: (a) heads in borehole between, above and below packers, (b) representation of packers in numerical model. Reproduced by permission of Geological Society, London, from Bliss and Rushton (1984)

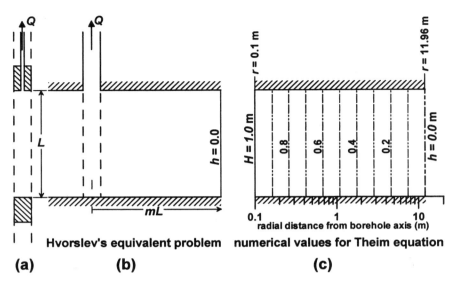

Figure 8.10   Hvorslev's method of analysis: (a) packers in borehole, (b) Hvorslev's equivalent problem, (c) numerical results for problem described in text

For mL at least ten times the borehole diameter, this expression reduces to

$$Q = 2\pi L K_r H / \ln[mL/r_w] \qquad (8.8)$$

This is of the same form as the Thiem equation, Eq. (5.2),

$$Q = 2\pi T s_w / \ln[R/r_w]$$

where s_w is the drawdown in the pumped borehole and R is the distance to an outer boundary where no drawdown occurs.

Since Eq. (8.8) is of identical form to the Thiem equation, it can be represented diagrammatically as shown in Figure 8.10b. In comparing the Thiem equation with Eq. (8.8),

T is identical to LK_r,

s_w is equivalent to H, the decrease in head compared to the unpumped conditions,

R, the distance to the outer boundary, is equivalent to mL.

When there is a single fissure in a packered section the following equation, derived by Barker (1981), can be used:

$$Q = \frac{2\pi KbH}{\ln\left[Kb\big/\exp(0.5772)r_w(K_rK_z)^{1/2}\right]} \qquad (8.9)$$

In Eq. (8.9) the fissure has a hydraulic conductivity K and thickness b, with background radial and vertical hydraulic conductivities K_r and K_z. In deriving this expression, Barker assumes that all the flow into the test section occurs through the fissure. Approximations are introduced in the mathematical derivation so that the equation does not involve Bessel functions.

8.3.3 Do the analytical solutions provide a reasonable approximation to hydraulic conductivity variations?

When packer tests are carried out in aquifers containing both higher and lower hydraulic conductivity layers, can the relatively simple expressions of Eqs (8.7)–(8.9) represent the differing aquifer properties with depth? This question is explored by developing a numerical model (time-instant r, z) which represents detailed information obtained from an intensive field investigation of a sandstone aquifer. Information about the field investigation can be found in Bliss and Rushton (1984), Walthall and Ingram (1984) and Brassington and Walthall (1985).

Steps in the investigation are as follows:

1. Extensive field investigations in a single borehole, using cores tested in a permeameter, down-hole geophysical investigations and packer testing, led to a detailed understanding of the sandstone aquifer (e.g. estimates of the hydraulic conductivities K_r and K_z for every 0.25 m over the full 200 m depth of the aquifer). There are five sandstone rock types. From the permeameter tests on horizontal and vertical core samples it was found that typically K_z is 50 per cent of K_r. In addition, there are very fine micaceous sandstone bands with an average vertical spacing of about 6 m. From packer testing, calliper readings, down-hole television and vertical flow monitoring, sixteen locations were identified with significant fissures, twelve of which were at depths of less than 80 m.

2. Sections were identified where either typical or unusual packer test results were obtained; these sections are studied in detail. For each of these 3.25 m sections, Eqs (8.7) and (8.9) (if applicable) were used to estimate the discharge Q required to cause a drawdown of 1.0 m.

3. A time-instant r-z numerical model was prepared using the properties identified in (1) above. In sections where there are many fissures the vertical mesh spacing is 0.25 m, a logarithmic radial mesh spacing is used. The general arrangement of the computational model, including the boundary conditions, is indicated in Figure 8.9a; Figure 8.9b shows how packers are represented in the numerical model. For different locations of packers, the numerical model is used to estimate the pumped discharge required to achieve a fall in groundwater head in the test section of 1.0 m.

Results for three representative sections are summarised in Table 8.3; further results can be found in Bliss and Rushton (1984). The column *Details of section* describes the nature of the aquifer in the test section, including the presence of beds of low permeability or fissures. Reference is also made to aquifer properties immediately above or below the test section. Two alternative calculated flows are listed under the column headed *Hvorslev*; for the first estimate isotropic conditions are assumed with m in Eq. (8.8) set equal to 1.0; for the second estimate a value of m is assumed. Barker's approach, Eq. (8.9), is only used when fissures are present. The last but one column lists the flows from the numerical model simulation, while the final column records the flow determined from the field results for a drawdown of 1.0 m.

For test section A the flows, deduced using Hvorslev's equation with $m = 1$ and using the numerical model, are virtually identical to the flow measured in the field. Test section B contains a major fissure; the flow according to Barker's analysis and that obtained using the numerical model are close to the field value; Hvorslev's anisotropic equation marginally overestimates the flow. It is for test section C that a substantial difference occurs between the measured flow in

Table 8.3 Study of three representative test sections; the table contains descriptions of the section and outflows in m³/d; Hvorslev's formula, Barker's formula and numerical model results are compared with field readings

No.	Details of Section	Discharge from section (m³/d)				
		Hvorslev		Barker	Model	Field
		Isotrop.	*Aniso.*			
A	*Test section*: homogeneous sandstone formation *Surrounding section*: bed of low permeability adjacent to top packer	1.8	2.6	–	1.8	1.9
B	*Test section*: fissure K = 24.45 m/d, 0.25 m thick bed of low K = 0.03 m/d, 1.0 m thick *Surrounding section*: bed of low permeability 1.0 m above top packer	–	9.2	7.6	8.5	8.0
C	*Test section*: bed of low permeability 0.25 m thick *Surrounding section*: large fissure 0.25 m below test section adjacent to packer	2.4	2.2	–	2.3	35.2

Notes: When there is a fissure present, Hvorslev's isotropic formula *(m = 1)* is not used: when there is no fissure, Barker's formula is not used.

the field and the Hvorslev and numerical model predictions. The large fissure immediately below the test section and supposedly isolated by the packer supplies a high flow into the test section. In the numerical model a connection was simulated between the test section and the fissure, the predicted flow became close to that measured in the field.

Apart from test section C, the packer test formula and the numerical model predict flows which are close to those actually measured. This indicates that packer tests can be used with confidence to estimate effective horizontal hydraulic conductivities in complex aquifer systems. This finding is reinforced by Figure 8.11a which contains detailed numerical model results for test section B. The figure shows the location of the fissure and low permeability bed with the groundwater head contours plotted from numerical model results. Values for the Hvorslev assumption, transferred from Figure 8.10c, are also included in the figure as broken lines. Although there are detailed differences between the two sets of groundwater head contours, there are strong similarities even though the contours deduced from the numerical model have complex shapes due to the influence of the fissure and the low permeability bed. The groundwater head in the column of water above the upper packer is 0.047 m (compared to the drawdown in the test section of 1.0 m). This indicates that some water is drawn from this column of water, through the aquifer and then into the test section. For the water below the lower packer, the groundwater head is 0.017 m.

8.3.4 Effectiveness of fissures in collecting water from the aquifer

When carrying out the packer testing, the presence of fissures was found to be very important. However, when a borehole was constructed about 100 m from the one discussed in Section 8.3.3, fissures were identified but not at the same elevations. Therefore the fissures in the sandstone aquifer are not continuous. If the fissures are discontinuous, does this influence their effectiveness in drawing water to the test section?

The numerical model was used to examine the impact of the length of a fissure on the quantity of water entering a packered section. Table 8.4 indicates that as the length of the fissure increases, the quantity of water entering the test section also increases. However, there is a negligible increase in the inflow for fissures with a length greater than 30 m. This occurs because much of the water which enters the fissure does so within 10 m of the borehole.

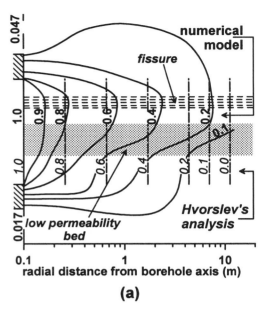

(a)

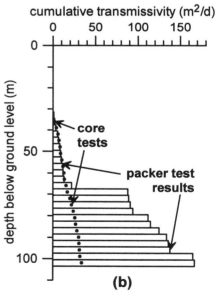

(b)

Figure 8.11 Results from packer testing: (a) groundwater potentials for test section B compared to values from Hvorslev's solution, (b) plot of cumulative transmissivity with depth from packer tests (bar diagram) and from core measurements (solid circles). Reproduced by permission of Geological Society, London, from Bliss and Rushton (1984)

Table 8.4 Effect of increased length of fissure on flow into test section

Fissure length (m)	Inflow m³/d
0.0	1.8
0.3	2.3
1.0	3.3
3.2	5.2
10.0	7.6
31.6	8.3
100	8.4
1000	8.4
3162	8.4

8.3.5 Comparison of properties of sandstone aquifers based on cores, packer testing and pumping tests

The usefulness of packer tests in estimating transmissivity can be assessed by comparing transmissivity estimates from the hydraulic conductivities of core samples, from packer test results over the full depth of the borehole and from pumping tests (Brassington and Walthall 1985).

Samples were prepared from the drilled core and tested in a permeameter. Average values of the horizontal hydraulic conductivity K_h for each 3.0 m interval below the water table (31.5 m below ground level) were calculated and the cumulative transmissivity from the water table to the bottom of the borehole calculated from the expression

$$T = \sum 3.0 \times K_h$$

Values of the cumulative transmissivity based on the hydraulic conductivities of the cores are plotted in Figure 8.11b as solid circles. The transmissivity over the full saturated thickness based on core samples totals only 34 m²/d.

Transmissivity values can also be deduced from packer test results. The packers were set 3.25 m apart to allow a small overlap between the sections. Cumulative transmissivity values from the packer test results are represented in the bar chart in Figure 8.11b. The cumulative values of transmissivity over the saturated thickness calculated from the packer tests is 167 m²/d.

Pumping tests have also been conducted; the transmissivity lies in the range 120–200 m²/d; this is consistent with the packer test results. However, the transmissivity deduced from the core samples is much smaller; this difference is due to the fissures. Major

fissure zones can be identified from Figure 8.11b as locations where the transmissivities from the packer tests show a significant increase. The core-tested cumulative transmissivity is 20 per cent of the cumulative transmissivity from packer testing; this indicates that 80 per cent of the transmissivity arises from flows through fissures. Similar findings for both chalk and sandstone aquifers are reported by Price *et al.* (1982).

8.3.6 Slug tests

Slug tests in single boreholes can be used to estimate the hydraulic conductivity of the region close to the borehole. There are many similarities with the packer tests described above. However, in a slug test measurements are made of the *change* in water level in the borehole with *time*; this avoids the need to measure the flow from the test section. The test starts with the water level depressed; the rise in water level is monitored. Alternatively, a 'slug' of water can be added to a borehole and the fall in borehole water level monitored. The following discussion is a brief resume of Bouwer (1989) which contains an update on the Bouwer and Rice slug test.

Flows into the test section are estimated in two ways.

1. Flows through the aquifer into the test section are calculated using a form of the Thiem equation. Rather than using Eq. (8.8), where the effective outer radius equals mL, the outer radius R is estimated as a function of the four dimensions r_w, L, L_w and H which are defined in Figure 8.12a. This approach allows corrections to be made for the dimensions of the open section and the distances to the base of the aquifer and the water table. Note that the effective borehole radius r_w is often greater than the radius of the casing r_c.
2. Water entering the borehole causes a rise in the water level in the casing. The initial depression of the borehole water level is y_0, and the volume of water in the borehole at the start of the test is shaded in Figure 8.12a. After time t, the water level rises so that the depression is y_t. The rate of rise in the water level multiplied by πr_c^2 (where r_c is the casing radius) provides a second expression for inflow into the borehole.

The inflows into the borehole from (1) and (2) above are equal; when the resulting equation is integrated the following expression for the hydraulic conductivity can be deduced:

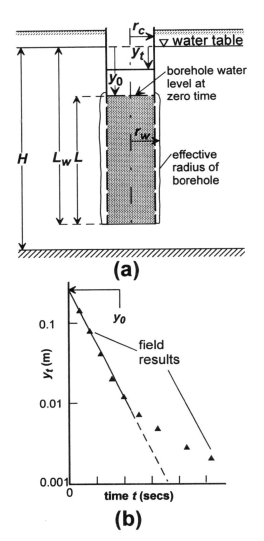

(a)

(b)

Figure 8.12 Slug tests: (a) parameters and dimensions included in the analysis, (b) representative plot of water level depression against time

$$K = \frac{r_c^2 \ln(R/r_w)}{2L} \frac{1}{t} \ln \frac{y_0}{y_t} \qquad (8.10)$$

According to Eq. (8.10), a plot $\ln y_t$ against time should be a straight line. Figure 8.12b indicates how values of the water level elevation y_t change with time; initially the values lie on a straight line but as the effect of the change in borehole water level spreads further into the aquifer, the values deviate from the straight line.

From the slope of the straight line a value for $(1/t)\ln(y_0/y_t)$ can be calculated which, when substituted in Eq. (8.10), provides an estimate for the hydraulic conductivity.

Further questions and issues related to the use of slug tests are discussed by Bouwer (1989).

8.4 INFORMATION ABOUT GROUNDWATER HEADS IN THE VICINITY OF PRODUCTION BOREHOLES

8.4.1 Background

For many pumping stations, there are no observation boreholes specifically constructed to monitor the elevation of the water table. Instead, records are kept of pumping and 'non-pumping' water levels. These recorded non-pumping water levels can often provide valuable insights into the approximate elevation of the water table. It is the water table elevation which indicates whether dewatering of the aquifer is occurring in the vicinity of the pumping station.

Figure 8.13 illustrates different conditions associated with a pumped borehole. Figure 8.13a shows the pump operating with the borehole water level substantially below the water table. Partial recovery after about 15 minutes is illustrated in Figure 8.13b; significant recovery of the water level in the borehole has occurred but the water level in the borehole may still be several metres below the water table elevation. Figure 8.13c shows complete recovery with the water level in the borehole at the same elevation as the water table; several hours or even days may elapse before this condition is achieved.

Information gained from non-pumping water levels proved to be important in identifying whether excessive pumping could be occurring at Trent Valley pumping station near Lichfield, England. In the sandstone aquifer there are four production boreholes, each 155 m deep, roughly in line over a total distance of 34 m. The boreholes are grouped in pairs about 7 m apart. Generally three boreholes are pumping; the continuous station output is 15.5 Ml/d. Due to the close spacing of the boreholes, water levels in non-pumped boreholes are dominated by the pumping boreholes so that the condition illustrated in Figures 8.13b or c did not apply. The pumped water levels are in the range 42 to 61 m below ground level. It was not possible to take the station out of supply, but boreholes were shut down in turn and for about 15 minutes there was no pumping; the highest recovery was to 36 m below ground level.

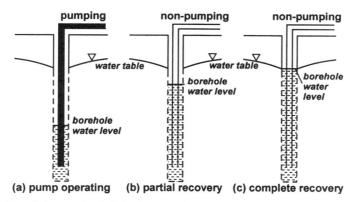

Figure 8.13 Recovery of borehole water levels following cessation of pumping

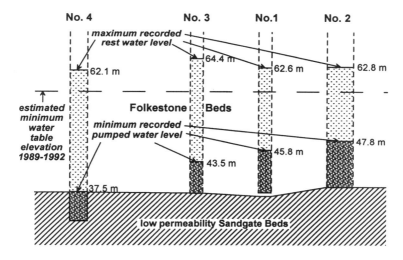

Figure 8.14 Features of four boreholes at Charing PS; numerical values indicate maximum and minimum borehole water levels

With uncertainty about the reliability of the yield of the pumping station, an observation borehole to monitor the water table was drilled at a distance of 40 m from the nearest production borehole. The required depth of drilling was difficult to estimate; if the borehole penetrated too far below the water table, its response would be dominated by the pumped water levels. Was the maximum recovery level, during the short period when there was no pumping, a reasonable estimate of the water table elevation? Since no additional information was available, the observation borehole was drilled to a depth of 40 m compared to the maximum recovery level of 36 m below ground level (this should be compared to the depth of the produc-

tion boreholes of 155 m). When drilled, the water table in the observation borehole was 35 m below ground level. During a year of careful monitoring the observation borehole water level varied by less than 0.8 m, compared to fluctuations in the pumped boreholes of up to 20 m. This confirms that this relatively shallow observation borehole does monitor the water table elevation.

8.4.2 Case study: identifying the water table elevation

Charing Pumping Station in Kent, the UK, which takes water from the Folkestone Beds aquifer, is used as a further illustration of information which can be gained

Table 8.5 Information about four boreholes at Charing pumping station

Borehole no.	1	2	3	4
Base of borehole (mAOD)	37.3	38.4	37.2	31.6*
Base of Folkestone Beds (mAOD)	36.6	38.4	37.2	37.4
Minimum pumped level (mAOD)	45.8	47.8	43.5	37.5
Max. recorded non-pumped level (mAOD)	62.6	62.8	64.4	62.1
Diameter of slotted screen (m)	0.38	0.76	0.38	0.51 max.
Pump speed	fixed	fixed	fixed	Variable

* Borehole extends 5.8 m below base of Folkestone Beds with plain casing.

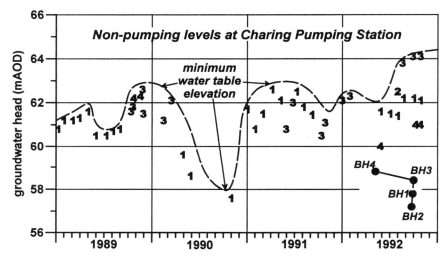

Figure 8.15 Non-pumped water levels from 1989 to 1992; numbers indicate the borehole in which the non-pumping water levels are recorded

from records of pumping and non-pumping water levels. The approximate layout of the four boreholes is shown in the small diagram in the bottom right hand corner of Figure 8.15; the distance between boreholes 3 and 4 is approximately 100 m. There are no observation boreholes.

Records of water levels in the supply boreholes are available; some of the readings are taken when pumping is occurring, others are non-pumped levels but, with the non-pumped levels, it is unlikely that complete recovery occurs. A summary of important information about the four boreholes is given in Table 8.5; some of this information is also presented in graphical form in Figure 8.14.

From this information and a test of individual boreholes it is possible to identify the responses of each borehole.

- *Borehole No. 1* is used occasionally; due to poor connection with the aquifer it is only able to pump about 0.1 Ml/d.
- *Borehole No. 2* is pumped almost continuously, the minimum pumped level is 9.4 m above the base of the borehole, which suggests that the capacity of the pump could be increased.
- *Borehole No. 3* has a larger drawdown than No. 2; the pumping level is often controlled by a cut-off which ensures that the column of water in the borehole is never less than 6 m.
- *Borehole No. 4* has a variable speed pump. The minimum pumped level is close to the base of the Folkestone Beds; this large drawdown can be achieved because the borehole extends almost 6 m below the base of the Folkestone Beds. The borehole operates almost continuously.

Concern was expressed about the ability of these bore-holes to supply water during a drought. Nineteen-ninety was a year with poor recharge and depressed pumping levels. Pumping levels in Nos 3 and 4 bore-holes were often in the range 47 to 41 m AOD; this is close to the base of the Folkestone Beds at 37.2 to 37.4 m. Was there a risk that the whole aquifer could become dewatered? Much can be learnt from a plot of the non-pumping levels; see Figure 8.15. For none of these readings is information available about how long the pump had been switched off. Most of the boreholes recover fairly rapidly; for Nos 2, 3 and 4 more than 90 per cent recovery occurs in 20 minutes, while No. 1 borehole takes more than an hour to recover.

Despite the uncertainty about the rate of recovery, a reasonable estimate of the *water table elevation* can be made from the discrete results of Figure 8.15. Reference to Figure 8.13 shows that the water table in the surrounding aquifer will always be *above* the non-pumping water level. Therefore it is possible to estimate the general elevation of the water table at the pumping station from the non-pumping levels. In Figure 8.15 a broken line is drawn just above the non-pumping levels to represent the estimated minimum water table elevation. Taking the lowest elevation of this line, the water table at the pumping station never fell below 57.8 m during the drought of 1989–92 (this estimated elevation is plotted on Figure 8.14). Consequently there was always a saturated depth of at least 20 m.

8.5 REALISTIC YIELD FROM AQUIFER SYSTEMS

8.5.1 Introduction

In semi-arid and arid environments when groundwater is used for irrigation, there is often the problem of a continuing fall of both pumped water levels and the water table. Frequently wells and boreholes are deep-ened or they are abandoned. Consequently a critical issue is the assessment of a realistic yield from an aquifer system.

In the state of Andhra Pradesh in Southern India during the period 1970–85, there was a substantial increase in the number of new groundwater structures due to faster drilling rigs, the availability of more drilling equipment and the provision of institutional finance for well construction. Yet the number of wells operating in 1985 was 55 000 compared to 48 000 in 1970, an increase far smaller than the number of new

tubewells constructed. The total area actually irrigated by groundwater did not increase. However, the crops grown have changed from a predominance of rice to crops such as groundnuts which require less water (Narasimha Reddy and Prakasam 1989). This example demonstrates that a lot of money can be spent in drilling additional tubewells without increasing the total volume of water abstracted.

The continuing construction of deeper wells is illus-trated by the Mehsana aquifer in Western India. Figure 8.16 shows how dug wells were deepened and then tubewells constructed to penetrate through low-permeability strata to tap deeper zones of the aquifer system (further information about the Mehsana aquifer can be found in Section 10.2). The deeper bore-holes now suffer from falling pumping levels and water tables due to over-exploitation and interference from neighbouring tubewells.

Estimation of the realistic long-term yield of an aquifer system depends on many features including the hydrogeology of the aquifer system and the recharge, together with the number, depth and method of oper-ation of the boreholes and wells. Identifying the quan-tity of water that can be abstracted from an aquifer system over the long term requires initial testing, devel-opment of conceptual and perhaps numerical models of the aquifer response and careful monitoring to revise, if necessary, the predicted yield. Three case studies are considered here; for the first two, the yield predicted from conventional analysis of pumping tests

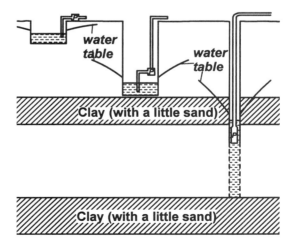

Figure 8.16 Deepening of wells in the Mehsana aquifer to maintain yields

is greater than the long-term yield of the aquifer system. In the first case study of a weathered-fractured aquifer (first considered in Section 7.4) comparisons are made between dug wells, dug-cum-bore wells and borewells. The site of the second study is an alluvial aquifer in Bangladesh; pumping test analysis suggests that boreholes can provide high yields but consideration of recharge to the aquifer system and the restricted flow through the uppermost clay layer indicates that the long-term yield of the aquifer system may be limited. For the third case study from Yemen, the ability of boreholes to collect water from the aquifer system limits the potential yield. For each case study, the two-zone radial flow model is important in understanding and interpreting the aquifer response.

8.5.2 Weathered-fractured aquifer

Large diameter wells are used on the Deccan Trap in India for supplemental irrigation of the rain-fed crop and for dry season irrigation of about 10 per cent of the available area. Since these wells are constructed in the low-permeability weathered zone, pumping can only continue for about two hours each day, otherwise daily recovery is inadequate, leaving insufficient water for the next day. The permeability of the weathered zone in which the wells are constructed is so low that the water levels recover by less than 0.5 m in twenty-four hours. Consequently the farmers are careful to ensure that the wells do not become dry. Section 6.2 provides further information about the manner in which large diameter wells draw water from the aquifer after pumping from the well ceases; yields of about 15 m³ each day can be maintained during the growing season.

In the 1970s, the availability of water in the underlying fractured zone was identified. Exploratory drilling into the underlying fractured zone, with boreholes open at depths from about 13 m to 30 m, provided greatly improved yields of up to 500 m³ in one day. These excellent yields suggested that a new aquifer had been discovered. This led to the drilling of many tube-wells into the underlying fractured zone; when submersible pumps were installed there was sufficient water for crops such as grapes. The farmers with dug wells were told that they had nothing to fear if a tube-well was drilled on their land since another deeper aquifer was being exploited.

These assurances proved to be unfounded. Figure 8.17a illustrates the operation of a dug well in the weathered zone; water can only be collected from close to the dug well, causing a small lowering of the aquifer

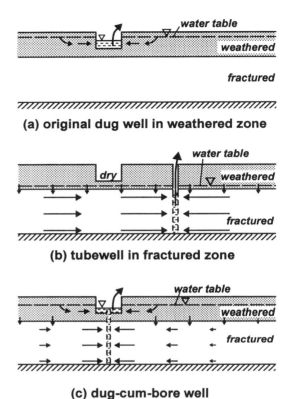

(a) original dug well in weathered zone

(b) tubewell in fractured zone

(c) dug-cum-bore well

Figure 8.17 Conceptual diagrams of flow processes to alternative types of wells in a weathered-fractured aquifer

water table. When tubewells are constructed to tap the fractured zone (Figure 8.17b), yields are far higher with significant flows through the fractured zone into the tubewell. However, the hydrogeologists who recommended that tubewells could be drilled in the vicinity of dug wells, failed to recognise that, although water is transmitted through the fractured zone towards the tubewell, the actual source of the water is the overlying weathered zone as shown by the small vertical arrows across the interface of the weathered and fractured zones in Figure 8.17b. The vertical flows result in a substantial fall in the water table to below the base of most dug wells. Not only do the dug wells dry up but also the quantity of water pumped from the fractured zone far exceeds the recharge to the weathered zone, with the result that all the tubewells fail within a few years.

The discussion in Section 7.4 demonstrates that water pumped from the fractured zone originates from the weathered zone; this is illustrated by Figure 8.18.

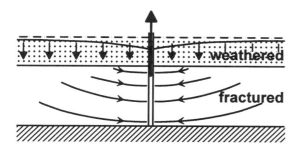

Figure 8.18 Idealised diagram of flow to a pumped borehole tapping the fractured zone

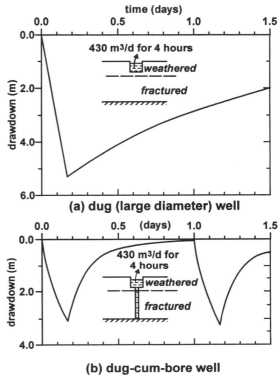

Figure 8.19 Differing responses to pumping from a dug well or a dug-cum-bore well. Reprinted from *Journal of Hydrology* **80,** Rushton and Weller, Response to pumping of a weathered-fractured granite aquifer, pp. 299–309, copyright (1985) with permission from Elsevier Science

Powerful submersible pumps in tubewells in the fractured zone withdraw too much water from the weathered zone. Does this mean that pumping from the fractured zone should never be allowed? The fractured zone is efficient at collecting water from the total aquifer system; however, a methodology must be introduced which does not allow powerful submersible pumps to suck the aquifer system dry. This can be achieved using dug-cum-bore wells in which a bore is drilled in the bottom of a large diameter well (Figure 8.17c). Preliminary field trials indicate that a dug-cum-bore well is a promising alternative provided that a suction pump is used to take water from the dug well, thereby preventing excessive drawdowns.

The operation of dug wells and dug-cum-bore wells is compared in Figure 8.19. This figure shows a dug well pumping at a rate of 430 m³/d for four hours; the pumped drawdown is more than 5.3 m. Twenty hours after the cessation of pumping, the recovery of the water level in the well is less than 50 per cent complete. However, if a dug-cum-bore well is used with the bore penetrating into the fractured zone (Figure 8.19b), the drawdown in the well after pumping at 430 m³/d for four hours is just over 3 m. A non-pumping period of 20 hours is sufficient for almost complete recovery to occur. The reason for these very different responses is apparent from the flow arrows in Figure 8.17c. The ability of the dug-cum-bore well to draw water, stored in the weathered zone, into the fractured zone for onward transmission to the bore well, is the key to the effectiveness of dug-cum-bore wells.

8.5.3 Alluvial aquifers with an uppermost layer of low hydraulic conductivity

Alluvial aquifers are often used to supply water for irrigated agriculture; however, there are situations where an upper silty-clay layer overlies the main alluvial aquifer, restricting vertical flows. This is a common situation in Bangladesh; even though pumping tests indicate the potential for a high yield, the available resource is limited by overlying lower permeability layers. This section considers the important flow mechanisms which occur when pumping from an alluvial aquifer overlain by low conductivity layers.

Studies of field conditions in areas with overlying clay layers have indicated that the clay layer restricts recharge to the main aquifer system. Ahmad (1974) describes early groundwater exploration studies in Bangladesh which show that in many areas there is in excess of 30 m of clay between the ground surface and the main aquifers. Banerji (1983) indicates that in the Calcutta region of the Bengal Basin in India, a thick clay layer overlies the main aquifer which consists of coarse to medium grained sand. In the Madras aquifer in Southern India, Krishnasamy and Sakthivadivel (1986) show that pumping from deeper aquifer zones has exceeded the rate at which water can pass through the overlying clay layers. Consequently, unconfined conditions have developed in the underlying aquifers resulting in the need to further deepen boreholes to draw water from lower aquifer zones which still contain water. Ramnarong (1983) describes how the blue-grey-yellowish marine clay in Bangkok, which is about 25 m thick and overlies the uppermost permeable aquifer layer, was thought to be impermeable. However, ground settlement in Bangkok indicates that some water has drained out of this overlying clay.

Introduction to case study in Bangladesh

The case study refers to the Madhupur aquifer in Kapasia, Bangladesh. Extensive drilling indicates that in the upper 120 m of the aquifer system, 40 to 90 m of the vertical section is sufficiently permeable to warrant the provision of a screened section in boreholes. An examination of the available borehole logs shows that the aquifer is a complex mixture of sand, silt and clay. A clay layer (containing some sand and silt) overlies the aquifer; its thickness ranges from 5 to 30 m.

A test at tubewell, KAP/13, is used to illustrate the aquifer response. As indicated in Figure 8.20a, the tubewell penetrates 89.3 m below ground level with the rest water level 6.6 m below ground level. The upper 12.8 m is a clay layer containing some silt and sand, with the remainder of the aquifer primarily of sand but with some clay zones; the slotted casing extends from 29.6 to 87.8 m below ground level. Pumping at a rate of 5141 m³/d lasted for 3.0 days, with recovery monitored for a further 2.0 days. Drawdowns during the pumping and recovery phases were monitored in the pumping well and in four observation piezometers; see Figure 8.20a. Some of the field drawdowns are plotted in Figures 8.20b and c; further information can be found in Miah and Rushton (1997). The field results are indicated by the discrete symbols; the lines represent numerical model results which are considered later. Only small differences occur between the shallow and deep piezometers; the differences are probably due to the properties of the aquifer in the vicinity of the piezometers.

Representation of moving water table in overlying layer

The two-zone radial flow model is ideally suited for analysing this problem; however, the model described in Section 7.4 does not include the effect of a moving

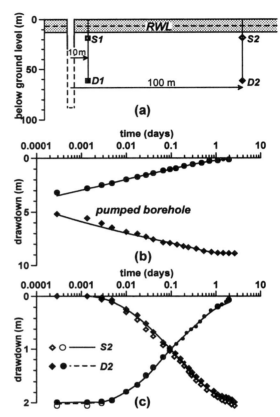

Figure 8.20 Details of pumping test, KAP/13: (a) layout of test site, (b) drawdown and recovery in pumped borehole, (c) drawdown and recovery in observation piezometers at 100 m. Reproduced by permission of IAHS Press from Miah and Rushton (1997)

water table in the overlying layer. For the test in the tubewell KAP/13, the overlying layer contains and sustains a water table which rises due to recharge and falls due to water moving vertically into the underlying aquifer system. In Figure 8.21a is a schematic flow diagram of the aquifer system. Flows in the Overlying low permeability layer depend on recharge and the release of water from storage at the water table, as indicated in Figure 8.21b. Conditions are more complex than classical leaky aquitard theory, and consequently all the conditions which can occur in the overlying aquifer must be considered. Three possible conditions are sketched in Figure 8.22 a–c.

(a) The water table in the Overlying low permeability layer, h_{wt} is towards the top of the aquifer while the

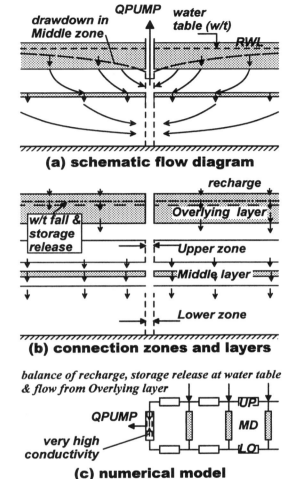

Figure 8.21 Idealisation of the aquifer system with overlying clay layer

underlying aquifer is under confined conditions so that the groundwater (piezometric) head h_{aq} lies within the Overlying layer. Additionally, there is recharge to the overlying aquifer. Two quantities can be calculated:

— the inflows from the base of the Overlying layer can be estimated using Darcy's Law; the calculation is shown on the diagram. The head difference is $h_{wt} - h_{aq}$ which applies over the vertical distance from the water table to the base of the layer, $h_{wt} - base$.

— the fall in the water table Δh_{wt} (which is defined as positive if the water table moves vertically

downwards) during a time increment Δt depends on the difference between the inflow and the recharge divided by the specific yield of the Overlying layer.

(b) For the situation in Figure 8.22b, the water table in the Overlying layer is again towards the top of the aquifer, but the underlying aquifer is unconfined, with the groundwater head below the base of the Overlying layer. Therefore, *base* (where the pressure is atmospheric) replaces h_{aq} in the equation for the head difference.

(c) For the third diagram, the water table is close to the bottom of the Overlying low permeability layer. Two possibilities must be considered; either the recharge is less than the vertical hydraulic conductivity, in which case the recharge moves through to the underlying aquifer, or if the recharge exceeds the vertical hydraulic conductivity there is a rise in the water table.

In the calculations for conditions (a) to (c) the water table is not allowed to rise above the top of the Overlying layer.

Results from two-zone model

These mechanisms have been incorporated in the two-zone model as indicated in Figure 8.21c. The analysis is described briefly below; full details can be found in Miah and Rushton (1997). Aquifer parameters are modified until a reasonable match is achieved between field and modelled groundwater heads during both the pumping and the recovery phases. The match of Figure 8.20b shows good agreement for the pumped borehole; it is necessary to include well loss factors for both the Upper and Lower zones. For the observation piezometers at 100 m from the axis of the pumped well, the difference between the shallow and deep piezometers is small, consequently the Middle layer has a nominal 1.0 m thickness with a hydraulic conductivity of 0.01 m/d. The modelled values for the observation piezometers in Figure 8.20b are generally close to or between the field values.

Aquifer values deduced from model refinement are listed in Table 8.6; note that the vertical hydraulic conductivity of the Overlying layer is estimated to be 0.007 m/d. From this analysis the specific yield of the Upper zone could not be deduced; the value quoted in the table is based on tests in other similar locations. The well loss factor represents the restriction on the movement of water through gravel packs and the well screen into the borehole; the hydraulic conductivity adjacent

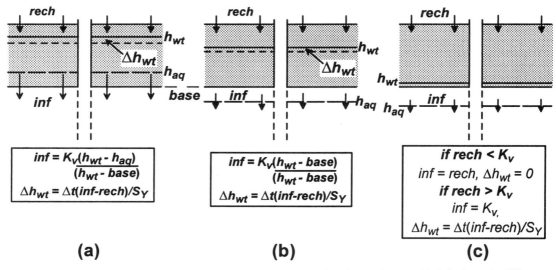

Figure 8.22 Conditions in the Overlying layer; method of calculating infiltration and water table fall Δh_{wt} under different conditions: (a) piezometric head h_{aq} within Overlying layer, (b) h_{aq} below Overlying layer, (c) h_{aq} below Overlying layer and water table at base of Overlying layer. Reproduced by permission of IAHS Press from Miah and Rushton (1997)

Table 8.6   Parameter values for test at tubewell KAP/13 (units m or m/d or non-dimensional)

Zone or layer	Thickness	Hydraulic conductivity		Storage coefficient		Well loss factor
		Radial	Vertical	Confined	Unconfined	
Overlying	6.2	–	0.007	–	0.03	
Upper	35.0	11.5	1.2	0.001	0.15*	4.0
Middle	1.0	–	0.01	–	–	
Lower	35.0	11.5	1.2	0.001	–	3.0

* Not available from test, deduced from other similar situations.

to the borehole equals the standard value divided by this factor. Although reliable values of certain aquifer parameters have been derived by matching the model to the field results of pumping and recovery, the analysis would have been enhanced if piezometers had also been provided in the Overlying lower-permeability layer.

Representation of long-term response

The next stage is to consider the response when this aquifer system is used to provide water for irrigation. The following assumptions are made which allow the specification of a problem which represents typical years. The average annual rainfall in the study area is 2370 mm; about 80 per cent of this rainfall occurs

during the monsoon months of June to October. For each year of the simulation (the year starts at the beginning of December) the recharge (for 10-day periods) and abstraction patterns are represented as follows:

Dates	Dec–Apr	May	June–Sep	Oct–Nov
Rech	Nil	Nil	0, 30, 0, 15, 0, 30, 30, 0, 30, 30, 15, 0+	Nil
Abst	2450 m³/d*	Nil	Nil	Nil

+ *recharge* values are in mm for periods of 10 days (annual recharge = 180 mm).
* *discharge rate*: 4900 m³/d for 12 hours, recovery for 12 hours each day.

The intensity of irrigation is assumed to be 5 mm/d; this is suitable for growing rice (total 750 mm in 150 days).

If the spacing of the boreholes is 700 m by 700 m (equivalent outer radius with no-flow condition at 395 m), the required pumping rate for twelve out of twenty-four hours should be $2.0 \times 700 \times 700 \times 0.005 = 4900 \, m^3/d$ for 0.5 day. In practice there are variations in water demand during the land preparation and growing seasons.

For *Example A* the drawdowns in the pumped tube-well during the first year are shown in Figure 8.23. During the first five months when abstraction occurs, the lower line indicates the maximum pumped draw-downs, and the upper line represents the recovery. The change in the pumping level between successive pumping and recovery phases is about 7.0 m; the pumped drawdown after 150 days is 17.2 m and the non-pumping level is 10.2 m. When the non-pumping drawdown reaches about 6 m there is a change in slope in the curves; this is due to conditions changing in the

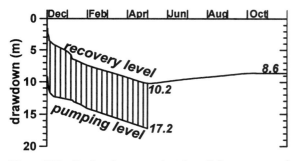

Figure 8.23 Predicted response in tubewell for one year of operation. Reproduced by permission of IAHS Press from Miah and Rushton (1997)

overlying layer from condition (a) to condition (b) of Figure 8.22.

During the recharge period of June to the end of September, drawdowns recover to 8.6 m and remain at this value during October and November when there is neither recharge nor abstraction. Predicted responses for five years of pumping are summarised in Table 8.7. The maximum pumped drawdown (month 5) and final recovery level at the end of each year (month 12) are recorded in the column headed *Example A* beneath the heading *drawdowns*. During Year 5 the pumped drawdown reaches 34.3 m with recovery to 23.8 m at the end of the recharge period. This rapid fall in ground-water levels, which averages more than 4.7 m per year, could result in the aquifer resources being exhausted within a decade. The fall in the pumped levels to 34.3 m during Year 5 is also likely to result in a deteri-oration in tubewell performance since the pumped levels are then within the perforated section of the borehole and a seepage face will form. These results indicate that, despite an initial pumping test at $5141 \, m^3/d$ for 3 days, a regular withdrawal of water of $2450 \, m^3$ each day for 150 days is higher than the aquifer system can sustain.

Two further situations are considered. For *Example B* the spacing between the tubewells is increased to 900 m with the discharge rate remaining at $4900 \, m^3/d$ for twelve hours. The average intensity of irrigation becomes 3.0 mm/d to compensate for the increased area from which the tubewell withdraws water. Even with this increased spacing with less interference between tubewells, the maximum pumped drawdown after five years is 23.0 m compared to 34.3 m with the standard spacing. The average groundwater head

Table 8.7 Summary of results for long-term pumping; quoted drawdowns (in metres) are for the pumped borehole

Example	A		B		C	
details of problem:						
well spacing, m	700		900		700	
irrigation intensity, mm	5.0		3.0		5.0	
discharge rate, m³/d	4900		4900		4900	
duration per day, hours	12		12		12	
annual recharge, mm	180		180		300	
drawdowns (m) for month:	*5*	*12*	*5*	*12*	*5*	*12*
Year 1	17.2	8.6	14.9	6.4	17.2	7.8
Year 2	21.4	12.4	16.9	8.3	20.5	10.8
Year 3	25.6	16.2	18.9	10.1	23.9	13.8
Year 4	29.8	20.0	21.0	11.9	27.3	16.9
Year 5	34.3	23.8	23.0	13.7	30.7	19.9

falls a total of 13.7 m which is equivalent to 2.7 m each year.

For *Example C* the annual recharge is increased from 180 to 300 mm/year; other parameters remain the same as for Example A. Summary results in Table 8.7 predict a fall in groundwater head during the five-year period of 19.9 m which is equivalent to 4.0 m per year. Although the annual rainfall is in excess of 2000 mm in a year, there is often a high runoff so that a recharge rate of 300 mm/year is unlikely to be exceeded.

This example demonstrates that initial pumping tests may not be a good guide to the long-term yield of an aquifer system. Instead, it is essential to consider the potential recharge and the restriction of infiltration to the permeable aquifer system caused by any Overlying layers of low vertical hydraulic conductivity.

8.5.4 Response of an alluvium-sandstone aquifer system

Extensive areas in Yemen in the Arabian Peninsular consist of alluvial fans with wadi deposits overlying sandstone with an intervening layer of conglomerate. In reviewing pumping test analyses in these aquifer systems, Sutton (1985) demonstrates the importance of the layering. When conventional methods of pumping test analysis are used, especially when they are based on drawdowns for later times, the transmissivity may be over-estimated by a factor of up to four. Sutton also explains how the two-layer model can be used to understand the responses of the different aquifer zones. Even if the only information relates to the drawdown and recovery in the pumped borehole, approximate estimates of the properties of the individual aquifers can be made.

Grout and Rushton (1990) studied in detail a particular location in Yemen. The aquifer system consists of a shallower unconfined unit of alluvial fan or wadi deposits of recent age underlain by sandstone of Cretaceous age which contains some fractures. Between the two units there is a well-cemented conglomerate layer which is typically three metres thick. Three boreholes and two piezometers were drilled at the test site; details can be found in Figure 8.24 and Table 8.8. The three boreholes were designed to operate in different ways: borehole B1 can collect water from both the alluvium and the sandstone, borehole B2 can only collect water from the alluvium while borehole B3 is cased through the alluvium and is an open hole in the sandstone. Piezometer P1 is in the gravel pack of borehole B1, piezometer P2 is adjacent to B2 but penetrates to just below the conglomerate.

This discussion will concentrate on a test in which water was pumped from borehole B1 at a rate of 1050 m³/d for two days (there was some difficulty in achieving a steady pumping regime) with recovery monitored for only a further 0.21 day. It was anticipated that the aquifer system would allow higher discharge rates. This assumption was based on aquifer properties deduced from an initial pumping test but when that test was re-examined it was found that the

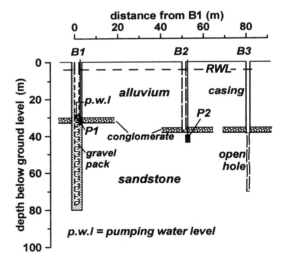

Figure 8.24 Details of the boreholes in the alluvial-sandstone aquifer system

Table 8.8   Details of boreholes and piezometers

Ref.	Diameter (mm)	Distance from B1 (m)	Depth (m)	Purpose
B1	322	–	75	Pumping from sandstone and alluvium
P1	25	0.01	30	Monitoring gravel pack
B2	304	50	37	Pumping from alluvium alone
P2	25	50	45	Monitoring sandstone below B2
B3	304	80	70	Pumping from sandstone alone

observation well was a village well which had been in use during the test! Even with a pumping rate of 1050 m³/d, the pumped drawdown approached 28 m which, with a rest water level of 3 m below ground level, meant that the pumping water level coincided with the elevation of the conglomerate layer.

Detailed results of the pumping test are contained in Figure 8.25; field results are indicated by discrete symbols. There are four separate plots:

(a) drawdowns due to pumping and recovery in the pumped borehole,
(b) results for the piezometer in the gravel pack; these are rather erratic but are still informative,
(c) drawdowns at 50 m in the borehole in the alluvium and in the piezometer which penetrates to the sandstone, and
(d) drawdowns in the borehole at 80 m.

Conceptual and numerical models

The key to understanding the causes of the restricted pumping rate can be gained from a careful examination of conditions in the pumped borehole and the gravel pack piezometer. Figure 8.26 shows the pumping water level and the water level in the piezometer. Because the piezometer is in the gravel pack, it will reflect both the aquifer water table and conditions in the borehole. The piezometer reading is higher than the pumped water level, consequently the *water table* in the aquifer is higher than the *water level* in the piezometer; an approximate water table elevation is indicated in Figure 8.26. Therefore there is a substantial seepage face between the water table in the alluvium and the pumping water level; water from the alluvium seeps or drains under gravity into the borehole. Consequently, as indicated in the schematic diagram, the flows from the alluvium into the borehole are smaller than the flows from the sandstone. With this understanding of the manner in which the borehole operates, it is possible to consider how the two-zone model can be adapted to simulate this test.

Detailed descriptions of the use of the two-zone model for this type of problem can be found in Section 7.4; parameter values used for this particular example are quoted in Table 7.6. The only new feature is the representation of the seepage face. Because a single zone is used to represent the alluvium, it is not possible

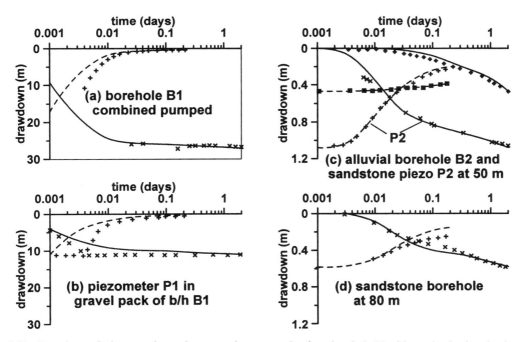

Figure 8.25 Drawdowns during pumping and recovery due to pumping from borehole B1 with monitoring in other boreholes and piezometers

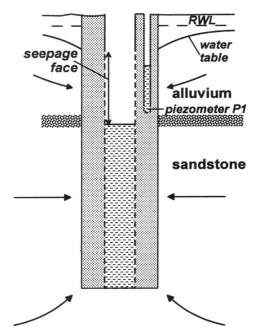

Figure 8.26 Schematic diagram of conditions in the vicinity of the pumped borehole in an alluvial-sandstone aquifer system

to include the varying groundwater heads with depth on the seepage face. However, the limited flow from the seepage face in the alluvium can be simulated by assuming that only 30 per cent of the total abstraction comes from the alluvium, even though the alluvium has a higher hydraulic conductivity than the sandstone.

Results from the analysis using the two-zone model are presented as continuous lines in Figure 8.25. Differences between field and modelled results do occur, especially during the early stages of the test when difficulties were encountered in achieving a stable pumping rate. Representation of conditions in the gravel pack is also difficult. Nevertheless the excellent agreement at 50 m (Figure 8.25c), and the satisfactory agreement elsewhere between field and modelled results, suggests that the model does represent the general aquifer response.

Difficulties occur in collecting water from the aquifer due to the limited effectiveness of the seepage face. Although there may be water available in the alluvium, the presence of the seepage face severely limits the quantity of water that can be drawn into the borehole. This was confirmed by a test in borehole B2, which draws water only from the alluvium. The maximum

yield was 500 m³/d; even this pumping rate could not be sustained over a long period.

8.6 INJECTION WELLS AND WELL CLOGGING

8.6.1 Introduction

Artificial recharge using injection wells has proved to be a viable technique, especially when the aquifer to be recharged is at some depth below ground surface and overlain by low-permeability strata (see Chapter 8 of Huisman and Olsthoorn 1983). In a highly transmissive aquifer, a single injection well is adequate but in other field situations, lines or circular arrays of wells are required. The most significant drawback of injection wells is their tendency to clog; this can occur due to the quality of the injected water or forcing fines back into the aquifer (Driscoll 1986). Cleaning of the injection wells, including the well screen and gravel pack, must be carried out at regular intervals. A number of successful injection schemes are described in the literature; for instance, Bush (1977) describes the injection of water into a deep, high-transmissivity limestone aquifer using water from a shallow sand aquifer. Other schemes are described by individual contributors to Asano (1985).

Two artificial recharge schemes will be considered in this discussion. The first is a pilot recharge project in an alluvial aquifer in India; serious problems occurred due to well clogging. The second scheme relates to chalk and overlying unconsolidated tertiary strata in North London, the UK.

8.6.2 Alluvial aquifer in India

To appreciate the problems involved in artificial recharge together with the associated data collection and analysis, a pilot artificial recharge project scheme in India is considered. An injection well of 350 mm diameter uses water from a shallow floodplain aquifer to recharge a deep alluvial aquifer. In India the feasibility of artificial recharge using injection wells was demonstrated by a short-term injection test in an alluvial aquifer in Western India (Desai *et al.* 1978); a recharge rate of 850 m³/d was achieved in an aquifer with a transmissivity of about 1400 m²/d. This successful short-term injection test was followed by an investigation in the floodplain of the Saraswati River, Gujarat, India. Available analytical solutions (Todd 1980, Huisman and Olsthoorn 1983) proved to be unsuitable for analysing the actual test. However, the two-zone model (Section 7.3) was used to represent a short-term pumping test and the long-term artificial recharge experiment. Clogging of the injection well proved to be a severe limitation of the scheme (Rushton and Srivastava 1988).

The Saraswati River floodplain was selected as the injection well site since there is a year-round supply of water; each monsoon season the water table in the floodplain unit approaches ground level (see the groundwater head hydrograph for A1 in Figure 8.27(b)). There are extensive clay and sand lenses; from the lithology of five boreholes it is possible to identify three aquifer units separated by less permeable layers. The first aquifer unit, A1, extends from the water table to a depth of about 25 m; the water table fluctuates between the ground surface and a depth of about 3 m. The second permeable aquifer unit, A2, extends from 45 to 85 m below ground level (BGL); the seasonal piezometric head fluctuation is within the range 10 to 20 m BGL. In the third aquifer unit, which is between about 110 and 145 m BGL, there is a larger fluctuation of between 20 and 35 m BGL.

This idealisation of the alluvial aquifer system is sketched in Figure 8.27a. The figure also provides information about the boreholes used for the pilot scheme, including the location of the source well in the water table aquifer A1, the injection well I screened in aquifers A2 and A3 plus piezometers P2 and P3 and observation boreholes O2 and O3 in aquifer unit A2 and A3. The piezometers are approximately 5 m from the injection well, while the distance to the observation wells is about 150 m. Further information about the wells and piezometers can be found in Table 8.9.

The groundwater hydrographs for aquifer units A1, A2 and A3 are shown in Figure 8.27b. There are two important points to note. First, aquifer unit A1 responds to rainfall recharge reaching a maximum towards the end of the monsoon rainfall, then declining steadily during the remainder of the year. However, the groundwater head hydrographs for aquifers A2 and A3 are dominated by the pumping. Pumping for irrigation starts at the end of the monsoon season and

continues until February or March when the temperature is high and the second crop has been harvested. Recovery due to the cessation of pumping occurs from March to August or September. The second important point is the substantial vertical hydraulic gradient with head differences across the low permeability strata between aquifers units A1 and A2 averaging about 14 m and a further head difference of about 15 m between aquifers units A2 and A3. Since the effective vertical thickness of these low permeability strata are approximately 20 to 25 m, the vertical hydraulic gradient is about 0.6, leading to substantial flows from

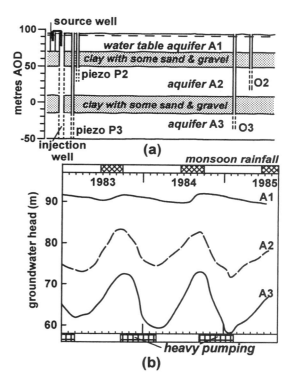

Figure 8.27 Groundwater head hydrographs for observation piezometers at different depths in an alluvial aquifer system

Table 8.9 Details of injection, source and observation wells and piezometers

Well	Dia. (mm)	Aquifer unit	Distance from I (m)
Injection well I	350	A2–A3	–
Source well	200	A1	5
Observation well O2	200	A2	164
Piezometer P2	37	A2	6
Piezometer P3	37	A3	6
Observation well O3	200	A3	150

aquifer unit A1 to unit A2 and then to A3. These substantial flows are required to meet the high abstraction demands from aquifer units A2 and A3; for further information about the regional groundwater response of the Mehsana aquifer see Section 10.2.

8.6.3 Initial pumping test

To estimate the aquifer parameters, a test was conducted with $1730\,m^3/d$ pumped from the injection well for 5000 minutes; monitoring of groundwater heads occurred in the pumped well, the two piezometers and the two observation wells. The results plotted to a *logarithmic* scale are shown in Figure 8.28b; the symbols represent a selection from the actual field readings. The figure also shows the groundwater heads before the start of the test in the injection well I and observation wells O2 and O3. The following list presents the groundwater heads above Ordnance Datum before the start of the test in descending order:

Observation well O2, aquifer A2	76.4 m
Piezometer P2, aquifer A2	72.7 m
Pumped well I between aquifers A2 & A3	68.0 m
Piezometer P3, aquifer A3	66.8 m
Observation well O3, aquifer A3	66.4 m

These readings indicate that water flows *from* aquifer A2, through the injection well and *into* aquifer A3 as sketched in Figure 8.28a. Further information about ambient flows in monitoring wells is presented by Elci *et al.* (2001). This condition of flow in the aquifer system violates the required initial conditions of most techniques for pumping test analysis since they assume zero-flow conditions at the start of the analysis.

Simulation of this pumping test is possible using the two-zone model of Section 7.4. Before proceeding with the analysis, it is helpful to look more carefully at the individual groundwater heads. Although the flow from A2 into the injection well equals the flow from the injection well into A3, the head difference between O2 and P2 is 3.7 m, whereas the head difference between P3 and O3 is only 0.4 m. This suggests that the transmissivity of aquifer unit A2 is less than A3. Considering next the differences between the piezometers and the injection well, the difference between P2 and I is 4.7 m whereas the difference between I and P3 is 1.2 m. This may be due to the different transmissivities of the two aquifers, but it may also be a function of well loss factors.

Considering the two-zone model, the aquifer zones and layers are similar to the example in Figure 7.11.

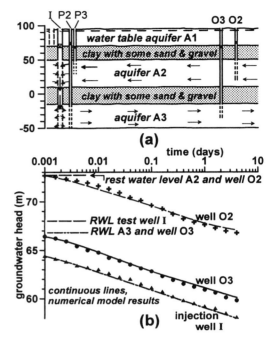

Figure 8.28 Pumping test in the injection well: (a) flows from aquifer A2 to aquifer A3 before the commencement of the test, (b) drawdowns in the test well and in observation wells with numerical model results indicated by continuous lines. Reprinted from *Journal of Hydrology* **99**, Rushton and Srivastava, Interpreting injection well tests in an alluvial aquifer, pp. 49–60, copyright (1988) with permission from Elsevier Science

However, the current problem requires that the groundwater heads at some distance from the test site are set at suitable values to achieve the initial conditions of flow from aquifer A2 through the injection well to aquifer A3; this is achieved by setting specified heads at 1000 m from the injection well at 78.2 and 65.8 m in the Upper and Lower zones (aquifers A2 and A3) respectively.

The following parameter values were deduced following model refinement:

Overlying layer between aquifers A1 and A2, thickness 20 m, $K_V = 0.0004\,m/d$

Upper zone, aquifer A2, $T = 50\,m^2/d$, $S_C = 0.001$, well loss factor = 2.0

Middle layer between aquifers A2 and A3, thickness 25 m, $K_V = 0.0002\,m\,/d$

Lower zone, aquifer A3, $T = 130\,m^2/d$, $S_C = 0.0045$, well loss factor = 6.0

Agreement between field readings and the continuous lines representing the numerical model results is good; with these aquifer parameters, the flow from aquifer A2 to aquifer A3 *before pumping commences* is estimated to be 292 m³/d.

8.6.4 Artificial recharge results and interpretation

A long-term artificial recharge test of the injection well system lasted for 250 days at a rate of 225 m³/d. The source well was constructed in the shallow floodplain aquifer; using a siphon pipe, water drawn from the shallow well was injected into the 350 mm injection well. There were two main reasons for using this procedure; the first was that the shallow aquifer provided a reliable source of water of the same quality as that in the deeper aquifers; second, the system did not require any power to operate. Frequent measurements were made of the groundwater heads in the injection well, in the piezometers and in the observation wells. The results, including those for fifteen days before the start of the test, are presented in Figure 8.29. To assist in identifying the changes in groundwater head, lines are drawn between the field points; the injection well is indicated by the unbroken line, the groundwater heads in aquifer unit A2 by short broken lines and those for A3 by chain-dotted lines.

An examination of the field results shows that:

1. from the cyclic nature of all the plots, the impact of the injection is superimposed on seasonal fluctuations,
2. aquifer unit A3 with observations in P3 and O3 is only slightly influenced by the injection; the hydrographs reflect the seasonal fluctuations and O3, the more distant measuring point, remains at about one metre below P3 suggesting a continuing inflow from the injection well,
3. as soon as injection starts the injection well I undergoes a considerable increase in groundwater head, the difference between P3 and I continues to increase throughout the test,
4. rapid changes also occur in aquifer unit A2, with increases in both O2 and P2; at 50 days the heads in O2, P2 and I are close together and from then onwards there is little difference between P2 and the injection well I.

To simulate the injection well test, the two-zone model is used with parameter values as listed above. For both the Upper and Lower zones the groundwater heads at 1000 m are varied to represent the seasonal ground-

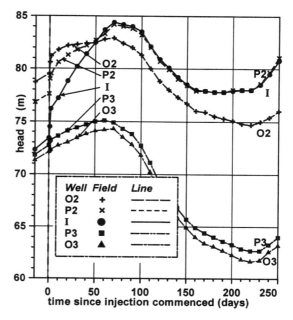

Figure 8.29 Field readings during the 250-day injection experiment. For location of observation wells and piezometers see Figure 8.28a

water head hydrographs. However, the most important feature is the inclusion of well losses. Figure 8.30a shows how the well loss factor is modified with time for aquifer unit A3; the well loss factor is multiplied by the hydraulic resistance between nodes 2 and 3 which are at radial distances of 0.175 m to 0.257 m. Although the model represents the increase in hydraulic resistance as being between 0.175 m to 0.257 m, deteriorating hydraulic conditions may spread further. Note that the well loss factor increases from 6.0 at the start of the test to 120 at 60 days (the time at which the groundwater head in the injection well is the same as the groundwater head in piezometer P2) with a reduced rate of increase to 150 at 250 days. It is not suggested that this is a precise representation of the clogging. Nevertheless, without this method of representing a serious deterioration in the contact between the injection well and the aquifers, it is not possible to reproduce the field response in the injection well and the four observation piezometers and wells.

A comparison between the field and modelled groundwater heads is found in Figure 8.30b. Agreement is encouraging, considering the complexity of the flow processes. From the model it is possible to calculate the rate at which water flows into aquifer units A2 and A3.

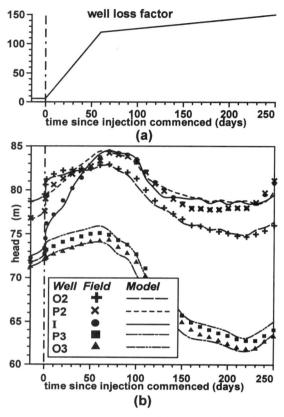

Figure 8.30 Numerical model of injection well experiment: (a) assumed variation in well loss for Lower zone, (b) model results compared with field results. Reprinted from *Journal of Hydrology* **99**, Rushton and Srivastava, Interpreting injection well tests in an alluvial aquifer, pp. 49–60, copyright (1988) with permission from Elsevier Science

Before the test starts there is a flow of $292\,m^3/d$ from aquifer unit A2 through the injection well to unit A3. As soon as artificial recharge commences at a rate of $225\,m^3/d$, flow from aquifer unit A2 reduces to $120\,m^3/d$. This reduction in flow from aquifer A2 to the injection well continues so that by day 50 the flow becomes effectively zero. The flow into aquifer A3 falls from a maximum of about $345\,m^3/d$ to a steady rate of $225\,m^3/d$. Clearly, serious clogging has occurred.

Step pumping tests, before the injection test, following the test and after the injection well was chemically treated and thoroughly cleaned, are reported in example (b) of Section 8.2.6 and Figure 8.8. Values of the coefficient B were $4.7 \times 10^{-3}\,d/m^2$, $46 \times 10^{-3}\,d/m^2$, and $2.9 \times 10^{-3}\,d/m^2$ respectively. As with most artificial recharge schemes, clogging caused serious problems

(an increase in B by a factor of ten) although by thorough cleaning the system was revitalised.

Finally, it is instructive to compare the injection rate with the quantity of water moving between the aquifer units due to the vertical gradient caused by pumping. Taking an effective vertical hydraulic conductivity of the clay layers of $0.0003\,m/d$, an average thickness of $22.5\,m$ with a typical groundwater head difference between layers of $14\,m$, the downward flow per square metre plan area equals $0.0003 \times (14/22.5) = 0.00019\,m^3/d/m^2$. When the artificial recharge rate of $225\,m^3/d$ is divided by this downward flow, the equivalent area is $1.2 \times 10^6\,m^2$ or $1.2\,km^2$. Therefore this injection experiment supplied as much water to the confined aquifers as occurs naturally by vertical flow over an area of about $1\,km^2$.

8.6.5 North London Artificial Recharge Scheme

In the North London Artificial Recharge Scheme, the Chalk/Lower London Tertiaries system is recharged to provide water for a 1 in 8 year drought, with the abstraction continuing for up to 200 days (O'Shea *et al.* 1995). The first attempt at artificial recharge was in the 1890s; current studies commenced in the 1950s. In 1977 an average of $48\,Ml/d$ was recharged for 6 months through 13 boreholes.

The primary aim is to develop the large storage potential of the Lower London Tertiaries which overlie the Chalk. The strata are similar to those in the Braintree area (Figure 5.30), but the water table is either in the Chalk or the Lower London Tertiaries due to long-term abstraction from the chalk aquifer. Typically the Lower London Tertiaries are 20–30 m thick; the middle and lower beds contain fine-grained sand, silty sand and some sandy clay. In assessing the suitability of artificial recharge schemes, specific yields are critical parameters; from the discussion in Section 5.16, the specific yields of the Chalk and the Lower London Tertiaries are approximately 0.01 and 0.05 respectively.

Each borehole can be used for injection and abstraction. Boreholes are of about 750 mm diameter; the solid casing extends 5 m into the Chalk with an open hole for at least a further 50 m. Test pumping indicates that boreholes can be classified into three groups: high-yielding boreholes with discharges of more than $7\,Ml/d$ for pumped drawdowns of no more than 8 m, medium-yielding boreholes of $5\,Ml/d$ but pumped drawdowns of 10–45 m, and low-yielding boreholes with yields of $3\,Ml/d$ or less and pumped drawdowns in excess of 25 m. For medium- and low-yielding boreholes there are substantial non-linear well losses which occur for reasons outlined in Section 8.2.5. Recharge rates into individual boreholes are typically 1 Ml/d using treated mains water. Such low recharge rates are sufficient since the aquifer is only heavily exploited during drought periods. The key to successful artificial recharge is that the water enters the aquifer system through the chalk aquifer, but is stored in the less permeable overlying Lower London Tertiaries.

Prior to 1860 the Chalk and Lower London Tertiaries were fully saturated and confined by the London Clay. For the next century there was a steady decline in groundwater heads, with the water table often falling to lie within the Chalk. During the period 1973 to 1993 the water table rose by 5–12 m, due to reduced abstraction and artificial recharge. Detailed hydrochemical monitoring has not shown any detrimental effects due to the artificial recharge; the aquifer system is protected from surface pollution by the overlying London Clay.

8.7 VARIABLE HYDRAULIC CONDUCTIVITY WITH DEPTH IN CHALK AND LIMESTONE

8.7.1 Introduction

It is generally recognised that the hydraulic conductivity of chalk and limestone aquifers varies with the saturated depth. This proved to be important for the Thames Groundwater Scheme, a major conjunctive use scheme in the Chalk of Southern England, in which groundwater is used during drought periods to supplement flow in the River Thames during drought years (see Section 11.2). The initial estimate of additional river flow due to pumping from the Chalk was 345 Ml/d from 38 boreholes during a six-month drought period. When the actual scheme was tested in a drought year, the sustainable yield was only 59 Ml/d. The major cause was the low transmissivity and storage coefficient of the chalk aquifer under drought conditions. Similar reduced yields during periods with low water tables have been observed in limestone aquifers; the Miliolite Limestone in Western India has already been considered (Section 6.6.3). Examples in the UK include the Southern Lincolnshire Limestone (Rushton *et al.* 1982) and the Cotswold Limestone (Rushton *et al.* 1992).

How can these changing hydraulic conductivities with saturated depth be identified and approximately quantified? An early study by Foster and Milton (1974) used laboratory estimates of intergranular properties, geophysical logging to identify borehole inflow levels and volumes, together with pumping tests, to demonstrate that permeability development is concentrated in the zone of seasonal water table fluctuation. Foster and Crease (1975) examined the yield of a particular borehole at two different water table levels; from

these tests they developed yield-drawdown curves which show that, for lower water tables, the yield falls off rapidly. A study in the East Kent chalk aquifer (Cross *et al.* 1995) of a well with adits showed that the yield could approach 16 Ml/d with very high water levels but during drought periods when the water table was 20 m lower, it was difficult to achieve a yield of 6 Ml/d.

8.7.2 Case study in Berkshire Downs, the UK

The case study selected to explore the change in hydraulic conductivity with depth refers to pumping tests in a borehole drilled as part of the Thames Groundwater Scheme. The pumping test site is situated at the top of a dry valley in the Berkshire Downs. Fortunately, two pumping tests were conducted in the same borehole at *different rest water levels*; the second test was three months after the first test with the rest water level 7.5 m lower. The abstraction borehole is over 100 m deep with a nominal diameter of 0.76 m; an observation well is positioned 220 m from the abstraction borehole (Rushton and Chan 1976).

In early May the first test was carried out with an abstraction rate of 6450 m³/d. The test continued for fourteen days; results for the pumping and recovery phases in both the pumped and observation boreholes are plotted in Figure 8.31. The second test took place at the end of July when the rest water level was 7.5 m lower; the pumping rate was 4650 m³/d. The drawdowns became excessive (43 m) after six days; the pump automatically switched off so that recovery data were only obtained for the observation well. All the drawdowns are plotted in Figure 8.31 relative to the rest water level of the test, which commenced early in May.

Before attempting a quantitative analysis of the test, a careful visual examination of the results is essential.

1. For each test the plot is of drawdown against *log* of time; this is the most useful way of examining the field data. If a *log-log* graph is used, the drawdowns at early times are given undue emphasis.
2. Considering the pumped water level with an abstraction rate of 6450 m³/d (the top left hand diagram of Figure 8.31), the shape of the curve is different from that normally observed in practice. In most field situations there is a rapid initial fall followed by a levelling-off (see, for example, Figure 5.2; Kenyon Junction). The increasing drawdown with time suggests that the borehole finds it increasingly difficult to draw in water.
3. The second test starts at the lower water table elevation. Even though the abstraction from the

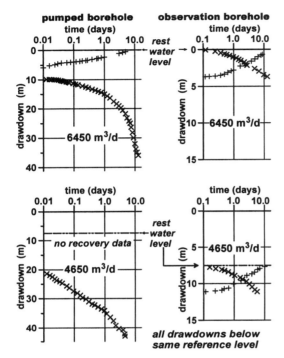

Figure 8.31 Pumping test in chalk aquifer with different rest water levels

Between 0.1 and 1.0 day the discharges per unit drawdown are 6450 [3580] m³/d/m; while between 1.0 and 10 days the values are 2690 [1110] m³/d/m. This suggests that the transmissivity for the second test is roughly half that for the first test.

Conventional pumping tests: methods of analysis were used for initial estimates of transmissivities and storage coefficients; data for only part of the test could be matched with a Theis type-curve. For the first test using observation well data, the transmissivities were estimated as 265–345 m²/d with a storage coefficient (specific yield) of about 0.01. For the second test the transmissivity was estimated to be 200–210 m²/d with a specific yield of 0.009. For the pumped borehole the transmissivities for the two tests were estimated as 190 and 120 m²/d. These results fail to highlight the true nature of the aquifer system.

Consequently an analysis was carried out using the *two-zone numerical model*. Three variations of hydraulic conductivity with depth are investigated (Rushton and Chan 1976); the same model with the same parameter values is used for the pumping and recovery phases of both tests.

1. For the first analysis a uniform value of hydraulic conductivity with depth is used, and the effective depth of the aquifer is taken as 65 m. In the numerical model, the saturated depth of the Upper layer of the two-zone model decreases with increasing drawdowns. However, the resultant decrease in transmissivity is insufficient to represent the rapid fall in pumped water levels, although the general shapes of the drawdown and recovery curves in the *observation well* are reproduced. This confirms that the hydraulic conductivity must decrease with saturated depth. Furthermore, an analysis based on a relatively distant observation well is unlikely to provide information of the significant changes of aquifer parameters with depth.

2. For the second attempt, information from geophysical logging of the production borehole was used. Major fissures were identified during pumping at 10, 22, 26 and 44 m with minor fissures at 55 and 65 m below the rest water level of the first test. If higher values of hydraulic conductivity are associated with the fissures, the transmissivity decreases as fissures are dewatered. Using this approach, the drawdowns derived from the numerical model show little resemblance to the field drawdowns. This is because acidising of the production borehole had taken place; this enhanced the permeability of the lower parts of

pumped borehole is 4650 m³/d (about 70 per cent of the abstraction rate for the first test), the rate at which the drawdown increases is substantially higher than for the first test. After one day, the drawdown relative to the rest level for this test is 27 m compared to 15 m for the test with the higher rest water level and higher abstraction rate.

4. The rapidly falling pumping levels may be partly due to limitations of the borehole such as the presence of a seepage face, therefore it is important to consider the observation borehole which reflects the response of the aquifer. Comparing drawdowns in the observation borehole between 0.1 and 1.0 day and between 1.0 and 10 days:

	0.1 to 1.0 d	*1.0 to 10.0 d*
change in drawdown for 6450 m³/d	1.0 m	2.4 m
change in drawdown for 4650 m³/d	1.3 m	4.2 m (estimate)

These drawdowns are used to calculate a discharge per unit drawdown at the observation boreholes, values for 4650 m³/d are shown in square brackets [].

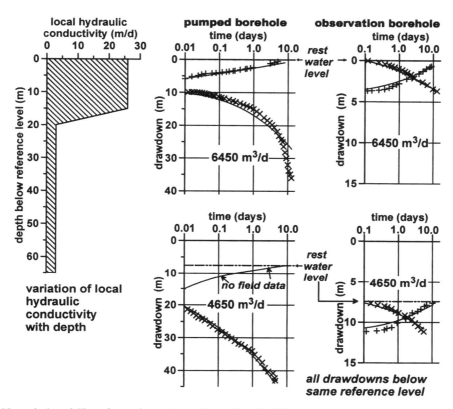

Figure 8.32 Numerical modelling of pumping test in chalk aquifer with different rest water levels; the numerical model results are indicated by continuous lines

the aquifer resulting in properties different from the undisturbed aquifer.

3. Finally, a trial and error method is used to determine an appropriate variation of hydraulic conductivity with depth so that the numerical model is able to reproduce the observed behaviour in both the pumping and observation well during pumping and recovery for each of the two tests. The final hydraulic conductivity variation is shown on the left of Figure 8.32. With this form of distribution of hydraulic conductivity, the modelled drawdowns in both pumped and observation boreholes are close to field values. The least satisfactory agreement is for the pumped borehole during the first test; the precise shape of the drawdown curve is not reproduced because the drawdowns depend on the detailed hydraulic conductivity variation close to the borehole. Nevertheless, there is good agreement for the recovery in the pumped borehole.

8.7.3 Consequences of variation in hydraulic conductivity

The precise details of the variation in hydraulic conductivity with depth are not critical; what is significant is that higher hydraulic conductivities occur in the zone of water table fluctuation which is partially or totally dewatered during drought periods. Connorton and Reed (1978) extended the analysis described above to include the spatial distribution of aquifer parameters, variable pumping rates, natural recessions and well field interference. They also suggest that variable pumping rates due to the pump characteristics can lead to a reduction in discharge of 30–40 per cent during drought conditions when the water level in the pumped borehole may be 40–60 m below maximum non-pumped water levels. A further feature included in their revised model is that the specific yield reduces with a decreasing saturated depth; extensive field studies suggest that specific yields as high as 3 per cent may occur at maximum water table elevation, falling to 0.8 per cent for minimum water table elevations. Having tested the ability of the model to reproduce pumping tests, it was used to examine the response of boreholes over periods as long as 240 days. Although their study is based on radial flow models, by incorporating additional physical processes, valuable insights were gained concerning the long-term yield of boreholes in the Thames Groundwater Scheme.

8.8 HORIZONTAL WELLS

Horizontal wells usually fall into one of two categories, either collector wells which are horizontal pipes radiating out from a central cylinder (caisson) or horizontal wells in shallow aquifers. The remainder of the discussion of horizontal wells is concerned with two issues. First, there are reviews of mathematical expressions for conditions in horizontal wells; alternative expressions are required for different times since the start of pumping. Second, a field study of a horizontal well is described; this field study provided sufficient evidence to develop a numerical model which represents both hydraulic conditions in the horizontal well and the interaction between the well and regional groundwater flows in the aquifer.

8.8.1 Collector wells

Collector wells, consisting of a central cylinder with perforated pipes radiating outwards, are often constructed in alluvial riverbed aquifers. They induce water from the surface water body and usually provide high yields of the order of 15 Ml/d. Raghunath (1987) provides detailed practical information about the construction and operation of collector wells in India. Frequently the collectors are drilled horizontally through the aquifer under a river (Figure 8.33).

Detailed analysis of collector well systems is difficult. McWhorter and Sunada (1977) suggest that a collector system can be represented as an ordinary well with an equivalent radius of 0.618 times the length of the individual collectors. Their analysis is based on steady-state flow in a confined aquifer. Time-variant methods of analysis have been developed by Hantush and Papadopulos (1962); they consider the drawdowns due to steadily discharging collector wells in unconfined and confined aquifers. One assumption, which is likely to limit the validity of their analysis, is that the yield of each lateral is uniformly distributed along its length. Milojevic (1963) provides a wealth of informa-

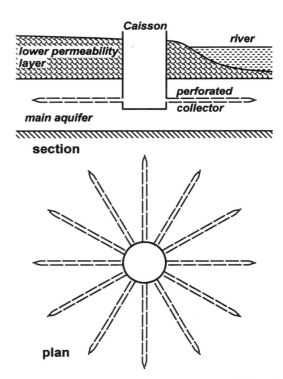

Figure 8.33 Diagram of a collector well in the vicinity of a river

tion about collector wells including a review of analytical and experimental studies; in addition an investigation into the effect of rivers on collector well responses is described.

Ball and Herbert (1992) describe field experiments where horizontal collector wells were drilled in large diameter wells in Sri Lanka; these horizontal wells are constructed in the regolith towards the bottom of the weathered zone immediately above the slightly weathered bedrock. The horizontal wells are of 90 mm diameter and extend from the large diameter well for distances up to 30 m. In the majority of the wells with radial collectors, improved yields were identified. This improvement was shown most clearly by a more rapid rate of recovery after the pump was switched off; a doubling of the rate of recovery was often achieved. With these horizontal collectors, water is drawn from greater distances in the weathered zone; the response may be similar to the dug-cum-bore well (see Section 8.5.2 and especially Figure 8.18, where the transmissivity of the fractured zone assists in drawing water from more distant parts of the weathered zone).

8.8.2 Mathematical expressions for horizontal wells

Huisman and Olsthoorn (1983) describe studies of horizontal wells using analytical methods. They use the word *gallery* to describe a horizontal well; the main focus of their work is the use of horizontal wells for artificial recharge. They argue that the vertical components of flow to a horizontal well are only significant very close to the well; consequently, their analyses include only horizontal flow components. Methods of correcting for vertical flow components in the vicinity of the well are included as an additional drawdown. Both steady-state and time-variant flow are considered.

Kawecki (2000) presents analyses of the variation of drawdown with time for horizontal wells in confined and unconfined aquifers. His solutions for confined aquifers are taken from the petroleum literature; he has prepared a useful table which defines corresponding parameters in oilfield and groundwater equations and relationships between the two sets of parameters.

There are a number of important steps in deriving the relevant equations:

1. *Uniform head or uniform flow*: in a horizontal well there are two conditions which can be included in an analytical solution, either a uniform flux along the well or a uniform groundwater head. With analytical solutions for horizontal wells it is more convenient to work in terms of uniform flux even though the physical situation is more likely to correspond to a constant head. Further discussion about the actual hydraulic conditions in a horizontal well can be found in the case study in Section 8.8.3.
2. *Skin factor*: a skin factor is similar to well loss as described in Section 5.5 in that it leads to additional drawdowns. The additional drawdown Δs is defined as a function of the discharge rate Q, the hydraulic conductivity K, the thickness of the aquifer m and the skin factor *skfac*:

$$\Delta s = \frac{Q}{2\pi K m} skfac \tag{8.11}$$

3. *Variables used in equations*: Figure 8.34 illustrates the co-ordinate axes and dimensions used in the analysis of a horizontal well in a confined aquifer.
4. *Flow regimes at different times*: the flow patterns around a horizontal well in an extensive aquifer change significantly with time. Three regimes can be identified; different analytical solutions are devised for these regimes. This approach has similarities

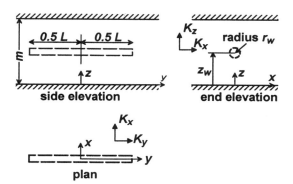

Figure 8.34 Geometry of horizontal well in confined aquifer (Kawecki 2000). Reprinted from *Ground Water* with permission of the National Ground Water Association. Copyright 2000

with the case study in Section 5.1 where different stages are identified in the development of flow patterns associated with a pumped vertical borehole in an unconfined aquifer. The three regimes for a horizontal well are illustrated in the schematic diagrams of Figure 8.35; the times are described as follows.

(a) *Early radial flow*: the well draws on confined (or elastic) storage close to the well; all the flows are perpendicular to the axis of the well; the sketches refer to side and end elevations.

(b) *Early linear flow*: the upper and lower boundaries begin to have an effect but all the flows are perpendicular to the axis of the well with no flow components in the y direction. Early linear flow occurs immediately after early radial flow.

(c) *Late pseudo-radial flow*: a significant time will elapse before these flow patterns are reached; the plan view is the most informative since it shows that flows in the y direction are now as important as flows in the x direction.

5. *Relevant equations and times when they are appropriate*: Table 8.10 lists an equation and a time which is applicable for each of the three types of flow. Frequently Kawecki presents alternative equations, only one of which is quoted in this table. The times for the end or start of the period are empirical approximations.

6. *Modifications for unconfined conditions*: Kawecki (2000) describes modifications to these equations which allow for the decreasing saturated thickness in an unconfined aquifer and also vertical flow components from the water table.

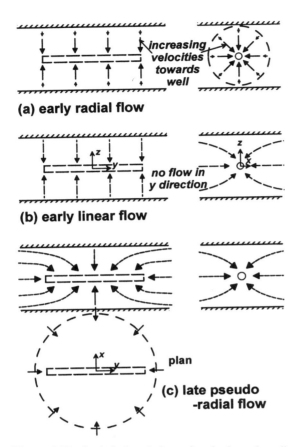

(a) early radial flow

(b) early linear flow

(c) late pseudo -radial flow

Figure 8.35 Analytical solutions for horizontal wells showing the three time periods. (Kawecki 2000). Reprinted from *Ground Water* with permission of the National Ground Water Association. Copyright 2000

A computational model for horizontal wells (or adits) feeding into vertical wells is described by Zhang and Lerner (2000). It uses a modelling code MOD-BRANCH, which combines MODFLOW with a code BRANCH for open channel flow. Modifications are required to represent the perimeter of the horizontal well instead of the wetted perimeter of a channel.

8.8.3 Horizontal well in a shallow coastal aquifer

This case study of a horizontal well relates to a shallow coastal aquifer at Loba in West Sarawak, Malaysia (Mailvaganam *et al.* 1993). A shallow sand aquifer lies above clay which is typically 4 m below mean sea level; the average groundwater elevation is 1.5 m above mean sea level. The quality of the water is good; the antici-

pated long-term requirement for water is 240 m³/d to cater for tourist development. Following a preliminary examination of the potential yield of vertical and horizontal wells it was concluded that a single horizontal well feeding to a central caisson could provide the required yield even during a period of up to 100 days without recharge. The alternative would be for several groups of vertical wells, each with a pump; the full supply cannot be guaranteed during extended dry periods if vertical wells are used.

Field information and construction of horizontal well

Pumping tests in the shallow sand aquifer indicated that the transmissivity is in the range 20–40 m²/d. The horizontal well consists of a central caisson 1.2 m in diameter with a pair of 150-mm diameter slotted PVC pipes linked to the caisson on two sides; Figure 8.36. The location of the well is at a groundwater mound. Datum is taken as the top of the clay; mean sea level is 2.8 m above datum, with the rest water table 2.2 m higher at 5.0 m above datum. The slotted horizontal pipes were placed directly above the underlying clay with a filter sand pack of about 60–80 mm. Vertical tubes are located along the length of the horizontal well to permit cleaning of the well, they also allow measurements of hydraulic heads in the well during pumping. Dewatering of the aquifer was necessary during construction of the horizontal well.

Several pumping tests were carried out on the horizontal well; this presentation will concentrate on a fourteen-day test. Important information gathered from Day 6 of the test pumping is included in Figure 8.37.

(a) Water table elevations perpendicular to the horizontal well are compared with the hydraulic head in a piezometer connected to the horizontal well as shown in Figure 8.37a; note that the water table is well above the hydraulic head in the horizontal well.

(b) Water table elevations for an extended line above the horizontal well and the hydraulic heads in the horizontal well are plotted in Figure 8.37b. The water table is drawn down below the original rest water level but it is 3.5–4 m above the horizontal well; the hydraulic head in the well is 0.1–0.22 m below the overlying water table.

(c) A more detailed plot of the hydraulic heads in the well is recorded in Figure 8.37c; the gradient increases towards the centre of the well where the caisson is located.

Table 8.10 Equations for horizontal well drawdowns and times when the equations are applicable

Condition and equation	Times
Early radial flow	*End time*
$$s_w = \frac{Q}{2\pi L\sqrt{K_x K_z}}\left\{\ln\left(\frac{1.5\sqrt{(K_x K_z)^{1/2}\,Lt/S_S m}}{r_w}\right) + skfac\right\}$$	$$t_e = \frac{0.475 d_z^2 S_s}{K_z}$$
where d_z is the shorter of the vertical distances from the well to the upper or lower boundaries of the aquifer	
Early linear flow	*End time*
$$s_w = \frac{1}{\sqrt{\pi}} \times \frac{Q}{LT_x}\sqrt{\frac{T_x t}{S}} + \frac{Q}{2\pi L\sqrt{K_x K_z}}(f_2 + skfac)$$ $$f_2 = \ln(b/2\pi r_w) + 0.25\ln(K_x/K_z) - \ln(\sin(\pi z_w/m))$$	$$t_e = \frac{0.042 L^2 S_s}{K_y}$$
Late pseudo-radial flow	*Start time*
$$s_w = \frac{Q}{2\pi\sqrt{T_x T_y}}\ln\left(\frac{1.5\sqrt{T_y t/S}}{0.25L}\right) + \frac{Q}{2\pi L\sqrt{K_x K_z}}(f_2 + skfac)$$	$$t_s = \frac{0.325 L^2 S_s}{K_y}$$

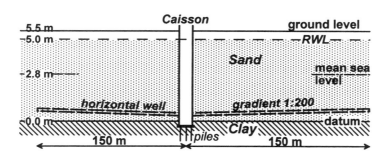

Figure 8.36 Details of horizontal well at Loba

These figures provide important information about the conditions associated with the horizontal well; any method of analysing this problem must allow for a variation in hydraulic heads along the well.

Quantified conceptual model

In the analytical approach, horizontal wells can either be represented as a constant head or a constant inflow. Figure 8.37c indicates that there is a variation in hydraulic head along the well, consequently the assumption of a constant head is incorrect. Nor is it possible to measure the distribution of inflows to the well. However, the differences between the water table elevation and hydraulic head in the well, as shown in Figure 8.37b, indicate that the uniform inflow approach is unlikely to be correct. The conceptual model requires consideration of both the interaction between the well

and the aquifer and the hydraulic response within the well.

A representative cross-section through the horizontal well and the aquifer is shown in Figure 8.38a; the decrease in saturated depth in the vicinity of the well must be included in any solution. A linear relationship, similar to that for rivers (Section 4.3) or drains (Section 4.4), is used to calculate the flow Q_a *from the aquifer to the horizontal well* depending on the difference between the groundwater head in the aquifer h and the hydraulic head in the water-filled pipe of the horizontal well h_p,

$$Q_a = C_a(h - h_p) \tag{8.12}$$

The coefficient C_a depends on the geometry of the well, any filter material, the aquifer hydraulic conductivity and the length over which Q_a is calculated. In practice

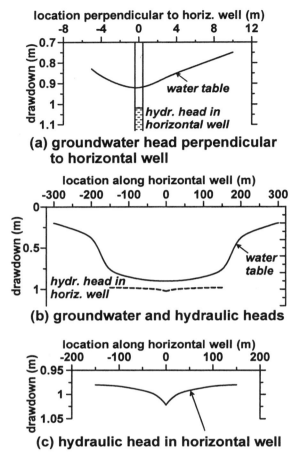

(a) groundwater head perpendicular to horizontal well

(b) groundwater and hydraulic heads

(c) hydraulic head in horizontal well

Figure 8.37   Groundwater heads and hydraulic heads in the well on Day 6

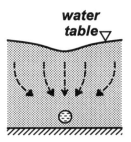

(a) flow, aquifer to horizontal well

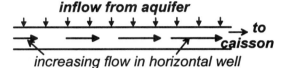

(b) conditions in horizontal well

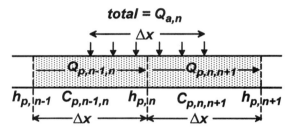

(c) flow balance in horizontal pipe

Figure 8.38   Development of conceptual models for horizontal well

the coefficient, which is initially based on analytical expressions or previous experience, is adjusted during model refinement.

Hydraulic conditions in the horizontal well are sketched in Figure 8.38b. Groundwater enters the pipe along the length of the well (for convenience, inflow is sketched as occurring at the top of the pipe) so that there is an increasing flow along the pipe towards the caisson. The flow within the pipe can be described using pipe flow theory such as the Hazen–Williams formula. However, conditions are different to those in most pipe flow problems in that there is an increase in flow towards the caisson. The Hazen–Williams formula states that the frictional *head loss* in a pipe h_f can be calculated from the expression

$$h_f = RQ^{1.85} \tag{8.13}$$

where Q is the rate of flow in the pipe. R is the coefficient of pipe resistance,

$$R = \frac{3.2 \times 10^6 L}{C_n^{1.85} d^{4.87}} \tag{8.14}$$

in which L is the length of the pipe, C_n is the Hazen–Williams discharge coefficient and d is the diameter of the pipe.

For computational purposes it is helpful to rewrite Eq. (8.13) as

$$Q = h_f \left[RQ^{0.85} \right]^{-1} = h_f C_p \tag{8.15}$$

The term in square brackets is treated as a coefficient, C_p, which is updated during iterative calculations.

Incorporation in numerical model

The numerical model has three components. First, there is a single-layer variable saturated depth regional groundwater model. Second, a numerical representation of the hydraulic response of the horizontal well, Figure 8.38c indicates that the component flows in the horizontal pipe between nodes can be calculated from the hydraulic heads and pipe coefficients. Considering the flow in the pipe between node $n - 1$ and n,

$$Q_{p,n-1,n} = C_{p,n-1,n}(h_{p,n-1} - h_{p,n}) \qquad (8.16)$$

where $h_{p,n-1}$ and $h_{p,n}$ are the hydraulic heads, while the pipe coefficient $C_{p,n-1,n}$ is given by the equation

$$C_{p,n-1,n} = \left[\frac{3.2 \times 10^6 \Delta x}{C_n^{1.85} d^{4.87}} Q_{p,n-1,n}^{0.85}\right]^{-1} \qquad (8.17)$$

In this equation Δx is the distance between the nodes; the coefficient $C_{p,n-1,n}$ depends on the unknown flow $Q_{p,n-1,n}$ which means that some form of iterative technique must be used,

also $\quad Q_{p,n,n+1} = C_{p,n,n+1}(h_{p,n} - h_{p,n+1}) \qquad (8.18)$

Third, considering the interaction between groundwater heads along the line of the horizontal well and the corresponding hydraulic heads in the aquifer, the inflow into the pipe from the mid-point between nodes $n - 1$ and n to the mid-point between n and $n + 1$ is calculated using Eq. (8.12),

$$Q_{a,n} = C_{a,n}(h_n - h_{p,n}) \qquad (8.19)$$

Continuity along the pipe (see Figure 8.38c), is satisfied by

$$Q_{p,n-1,n} + Q_{a,n} = Q_{p,n,n+1} \qquad (8.20)$$

Substituting from Eqs (8.16), (8.18) and (8.19)

$$C_{p,n-1,n} h_{p,n-1} - (C_{p,n-1,n} + C_{a,n} + C_{p,n+1,n})h_{p,n}$$
$$+ C_{p,n+1,n}h_{p,n+1} = -C_{a,n}h_n \qquad (8.21)$$

Equations for each of the thirteen nodal points along the horizontal well at Loba are written; these equations

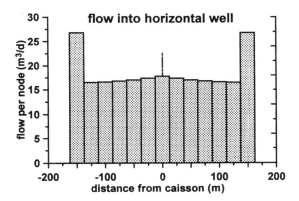

Figure 8.39 Bar chart of numerical model flows at Day 1 into horizontal well for each 25 m length of the well, pumping rate 240 m³/d

are solved iteratively with the equations for the regional groundwater model. Solutions are obtained alternately for the aquifer and the well until both the out-of-balance at all nodal points in the aquifer are less than 0.01 mm/d and also changes in hydraulic head in the horizontal well all less than 0.001 m. First estimates of the coefficients $C_{p,n-1,n}$ etc. can be determined from Eq. (8.17); adjustments are made during model refinement. The coefficients at either end of the horizontal wells are increased by a factor of 1.3 to represent the more spherical flow towards the ends of the pipes.

Detailed comparisons between field and modelled results can be found in Azuhan (1997); the one result included in this discussion is the calculated distribution of inflows to the horizontal well. Figure 8.39 indicates that the maximum inflows are at the ends of the pipes; there is also an increase in flow towards the pipe centre due to the greater difference between aquifer and hydraulic heads.

8.9 CONCLUDING REMARKS

Chapter 8 considers a range of practical issues related to flow to boreholes; for most of the examples the analysis is based on methodologies introduced in Chapter 7. There are two major themes in this chapter. The first three sections describe techniques for identifying the well losses that occur as water is drawn into a pumped borehole, how the variation of hydraulic conductivity with depth can be identified and quantified and how the water table elevation can be estimated from water level readings in supply boreholes when they have ceased pumping. The remaining four sections consider the impact of aquifer conditions on the yield of vertical boreholes and horizontal wells and also the effectiveness of injection boreholes.

Step-pumping tests and packer tests are widely used in groundwater investigations, but there are questions about the reliability of the methods of analysis. The study of alternative conditions in the vicinity of pumped boreholes using a time-instant radial-vertical numerical models shows that distinctively different responses are obtained when specific drawdown-discharge plots are prepared. These findings are used to interpret field step-pumping test results for four different aquifer systems. Packer tests can be used to estimate the aquifer properties in the vicinity of boreholes but there are questions about the validity of the simple analytical expressions used to estimate aquifer properties for complex layered systems. A radial-vertical numerical model is used to represent a sandstone aquifer system which contains both low-permeability layers and fissures. The standard formulae for analysing the tests provide similar estimates to the numerical model apart from one location where the packer was in contact with a fissure, but failed to totally seal the fissure. A further case study shows how non-pumping levels in supply boreholes can be used to identify the water table elevation at multi-borehole pumping stations.

Reliable yields of aquifer systems are investigated by preparing conceptual models and using the two-zone time-variant radial flow model first to represent field results and then to predict the consequences of long-term pumping. Explanations are presented for aquifer yields which are lower than originally estimated. Reduced yields with lower initial water table elevations in chalk aquifers are also explained. Artificial recharge using injection wells is studied. The impact of well clogging is examined and the success of the North London Artificial Recharge Scheme is explained in terms of the transmissive properties of the chalk and the storage properties of the overlying strata. Finally, horizontal wells are considered. Mathematical expressions are presented for various types of horizontal well system. A numerical model for a shallow coastal aquifer is described which represents both groundwater flows in the aquifer and hydraulic flows in the horizontal well pipe.

Part II introduces a wide range of situations and conditions which occur when water is withdrawn from aquifers through boreholes and wells. Practical issues are highlighted which limit the effectiveness of the wells and boreholes in collecting water from an aquifer and the ability of the aquifer system to supply water. Numerical models are especially useful in understanding flow processes and predicting future aquifer responses.

PART III: REGIONAL GROUNDWATER FLOW

Most regional aquifer systems involve complicated processes such as a combination of groundwater flow through aquifers and aquitards, complex recharge mechanisms, poorly defined boundaries to the aquifer system, sensitive interaction with surface water bodies and abstractions changing over long time periods. The objectives of regional groundwater studies are also varied; they include examining the risk of over-exploitation of the aquifer resources, determining whether low river flows are due to groundwater abstraction, and considering quality issues. At the outset it is essential to identify the objectives of a study; different objectives may lead to different emphases in data collection, data interpretation, conceptual model development and numerical model simulations.

The discussion of regional groundwater flow is presented in four chapters:

Chapter 9: studies where the transmissivity remains effectively constant.

Chapter 10: studies where vertical flows through lower permeability strata are significant.

Chapter 11: studies where the hydraulic conductivity varies with saturated depth.

Chapter 12: insights into selected modelling features and techniques.

In each chapter, case studies are introduced to illustrate the important concepts.

For each case study the focus is the development of conceptual understanding. This requires an examination and interpretation of all available sources of field evidence. For every case study, quantified conceptual models are prepared, usually followed by the development of numerical groundwater models. Each case study focuses on specific issues as indicated in Table III.1. In addition, a summary is provided in Table III.2 of topics considered in these chapters with reference to the sections where information can be found.

Table III.1 List of major case studies and important insights

Section	Case Study	Selected issues
9.2	Nottinghamshire sandstone	Sources of recharge include rainfall, minor aquifers and leakage through low permeability strata, river coefficients based on field information, flow balances for different time intervals deduced from groundwater model
9.3	Doncaster sandstone	Inflows to sandstone aquifer through extensive drift deposits, recharge from drainage channels, groundwater model reproduces slow decline in groundwater heads, comparison between field and modelled groundwater heads
9.4	Lower Mersey sandstone	Recharge from urban areas through drift, aquifer response studied for 150 years, old saline water at depth and modern saline water drawn into aquifer, use records from pumped borehole to verify model, predictions of responses to reduced abstractions
9.5	Barind alluvial	Aquifer of limited thickness, high recharge through ricefields, overlying clay of sufficiently high permeability to allow recharge to pass through, inverted well design excellent for aquifers of limited thickness
10.2	Mehsana alluvial	Important information from observation wells in shallow and deeper aquifers, most water pumped from deeper aquifer is drawn from shallow aquifers, overexploited aquifer, vertical slice model and model of limited area
10.3	Vanatavillu	Aquifer system in Western Sri Lanka, alluvium over limestone, incorrect estimates of deep limestone aquifer properties due to mistaken leaky aquifer pumping test analysis, limited resources of limestone aquifer

Table III.1   *Continued*

Section	Case Study	Selected issues
10.4	SLP	San Luis Potosi in Mexico, granular and clay overlying metamorphic strata, inputs to deep aquifer due to runoff from overlying aquifer plus contribution from deep thermal water, use temperature and chemical changes to identify and approximately quantify inflows
10.5	Bromsgrove sandstone	Importance of head loss across thin low-permeability layers within sandstone aquifer, water mills from 100 years ago provide insights into historical aquifer response, differences in detailed responses of shallow and deep piezometers
11.2	Berkshire Downs chalk	Chalk aquifer used for river augmentation scheme, location of boreholes important for good net gain, fivefold seasonal variation in transmissivity at certain locations, good agreement between field and modelled heads and flows when variable hydraulic conductivity with depth is included in numerical model
11.3	S. Lincolnshire limestone	Most of limestone aquifer covered by drift, half of recharge due to runoff from less permeable areas, reduced hydraulic conductivity with depth tends to isolate unconfined area during droughts, complex river–aquifer interaction
11.4	Miliolite Limestone	Limestone aquifer in western India, contrast of transmissivity between post- and pre-monsoon, groundwater abstraction from a large number of wells, unsuccessful attempt to displace saline water using artificial recharge
11.5	Gipping Chalk	Importance of groundwater hydrographs in identifying recharge mechanisms, immediate and delayed runoff from boulder clay enters chalk aquifer on boulder clay margins, evidence of flow processes from hydrochemistry

Table III.2   Key topics and section numbers where information can be found

Topic	Sections where topic is discussed
Nature of aquifer system	
Regional distribution of transmissivity	10.3.3, 11.2.2, 11.3.6, 11.5.1
Variable K with depth	11.2.2, 11.3.5, 11.4, 11.6.2, 12.5.3
Flow from shallow to deep aquifers	10.2.4, 10.2.6, 10.2.7, 10.4.2, 10.6.1
Leakage through low K layers	9.3.2, 9.5.2, 10.3.2, 10.6.2, 10.6.3, 10.6.4, 11.3.6, 11.5.1, 12.6.1, 12.7.2
Conditions and locations of boundaries	9.2.3, 10.5.3, 11.3.1, 11.3.7, 12.6.1, 12.6.3
Coastal boundaries	9.4.3, 11.4
Flows from adjacent low K strata	9.2.3, 9.4.1, 10.6.4
River–aquifer interaction	9.2.3, 11.2.1, 11.3.3, 11.3.8, 11.6.2, 12.2.1, 12.2.2, 12.6.7
Springs	9.3.2, 9.4.1, 9.4.5, 10.2.1, 10.5.1, 10.5.3, 11.3.3, 11.3.8
Location and yield of boreholes	9.4.3, 9.5.2, 10.2.2, 10.2.3, 10.2.5, 10.5.1, 10.6.2, 11.2.1
Inflows to aquifer system	
Rainfall recharge	9.2.3, 9.3.2, 9.4.2, 10.5.1, 11.3.2, 11.4, 11.5.1, 12.4.3
Runoff recharge	10.4.2, 11.3.2, 11.5.1
Recharge through drift	9.2.1, 9.4.2, 9.5.2, 10.6.3, 10.6.4, 11.3.2, 11.5.1
Urban recharge	9.2.1, 9.4.2, 10.4.1, 10.5.1, 10.6.2
Impact of release from storage	9.2.1, 9.4.5, 10.2.5, 11.6.1, 11.6.2, 12.4.3, 12.5.2
Time-related issues	
Duration of study and impacts	9.2.6, 9.4.1, 10.5.5, 12.6.2, 12.7.1
Initial and starting conditions	9.2.3, 10.5.3, 12.5.2, 12.5.3
Time-instant solutions	10.2.5, 12.4.2, 12.4.3
Outputs and techniques for groundwater studies	
Refinement of groundwater models	9.2.1, 10.6.4, 11.3, 11.5, 11.6.1, 12.6.5
Comparison of field and model heads	9.2.5, 9.3.3, 9.4.4, 10.2.5, 10.2.7, 11.2.2, 11.3.8, 12.7.2
Comparison of field and model flows	9.2.5, 9.4.4, 11.2.2, 11.3.8, 12.7.2
Water balances	9.2.1, 9.2.6, 9.3.3, 9.4.5, 10.4.2, 11.6.3
Predictive simulations	9.4.5, 11.3.8, 12.7.1, 12.7.2

Regional Groundwater Studies in which Transmissivity is Effectively Constant

9.1 INTRODUCTION

This chapter focuses on regional groundwater flow situations when the transmissivities remain effectively constant with time. Significant features are introduced using four case studies in which the aquifer systems consist of a single main aquifer. Identifying and estimating the magnitude of recharge to the aquifer system is an important task; frequently lower permeability strata, which overlie the main aquifer, modify the recharge. In most of the studies, interactions of aquifers with lakes, rivers and springs are essential components of the overall aquifer water balance. Many of these aquifer systems have the ability to store substantial quantities of water; during high recharge years water is taken into storage, and during low recharge periods water is released from storage to flow through the aquifer.

For the first case study there is a detailed presentation of the steps necessary to develop and refine the conceptual model; this is followed by a description of the estimation of aquifer and other parameters in preparation for numerical model simulations. An assessment is made of the adequacy of the numerical models to reproduce the observed field responses. Finally, outputs from the numerical model are presented as flow balances. In the later studies the discussion focuses on additional issues which are crucial to the development of an adequate understanding and representation of regional groundwater flow.

9.2 NOTTINGHAMSHIRE SHERWOOD SANDSTONE AQUIFER

Objectives of the study: identify the groundwater resources available for exploitation; examine the impact of groundwater abstraction on rivers.

Key issues: recharge through less permeability strata, leakage through Mercia Mudstone, field measurements to estimate river coefficients, flow balances at different timescales.

The first step in all regional groundwater studies is to identify the essential features of the conceptual models. This is followed by the development of idealised models together with the estimation of numerical values for all the parameters. From the quantified idealised models, numerical groundwater models are prepared; numerical solutions are obtained using appropriate software. The adequacy of the numerical model simulation is tested against field data of groundwater heads and flows between the aquifer and rivers, springs etc. A wide variety of alternative outputs can be obtained including the distribution of groundwater heads and flows in space and time. For this presentation, the aquifer response is summarised using annual and monthly water balances which provide insights into the way in which the aquifer system functions and in particular the impact of abstraction on river–aquifer interaction.

9.2.1 Identifying the conceptual model, focus on recharge components

The Nottingham Sherwood Sandstone aquifer in the English Midlands has been an important source of groundwater for many decades. Figure 9.1 is a plan view of the aquifer system; Figure 9.2 contains a representative west–east cross-section. There are more than forty principal pumping stations; some are located in the unconfined area, others in the confined area are located close to their supply areas.

The initial conceptual model of the aquifer system, Figure 9.2a, assumed that the only source of recharge is rainfall recharge on the outcrop of the Sherwood

Groundwater Hydrology: Conceptual and Computational Models. K.R. Rushton
© 2003 John Wiley & Sons Ltd ISBN: 0-470-85004-3

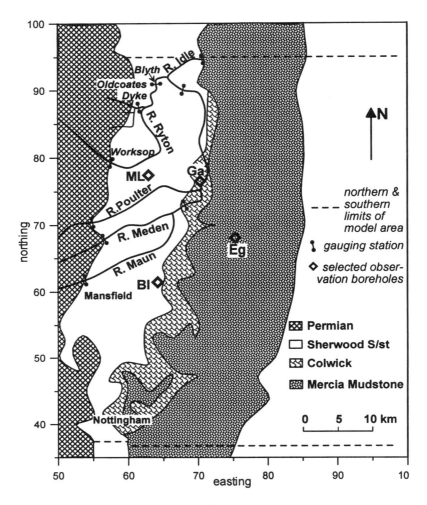

Figure 9.1   Plan view of the Nottinghamshire sandstone aquifer system

Sandstone; other strata overlying the sandstone aquifer were assumed to be effectively impermeable. Based on this assumption, the long-term water balance for the period 1974 to 1992 is as follows (water taken from the aquifer system such as abstraction is written as a negative quantity):

rainfall recharge on sandstone outcrop	230 Ml/d
abstraction	−285 Ml/d
groundwater flow to rivers	−40 Ml/d
estimated release of water from storage	25 Ml/d
lack of balance	*70 Ml/d*

Rainfall recharge for the outcrop area of 513 km² is calculated using methodologies described in Chapter 3.

Reliable abstraction records are available; in addition, there is sufficient river gauging to allow a reasonable estimate of the net groundwater flow to rivers as they cross the Sherwood Sandstone. Furthermore, the extensive groundwater head monitoring network allows estimates to be made of the quantity of water released due to the long-term decline in groundwater heads. The average decline is 0.12 m/yr; with an estimated specific yield $Sy = 0.15$, this decline is equivalent to an inflow to the aquifer system of 25 Ml/d. As indicated above, the overall lack of balance is 70 Ml/d, which is 30 per cent of the rainfall recharge.

The lack of balance is primarily due to an over-simplification of the aquifer system by failing to include all the sources of recharge. In the revised conceptual

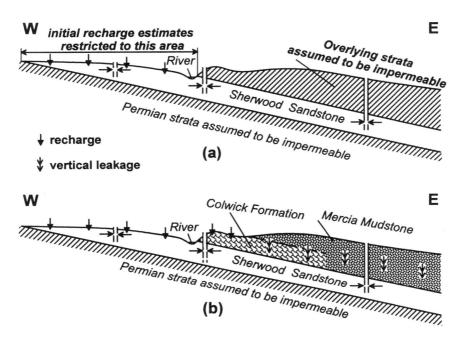

Figure 9.2   Conceptual models of the Nottinghamshire sandstone aquifer system: (a) original model, (b) revised model including additional recharge and leakage components

model, Figure 9.2b, the representative cross-section now includes the Mercia Mudstone and the Colwick Formation which overlie the Sherwood Sandstone. The Mercia Mudstone group is a predominantly argillaceous sequence; a leaky aquifer pumping test in the Mercia Mudstone (see Section 5.8) indicated a vertical hydraulic conductivity of 0.0001 m/d. Although the vertical hydraulic conductivity of the Mercia Mudstone is low, pumped boreholes can set up substantial vertical gradients so that water is drawn through the Mercia Mudstone into the underlying sandstone. In Figure 9.2b, vertical leakage is represented as arrows within the low permeability stratum while recharge is drawn as arrows crossing the top of the Sandstone.

On the western margins of the Mercia Mudstone, the lower strata are described as the Colwick Formation. This Formation is a transition facies between the arenaceous Sherwood Sandstone and the overlying argillaceous Mercia Mudstone. Pumped boreholes into the Sherwood Sandstone, which are located beneath the Colwick Formation, often produce high yields; this suggests that significant flows occur through the Colwick Formation. This possibility is supported by information such as private wells constructed in the

Colwick Formation which provide a reliable yield and by spring flows which occur from the Colwick Formation outcrop. Therefore, for the revised conceptual model of Figure 9.2b, in the transition zone of Sherwood Sandstone to Colwick Formation, the recharge is taken to be 75 per cent of the potential rainfall recharge. Where the Colwick Formation becomes thicker, the effective vertical permeability is assumed to vary from 0.002 to 0.0005 m/d. Pumping from the confined sandstone aquifer results in vertical leakage through the Colwick Formation; this leakage (vertical flow) is estimated from the vertical head gradient across the Formation multiplied by the effective vertical hydraulic conductivity. The estimated contribution from rainfall recharge through the Colwick Formation is 23 Ml/d, while leakage through the Mercia Mudstone and Colwick Formations sums to 41 Ml/d. An additional source of recharge is leakage from water mains in urban areas on the Sandstone outcrop; from flow measurements of the losses from certain water mains, urban recharge is calculated to be 9 Ml/d.

Consequently, the revised estimated inflows to the sandstone aquifer are:

rainfall recharge on sandstone outcrop	230 Ml/d
rainfall recharge to Colwich Formation	23 Ml/d
leakage through MM and CF	41 Ml/d
release of water from storage	25 Ml/d
urban recharge	9 Ml/d
total inflow	*328 Ml/d*

This is of similar magnitude to the combined effect of abstraction, 285 Ml/d, plus groundwater flow rivers, 40 Ml/d.

9.2.2 Idealisations introduced in the regional groundwater model

Figure 9.1 and Figure 9.2b embody a three-dimensional time-variant conceptual understanding of the aquifer system. Before this can be represented using a numerical model, idealisations must be made; these idealisations are summarised in the cross-section of Figure 9.3. Within the sandstone aquifer, lateral flows predominate; there is no evidence of significant vertical head gradients or the presence of continuous low-permeability zones which isolate sections of the main aquifer. Consequently, it is appropriate to represent the flows in the main aquifer in terms of the co-ordinates (x, y, t). Assumptions inherent in this approach and the manner in which this formulation implicitly represents vertical flow components are discussed in Section 2.6. The flow processes can be summarised as follows:

1. recharge enters the aquifer system on the outcrop of the sandstone aquifer and through the thinner sections of the Colwick Formation,

2. vertical leakage occurs through the Mercia Mudstone and Colwick Formations,
3. water is pumped from the sandstone aquifer using boreholes,
4. there are interactions between rivers and the main aquifer,
5. a further feature which is not represented diagrammatically in the cross-section of Figure 9.3 is that water is taken into and released from storage.

9.2.3 Quantifying the parameters of the conceptual model

The next stage is to quantify the conceptual model boundaries and parameters. In most practical situations, only limited information is available. Nevertheless, using all the field information available for the study area together with insights from studies of similar areas, reasonable estimates can be made. The estimated parameter values are summarised in Table 9.1. Uncertainties in parameter values can be resolved by conducting sensitivity analyses during the model refinement (see Section 12.6.6).

Boundaries (see figure 9.1)

Southern boundary: this boundary coincides with the southern limit of the Nottinghamshire sandstone aquifer unit; it is assumed that no flow crosses this boundary.

Eastern boundary: there is little information about groundwater conditions to the east other than that the

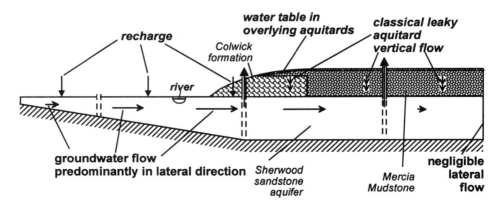

Figure 9.3 Cross-section of the Nottinghamshire sandstone aquifer system modified to be in a form suitable for transfer to a numerical groundwater model

Table 9.1 Summary of parameters used for the Nottingham Sherwood Sandstone aquifer model

Parameter	Range	Location and reasons
Transmissivity	500 to 1000 m²/d	Close to and under Colwick Formation; pumping test results and borehole yields
	260 to 30 m²/d	Outcrop, decrease to west with reduced sat. depth
	400 to 40 m²/d	Confined, decrease to east where yields are poor
Storage coefficient	0.15–0.10	Outcrop and under Colwick (from other studies)
	0.001	Under Mercia Mudstone, pumping test results
Vertical hydraulic conductivity	0.0001 m/d	Mercia Mudstone, from pumping test
	0.002–0.0005 m/d	Colwick, from ability to transmit water vertically
Abstractions	1.0 to ≈20 Ml/d	Actual historical values for individual pumping stations
River coefficients	0.5 to 2.0 Ml/d/km	Field accretion records (Section 4.3.4) and model refinement

transmissivity becomes lower and the salinity becomes higher as the aquifer dips deeper below the ground surface; therefore a no-flow condition is specified at about 25 km to the east of the sandstone outcrop.

Northern boundary: it is known that flows occur across the northern extension of this boundary; this is represented by extending the groundwater model further north as discussed in Section 9.3.

Western boundary: the sandstone aquifer becomes very thin, but small quantities of water could be transmitted from the low permeability Permian strata. Consequently, a small flow of 9 m³/d/km length of boundary is transferred across the western boundary of the Sherwood Sandstone. Alternative magnitudes for this flow can be explored during the model refinement.

Aquifer parameters

Since groundwater fluctuations are small compared to the aquifer thickness, it is appropriate to work in terms of constant transmissivities. More than sixty pumping tests have been analysed; trends can be identified but individual values show significant variations from the general trends. These general trends are confirmed by values of the specific discharge of boreholes (discharge per unit pumped drawdown). Reduced values indicate that the transmissivities are lower to the west where the saturated thickness decreases and lower to the east where the aquifer is confined beneath 200 m or more of Mercia Mudstone. Representative values of the transmissivity are listed in Table 9.1.

None of the pumping tests used both shallow and deep piezometers (see Section 5.1); consequently,

values of the specific yield estimated from pumping tests are not reliable. A value of specific yield of 0.15 for the sandstone outcrop is based on studies of similar sandstone aquifers. The confined storage coefficient is set at 0.001, based on a number of pumping tests.

Recharge

The potential recharge for the sandstone outcrop is estimated using the techniques described in Chapter 3. There is a major difference between the magnitude of average annual recharge in the south and north of the study area; values are 245 mm and 143 mm respectively. This occurs due to differences in average annual rainfall which is 725 mm in the south compared to 575 mm in the north. The average annual potential evapotranspiration is approximately 500 mm. The area is divided into seven rainfall zones with six different crop types.

Even though there is little drift cover on the outcrop of the sandstone, field observations during wet periods show that runoff does occur; this is confirmed by the presence of drainage ditches. The runoff is estimated from the daily rainfall intensity and the current soil moisture deficit. The basis of this approach is that for high intensity rainfall and for wet catchments (indicated by a low soil moisture deficit) the runoff is a higher proportion of the precipitation than for less intense rainfall on drier soil. The values used for the Nottinghamshire aquifer are similar to those quoted in Table 3.5.

Due to the thickness of the unsaturated zone between the base of the soil zone and the water table in the sandstone aquifer, delays do occur between the recharge event and water reaching the water table. The duration of the delays can be identified from

Table 9.2 Delays in potential recharge R reaching the water table dependent on the thickness of the unsaturated zone; M is the month when the potential recharge occurs

Thickness of unsaturated zone	Month					
	M	$M+1$	$M+2$	$M+3$	$M+4$	$M+5$
0 to 10 m	$0.9R$	$0.1R$	–	–	–	–
10 to 30 m	$0.3R$	$0.5R$	$0.2R$	–	–	–
30 to 50 m	$0.2R$	$0.3R$	$0.35R$	$0.1R$	$0.05R$	–
50 m+	$0.1R$	$0.15R$	$0.2R$	$0.3R$	$0.2R$	$0.05R$

comparisons between the time of the onset of major recharge events and the start of recovery in observation boreholes (or a decrease in the rate of recession). The delays are estimated for different thicknesses of the unsaturated zone as indicated in Table 9.2; monthly estimates of the potential recharge in month M are distributed in the subsequent months.

River–aquifer interaction

A detailed description in Section 4.3.3 indicates how river coefficients for the River Meden are estimated from flow gauging measurements. For the River Meden the coefficient is $1000\ \mathrm{m^3/d/km}$, which means if the groundwater head is 1.0 m above the river surface level, $1000\ \mathrm{m^3/d}$ will flow from the aquifer to the river over a river reach of 1.0 km. Coefficients for other rivers crossing the sandstone aquifer were deduced from the size of the river and from accretion diagrams. During numerical model refinement, modifications were made to some of the coefficients. Since most of the rivers originate on the Permian strata, there is a substantial flow as they cross onto sandstone. Even though there are some losses to the sandstone, the losses are never greater than the inflow at the Permian boundary, consequently none of the rivers become dry.

Initial conditions

For initial conditions, the input is the distributed recharge for a typical year. Abstraction corresponds to the values for 1970. This allows a time-instant calculation to be carried out (see Section 12.5.2), thereafter historical estimates of recharge and abstraction are used.

9.2.4 Numerical groundwater model

There are many regional groundwater computer codes which can be used to represent the aquifer system as

defined above. Particular care is needed in simulating the leakage through the overlying Mercia Mudstone and Colwick Formations; the leakage depends on the elevation of the overlying water table (which is usually close to the ground surface), the thickness of the overlying strata and its effective vertical permeability. Due to the slow changes which occur in this aquifer system with abstractions exhibiting no rapid changes, it is acceptable to use the conventional leaky aquifer response of Section 5.8 rather than including the effect of aquitard storage (Section 7.5).

The regional groundwater model code described by Rushton and Redshaw (1979) was used for this study; other standard regional groundwater models would be equally suitable. The model of Rushton and Redshaw uses a backward difference finite difference technique. Since this is a regional resources study, a mesh interval of 1.0 km is appropriate. There are two time steps per month.

9.2.5 Adequacy of model

The adequacy of the model can be assessed by comparing field and modelled values of both groundwater heads and river–aquifer interaction. At *most locations* differences between field and modelled groundwater heads are *less than two metres*. However, for this discussion groundwater head hydrographs are presented *where larger differences occur* to illustrate why a close match cannot always be obtained. Four groundwater head hydrographs are plotted in Figure 9.4, and the locations of the boreholes are shown on Figure 9.1.

Manton Lodge (ML) observation borehole is located on the Sherwood Sandstone aquifer on the watershed between the Rivers Ryton and Poulter. There are a number of major pumping stations at 5 km or more from this observation borehole. In terms of elevation, the modelled groundwater heads are 1 to 3 m higher than the field results. Since this observation borehole is

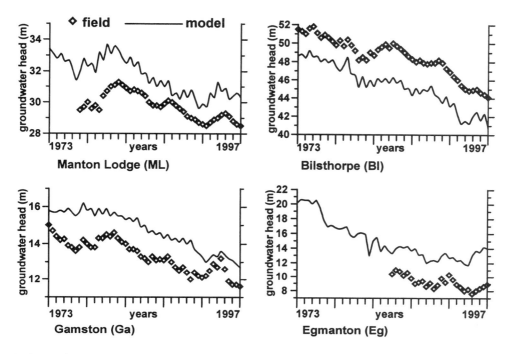

Figure 9.4 Comparison of field readings (indicated by diamonds) and model groundwater head hydrographs (full lines) for locations where the agreement is less satisfactory. For position of observation boreholes see Figure 9.1

on a watershed, there are no strong constraints on the groundwater heads; if rivers or lakes were nearer to the site of the observation borehole, closer agreement with field readings should occur. Field results show a rise in groundwater heads during 1975–80, then a fall until 1991, followed by a rise and a further fall in the 1990s. Each of these responses is reproduced by the groundwater model.

Bilsthorpe (Bl) observation borehole is on the eastern side of the sandstone outcrop, close to the Colwick Formation; consequently the agreement between field and modelled results provides information about the ability of the model to represent recharge to both the Sandstone and Colwick Formations. There are several pumping stations within a few kilometres; the continuing decline in groundwater heads reflects the high abstraction in that area. The model reproduces the historical trends *but* the absolute values are about three metres too low. There are no rivers in contact with the sandstone aquifer in the vicinity of Bilsthorpe, hence there is no major constraint on the groundwater heads.

Gamston (Ga) observation borehole monitors groundwater heads in the Sandstone beneath the

Colwick Formation. Other influences include a number of pumping stations at about 5 km from the observation well and partial contact between the River Idle and the aquifer. Both the model and the field results show a decline from 1970 to 1996, although there are detailed variations in the field results which are not reproduced by the model. Nevertheless, the agreement between modelled and field results is adequate for this study in which the objective is to quantify the groundwater resource available for exploitation.

Egmanton (Eg) observation borehole measures groundwater heads in the sandstone aquifer where it is overlain by Mercia Mudstone; the observation borehole is approximately 8 km to the east of the unconfined-confined boundary. A major public supply borehole at about 7 km has an influence on the aquifer response at Egmanton. The modelled groundwater heads are above the field values, but they do reproduce the general decline and subsequent increase in the late 1990s. This borehole is about 10 km from any location where there is contact between a river and the sandstone aquifer.

The above discussion indicates why differences between the field and modelled results occur; this examination of hydrographs where the agreement is

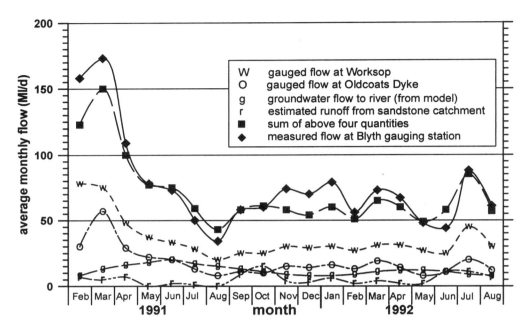

Figure 9.5   Components of river flows at Blyth gauging station

less satisfactory provides insights into the total aquifer response.

Changes in river flows arising from river–aquifer interaction also provide insights into the reliability of the groundwater model. In this study area, flows between river and aquifer are often small compared to total river flows; there are also uncertainties about the reliability of some of the gauging stations due to settlement associated with coal mining.

Information is presented for the gauging station at Blyth on the River Ryton; see Figure 9.1. Average flows for March 1991 are listed below:

river flows from Permian strata at Worksop	75 Ml/d
river flows from Permian strata at Oldcoates Dyke	57 Ml/d
runoff	5 Ml/d
gain from aquifer (numerical model)	13 Ml/d
Total	**150 Ml/d**
gauged flow at Blyth	173 Ml/d

Comparisons for all the flow components from February 1991 to August 1992 are included in Figure 9.5. When examining these results, account must be taken of the uncertainties in the reliabilities of the gauged flows at Worksop, Oldcoates Dyke and Blyth; for each

of these gauges the tolerance is of the order of ±15 per cent. Although the comparison between the total flows at Blyth does not provide entirely convincing evidence, it does suggest that the numerical model provides an acceptable representation of the river–aquifer interaction.

9.2.6 Flow balances

One of the *key issues* of this study is to use flow balances to understand the interactions between the different components of the aquifer system. Now that an acceptable numerical model has been devised, annual and monthly flow balances can be prepared (Rushton *et al.* 1995).

Annual flow balances between 1974 and 1992 are presented in Figure 9.6. Note that annual balances refer to water years, hence the values for 1974 are the average from October 1973 to September 1974. Certain of the components remain relatively constant including leakage through the Colwick Formation and Mercia Mudstone while the total abstraction (plotted as a negative quantity) exhibits long-term trends; the other components show considerable variations. For instance, the total estimated recharge was 61 Ml/d in 1976 but 540 Ml/d in 1977. Variations in the annual recharge are balanced by water drawn from or taken

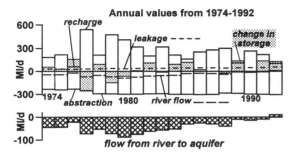

Figure 9.6 Annual flow balance for Nottinghamshire Sherwood Sandstone aquifer with additional detailed diagram of flow between river and the aquifer

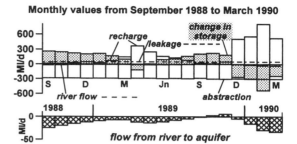

Figure 9.7 Monthly flow balance from September 1988 to March 1990 for Nottinghamshire Sherwood Sandstone aquifer

into storage. The shaded parts of the upper diagram represent the storage contributions; during 1976 the groundwater heads declined, providing a storage release of 161 Ml/d whereas during 1977 there was a rise in groundwater heads with an additional 251 Ml/d taken into storage. Therefore, a year with poor recharge has little effect on the ability of the aquifer to meet the abstraction demand.

The low recharge in 1976 did have an effect on aquifer to river flows. Because aquifer to river flows are small compared to the recharge and abstraction, an enlarged diagram in Figure 9.6 represents water from the aquifer which enters rivers. An examination of long-term flows to rivers shows that from 1980 there was an overall decrease in the total flow to rivers. By 1992 there was a reversal, with a net loss from the rivers to the aquifer. The decrease in aquifer to river flows is caused by increased abstractions throughout the study period. However, the effect of any major change in aquifer conditions only extends slowly in these high storage aquifers. Equation (12.8) provides an expression for t_{eq}, the time taken for the effect of changed conditions to spread through the aquifer. For this aquifer system with appropriate numerical values

$$t_{eq} = \frac{1.5L^2 S}{T} = \frac{1.5 \times 4000^2 \times 0.15}{500}$$
$$= 7200 \, \text{days} \approx 19.7 \, \text{years}$$

The dimension L refers to a representative flow path length; because of the network of rivers, water typically moves about 4 km before it reaches a river. Consequently the effect of changes in abstraction may be spread over twenty or more years. In the southern part of the study area where there are fewer rivers, a larger value of L should be used.

Monthly flow balances from September 1988 to March 1990 are plotted in Figure 9.7. This diagram allows a more detailed examination of the manner in which water, released from storage, can balance shortages in recharge. For fourteen of the fifteen months between September 1988 and November 1989, there was insufficient recharge to meet the abstraction demand, so that water was withdrawn from storage. This resulted in a considerable impact on the net flows from the sandstone aquifer to the river system. However, from December 1989 to March 1990, there was substantial recharge, the storage was partially replenished and the aquifer to river flows recovered.

The numerical model of the aquifer has also been used to identify source protection zones. Diagrams of flow vectors similar to Figure 2.23 are used to identify flow-paths into boreholes; alternatively, particle-tracking procedures can be used (Zheng and Bennett 2002). From this information areas can be delineated which supply water to an individual or group of boreholes. For a pumping station in the vicinity of a river, some of the water entering a borehole originates from the river; this complicates the definition of a source protection zone. In fact any part of the river catchment, upstream of a reach which loses water to the aquifer, can be a source of contamination.

9.3 NORTHERN EXTENSION OF NOTTINGHAMSHIRE SHERWOOD SANDSTONE AQUIFER

Objectives of the study: identify and quantify all the potential sources of recharge; determine what reductions in abstraction are required to prevent continuing declines in groundwater heads.

Key issues: inflows to the sandstone aquifer when drift is present, drainage channels recharge the aquifer.

This case study of the Doncaster sandstone aquifer unit relates to the northern extension of the Nottinghamshire Sandstone aquifer. Different approaches are required, since drift deposits cover most of the study area. A further distinction is that the topography is flat with much of the land surface close to or even below sea level. To obtain reliable estimates of all sources of recharge to the aquifer, information is required about the drift geology together with careful analysis of field observations.

9.3.1 Brief description of groundwater catchment

A plan of the study area and a representative cross-section are shown in Figures 9.8 and 9.9; in Figure 9.8 information is presented on a 1 km grid in a form

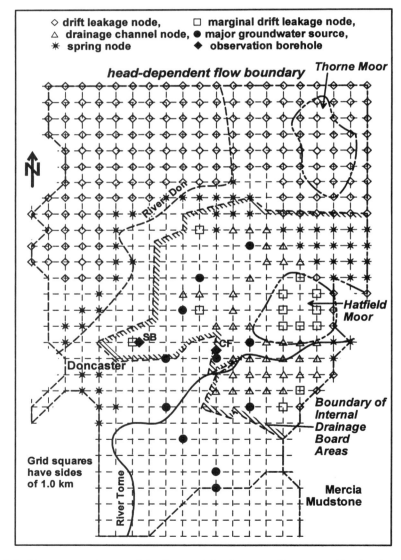

Figure 9.8 Plan of the Doncaster study area of the Sherwood Sandstone aquifer system; this plan shows how different features are allocated to a 1-km grid covering the study area

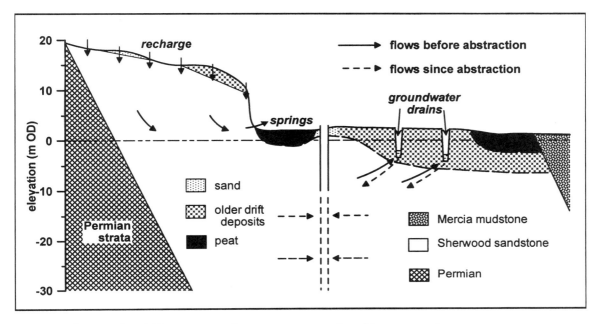

Figure 9.9 Conceptual model for a representative west–east cross-section of Doncaster study area

suitable for inclusion in a numerical model. The major aquifer system is similar to that to the south with the Sherwood Sandstone underlain by Permian strata and overlain by Mercia Mudstone to the east; the dip of the base of the sandstone is approximately 2 per cent. Apart from some higher ground to the west, which approaches 20 m, 95 per cent of the study area is below 10 m OD. In the west–east direction, the lateral distance from the edge of the Permian outcrop to the Mercia Mudstone is 15 to 20 km.

A critical feature is the thickness, nature and extent of the drift covering the solid geology. Deposits include clay tills, channel deposits, glacial and fluvioglacial sand and gravel with more recent lacustrine deposits, blown sands and peat. Figure 9.8 indicates the extent of the less permeable deposits; there are also extensive peat deposits on Hatfield and Thorne Moors.

The largest river in the study area is the Don; since this river flows over thick low-permeability alluvium, there is little interaction with the aquifer. A smaller river, the Torn, crosses the aquifer from west to east; the eastern section of the river has been canalised on an embankment to act as a high-level carrier for drainage water. The northern and eastern areas of the catchment were historically poorly drained with fenland and marsh to the east. In the early seventeenth

century, deep drains were constructed in the marsh; this drainage has been improved so that the drainage system is now used to take away excess water in winter and provide water for irrigation in summer (see Figure 9.9). A further feature of this area is the extensive workings for sand and gravel; in addition, peat has been milled on both Hatfield and Thorne Moors.

Major abstractions from the sandstone aquifer started in the 1920s. As more boreholes were drilled, a continuing steady decline occurred in observation well water levels with some groundwater heads drawn below Ordnance Datum; Figure 9.11.

9.3.2 Conceptual models

This discussion on conceptual models of the flow processes in the aquifer system is based on Figures 9.8 and 9.9. Various recharge and discharge processes occur.

Precipitation recharge: drilling records for the study area show that the sandstone is covered with drift in most locations and that the drift contains some layers of clay. However, over certain drift-covered areas there is no field drainage; during periods of heavy and continuous rainfall, no runoff or ponding is observed. Consequently, over these areas the potential recharge moves through the drift to become actual recharge to the sandstone aquifer. The route followed by the recharging water is not known in detail, but the *effective* vertical permeability of the drift does not limit the occurrence of recharge. Locations where potential recharge occurs are represented in Figure 9.8 as mesh intersection points with no symbol.

Classical leaky aquifer conditions: in the northern part of the study area thick lacustrine clay deposits occur which limit the vertical movement of water; there is also extensive field drainage. The effective vertical permeability of these clay deposits, deduced from laboratory tests of undisturbed samples, is about 0.0005 m/d. Since there is little abstraction in these areas, the small vertical groundwater flows are represented as a classical leaky aquifer response (Section 5.8); the vertical permeability and the vertical head gradient across the clay determine the magnitude of the vertical flow. The surface water head providing the source (or sink) for this leakage is set at 0.0 m OD. Locations of leakage are indicated on Figure 9.8 by open diamonds. To the east there are some areas where the thickness and nature of the deposits result in a higher effective vertical permeability, these locations are indicated by open squares.

Spring outflows: the representative section of Figure 9.9 indicates the presence of peat filled depressions; springs form immediately above the peat when the groundwater head is above the spring line. Other spring locations (shown on Figure 9.8 by stars) are on the edge of the alluvial deposits of river channels etc.

Groundwater drainage: elsewhere to the east of the Doncaster area, where the drift is more permeable, significant upwards flows occur between the sandstone aquifer and the extensive drainage ditches. Upwards flows also occur in areas covered mainly by sands and gravels. Reference has been made to deep drainage introduced to remove this groundwater; the water is pumped from the deep drainage channels, with water surface elevations of −0.5 to −1.5 m OD, into high-level carriers.

Recent pumping returns from the Internal Drainage Boards show a substantial decrease in pumped discharge with the annual discharge reduced to about one-third during the decade 1980 to 1990. Towards the end of that period during some summer months, no pumping was required. This, and other information, confirms that there has been a partial reversal of flows due to increased abstraction from the aquifer so that the drainage system, which was originally designed to take away excess groundwater, is now providing a source of groundwater at times of low groundwater

heads. This reversal of flows is represented in Figure 9.9 by the change from full flow arrows, showing pre-groundwater abstraction conditions, to broken flow arrows representing the reversal of flows due to groundwater abstraction.

Figure 9.10 provides further information about the flow processes in the drained areas. In Figure 9.10a the historical upwards flows are illustrated; these flows occur when the piezometric head in the sandstone aquifer is above the water table in the drift. A reversal of flows is illustrated in Figure 9.10b. If groundwater heads in the sandstone are drawn below the top of the aquifer, Figure 9.10c, flow through the drift is governed by the unit vertical hydraulic gradient across the drift. Consequently further reductions in groundwater head do not lead to increased flows. Nodes where these conditions hold are indicated by open triangles; the effective vertical permeability of the drift is taken to be 0.001 m/d, increasing to 0.01 m/d where the sand content is high.

9.3.3 Numerical groundwater model and flow balances

Features of the numerical groundwater model are recorded in Figure 9.8; this groundwater model is an extension of the Nottinghamshire regional groundwater model described in the previous section. However, additional features are required including leakage through low-permeability strata and the groundwater drainage described in Section 9.3.2.

Two observation well hydrographs are selected to illustrate the match between field and model results (Figure 9.11).

Sandall Beat (SB) observation borehole is about 4 km from the western limit of the Sherwood Sandstone; there are no major pumping stations in the near vicinity (Figure 9.8). Nevertheless, the field readings reflect the steady decline due to the heavy pumping from the aquifer, with a partial recovery since 1994. A comparison between the full line from the numerical model and the discrete symbols representing the field results (Figure 9.11), shows that the groundwater heads are reproduced accurately by the numerical groundwater model.

For *Cockwood Farm (CF)* there are major pumping stations exerting a strong influence on the observation borehole. In Figure 9.8, pumping stations are shown as a solid circles. The field measurements reflect the changing abstraction patterns; these variations are reproduced adequately by the numerical model.

This close agreement between field and modelled groundwater heads gives confidence in the reliability of

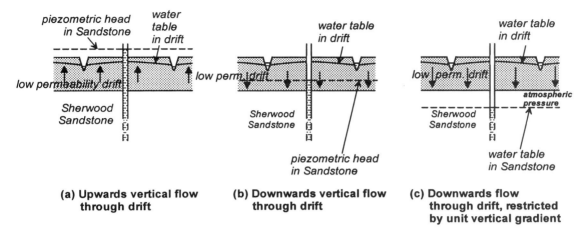

Figure 9.10   Alternative flow conditions in low permeability drift where extensive drainage ditches are provided

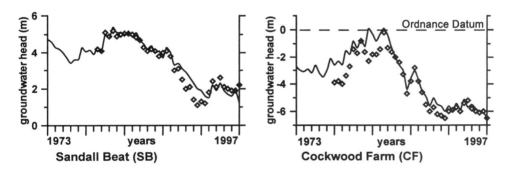

Figure 9.11   Groundwater head hydrographs in Doncaster study area; discrete symbols indicate field readings, continuous line numerical model results. For location of observation boreholes see Figure 9.8

the groundwater model. The success in reproducing the decline and levelling-off of the groundwater heads since 1993 demonstrates that the differing flow components are represented satisfactorily in the model. In fact, the agreement between field and modelled results may appear to be superior to that for the Nottinghamshire part of the aquifer system; Figure 9.4. However, the observation well hydrographs in the Doncaster area are strongly influenced by the large number of spring and leakage nodes at which the outflow elevations are specified; in the Nottinghamshire area the groundwater heads are not so well constrained.

Annual flow balances are determined from the groundwater model outputs, in a similar manner to those for the Nottinghamshire aquifer. To demonstrate how flow balance components can be presented to

highlight important aquifer responses, Figure 9.12 illustrates how the contributions to aquifer recharge from the drained areas (flow from drains) has changed with time. The upper diagram refers to the equivalent recharge (see Section 3.3) which equals the recharge due to precipitation and losses from water mains, plus water released from storage. This total is indicated by the unbroken lines of the bar chart. The annual average total groundwater abstraction is also plotted on the diagram: for comparisons with the equivalent recharge, abstraction is plotted in the positive direction. The difference between the equivalent recharge and the abstraction is shaded. Cross-hatching indicates that the equivalent recharge exceeds the abstraction; dotted shading indicates that the equivalent recharge is less than the abstraction.

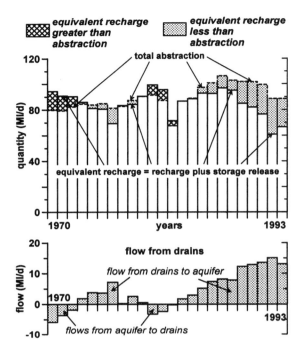

Figure 9.12   Balance of annual average flows for the sandstone aquifer system of Doncaster study area

The bar chart of the lower diagram in Figure 9.12 represents the flow between the aquifer and the drains. For the first three years of the record, the net response is that water flows from the aquifer to the drains. Most of the flow occurs during the winter; there may be a reversal of flows in summer. For each of the last nine years, abstraction exceeds the equivalent recharge; the only possible additional source of water is that drawn from the drains. Figure 9.12 shows that up to 15 Ml/d is drawn from the drains. Consequently groundwater abstraction in the Doncaster area has been supported by inflows from the drainage ditches. Since these drains are not intended to supply groundwater, the current situation is unsatisfactory. The numerical model is being used to examine the impact of reducing the net groundwater abstraction on both groundwater heads and contributions from drains.

9.4 LOWER MERSEY SANDSTONE AQUIFER

Objectives of the study: to identify how the aquifer system has supported abstractions significantly in excess of rainfall recharge, identify possible remedial action and the impact of future groundwater abstraction.

Key issues: important contribution to recharge through drift and from urban areas, old and recent saline waters, the need to consider aquifer behaviour for over 100 years, using data from pumped boreholes to confirm adequacy of numerical model, predictions of continuing inflow of poor quality water even if all pumping ceased.

Studies of the Sherwood Sandstone aquifers of the Lower Mersey Basin between Liverpool and Manchester were carried out in the early 1980s; the Lower Mersey Unit was considered first followed by the Liverpool Unit. There are many similarities with the Nottinghamshire and Doncaster aquifer systems described in the previous sub-sections. There are, however, a number of additional characteristics which are considered below.

9.4.1 Conceptual model

A plan of the Lower Mersey and Liverpool units of the Sherwood Sandstone is presented in Figure 9.13; a representative north–south cross-section for the Lower Mersey unit is included as Figure 9.14. Geological units of late Carboniferous, Permo-Triassic and Quaternary age occur in the study area. One feature of the aquifer system is the presence of Quaternary drift deposits which restrict recharge to the sandstone aquifer. Furthermore, poor quality water is drawn in from the Mersey Estuary and the Manchester Ship Canal. There is also old saline water at depth. There may be a small contribution of flows from the Carboniferous strata which are faulted against the sandstone aquifer in the north.

Groundwater has been a key to the commercial and industrial development of the Lower Mersey area over the last 150 years; during much of the history of exploitation, resource assessments were not carried out to check whether over-exploitation was taking place. However, increasing salinities of pumped water, especially in the Widnes area, led to the abandonment of many of the boreholes close to the Mersey Estuary. The substantial decline in the water table elevation has also been a cause of concern.

The thickness of the Sherwood Sandstone aquifer increases in the south to about 500 m. The hydraulic properties of the sandstone were investigated using borehole drilling, core sampling and packer testing. The permeabilities of horizontal and vertical core samples were measured; samples in the medium to coarse range had intrinsic permeabilities almost three

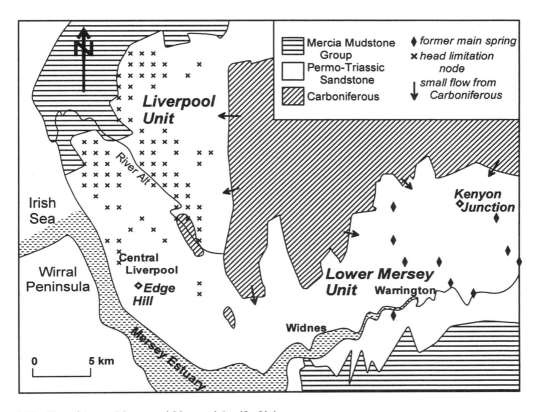

Figure 9.13 Plan of Lower Mersey and Liverpool Aquifer Units

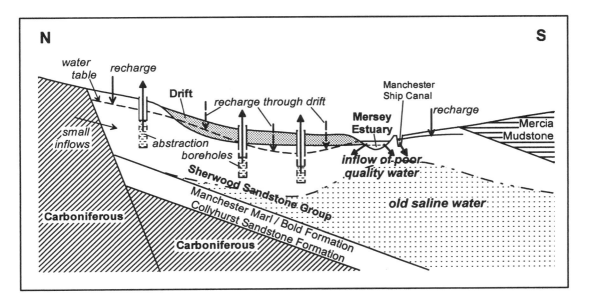

Figure 9.14 Representative north-south conceptual section of the Lower Mersey Unit

orders of magnitude greater than samples classified as fine-grained. For coarse-grained samples the average ratio of horizontal to vertical permeability is 1.5 to 1; for fine-grained samples the ratio is 7 to 1. However, pumping tests suggested higher transmissivities than those deduced by summing the core permeabilities. Consequently packer testing was carried out, this identified the importance of fissures. Figure 8.11b and the associated discussion refers to packer testing at Kenyon Junction; the presence of the fissures leads to a transmissivity five times the value from the core samples alone. The analysis of the pumping test at Kenyon Junction, described in Section 5.1, indicates a specific yield of 0.14.

There is extensive faulting in the Sherwood Sandstone; however, an examination of groundwater hydrographs suggests that most faults have little influence on regional groundwater flow. For two faults, there appears to be a small discontinuity in groundwater heads; these faults are represented by reducing the transmissivity to 0.05 of the standard value across a distance of 1.0 km. In addition, there is evidence of small flows from the Carboniferous strata to the north across to the Sherwood Sandstone; this is probably due to sandstone horizons in the Carboniferous. The connection due to the Sandstone horizons is estimated to cover an area of $250 \times 10^3 \, \text{m}^2$, the head gradient is approximately 20 m in 1.0 km. If the horizontal hydraulic conductivity is $K = 1.0$ m/d, the total flow equals 5 Ml/d. It is unlikely that the inflow from the Carboniferous strata exceeds this value.

One final stage in defining conceptual models is to consider the groundwater conditions in the 1850s when there was little abstraction. Present-day hydrochemical distributions are dominated by the regional aquifer behaviour of the pre-pumping history of the aquifer. The hydrochemistry indicates a slow but continuous through-flow from recharge areas to discharge areas. An examination of old records, including maps, shows that there were major springs; locations of these springs are indicated by the filled diamonds in Figure 9.13. If the spring outflows are not included in the regional groundwater model, the groundwater heads rise above ground level to the north. Consequently, the mid-1850s are used as starting conditions for the groundwater simulation.

9.4.2 Recharge through drift

Field observations during periods of heavy rainfall indicate that there are locations where significant runoff occurs. For other areas there is limited runoff,

and elsewhere there is no runoff due to the good natural drainage of the soil and underlying strata. The extent of areas with sub-surface drainage, the depths of drains or ditches and the size of culverts allowed approximate estimates to be made of the proportion of the potential recharge which would run off (Rushton *et al.* 1988). The nature of the runoff was compared with information from geotechnical investigations which provide information about the thickness of the drift and the percentage of the more permeable sandy clays (see the first two columns of Table 9.3). The geotechnical boreholes do not provide the detailed information required to construct drift cross-sections, yet there is sufficient information to develop a methodology whereby the water passing through the drift can be defined as a factor of the potential recharge. The third column of Table 9.3 lists factors by which the potential recharge is multiplied. These factors are based on:

- information about the thickness and percentage of sandy clay in the drift,
- observations of the drainage required to take away excess water following heavy rainfall,
- adjustments during numerical groundwater model refinement when the factors were modified to improve the representation of groundwater head hydrographs.

Figure 9.15 illustrates the way in which drift factors are used in estimating recharge.

An alternative method of estimating the actual recharge would be to use a leaky aquifer approach in a regional groundwater model. However, there is a serious risk with the leaky aquifer approach that the vertical flow through the drift exceeds the potential recharge. Although a leaky aquifer approach is not used directly, the vertical flow through the drift is modified

Table 9.3 Factors for estimating movement through the drift in the Liverpool Unit

Drift clay thickness	Description % sandy clay	Factor	Area km²	Area %
0	–	1.0	28.7	17.7
<2 m	50–90	0.7	22.6	13.9
>2 m	90–100	0.65	19.5	12.0
>2 m	50–90	0.1	10.1	6.2
>2 m	0–50	0.02	81.5	50.2

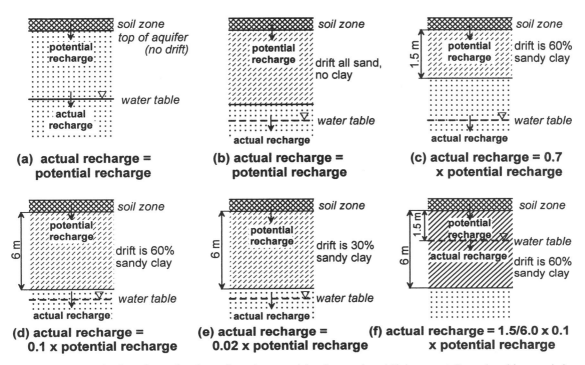

Figure 9.15 Determination of actual recharge from the potential recharge when drift is present. Reproduced by permission of CIWEM from Rushton *et al.* (1988)

to reflect the elevation of the groundwater head within the aquifer system. For Figure 9.15a to e, the water table is within the underlying aquifer. However, if the groundwater head is within the drift (Figure 9.15f), an additional factor is introduced to modify the actual recharge calculated from Figure 9.15d. In Figure 9.15f the groundwater head is within 1.5 m of the top of the drift which is 6.0 m thick, hence the groundwater elevation factor is 1.5/6.0 = 0.25; this is multiplied by the factor of 0.1 which refers to more than 2 m of drift with 50–90 per cent sandy clay. When the water table (groundwater head) is within the aquifer, and hence beneath the base of the drift, the groundwater elevation factor is 1.0. On the other hand, if the groundwater head is at the ground surface there is no vertical gradient since the aquifer is full; therefore the groundwater elevation factor is zero. As the water table within the drift is lowered, the actual recharge increases. In the Lower Mersey unit this mechanism is significant; calculations suggest that the average groundwater elevation factor increased from 59 per cent in the mid-1850s to 88 per cent in the late 1970s. This change is reflected in the flow balances of Figure 9.20.

For the Liverpool unit of the study area, rainfall recharge in drift free areas (indicated by the inclined hatching in Figure 9.16) plus recharge through the drift (the cross-hatching in Figure 9.16) is significantly less than the abstraction (represented by the broken line) for all but the later years. A careful examination of losses from sewers and water mains (shallow hydrochemical evidence indicates losses from sewers) demonstrates that these components are substantial in urban areas. Losses from water mains and sewers are treated as potential recharge; the same drift factors apply when estimating the actual recharge. From the results of Figure 9.16 it is apparent that in the 1950s and early 1960s the total actual recharge was of similar magnitude to the abstraction. Due to reduced abstractions from the mid-1960s onwards, the total actual recharge exceeds the abstraction, leading to a recovery in groundwater heads.

9.4.3 Saline water

In this aquifer system, hydrochemical evidence allows the identification of two types of saline water; old

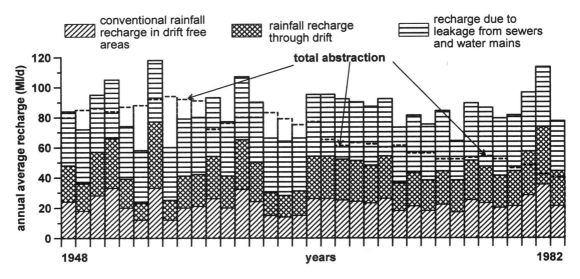

Figure 9.16   Components of recharge in the Liverpool unit. Reproduced by permission of CIWEN from Rushton *et al.* (1988)

saline water at depth and recent saline water from the Mersey Estuary (for detailed information, see Lloyd and Heathcote 1985). In Figure 9.14, old saline water occurs at varying depths. Inland, the saline water is at least 200 m below OD, whereas in the vicinity of the Mersey Estuary it is shallower. A variety of techniques have been used to locate and quantify the saline water including borehole conductivity logging, pore water sampling, resistivity logs and the chloride concentration from pumped boreholes; from this information maps of the highest elevation of saline water have been prepared. The old saline water has salinity two to three times that of sea water. Its origin is probably due to halite dissolution; direct contact between the Sherwood Sandstone Group and Northwich Halite formation occurs in the south-east of the study area. The final stage in aquifer evolution involved the flushing out of saline water from regions where groundwater flow was significant, leaving saline water where flows were small. The horizontal low conductivity zones restrict the deeper movement of fresh water.

Is there a risk that the saline water at depth will be drawn into pumped boreholes? Experience gained from the drilling of some of the earlier boreholes is instructive. During drilling, certain boreholes penetrated into the saline zone with the result that the abstracted water was very saline. However, when the lowest 10 m of the borehole was backfilled with impermeable material, the borehole operated successfully for many decades with

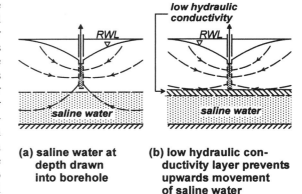

(a) saline water at depth drawn into borehole

(b) low hydraulic conductivity layer prevents upwards movement of saline water

Figure 9.17   Effect of the presence of a low conductivity layer restricting upward movement of saline water from depth

no salinity problems. The reason for the success of this backfilling is apparent from Figure 9.17. Within the Sherwood Sandstone aquifer there are many layers of lower hydraulic conductivity. Packer testing and the hydraulic conductivity of core samples show that the effective vertical hydraulic conductivities of the low permeability zones are typically one-thousandth of horizontal hydraulic conductivities. Figure 9.17a illustrates a situation with no layers of lower hydraulic conductivity; even though the borehole does not penetrate into the saline zone, saline water is drawn to the borehole. However, in Figure 9.17b there is a layer of low hydraulic conductivity between the bottom of the borehole and the saline water. For this situation, virtually all the water entering the borehole is fresh water. This explains the success of backfilling a borehole which was originally drilled into the saline zone. When the borehole terminates above a low permeability layer there is only a small upward movement of saline water.

The time-instant radial flow model in Section 7.3 was used to explore the situation further. The conditions of Figures 9.17 a and b were modelled; in Figure 9.17b the vertical hydraulic conductivity of the low permeability layer is set at 0.001 of the horizontal hydraulic conductivity of the more permeable layers. According to the radial flow model, the upward velocity below the layer of low hydraulic conductivity is 1 per cent of the velocity when the low permeability layer is missing. Consequently, the upward flow of saline water is small. Supporting evidence of small disturbances beneath a low permeability layer is provided by the lowest piezometer of the Kenyon Junction pumping test (Section 5.1).

Hydrochemical evidence demonstrates that close to the Mersey Estuary there is a different type of saline water. Dating using C^{14} shows that it is modern water; comparisons between the hydrochemistry of this groundwater and water in the Mersey Estuary confirm that the estuary is the source of this water.

9.4.4 Numerical groundwater model

Since this is a resource study, it is acceptable to represent the aquifer as a single layer numerical model with predefined transmissivities. Features and properties of the numerical groundwater model are similar to the aquifer systems described in Sections 9.2 and 9.3. However, for a study in which the primary purpose is an investigation of the movement of contaminants, a multi-layered model would be required with representation of the lower hydraulic conductivity zones within the sandstone aquifer.

A variety of sources of information are used to check the validity of this single-layer groundwater model. This discussion considers both groundwater heads and groundwater flows. No long-term records of groundwater heads from specially designed observation boreholes are available, but there are records from abstraction boreholes. Edge Hill (Figure 9.18) is a typical example. During the 1950s and 1960s abstraction took place from the Edge Hill boreholes to provide water for steam-powered railway engines; by 1969 abstraction had ceased. In Figure 9.18a, field values of

borehole water levels are plotted as discrete symbols for pumping and non-pumping conditions; for the non-pumping readings, the elapsed time since pumping stopped is unknown.

Comparisons between modelled groundwater heads at the nodal point corresponding to Edge Hill (the unbroken line of Figure 9.18a) and the measured borehole water levels, indicate that the field readings are always below the modelled results. This occurs because the nodal reading from the numerical model represents an average over the corresponding nodal area which, in this example, is 1.0 km^2. However, corrections can be made so that the numerical model results represent more closely the conditions in the pumped borehole. Using the approach of Eq. (5.36), the drawdown s_a at the edge of the pumped borehole of radius r_w relative to the calculated groundwater head at the regional model node is given by

$$s_a = \frac{Q}{2\pi T} \ln \frac{0.208 \Delta x}{r_w} \qquad (9.1)$$

where Δx is the mesh interval. As indicated in the small diagram of Figure 9.18a, there is a seepage face between the pumped water level and the intercept of the water table with the borehole sides. As a first approximation, the seepage face height is set at 50 per cent of s_a. For the parameters used in the model, the drawdown below the nodal reading from the groundwater model is 6.0 m for each 1 Ml/d of abstraction. When this additional drawdown is subtracted from the modelled groundwater heads, the corrected heads are represented by the broken line of Figure 9.18a. Figure 9.18b contains values of the average annual abstraction from Edge Hill which are used to estimate the additional drawdowns. A careful comparison between field and modelled results indicates that:

* the overall recovery from 1947 to 1982 is represented satisfactorily by the model; this gives confidence in the choice of the specific yield and the representation of the various sources of recharge.
* the correction for the pumped water levels is consistent with the recorded pumped levels which are indicated by the solid diamonds in Figure 9.18a. Since the correction is based on the average annual abstraction for each year, it may not represent the actual abstraction at the time when the field reading was taken.
* the non pumping levels, the crosses in Figure 9.18a, all lie between the numerical model results without and with the correction for pumping.

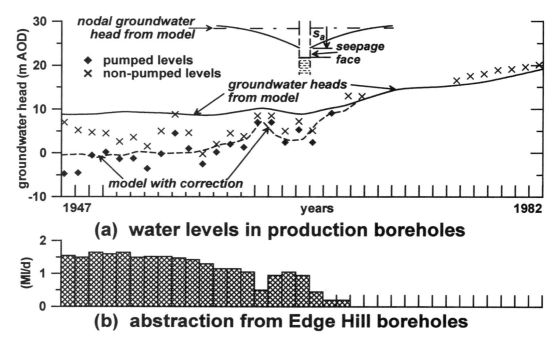

(a) water levels in production boreholes

(b) abstraction from Edge Hill boreholes

Figure 9.18 Edge Hill information: (a) pumped and non-pumped levels, unbroken line represents modelled results, broken line shows model results corrected for radial flow and seepage face, (b) abstractions from Edge Hill. Reproduced by permission of CIWEM from Rushton *et al.* (1988)

The second check on the validity of the numerical model is based on *groundwater flows*. There are no specific measurements of groundwater flow, but the extent of the movement of saline water from the Mersey Estuary into the aquifer in the Widnes area can be used to explore the ability of the model to represent groundwater flows. In Figure 9.19 there is a comparison between the extent of saline zone in the Widnes area in 1980 and model estimates of the total movement of saline water from the Mersey Estuary. Figure 9.19a plots the contours of salinity of the water (mg/l) due to pumping in the vicinity of the saline Mersey Estuary; areas of saline water identified from geophysical measurements are also included in the figure. Estimates from the groundwater model of the movement of saline water from the estuary, based on groundwater head gradients, the aquifer depth and the effective porosity, are plotted in Figure 9.19b. The predicted movement of saline water deduced from the model in Figure 9.19b is similar to the ingress of saline water based on field readings in Figure 9.19a.

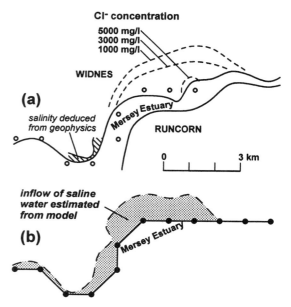

Figure 9.19 Movement of saline water from Mersey Estuary into aquifer: (a) field information, (b) prediction using regional groundwater model

9.4.5 Flow balances and predictions

Flow balances for the Lower Mersey Unit (Figure 9.20), provide important insights into the aquifer response; the balances cover decades from 1847 to 1947 followed by annual water balances until 1980. For clarity, the quantities are presented in two diagrams; the lower diagram contains information about recharge, abstraction and storage changes, while the upper diagram refers to interactions of groundwater with springs and rivers.

For the lower diagram of Figure 9.20, the following responses can be identified.

- *Abstraction*: at the start of the study period, groundwater abstractions were very small. There was a steady increase to a maximum in 1970 followed by a decline during the 1970s.
- *Recharge*: potential recharge is assumed to be constant during the first hundred years; however, due to the falling water table the actual recharge increases with time. For the detailed recharge estimates from 1948 onwards, there are large annual variations, with the highest value of actual recharge two and a half times the lowest.

- *Storage changes (shown shaded)*: water released from storage as the water table falls balances out the low recharge during certain years. In years with excess recharge, water is taken into storage; this is indicated by shaded bar charts below the zero line. Since 1948, there have been few years when water is taken into storage, indicating over-exploitation of the aquifer system.

In the upper diagram of Figure 9.20, it is possible to observe the impact of the changing abstraction patterns on spring and river flows.

- *Spring flows (shown hatched)*: at the start of the study period, when abstraction was small, the outflow from springs totalled 28 Ml/d. As abstractions increased, spring flows decreased to very small values becoming zero during the 1960s.
- *Groundwater interaction with the River Mersey (shown shaded)*: groundwater outflows to the northern side of the Mersey Estuary occurred in the early years of the study period. As abstraction increased, these outflows decreased. Unlike springs, flows at the Mersey Estuary can reverse with poor quality water

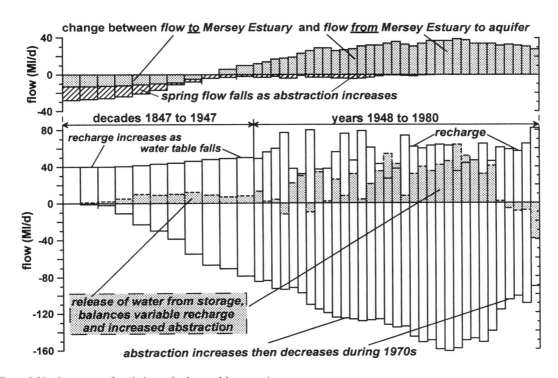

Figure 9.20 Long-term flow balance for Lower Mersey unit

supplied to the aquifer. Since the 1930s the Mersey Estuary has provided a net inflow to the aquifer system. During the 1960s and 1970s this contribution from the Mersey aquifer averaged about 35 Ml/d compared to an average recharge of around 60 Ml/d. Without the inflows from the Mersey Estuary there would have been more serious water table declines.

A number of *predictive simulations* have been devised; these simulations are designed to investigate whether the inflow of saline water from the Mersey Estuary in the Lower Mersey unit can be prevented. Three alternative predictive scenarios are considered,

A: maintain abstraction at 1980 rates (91 Ml/d),
B: stop all abstraction (0 Ml/d),
C: discontinue abstractions close to the Estuary but continue abstractions inland (62 Ml/d).

For the predictive simulations, the same regional groundwater model is used but with the groundwater heads at the start of the predictive simulation equal to those at the end of the historical simulation. This aquifer has a high storage capacity, therefore recharge for the predictive simulation is set equal to the long-term average; the predictive simulation continues for thirty years. Since the purpose of the predictions is to examine the effect of the abstraction scenarios on saline intrusion from the aquifer, two plots are prepared. Figure 9.21a refers to the groundwater heads at an observation borehole within 2 km of the estuary, and Figure 9.21b is the predicted net saline flow *from* the Estuary.

Figure 9.21a demonstrates that for *Scenario A*, with abstraction remaining at 1980 values, the groundwater heads in the vicinity of the estuary would change little; from Figure 9.21b the inflow of saline water from the Estuary would reduce during the thirty-year predictive period from 28 Ml/d to 20 Ml/d.

For *Scenario B*, all groundwater abstraction ceases; this is indicated by the shaded curve and bar chart. This is not a practical option, yet even if such a severe approach is followed, Figure 9.21 indicates that groundwater heads in the observation borehole would only rise above sea level after fifteen years and the net inflow from the Estuary would become a net outflow after nine years. Even though a net outflow from the aquifer is predicted, this would not preclude continuing inflows from the estuary in the vicinity of major abstraction sites.

For *Scenario C*, where the overall abstraction is reduced to 68 per cent by discontinuing abstraction in

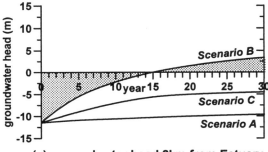

(a) groundwater head 2km from Estuary

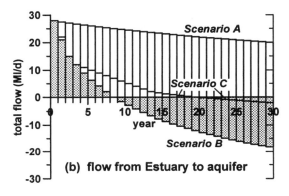

(b) flow from Estuary to aquifer

Figure 9.21   Predictive runs for Lower Mersey unit; Scenario A maintain abstraction at 91 Ml/d, Scenario B stop all abstraction, Scenario C discontinues abstractions close to the Estuary but continue abstractions inland (62 Ml/d)

areas of higher salinity, a rise in groundwater heads is predicted but the heads would still be 4.5 m below sea level after 30 years (Figure 9.21a). According to Figure 9.21b there would be a small net outflow of water from the aquifer to the Estuary. However, this refers to the total outflow along the Estuary in the Lower Mersey unit; this net outflow masks continuing saline inflows at several locations.

This preliminary predictive analysis demonstrates that the groundwater model can be used to explore the consequences of changed abstractions. However, care is required when examining the outputs, to ensure that the predictions are physically realistic.

9.5 BARIND AQUIFER IN BANGLADESH

Objectives of the study: determine whether the present high abstraction from the Barind aquifer is likely to be sustainable.

Key issues: high yield of inverted wells in aquifer of limited thickness, importance of recharge through ricefields, overlying clay has adequate vertical permeability to allow significant recharge to aquifer.

This case study from Bangladesh is selected because it demonstrates that insights can be gained from a detailed analysis of available field information allowing an assessment of the dependability of an existing groundwater scheme. Even though there is insufficient information to prepare a detailed numerical groundwater model of the system, by developing quantified conceptual models, conclusions can still be drawn about the long-term reliability of the aquifer system.

9.5.1 Background

Irrigation in north-west Bangladesh is almost entirely dependent on groundwater. The Barind Tract (Figure 9.22) is one of the driest parts of the country. Since the ground surface is above the adjacent rivers (the Ganges is the western boundary of the study area and the Jamuna is about 50 km to the east), surface water

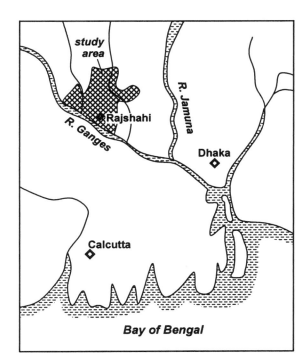

Figure 9.22 Location plan of the Rajshahi Barind

supplies are limited. Initial surveys indicated only limited availability of groundwater. However, due to the ingenuity of the local engineers, significant groundwater development has taken place. Successful tubewells have been constructed; the initial poor yield of certain tubewells has been improved by adopting an 'inverted well' construction (Asad-uz-Zaman and Rushton 1998).

During an initial survey by UNDP, the Barind aquifer system was found to consist of older alluvial deposits known as the Barind Tract; thick clay deposits were proved by test drilling which suggested that the main aquifer does not occur in the upper 300 m. Therefore, according to the UNDP study, groundwater potential is limited to development from relatively thin, fine-grained sand zones that occur within the clay sequence. The aquifer was classified as being suitable only for small domestic demands.

Despite this negative report, tubewell development took place. For instance, a tubewell was constructed in 1982 with a total depth of 56.4 m, the upper 24.7 m with a 0.36 m diameter solid casing connected through a reducer to 10.2 m of 0.15 m diameter slotted casing. With a rest water level at a depth of 14.2 m and a pumped water level of 22.0 m below reference level, a discharge rate of 3.0 Ml/d was achieved.

In 1988–89, 117 exploratory test boreholes of varying depth from 92 m to 152 m were drilled in the Rajshahi Barind to examine the lithology; twelve of these boreholes were geophysically logged. Seventy-seven new observation wells with depths ranging from 30 to 40 m were installed to measure groundwater head fluctuations. Twenty-nine pumping tests were used to estimate aquifer hydraulic properties. Data from these field investigations allowed reliable assessments of the location and areal extent of aquifers as well as realistic estimates of the aquifer properties. Figure 9.23 shows the frequency distribution of transmissivities. By 1988 more than 2000 deep tubewells were operating; in addition, there were many shallow tubewells. The total annual abstraction in the project area of almost 3000 km^2 was estimated to be 744 Mm^3.

9.5.2 Development of conceptual models

Currently the Rajshahi Barind aquifer is supplying sufficient water for extensive irrigation. However, to assess the sustainability of the abstraction rates, a thorough study is required of the nature of the aquifer system, the recharge processes and the design and operation of tubewells.

Nature of aquifer system: the geological cross-section of Figure 9.24 shows that the near surface clay layer is

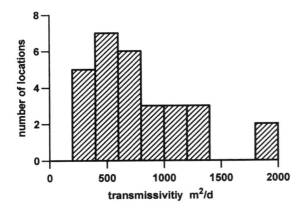

Figure 9.23 Frequency distribution of transmissivities in the Rajshahi Barind

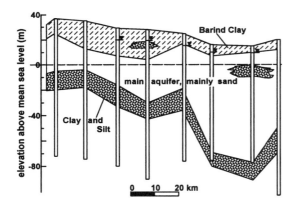

Figure 9.24 Geological cross-section of the Rajshahi Barind

underlain by sandy horizons which may contain clay zones. Beneath the sandy aquifer zone there are extensive thick clays. The hydraulic properties of the Barind Clay, which is directly below the soil zone, are important for assessing the possibility of movement of water from the ground surface into the underlying aquifer. The thickness of the clay varies from 2 to 20 m. Evidence from drilling suggests that the clay is not well consolidated, since boreholes may collapse unless a temporary casing is installed quickly. Furthermore, it is possible to construct wells by hand-digging through this overlying clay. This information indicates a classification as a poorly consolidated clay hence the effective vertical permeability may be as high as 0.01 m/d.

Studies by Alam (1993) of the stratigraphy and paleo-climate of the Barind Clay included a careful field examination of the undisturbed deposits; this was achieved by studying the lithology from within large diameter wells. The observations indicate that the Barind Clay was laid down during a series of periods with different climatic conditions. There is evidence that the decayed roots of the vegetation and other disturbances provide pathways for vertical flow. From all the above evidence, it is reasonable to assume that the overlying Barind Clay does not prevent potential recharge from moving through to the underlying aquifer.

The geological cross-section of Figure 9.24 also shows that the sandy aquifer varies in thickness; in some locations there are clay lenses or layers within the sand and sandy-gravel aquifers. In a review of 27 driller's logs, eleven were found to have intermediate clay layers with thickness varying from 2 to 6 m. Unless these clay layers are continuous over large areas they are unlikely to have a substantial impact on the effectiveness of the aquifer unit. The thickness of the sand or sand and gravel sequences vary from 18 to 80 m. All this information suggests that the aquifer system, though shallow, should provide a reliable source of water provided that there is sufficient recharge and provided that the wells are designed to collect the water efficiently.

Recharge to aquifer: over 70 per cent of the Rajshahi Barind area is covered by ricefields, consequently the soil moisture balance method of recharge estimation described in Chapter 3 is not appropriate. Instead information about losses from flooded ricefields is used to estimate the recharge. Section 4.5 indicates that the flow of water through the bunds of a flooded ricefield to the underlying aquifer is equivalent to a loss each day of a 10–20 mm depth of water distributed over the area of the ricefield. A number of field observations in the study area indicate that recharge through the bunds of ricefields could be significant:

- there is little surface drainage in the Barind area, suggesting that much of the monsoon rainfall is removed without the need for drainage channels,
- the bunds of the ricefields are high, often in the range 0.3–0.5 m; this suggests that the ricefields can store substantial quantities of water during the monsoon period which may become recharge,
- after the end of the monsoon season the depth of water in most of the ricefields falls quickly, indicating that much of the water infiltrates into the aquifer,
- lysimeter studies (similar to Figure 4.30) show that

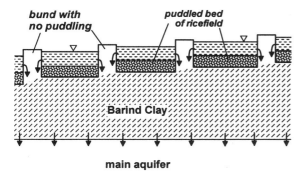

Figure 9.25 Movement of water through ricefields and the Barind Clay

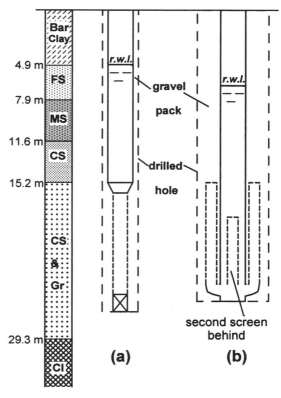

Figure 9.26 Details of tubewell design: (a) conventional design, (b) design with four inverted screens: Bar Cl is Barind Clay, FS, MS, CS are fine, medium and coarse sand, Gr is gravel

there are water losses through the bunds to the underlying aquifer.

This information indicates that a significant proportion of the monsoon rainfall is collected in the ricefields, subsequently infiltrating into the aquifer system through the bunds; a conceptual model of this process is shown in Figure 9.25. Recharge through the bunds also occurs during periods when groundwater is used to irrigate the ricefields; consequently recharge to the aquifer system occurs for much of the year. Indeed, the ricefields provide an efficient recharge system.

Appropriate well design: field evidence suggests that in many locations the transmissivities are moderate. Furthermore, the thickness of the aquifer is limited so that with conventional well design, the pumped drawdown is not sufficient to attract large quantities of water to the tubewell. Therefore an alternative well design has been devised, Figure 9.26b, in which the solid casing is taken to the full depth of the aquifer with four slotted screens connected to the bottom of the casing and extending upwards into the aquifer; this system is described as an *inverted well*.

A comparison between the construction and responses of a conventional well (Figure 9.26a) and an inverted well (Figure 9.26b) is made in Table 9.4. The aquifer system has a 4.9 m layer of clay directly beneath the ground surface; this is underlain by 24.4 m of water-bearing strata consisting of fine sand, medium sand, coarse sand and gravel. Beneath the water-bearing strata is thick clay which extends for at least a further 32 m.

For the initial well design in 1976, a solid casing was installed extending for 15.2 m with a further 9.1 m of 0.20 m diameter slotted screen (Figure 9.26a). The rest water level was at a depth of 4.8 m. To allow a sufficient depth of water over the pump (which is situated towards the bottom of the solid casing) the lowest safe pumping level is 13.4 m below reference level. Consequently the pumped drawdown must not exceed 8.6 m. During a pumping test with an abstraction rate of 2.45 Ml/d the pumped drawdown was 7.44 m; this is equivalent to a specific capacity of 330 m^3/d/m.

In 1986 a new tubewell was constructed within 30 m of the first tubewell; the aim of this new well was to provide an increased yield. The new tubewell was drilled at 1.02 m diameter; this is more than double the diameter of the original. The solid casing in this new tubewell extends for 22.6 m with four screens of 0.15 m diameter *projecting upwards* from the base of the solid casing as shown in Figure 9.26(b). The hole is

Table 9.4   Comparison of single-screen well and multiple inverted screen well

	Single screen (m)	Multiple screen inverted (m)
Lithology and drilling		
Date of drilling	12.01.76	14.07.86
Thickness of upper clay	4.9	4.9
Thickness of water bearing strata	24.4	24.4
Total depth of well	29.3	29.3
Drilled diameter	0.51	1.02
Well construction		
Length of solid casing	15.2	22.6
Diameter of solid casing	0.36	0.36
Length of screen	9.1	2×9.1
		2×6.1
Diameter of screen	0.20	0.15
Lowest possible pumping level*	13.4	20.7
Approx. gravel pack volume	$3.3 \, m^3$	$20.8 \, m^3$
Pumping Test		
Rest water level*	4.8	6.6
Pumped discharge	2.45 Ml/d	7.34 Ml/d
Pumped water level*	12.2	14.2
Pumped drawdown	7.44	7.65
Specific capacity	$330 \, m^3/d/m$	$960 \, m^3/d/m$

* Below datum.

backfilled with pea gravel so that the volume of gravel pack associated with this tubewell is approximately $20.8 \, m^3$ compared to $3.3 \, m^3$ for the original design. Even though the rest water level during the test was 1.8 m lower than for the original test, the yield is three times the original value at 7.34 Ml/d (Table 9.4).

There are several reasons for the higher yield of the inverted well system.

- The screen area is increased from $5.8 \, m^2$ to $14.6 \, m^2$. Despite the large increase in pumping rate the entry velocities are similar; for the original well the entry velocity was 0.0049 m/s and for the replacement well 0.0058 m/s. For these small values, substantial head losses as water flows through the screen are unlikely.
- A greater pumped drawdown is possible with the inverted well; even though the rest water level for the second test was 1.8 m lower, the greater length of solid casing allows the pump to be set 7.3 m lower, hence the maximum possible drawdown is increased.
- Doubling the effective well radius, which results from a doubling of the drilled diameter, and the significant increase in the volume of the gravel pack, ensure that

the borehole is more effective at attracting water; see Figure 9.27.

This increase in yield using the inverted well screens is remarkable; the specific capacity for the inverted well screen is 2.9 times the original design. A threefold increase in yield is unusual. Replacement tubewells using inverted screens have been installed in about 250 locations. For most wells the increase in specific capacity is between 1.4 to 1.7 times that of the conventional tubewell. However, in some locations there is little improvement in the specific capacity; this probably occurs because the limiting factor is the ability of the aquifer to supply water rather than the ability of the tubewell to abstract the water.

9.5.3 Can this rate of abstraction be maintained?

In the above discussion, the successful aspects of the exploitation of the Barind aquifer have been highlighted. However the question remains of whether this abstraction is sustainable. The clearest evidence is provided by observation well hydrographs. Figure 9.28

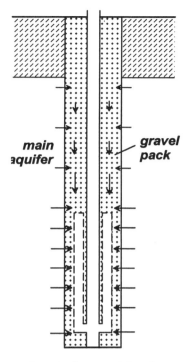

Figure 9.27 Schematic diagram of flow from main aquifer through gravel pack to inverted well

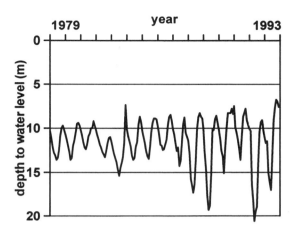

Figure 9.28 Hydrograph of observation well in Rajshahi Barind aquifer

contains a typical hydrograph of monthly readings of groundwater head. The construction of the observation well is a solid casing through the clay layer connected to an open section of 1.8 m. Prior to 1987 there was a relatively uniform seasonal fluctuation with the largest pre-monsoon depth to water of about 14 m and a recovery following the monsoon to depths of 10 to 8 m. In late 1987 three production wells started operating in the vicinity of this observation well; one production tubewell is at a distance of around 0.25 km while the other tubewells are about 0.5 km distant. The effect of these production tubewells is shown by greater drawdowns in the observation well for most years from 1987 onwards. Minimum water levels are almost 5 m lower, demonstrating that the influence of the production wells extends for up to 0.5 km. The notable feature of the hydrographs is that the recovery following monsoon rainfall is unaffected by the pumping. This is a clear demonstration that there is sufficient recharge to refill the aquifer following groundwater abstraction for irrigation purposes.

Hydrographs are available for a number of observation wells. From an examination of over thirty hydrographs, none show trends of declining recovery levels; this confirms that there is sufficient recharge for complete recovery after each pumping season. If incomplete recovery ever occurs, this would indicate that the aquifer is being over-exploited.

9.5.4 Possible provision of a regional groundwater model

If a groundwater model is to be constructed of the Barind aquifer, reliable information is required about the properties and extent of the aquifer system, including the Barind Clay, together with the timing and intensity of the recharge. No accurate assessment of the recharge can be made, but information such as the 'natural' drainage for excess monsoon rainfall and the ability of the ricefields to store and recharge water suggests that a significant proportion of the monsoon rainfall enters the aquifer system. The recharge is likely to be between 400–600 mm/year; this is equivalent to between 25–35 per cent of the annual rainfall of 1400–1700 mm/yr. This estimated recharge is of a similar order to the water requirement of an irrigated rice crop of about 700 mm multiplied by a cropping intensity of 70 per cent. The response of observation wells, with complete recovery after irrigation pumping, gives confidence that the development of groundwater in Barind is sustainable.

Would a groundwater model of the Barind aquifer system provide greater confidence in the sustainability of the aquifer system? As indicated above, reliable estimates of the recharge are not yet available; extensive fieldwork using lysimeters would be required to obtain accurate assessments. There is no quantitative information about the vertical hydraulic conductivity of the Barind Clay and only limited information is available concerning the transmissivity of the aquifer. Consequently, a numerical groundwater model would be speculative and would add little to the understanding of the aquifer system. Nevertheless, the quantified conceptual models, supported by observation well records, provide sufficient information to conclude that the current exploitation of the Barind aquifer is sustainable.

9.6 CONCLUDING REMARKS

Chapter 9 introduces regional groundwater studies in which the transmissivity is effectively constant; the presentation is focused on four case studies. The main emphasis for each case study is the identification of conceptual models which correspond to the important flow processes. The next step is to quantify the aquifer parameters and coefficients. Once the quantified conceptual models are incorporated in regional groundwater modelling packages, model refinement can proceed by comparing field information and data with model responses. Model outputs should be selected and presented in a form that is helpful in understanding the aquifer responses.

For the Nottingham Sherwood Sandstone aquifer the initial challenge was to identify and quantify all the sources of inflow to the aquifer system. Inflows through low-permeability overlying formations proved to be important; leakage from water mains is a further source of recharge. When estimating aquifer parameters and coefficients, one issue of particular significance is the magnitude of the coefficients describing river–aquifer interaction (river conductances); flow gauging of a river which is influenced by a major pumping station provides valuable information. Comparisons between field results and numerical model simulations for both river–aquifer flows and groundwater head hydrographs indicate that the numerical

model provides an acceptable representation of the flow processes. Flow balances with an annual or monthly time step are used to interpret the interactions between recharge, abstraction, aquifer storage changes and river–aquifer flows.

When the northern extension of the Nottinghamshire sandstone aquifer is studied, the dominant flow processes are different to the Nottinghamshire unit. Classical leaky aquifer conditions apply where there are extensive overlying clay deposits. To the east of the outcrop area, the overlying deposits are more permeable. To prevent waterlogging due to up-flowing groundwater, extensive drainage systems were constructed; these drains now provide a source of recharge due to the increased abstractions from the main aquifer.

In the Lower Mersey sandstone aquifer, recharge through the drift deposits and contributions from leaking water mains and sewers account for almost two-thirds of the recharge. In addition, poor quality water has been drawn in due to over-abstraction from the aquifer. Owing to the high storage of the aquifer system, the initial conditions must be represented satisfactorily; the simulation commences in the mid-1800s with larger time steps for the first hundred years.

The final case study refers to the Barind aquifer in Bangladesh. Important issues in the study include recharge from irrigated ricefields, movement of water through an overlying clay layer into the main aquifer, and an ingenious design of boreholes which is ideally suited to aquifers with a limited saturated thickness. Although a numerical model has not been prepared for this aquifer due to insufficient field information, the recovery of groundwater heads each year following the main abstraction period suggests that the current abstractions can be maintained.

Information about various aspects of regional groundwater modelling is presented in the case studies in Chapter 9. However, in the introduction to Part III, tables are prepared which indicate where specific information can be found.

Additional complexities arise for regional aquifers in which vertical flows are significant. This is the subject of Chapter 10.

10

Regional Groundwater Flow in Multi-Aquifer Systems

10.1 INTRODUCTION

In many regional groundwater flow systems there are several permeable aquifer zones separated by material of a lower permeability. Vertical flows through these low permeability zones are of crucial importance in quantifying the groundwater flows. In a number of earlier studies the assumption was made that, because low permeability zones have hydraulic conductivities three or more orders of magnitude less than the productive aquifer zones, vertical flows through these low permeability zones are negligible. However, more recent studies have recognised that vertical flows through low permeability zones are often significant. The availability of multi-layered groundwater models has assisted in the growing understanding of vertical components of flow.

One of the practical difficulties is determining the effective vertical hydraulic conductivity of low permeability zones. Leaky aquifer pumping test analysis can be used to estimate the vertical hydraulic conductivity of aquitards, but inaccuracies are likely to occur when the storage effects of the aquitard are ignored; see Section 7.5. Realistic first estimates of the vertical hydraulic conductivity can usually be deduced from borehole logs and permeameter tests on samples. An instructive example of detailed field investigations supported by groundwater flow modelling of a complex system of aquitards and thin aquifers with interbeds in Canada, is described by Gerber *et al.* (2001); the bulk vertical hydraulic conductivity of the aquitard is estimated as 0.0001 m/d.

Studies of multi-aquifer systems require field records of the groundwater head fluctuations in individual aquifers. In addition, all the available data and information must be collected to formulate conceptual models of the aquifer system. Only when conceptual models have been developed and acceptable idealisations have been devised should attention be turned to the selection of a numerical model. Models of the entire aquifer system are not always appropriate; vertical section models, radial flow models and models of a sub-area of the system are helpful in developing an understanding of the important flow processes in the aquifer system. An approach such as telescopic mesh refinement can be beneficial (Ward *et al.* 1987).

Four case studies, from different countries and with very different aquifer formations, are considered in detail in this chapter. For each example, the first step is an examination of all the available data and information to develop conceptual models. For two of the case studies there is insufficient information to proceed to mathematical modelling. Nevertheless, the preparation of quantified conceptual models assists in identifying the groundwater resources and provides guidance about safe exploitation. Brief reference is made in the final section to four further situations where vertical components of flow are important.

10.2 MEHSANA ALLUVIAL AQUIFER, INDIA

Objectives of the study: consider reasons for over-exploitation, identify the groundwater resource available for exploitation, develop a detailed understanding of a small area.

Key issues: water in deeper aquifers mainly originates from shallow aquifers; identify vertical permeability of less permeable zones; importance of shallow and deep observation wells; use of a vertical section model plus a model of a limited area.

Groundwater Hydrology: Conceptual and Computational Models. K.R. Rushton
© 2003 John Wiley & Sons Ltd   ISBN: 0-470-85004-3

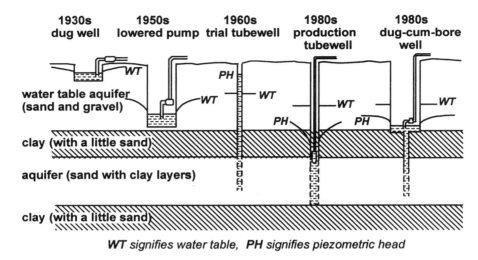

WT *signifies water table,* PH *signifies piezometric head*

Figure 10.1   History of exploitation of the Mehsana alluvial aquifer

10.2.1 Introduction

The Mehsana Alluvial aquifer in western India is selected to provide an introduction to the impact of groundwater development in multi-layered alluvial aquifers and to illustrate many of the issues which arise when studying aquifer systems in which vertical components of flow are significant. The history of exploitation of the Mehsana aquifer is summarised in the idealised diagram of Figure 10.1. For centuries water was withdrawn from large diameter dug wells in the shallow aquifer using animal power. Consequently the maximum drawdown in well water level was no more than 10 m; this self-regulation occurred due to the limited depth that a large 'bucket' could be lowered into the dug well thereby restricting the 'pumped drawdown'. When suction pumps were first used in dug wells (1930s in Figure 10.1), water levels fell so that many wells became dry. Consequently the wells were deepened with suction pumps set on platforms just above the water surface (1950s lowered pump) thereby continuing the decline in water levels. The next stage of development of the Mehsana aquifer was the use of deep drilling rigs to construct tubewells to depths of 100 m or deeper. Submersible pumps provided excellent yields. In the initial tests, the pumped drawdowns were small with yields of 3000 m³/d or more (1960s trial tubewell). At the time of the initial tests, rest water levels (piezometric heads) in the wells tapping the deeper aquifers were usually above the current water table elevation. This suggested the discovery of a 'new' deeper aquifer

fed from distant recharge sources. This 'discovery' led to the drilling of many production wells into the deeper aquifers. Initially excellent yields were obtained but, after several years, severe declines in both the pumped levels in the tubewells and in the water table elevations were observed (*1980s* diagrams in Figure 10.1).

The original idealised conceptual understanding of the aquifer system is shown on the schematic ENE–WSW cross-section of Figure 10.2. Recharge occurs over the whole of the aquifer. The higher ground, where the depth of the aquifer is limited and little clay is present, was defined as a 'common recharge zone'. It was assumed that water moves laterally from the common recharge zone on the higher ground to the confined zone beneath the clay layers. To overcome the declining groundwater heads in the confined zone, pilot artificial recharge investigations were initiated; information about spreading channel and injection well experiments can be found in Sections 4.2.7 and 8.6 and in Rushton and Phadtare (1989).

In this review of the Mehsana alluvial aquifer system, the first step is to introduce and examine available field data and information; this allows the development of quantified conceptual models. Two types of mathematical model are chosen to explore the aquifer response; the first model represents a vertical section, the second is a multi-layered time-variant model of a limited area. The fieldwork and modelling were carried out in India with limited resources and computing power. With careful interpretation of field data and the development

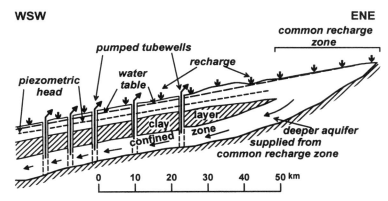

Figure 10.2   Initial conceptual understanding of the Mehsana aquifer

of simple mathematical models, an understanding of the flow processes was realised. This led to the development of aquifer management strategies.

10.2.2 Description of the aquifer system

The Mehsana alluvial aquifer is located in Gujarat, Western India; Figure 10.3. It extends over an area of approximately 3000 km² with a ground elevation of 180 m in the north-east falling to less than 70 m in the south-west. The study area is bordered by hills to the north and north-west and to the south-west by the Rann of Kutch. The boundaries of the study area include the perennial Sabamarti River in the east and the intermittent Saraswati River to the north-west. Towards the Rann of Kutch, groundwater salinities increase significantly; the study area is restricted to regions with good quality groundwater. Brief details of the lithology in the study area are presented in Table 10.1, and a more detailed account of the paelohydrology of the aquifer system is presented by Bradley and Phadtare (1989). Detailed lithological logging of boreholes in the Mehsana region allowed the construction of a representative vertical cross-section; Figure 10.4. Information is available from seven boreholes spread over a distance of 50 km. The cross-section of Figure 10.4 was drawn assuming that the clays form extensive lenses.

The long-term average annual rainfall in the south and west of the study area is 520 mm but in the 1980s there were three successive years with rainfall less than 200 mm. Over the higher ground to the north, the average rainfall is 900 mm. Recharge calculations for the Mehsana area are described in Section 3.4.11. However, the recharge is generally less than that plotted in Figure 3.20 since the results in that figure refer to areas under surface water irrigation. Usually ground-

water irrigation is based on crop needs, so that there is little recharge after the monsoon season.

Abstraction from tubewells is primarily for irrigation and public water supply; there are many private wells for which abstraction records are not available. Some indication of the magnitude of the withdrawal of water can be obtained from the number of hours that a well operates or from the electricity consumption. Kavalanekar *et al.* (1992) describe how monthly estimates of abstraction were obtained from a careful study of electricity consumption and crop water requirements.

10.2.3 Field records of groundwater head

Field records of groundwater heads are available for both pumped tubewells and observation wells. An examination of groundwater head changes provides insights into the flow processes.

Table 10.2 presents data from an individual pumped borehole over a period of ten years; these data permit an examination of changes in the specific capacity including its dependence on the intensity of the pumping. Each of the tests is discussed in turn.

- January 1980 was during the early years of the exploitation of the deeper aquifer system. The water table was 18 m below the ground elevation of 90.5 m with the rest piezometric head in the deeper aquifer almost 10 m lower. Pumping to test the potential yield of the borehole at a high discharge rate of 3636 m³/d resulted in a drawdown of 11.7 m after 300 minutes of pumping; the specific capacity (discharge per unit drawdown) is 310 m³/d/m.
- Almost eight years later in November 1987 the water table was 15 m lower and the non-pumping

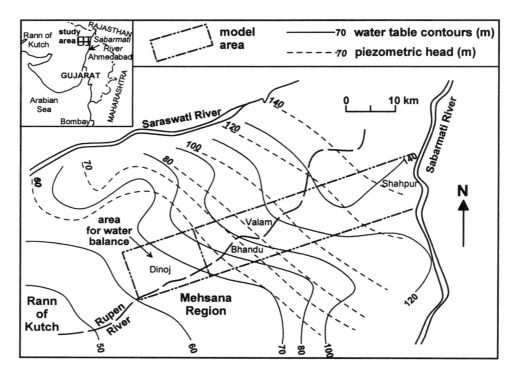

Figure 10.3   Plan of the Mehsana alluvial aquifer including contours of the water table and piezometric head

Table 10.1   Aquifer system in Mehsana study area, depths are below ground level

Formation	Depth range (m)	Lithology
Phreatic aquifer	3–40	Fine- to medium-grained pale yellowish brown sands with kanker. Gravel clay intercalations of 2 to 5 metres are common
1st aquitard	40–50	Clay yellowish silty with kanker
1st confined aquifer	50–125	Sand medium- to fine-grained with gravel and sandy clay lenses. Pebble boulder bed on eastern side. Gradual reduction in grain size from east to west
2nd aquitard	125–135	Clay yellowish, sometimes sticky with kanker
2nd confined aquifer	135–225	Same as first confined aquifer
3rd aquitard	225–265	Clay brownish, sticky with little kanker and fine sand

piezometric head more than 24 m lower. At a discharge rate of 1426 m³/d the pumped drawdown was 6.6 m. The reduced specific capacity of 214 m³/d/m, which is 70 per cent of the original value, results from interference due to neighbouring tubewells.

• The third test took place less than two years after the second test; the decline in the water table elevation continued but this test was carried out immediately before the monsoon when the water table tends to be lower. A slightly higher discharge rate was achieved

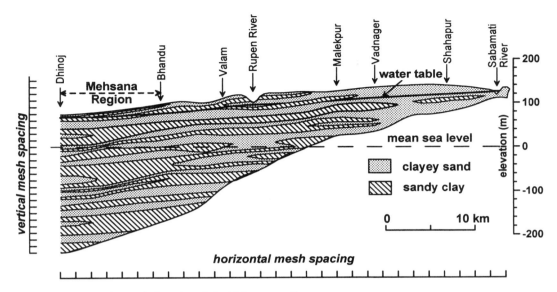

Figure 10.4 Representative vertical section of the Mehsana aquifer

Table 10.2 Borehole conditions and specific capacity in a single borehole; all levels are relative to Ordnance Datum

	Date of test			
	Jan. 1980	Nov. 1987	June 1989	March 1990
Water table elevation (m)	72.5	57.5	50.5	47.5
Non-pumping piezometric head (m)	62.4	38.1	27.6	22.6
Pumped level (m)	50.7	31.5	21.5	16.8
Pumping rate (m³/d)	3636	1426	1493	1150
Duration (m)	300	300	300	300
Drawdown (m)	11.7	6.6	6.1	5.8
Specific capacity (m³/d/m)	310	216	245	198

because only a limited number of tubewells were pumping; the specific capacity of 245 m³/d/m is higher than the second test but lower than the first test. This is an important result since it demonstrates that the specific capacity is influenced by pumping from surrounding tubewells.

• March 1990 was a period of substantial pumping. Further declines had occurred in the water table elevation, the rest piezometric level and the pumped discharge. The specific capacity of 198 m³/d/m demonstrates the increasing difficulty in obtaining an adequate discharge from the deeper aquifers.

This deterioration in the yield and specific capacity of tubewells is primarily due to the influence of the surrounding tubewells. A study was made of one of the original tubewells for which, even though the pump was operating, the discharge was almost zero. This very low yield occurred because the surrounding tubewells penetrated to greater depths. An experiment was conducted on this poor yielding tubewell to assess the impact of the neighbouring deeper tubewells. All of the tubewells were electrically driven; power is only available for part of the day. Therefore a generator was brought on site so that this low-yielding tubewell could operate when the other tubewells had no power and therefore were not operating. A good yield was obtained from this tubewell when pumping on its own. This leads to the concept of tubewells competing for water, with the deeper tubewells achieving higher yields. This is one item of evidence to be included in the conceptual model.

Records of groundwater heads in shallow and deep aquifers, based on monthly readings, are also available. Figure 10.5 shows the groundwater heads in shallow (S) and deep (D) piezometers at the same location. The shallow piezometer recovers each monsoon; unlike most shallow piezometers in the study area there is a complete recovery each year since this piezometer is located in the floodplain of the Saraswati river. The response of the deep piezometer is not in phase with the shallow piezometer; the deep piezometer does not respond to the monsoon recharge since there are extensive clay layers separating the deep aquifers from the ground surface. Instead the deep piezometer reflects the irrigation pumping from the deeper aquifers (indicated at the foot of the graph in Figure 10.5) which commences shortly after the end of the monsoon season and continues until the harvesting of the irrigated crops in February or March.

Careful examination of Figure 10.5 also shows that there is a *significant vertical head gradient* with the groundwater head in the deeper piezometer on average 25 m lower than the water table elevation. This hydraulic gradient is required to draw water from the overlying water table, vertically through the low permeability zones within the alluvial aquifer and into the deeper aquifers to meet the demand of the deeper tubewells.

10.2.4 Flow processes in aquifer system

From insights gained from the analysis of the field information, an alternative conceptual model is proposed; Figure 10.6 (Rushton 1986b). The two diagrams show changes to the flow system which have occurred between the original drilling and testing of the tubewells and the current extensive abstraction. Diagram (a) represents part of Figure 10.2, the initial conceptual model (Rushton 1986b). When isolated test wells were constructed into the deeper aquifer, there was

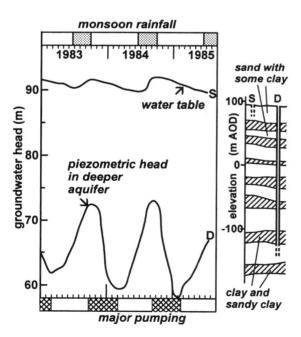

Figure 10.5 Groundwater heads at the water table and in a deep piezometer at Kamliwara in the flood plane of the Saraswati River; S = shallow piezometer, D = deep piezometer

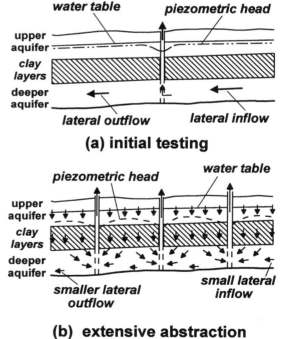

Figure 10.6 Conceptual models: (a) original conceptual model, (b) changed conditions due to major abstractions from the deeper aquifer (Rushton 1986b). Reprinted from *Ground Water* with permission of the National Ground Water Association. Copyright 1986

sufficient lateral flow from the common recharge zone to meet the abstraction demand of the small number of deep tubewells. However, once extensive abstraction occurred (diagram (b)), lateral flows proved to be insufficient to meet the demand. The total lateral flow remains largely unchanged because its magnitude depends on the overall groundwater gradient from the common recharge zone to the main abstraction zone; this groundwater gradient follows the topographical gradient. Since the lateral inflow is insufficient (see the small lateral arrows in Figure 10.6b) the borehole pumps have to work much harder to draw water vertically from the overlying water table through the clay layers; the plan area from which they can draw water is restricted by interference from surrounding boreholes.

Information presented in the previous section, *Field records of groundwater head*, confirms this revised conceptual model.

- As the number of boreholes increased, the yield of individual boreholes decreased (Table 10.2) resulting in a decrease in the specific capacity.
- There has also been an increase in the difference between water table elevations and the non-pumping piezometric heads in the deeper aquifer; the difference has increased from 10.1 m in 1980 to 24.9 m in 1990 (see Table 10.2). This change has occurred because, as more tubewells were drilled, the plan area of the water table associated with an individual tubewell decreased. In an attempt to maintain the tubewell yields, greater differences between pumping levels and water table elevations are required to draw sufficient water into the tubewells.
- In Figure 10.5 seasonal differences between the water table elevation and the non-pumping piezometric head in the deeper aquifer can be identified; before the start of pumping for irrigation in October the difference is about 17 m, increasing to more than 31 m at the end of the irrigation pumping period.

10.2.5 Mathematical model of a vertical section

A vertical section mathematical model is used to explore the balance between water drawn laterally and vertically into the deeper aquifers. Is lateral flow from the common recharge zone the major source of water or is a significant proportion of the pumped water due to vertical flow from the water table through the less permeable clay lenses? The conceptual model has identified both of these processes, but a numerical model is required to quantify the flow components. This is achieved by developing a computational model to represent flows through the vertical section of Figure 10.4; Rushton and Tiwari (1989). Since the information required is the average contributions from lateral and vertical flow, a time-instant approach is adopted. The model corresponds to conditions in the area indicated on Figure 10.2, which is 58 km long by 10 km wide.

A finite difference model is used to represent the vertical section of Figure 10.4. A rectangular grid is chosen; the sub-division of the grid is shown to the left and below the cross-section. The mesh is designed for a similar number of mesh intervals in the horizontal and vertical directions, hence $\Delta x = 2000$ m and $\Delta z = 15$ m. Even though a rectangular grid is used, it is possible to represent the complex distribution of the sand and clay lenses. A detail of the sand-clay distribution is shown in Figure 10.7; the method of calculating equivalent conductivities in the horizontal and vertical directions for the mesh spacing of 2000 m by 15 m is explained on the figure. Between *a-b* in the horizontal direction the equivalent hydraulic conductivity is 3.10 m/d; between *b-c* the equivalent vertical hydraulic conductivity is 0.00208 m/d.

A time-instant approach is appropriate for this study of average conditions. Because this model corresponds to an average response over a period of one year, the water table is represented as a known averaged groundwater head distribution; see the unbroken line in Figure 10.4. Although the water table is represented on the finite difference grid as specified groundwater heads, the flow from the water table is not predetermined. Once nodal groundwater heads throughout the aquifer have been determined from the numerical model, the quantity of water being drawn from the water table can be calculated. These calculated flows represent the sum of the estimated recharge to the water table and the quantity of water released from unconfined storage due to changes in water table elevation.

Tubewells are represented in the mathematical model as discharges from the appropriate nodal points. However, it is essential to consider how tubewells take water from the aquifer. Tubewells have a solid casing through the upper part of the borehole. At greater depths, slotted casing is provided adjacent to major permeable layers; in many boreholes there are five or more slotted sections of the borehole assembly. Gravel packs are provided between the casing and the formation. The groundwater head takes the same value at each location where there is contact between the tubewell and the aquifer because the column of water in the borehole is a line of constant groundwater head. A detailed discussion on this topic can be found in Section 12.3.

The adequacy of the mathematical model can be

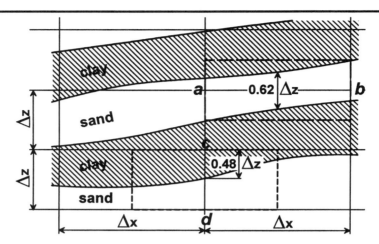

(a) equivalent *horizontal hydraulic conductivity* between *a-b*

The general formula for the equivalent horizontal hydraulic conductivity with a series of layers of thickness t_i and hydraulic conductivity K_i is

$$K_x = \sum_{i=1}^{n} K_i t_i \bigg/ \sum_{i=1}^{n} t_i$$

For this example the thickness and hydraulic conductivity of the two layers between *a* and *b* are sand, thickness 0.62 Δz, $K = 5.0$ m/d and clay, thickness 0.38 Δz , $K = 0.001$ m/d. Substituting in the above formula gives $K_x = 3.10$ **m/d**.

(b) equivalent *vertical hydraulic conductivity* between *c-d*

The general formula for the equivalent vertical hydraulic conductivity is

$$K_z = \sum_{i=1}^{n} t_i \bigg/ \sum_{i=1}^{n} (t_i / K_i)$$

For this example the thickness of the two layers between *c* and *d* are clay, $t = 0.48 \Delta z$ and sand, $t = 0.52 \Delta z$. Substituting in the above formula gives $K_z = 0.00208$ **m/d**.

Figure 10.7 Estimation of conductivities in vertical section finite difference model

assessed by considering groundwater head distributions and conditions at the water table. Field values of groundwater head are available in two piezometer nests, one at Bhandu and the second at Valam (for locations see Figure 10.3). At Bhandu the water table elevation is 83 m AOD and the average groundwater head at the four piezometers located from +20 to –140 m AOD is 51 m. Groundwater heads from the mathematical model, for locations corresponding to the piezometers, average 54 m.

At Valam the water table is at 104 m AOD, the average reading of three piezometers is 76 m and the model average is 73 m (for more detailed information see Rushton and Tiwari 1989). These differences between field and modelled groundwater heads of about 3 m are certain to occur with such a complex aquifer system. In the groundwater model the vertical hydraulic conductivity of clay is set at 0.001 m/d. A sensitivity analysis of the model with the vertical hydraulic conductivity of the clay set at 0.0003 m/d or

0.003 m/d leads to serious errors in the representation of groundwater heads in the piezometers.

With a time-instant simulation, the quantity of water drawn from the water table can be estimated from the vertical hydraulic gradients. In the Mehsana region (the location of this region is indicated in Figures 10.3 and 10.8) the quantity drawn from the water table is shown in Figure 10.8b; the average is 0.65 mm/d. Considering a typical year, the annual recharge is unlikely to exceed 73 mm or 0.2 mm/d. The quantity of water released from storage is estimated from the long-term decline in the water table elevation of at least 1.0 m per year and the specific yield due to slow drainage, which is estimated to equal 0.15 for sand. Hence

storage release at water table
= 1.0 × 0.15/365 = 0.00041 m/d

Thus the estimated recharge plus water released from storage totals 0.61 mm/d, which is similar to the modelled value of water drawn from the water table of 0.65 mm/d.

10.2.6 Origin of flows as determined from vertical section model

The vertical section model can be used to identify the sources of water pumped from the deeper aquifers. Results for the Mehsana region are presented in Figure 10.8b where the vertical velocity in mm/d from the water table is plotted for each 2 km mesh interval of the Mehsana region. Multiplying each of these velocities by 2 km × 10 km (since the plan area represented by the vertical section model extends 10 km perpendicular to the line of the vertical cross-section), the total flow from the water table is 91 Ml/d. The modelled deep aquifer abstraction for this area is 98 Ml/d; consequently, almost 93 per cent of the abstracted water is drawn from the overlying water table through the low permeability zones with only 7 per cent due to lateral flow. A re-examination of field information of Section 10.2.3 supports the conclusion that a high proportion of the abstraction demand is met from vertical flows.

- If the cross-section of Figure 10.2 is drawn to true scale, the horizontal dimension needs to be multiplied by a factor of eighty. The lateral pathways then become very long and the proximity of the overlying water table becomes more significant despite the intervening low-permeability layer.
- Considering the equivalent hydraulic conductivities of the finite difference mesh (Figure 10.7), equivalent

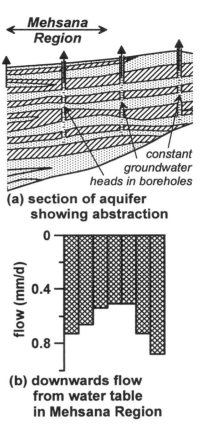

(a) section of aquifer showing abstraction

(b) downwards flow from water table in Mehsana Region

Figure 10.8   Vertical flows from water table into the deeper aquifer system as determined from the vertical section model

hydraulic resistances can be calculated. They are proportional to the distance between the nodal points divided by the width and equivalent hydraulic conductivity of the flow zone. With $\Delta x = 2.0$ km and $\Delta z = 15$ m, the ratio of the horizontal to vertical hydraulic *resistances* is 2000/(3.1 × 15) to 15/(0.00208 × 2000) or 12 to 1. With similar numbers of mesh intervals in the horizontal and vertical directions, there is a *lower* resistance to vertical flow in the aquifer system despite the presence of low permeability clays.
- Since vertical hydraulic conductivites of the clay are low at 0.001 m/d, substantial vertical groundwater head gradients of about 0.3 occur across the clay zones. These vertical gradients result from the lowering of piezometric heads due to abstraction. Furthermore, the plan area of the Mehsana region represented by the model of 140 km² is large compared to a lateral cross-section of 10 km × 200 m = 2 km².

Consequently significant quantities of water can be drawn vertically through the overlying clays.

10.2.7 More detailed study of smaller area

Since the mathematical model described above is a time-instant (pseudo-steady state) representation of a vertical section of the aquifer system, it does not provide information about the time-variant response of the deeper aquifer system or the water table decline. There is insufficient information about the lithology, abstraction, recharge etc. to justify a three-dimensional time-variant numerical model for the whole aquifer system. Therefore a smaller area of 16 km^2 within the Mehsana region is studied in detail (Kavalanekar *et al.* 1992). The quantified conceptual and mathematical models used in this study can be classified as being straightforward; nevertheless they are sufficiently detailed to represent the key features of the aquifer system.

Field information for detailed study area

Field information includes the following.

- The lithology for five boreholes is available; the boreholes indicate that the aquifer is a complex mixture of sands, clayey sands, silts and sandy clays with kankar in certain zones. Although the thickness of sand and clay is roughly the same in each borehole, no continuous layering can be identified across the study area.
- The study period extends from the start of the monsoon season in 1984 to the end of the dry season in 1989. In a typical year it is possible to grow a first rain-fed crop followed by a second crop using water from tubewells. For the middle three of the five years, rainfall was below 200 mm with the result that there was no recharge. Recharge for the first and fifth year is set at 1 mm/d for the three months of August, September and October; this is consistent with water balance estimates described in Section 3.4.11.
- Within the study area there are three government tubewells and 54 private tubewells. Estimates of the abstraction are based on power consumption of the government tubewells and the crop water requirements for the private tubewells. During 1988–89 the average abstraction was 11 Ml/d. The peak monthly abstraction was 20.2 Ml/d; for a cropping intensity of 50 per cent of the total area of 16 km^2, this is equivalent to 2.5 mm/d.
- Groundwater head data for the study area is presented and discussed in Section 2.8.2 and Figure

2.33; the data are also plotted in Figure 10.10. Water table information was collected from an abandoned dug-cum-bore well but a consequence of the rapid fall in the water table is that no readings are available during the latter part of the study period. Piezometric head readings were obtained from a production borehole with readings always taken at least five hours after pumping stopped. A detailed study of the recovery shows that five hours after the pump is switched off the recovery is 98 per cent complete.
- Pumping tests using conventional analysis suggested that the regional transmissivity of the deeper aquifer system is about 450 m^2/d.

Conceptual model

In developing a conceptual model of this restricted study area, the *main characteristics*, which have a strong influence on the flow processes within the aquifer system, must be identified. Figure 10.9a illustrates the flow processes.

1. Recharge occurs at the water table and water is released from storage as the water table falls; the water table fall is indicated by the difference between the unbroken and broken lines.
2. Vertical flows occur *through* the clays and sands above the main aquifer zones; these flows are indicated by double arrows.
3. Mainly horizontal flows occur through the permeable horizons of the main aquifer zones towards the slotted portions of the tubewells; it is also necessary to represent the confined storage coefficient of the main aquifer zones.
4. Tubewells are used to abstract water from the main aquifer.

The next step is to idealise the conceptual model and estimate aquifer parameters; the idealised conceptual model is shown in Figure 10.9b. The four *main characteristics* are idealised and quantified as described below.

1. Recharge to the water table together with water released from storage due to a fall in the water table is the source of water for the deeper aquifer system; water is drawn into the deeper aquifer system to meet the abstraction demand. The specific yield is set at 0.12 but reduces to 0.04 when the water table falls below 63.5 m to reflect the higher clay content of the lenses of lower hydraulic conductivity.
2. The low permeability material between the water table and the main aquifer is described as the *less permeable zone* and is represented as a single con-

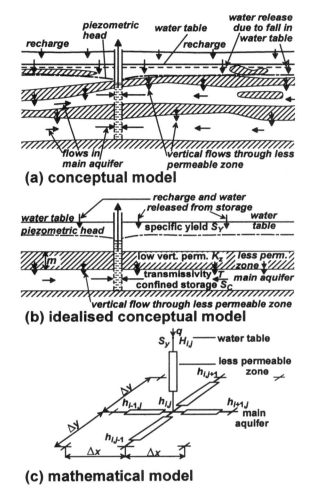

Figure 10.9 Conceptual and numerical models of an area of 16 km² in central Mehsana

tinuous layer of thickness $m = 28$ m and effective vertical hydraulic conductivity $K_z = 0.0009$ m/d (see the last paragraph of the *Mathematical model* section below).

3. All the permeable zones from which water is pumped are combined as a single layer, described as the *main aquifer*. From pumping tests, the transmissivity is set at $T = 450$ m²/d with a confined storage coefficient $S_c = 0.005$. This relatively high confined storage coefficient is appropriate for unconsolidated material (Rushton and Srivastava 1988).

4. Since the permeable zones of the main aquifer are represented as a single layer, all the abstracted water is taken from this layer; the monthly values of

abstraction for the study area lie in the range 2.0 and 22.0 Ml/d.

Mathematical model

The mathematical model, which represents the conceptual model defined above, is illustrated in Figure 10.9c. A finite difference approximation is used with mesh spacings of Δx and $\Delta y = 250$ m; nodal positions are defined by i, j etc., with a time step of $\Delta t = 10$ days (adjusted as necessary for the third time step of each month). The inclusion in the mathematical model of the four *main characteristics* is described below.

1. The balance at the water table has three components: the water lost vertically through the less permeable zone, the recharge and the water released from storage. The appropriate equation is

$$-[K_z\Delta x\Delta y(H_{i,j} - h_{i,j})]/m + q\Delta x\Delta y$$
$$-[S_Y\Delta x\Delta y(H_{i,j} - H_{i,j}^*)]/\Delta t = 0 \qquad (10.1)$$

where $H_{i,j}$ and $h_{i,j}$ are the water table elevation and piezometric head; * signifies the value at the previous time step.

2. Vertical flow through the less permeable zone depends on the difference between the water table elevation and the piezometric head; this flow is represented by the first term of Eq. (10.1) and the term immediately before the equals sign in Eq. (10.2).

3. The three contributions to the water balance at a node in the confined main aquifer are the lateral flow components, the vertical inflow from the less permeable zone and the release of water from confined storage as the piezometric head changes with time. The relevant equation is

$$[T_x\Delta y(h_{i-1,j} - 2h_{i,j} + h_{i+1,j})]/\Delta x$$
$$+ [T_y\Delta x(h_{i,j-1} - 2h_{i,j} + h_{i,j+1})]/\Delta y$$
$$+ [K_z\Delta x\Delta y(H_{i,j} - h_{i,j})]/m$$
$$= [S_C\Delta x\Delta y(h_{i,j} - h_{i,j}^*)]/\Delta t \qquad (10.2)$$

4. Abstraction from the main aquifer is represented at fifteen nodes; these abstractions are included as extra terms on the right-hand side of Eq. (10.2).

Also, it is assumed that the study area is isolated from other parts of the aquifer system. This assumption is acceptable since the analysis using the vertical section model showed that, in the Mehsana region, lateral flows account for only 7 per cent of the abstraction.

For the numerical solution, initial conditions must be specified. The historical field value for the water table in July 1993 was 67.0 m, hence for initial conditions all the water table heads are set to this value. For the main aquifer the average abstraction for the year 1982–83 is withdrawn from the modelled tubewells with the vertical hydraulic conductivity adjusted so that the piezometric heads average 51 m. This requires a vertical hydraulic conductivity of 0.0009 m/d which is close to the value of 0.001 m/d used in the vertical section model. Numerical solutions to the problem described above can be obtained using most standard regional groundwater flow packages.

Comparison with field data

The field data for this study area are presented and considered in detail in Section 2.8.2 and Figure 2.33. Comparisons between the values from the mathematical model and the field information, Figure 10.10, show acceptable agreement. For instance:

- the general decline in the water table is consistent with the available information,
- the long-term decline in the piezometric heads is also represented satisfactorily,
- the modelled fluctuations in the piezometric heads, which are mainly due to pumping, show similar trends to the field values but there are differences of up to 4 m. However, in the vicinity of pumped boreholes there are local depressions of 2 to 5 m; consequently differences between field and modelled values of this order are likely to occur.

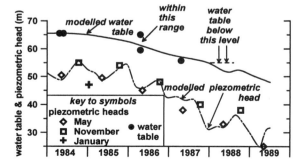

Figure 10.10 Comparison between field readings and model results for detailed study area. Reproduced by permission of IAHS Press from Kavalanekar *et al.* (1992)

10.2.8 Concluding discussion

Two important questions remain; first, during the initial field testing in the 1960s, was there sufficient information to deduce the nature of the aquifer system and identify the importance of vertical flows? Second, how can the aquifer be managed to prevent further reductions in the water table level and further deepening of the tubewells? Figure 10.6a indicates that trial tubewells had no difficulty in collecting water flowing laterally through the deeper aquifer. There was no indication that there would be vertical flows or large pumped drawdowns. Therefore individual widely spaced deep test wells in an un-exploited aquifer are unlikely to identify the flow processes which will occur under heavy exploitation. In practice, the only reliable method of ensuring that the yields can be maintained is to carry out a pilot scheme in a restricted area with a close spacing of tubewells, pumping hard for at least one dry season. If the pilot scheme covered an area of about 4 km^2, with sufficient water pumped out to meet crop water requirements, any limitations of the resource should become apparent. However, the implementation of this kind of pilot scheme is expensive.

Simple realistic water balance calculations can indicate whether there is likely to be sufficient water. In most relatively flat alluvial aquifers (the overall topographical and water table gradients of the Mehsana aquifer are about 1 : 1000) lateral flows are small, hence most of the abstracted water must come from the overlying water table. With an average annual rainfall in the range of 600 to 700 mm, it is unlikely that the rainfall recharge will exceed 100 mm in a year. Return flows from irrigation are small because farmers only irrigate when it is essential to do so. Therefore, for a crop water requirement in excess of 400 mm, each crop requires the equivalent of at least four years' recharge. This straightforward calculation indicates that mining of groundwater is likely to occur.

As a means of mitigating the over-exploitation of the aquifer system, pilot artificial recharge experiments were carried out with UNDP support. Artificial recharge using injection wells is described in Section 8.6 and artificial recharge using spreading basins is considered in Section 4.2.7. Although the field experiments were moderately successful, the schemes were not viable economically (Rushton and Phadtare 1989).

The second question is concerned with the management of the aquifer system. Once the serious nature of the over-exploitation was recognised attempts were made to persuade the farmers to reduce abstraction. Legislation, which has not been fully implemented, was drawn up to prevent the drilling of new tubewells or the deepening of existing tubewells. One positive outcome is that farmers are changing to crops which have a lower water requirement.

This phenomenon of the over-exploitation (or mining) of aquifers in semi-arid areas, when the water is used for irrigation, is a serious problem in many parts of the world. The Mehsana aquifer study, which is based on the careful interpretation of data and information, the development of quantified conceptual models and the preparation of two basic but appropriate numerical groundwater models, has led to an understanding of the aquifer system which allows both groundwater experts and decision makers to be aware of the outcomes of continuing over-exploitation.

10.3 VANATHAVILLU AQUIFER SYSTEM, SRI LANKA

Objective of the study: examine the potential resources of a deep limestone aquifer.

Key issues: validity of leaky aquifer pumping-test analysis to estimate the properties of the deeper aquifer; insights gained from an overflowing borehole.

10.3.1 Introduction

The Vanathavillu aquifer system in north-west Sri Lanka is chosen as a case study to illustrate how misleading conclusions can be drawn when too much reliance is placed on the analysis of pumping tests. In the Vanathavillu study area there are two significant formations, a deeper limestone aquifer (which elsewhere in Sri Lanka is an important source of water), overlain by a formation consisting of clays, silts and sands. In detailed studies of the limestone aquifer, resource assessment of the aquifer system relied primarily on leaky aquifer pumping test analysis. Information presented below is based on a paper by Lawrence and Dharmagunwardena (1983), earlier project reports and visits to the study area.

Figures 10.11 and 10.12 contain a plan and a representative west–east cross-section of the aquifer system of the study area. Figure 10.12 indicates that there are five distinct stratigraphic units:

Basement Complex: this is of Precambrian age and dips to the west; it has no influence on the groundwater flow in the Vanathavillu Basin.

Mannar Sandstone: this stratum lies unconformably on the basement rocks. Where it is beneath the limestone, there is little significant groundwater flow.

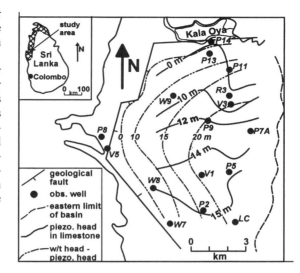

Figure 10.11 Plan of Vanathavillu study area; contours of piezometric head in limestone aquifer (unbroken line) and water table elevations minus piezometric heads (chain-dotted line). Reprinted from *Journal of Hydrology* **63**, Lawrence and Dharmagunwardena, vertical recharge to a confined limestone in northwest Sri Lanka, pp. 287–97, copyright (1983) with permission from Elsevier Science

However, to the east where the limestone wedges out, inter-granular flow in the sandstone becomes more important.

Vanathavillu Limestone: two units can be identified; an upper sequence consisting mostly of inter-bedded highly-porous reef limestone and massive partly-crystalline limestone with hard siliceous limestone and clay as minor members. The lower sequence consists of arenaceous limestone which has a gradational boundary with the underlying Mannar Sandstone. During drilling of the upper unit, loss of circulation indicated zones of major fissuring. Boreholes drilled into the Vanathavillu Limestone support yields of up to 2 Ml/d.

Moongil Aru Formation: this formation consists of a series of clays, sands and silts with occasional thin argillaceous limestones; it thickens to more than 150 m in the west. In the eastern part of the basin, the formation consists of inter-bedded thin silty sands and gravels with clays; it is probably a more permeable sequence than in the west.

Soil cover: the soils are mostly sandy and moderately permeable. Alluvium is present along the Kala Oya Delta in the north.

Faulting occurs in the study area; the probable locations of the faults are indicated in Figures 10.11 and

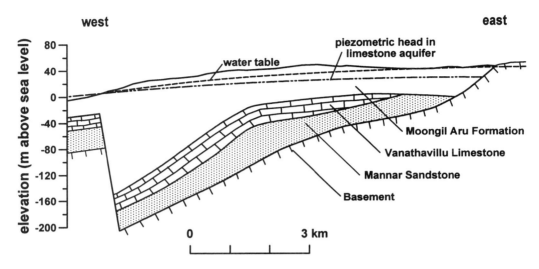

Figure 10.12   Representative cross-section through Vanathavillu study area. Reprinted from *Journal of Hydrology* **63**, Lawrence and Dharmagunwardena, vertical recharge to a confined limestone in northwest Sri Lanka, pp. 287–97, copyright (1983) with permission from Elsevier Science

Table 10.3   Details of pumping tests; methods of analysis, H, Hantush (1956), W, Walton (1962) and Th, Theis (1935). Reprinted from *Journal of Hydrology* **63**, Lawrence and Dharmagunwardena vertical recharge to a confined limestone in northwest Sri Lanka, 287–97, copyright (1983) with permission from Elsevier Science

Site	Obs well	Method	Trans (m²/d)	Storage	B (m)	K' (m/d)	Remarks
V1	V14	H	727	0.00006	1005	0.04	Only small drawdown in
		W	737	0.00009	784	0.07	water table, 3-day test
V3	V3-1	H	1735	0.0003	1592	0.027	3-day test, further results
		W	1639	0.00035	1114	0.053	in Table 10.4
P2	P2a	Th	483	0.000057			420-min pumping only
P9	A3	Th	734	0.0037			24-hr pumping
P13	P13a	W	2590	0.00026		0.066	350-min pumping

10.12. The throw of the fault in the west of more than 120 m means that the permeable limestone is not continuous across this fault.

10.3.2 Aquifer parameters

Aquifer parameters were determined from pumping tests. Information about the pumping tests can be found in Table 10.3, these parameter values are based on the analyses of Lawrence and Dharmagunwardena (1983). Table 10.4 provides more detailed information about one of the pumping tests. Most of the analyses are based on leaky aquifer curve matching procedures either of Hantush (1956) or Walton (1962). From these

analyses, together with specific capacity tests, a transmissivity distribution was derived with values of more than 1500 m²/d in the north and north-east to 100–500 m²/d in the west and south.

Leaky aquifer pumping tests (see Section 5.8) also provide estimates of the leakage coefficient B which is defined as

$$B = \sqrt{Tm'/K'} \qquad (5.22)$$

where T is the aquifer transmissivity, with m' and K' the thickness and vertical hydraulic conductivity of the aquitard. Consequently estimates of the vertical hydraulic conductivities can be derived from the expression

Table 10.4 More detailed examination of pumping test at borehole V3

Obs. well	Distance (m)	Method	Trans (m^2/d)	Storage	B (m)	K' (m/d)	Duration of match
V3-1	78	H	1735	0.0003	1592	0.027	4 to 120 min
		W	1639	0.00035	1114	0.053	8 to 150 min
A3	576	H	2565	0.00013	3194	0.0075	18 to 180 min
		W	2786	0.00014	5273	0.006	12 to 180 min
R3	562	H	2407	0.00009	4323	0.004	15 to 180 min
		W	2857	0.00009	5620	0.003	13 to 180 min

$$K' = Tm'/B^2 \qquad (10.3)$$

When assessing the results from pumping test analysis it is essential to look carefully to see whether there are inconsistencies. The results of Table 10.3 appear to be plausible with consistent estimates from the alternative methods of Hantush or Walton. Values of the transmissivity are high, often exceeding 1000 m^2/d. Unless an aquifer has major continuous fissures, a transmissivity of 500 m^2/d is a typical maximum for aquifers with a thickness of 20 to 40 m. However, it is the values of the vertical hydraulic conductivity which are not consistent with the field description of the Moongil Aru Formation as a series of mainly clays with sands and silts. The vertical hydraulic conductivities in Table 10.3 vary from 0.027 to 0.07 m/d; hydraulic conductivities of this magnitude usually apply to a sand containing some clay. If these are valid vertical hydraulic conductivity values, recharge could move without difficulty to the underlying limestone aquifer so that the difference in groundwater head between the Moongil Aru Formation and the Vanathavillu Limestone would be no more than a fraction of a metre compared to the actual difference of 10 to 20 m (Figure 10.11).

Consequently, a more detailed examination of the actual analyses of the pumping test at borehole V3 was conducted; the findings are summarised in Table 10.4. The duration of the test at V3 was 3 days, drawdowns are available at three observation boreholes. As indicated in the final column of Table 10.4, a review of the actual curve matching showed that, although a match was obtained for the earlier part of the curves, beyond 180 minutes (out of a test lasting 4320 minutes) no match was possible. There are two contributing factors:

- Before the start of the test the piezometric head in the limestone aquifer was 20 m below the water table in the overlying aquifer. At the nearest observation well, V3-1, the pumping leads to a difference between the water table and piezometric head of only 0.4 m, while at the other observation wells the head change

due to pumping is much smaller. However, classical leaky aquifer theory assumes that the piezometric head before pumping is identical to that of the overlying water table. Therefore one of the assumptions of classical leaky aquifer analysis is violated.

- In the analysis, no account is taken of the aquitard storage. The significant impact of the storage of aquitards is considered in Section 7.5; one of the findings is that *drawdowns during the early stages of pumping when aquitard storage is included can be as little as one tenth of the value when aquitard storage is ignored.* Consequently, when the field drawdowns, which are influenced (indeed almost dominated) by aquitard storage, are analysed using a method which ignores aquitard storage, the transmissivity and vertical hydraulic conductivity of the aquitard will appear to be much higher than the true field values.

Due to these two contributing factors, the values of transmissivity and vertical hydraulic conductivity quoted in Tables 10.3 and 10.4 are too high and should not be used for groundwater resource assessment.

10.3.3 Groundwater head variations and estimates of aquifer resources

The relationship between rainfall recharge and groundwater head fluctuations can provide insights into the flow processes within the aquifer. Water table fluctuations in the Moongil Aru Formation and piezometric head variations in the Vanathavillu limestone for a sixteen-month period are plotted in Figure 10.13. A number of insights can be gathered from these plots.

- Water table fluctuations indicate that recharge commences one or two months after the start of the rainy season, i.e. in October/November of the first year and September/October of the second year. Recoveries in the *water table* elevation due to recharge are 4 to 6 m.
- Groundwater (piezometric) heads in the deeper

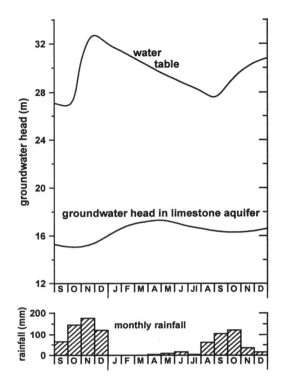

Figure 10.13 Variation with time of water table elevation and groundwater head in limestone aquifer for a sixteen-month period. Reprinted from *Journal of Hydrology* **63**, Lawrence and Dharmagunwardena, vertical recharge to a confined limestone in northwest Sri Lanka, pp. 287–97, copyright (1983) with permission from Elsevier Science

limestone aquifer do not appear to respond to recharge. Following the discussion in Section 10.2.3 concerning the Mehsana aquifer, the limestone aquifer, which is overlain by a low-permeability zone typically 50 m thick, responds to changes in pumping. From the hydrograph in Figure 10.13, there is no clear evidence of significant seasonal pumping from the limestone aquifer.
• Groundwater heads in the deeper limestone aquifer at this location are 12 m lower than the water table elevation.

Groundwater head contours are presented in Figure 10.11. The full lines refer to piezometric heads in the limestone aquifer; the groundwater gradient increases from about 0.3 m per km in the south to 1.0 m per km in the north. The broken line in Fig. 10.11 shows the *height* of the water table *above* the piezometric heads in the limestone aquifer. Differences in heads are also plotted on the east-west cross-section of Fig. 10.12 and the south-north schematic section of Fig. 10.14. Note that the water table heads in the Moongil Aru formation are primarily related to ground level with a seasonal fluctuation of up to 6 m.

The differences between the water table elevation and the piezometric heads in the limestone aquifer were used to postulate the flow mechanisms sketched in Fig. 10.14. To the south, where the water table is *above* the piezometric head, a 'leakage' through the Moongil Aru Formation to the limestone aquifer is assumed. This water then moves laterally through the limestone in a

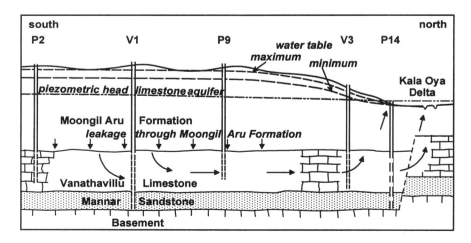

Figure 10.14 Schematic south–north cross-section showing flow mechanisms postulated by Lawrence and Dharmagunwardena (1983)

northerly direction to locations where the water table is *below* the piezometric head. In the Kala Oya Delta, the piezometric head is above the ground surface so that artesian conditions occur. Field visits showed that there is a small artesian outflow from borehole P14 in the Kala Oya Delta, but the magnitude of this flow is not consistent with the concept of a highly transmissive limestone aquifer with substantial upward flows to the Delta as indicated in Figure 10.14. Although the suggested flow mechanisms of Figure 10.14 are probably valid, the quantities of water moving through the aquifer system are lower than indicated in the figure.

Assuming that the transmissivities and vertical hydraulic conductivities have been over-estimated by a factor of five, the flow through the limestone aquifer is probably adequate to meet the current abstraction of about 4 Ml/d but a higher total abstraction is unlikely to be sustainable. Conditions in the Mehsana aquifer can provide useful insights into how the deep aquifer is likely to respond to increased abstractions. With an abstraction from the Vanathavillu limestone less than the natural aquifer through-flow, conditions are similar to Figure 10.6a. However, an increase in abstraction to, say, double the current value would require increased vertical gradients across the overlying stratum to draw water downwards; Figure 10.6b. Attempts to increase the abstraction from the Vanathavillu Limestone have met with limited success; increased pumped drawdowns have been one negative feature.

Summarising this study, the initial concept was that the important aquifer for groundwater exploitation is the Vanathavillu Limestone aquifer. Detailed field investigations were used to develop a conceptual model of the flow processes; the flows were quantified using parameters deduced from pumping tests. Unfortunately, conventional leaky aquifer test analysis, with no allowance for aquitard storage, was used. In practice the storage properties in the aquitard are significant. Consequently the transmissivity and vertical hydraulic conductivity values are over-estimates; this leads to an over-estimate of the available resources of the limestone aquifer. Although the resources of the deep limestone aquifer are limited, the overlying lower permeability aquifer stores much of the recharge. Exploitation of this shallower aquifer using large diameter wells has proved to be successful.

10.4 SAN LUIS POTOSI AQUIFER SYSTEM, MEXICO

Objectives of the study: identify the sources of water in the deeper aquifer; consider possible contributions due to runoff from the valley sides and from deeper thermal water.

Key issues: temperature and chemical composition of water indicate where thermal water may be contributing to the aquifer system; development of a plausible water balance for the aquifer system.

The San Luis Potosi aquifer system in Mexico is selected as a case study to illustrate how, even though the available field information is limited, it is possible to develop conceptual models. These conceptual models are used for preliminary resource estimation.

The groundwater system in San Luis Potosi, Mexico, has a number of noteworthy features. The aquifer is multi-layered with certain strata acting as aquifers and others as aquitards as shown in the schematic east–west cross-section in Figure 10.15. Abstraction occurs from both shallow and deep aquifers. There are several possible sources of water feeding the aquifer system; the hydrochemistry and the temperature of the water indicate that some water originates from deep sources. Much of the field information presented below is taken from the PhD thesis of Carrillo-Rivera (1992); insights into similar situations of mountainous terrain can be found in Forster and Smith (1988).

In the representative west–east cross-section of Figure 10.15 the geology is simplified as follows.

- Underlying formations include Rhyolites, Ignimbrites, Latites and Intrusives; the irregular surface is due to extensive faulting.
- In the valley there is a granular unit which contains mainly volcanic debris, gravels, sands and silts in various degrees of consolidation with an irregular distribution; the base of the granular unit can be more than 300 m below ground level.
- Within the granular sequence in the middle of the valley there is a thick bed of fine-grained compacted sand; the thickness of this low permeability bed varies between 70 and 100 m.
- On the valley sides there are piedmont sediments consisting of conglomerates, sands, gravels and residual soils.

Figure 10.15 also contains information about the groundwater heads in the aquifer system. There is a shallow water table overlying the low permeability bed with a deeper piezometric head associated with the deeper aquifer below the low permeability bed. The groundwater heads in the deeper aquifer are influenced by major abstractions. These two aquifer systems are considered individually below.

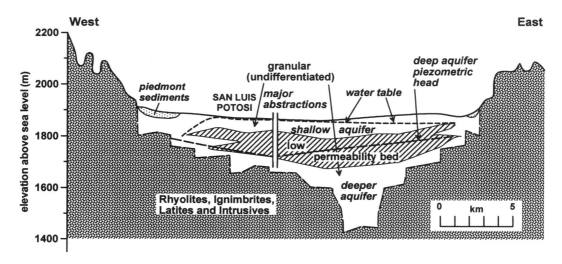

Figure 10.15   Schematic west-east cross-section through San Luis Potosi aquifer system

10.4.1 Shallow aquifer system

The shallow aquifer system consists of granular material above the low permeability bed. Recharge occurs due to precipitation, return flow from irrigation and wastewater disposal. Abstraction occurs from many shallow wells. Water table fluctuations are small with no clear evidence of a decline in the water table elevation. Over much of the area the depth to water table is less than 10 m but to the west it becomes deeper due to the ending of the low permeability bed.

The temperature of the shallow groundwater is 20.9 ± 1.0°C compared to a summer rainy season air temperature of 20.5°C. It is possible that some water moves from the shallow aquifer into the deeper aquifer through the low permeability bed; this flow must be small otherwise there would be a continuing decline in the water table. On the edges of the low permeability bed, where it becomes thin, the water table is lower; this indicates that there is a flow from the shallow to the deeper aquifer.

10.4.2 Deeper aquifer system

Significant abstractions occur from the deeper aquifer system; total abstraction increased from 75 Ml/d in 1960 to 163 Ml/d in 1977 and to 225 Ml/d in 1987. Of this abstraction of 225 Ml/d, 144 Ml/d is for public water supply, 55 Ml/d for agriculture and 26 Ml/d for industry. Abstractions are concentrated in the vicinity,

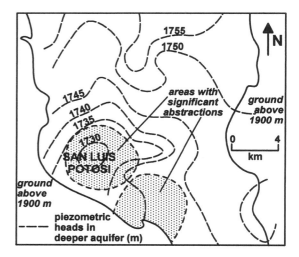

Figure 10.16   Plan of San Luis Potosi aquifer system; high ground to the west and east, contours refer to piezometric heads in the deeper granular aquifer

and to the south-east, of San Luis Potosi; see Figure 10.16.

Groundwater heads in the deeper aquifer are more than 100 m below ground level over much of the study area, increasing to 140 m below ground level in the high abstraction area of San Luis Potosi (Carrillo-Rivera *et al.* 1996). Figure 10.16 contains contours of the

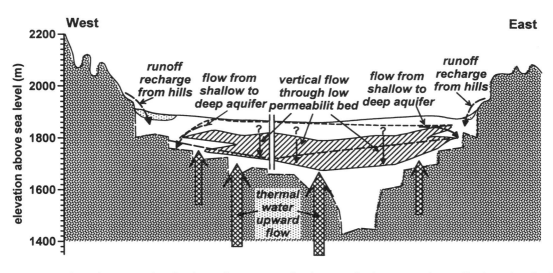

Figure 10.17 Schematic cross-section showing various sources of recharge to the deeper granular aquifer (note that the size of the arrows does not signify the magnitude of the flows)

piezometric head; piezometric heads are also shown on the cross-section of Figure 10.15. A continuing decline in piezometric heads has occurred; during the period 1977 to 1987, the decline varied from 10 m to 30 m in areas of significant abstraction.

There are several sources of inflow to the deep aquifer; see Figure 10.17.

(i) Runoff from the hills can enter directly into the aquifer system (for a methodology to estimate recharge due to runoff from hills, see Stone *et al.* 2001). If runoff occurs for 60 days at an average rate of 10 m^3/d per metre length at the base of the hill-slope (equivalent to 0.12 litres/sec per metre length), when averaged over the year this becomes 1.7 m^3/d/m (this is an order of magnitude estimate).

(ii) Also, because the low permeability bed does not extend to the edges of the granular aquifer, some flow can occur from the shallow to the deeper aquifer; a nominal figure of 10 Ml/d is chosen for this component.

(iii) Another possible route for recharge is vertical flow through the low permeability bed; this is shown in Figure 10.17 by vertical arrows with question marks. There is a substantial vertical hydraulic gradient between the water table and the piezometric head of the deeper aquifer. If the vertical

hydraulic conductivity is 0.0002 m/d with the hydraulic gradient across the low permeability bed between 0.5 and 1.0 (see the difference between the water table and piezometric heads in Figure 10.15), the vertical flow is 0.1 to 0.2 mm/d. This is equivalent to between 36 and 73 mm per year; this estimate is consistent with recharge less abstraction from the shallow aquifer. Consequently the vertical flow through the low permeability bed is taken as a maximum of 0.15 mm/d.

(iv) Due to the decline in groundwater head in the lower aquifer, and the fact that unconfined conditions develop over part of the study area as the water table falls, the release of water from storage is an important component. The unconfined area beneath the low permeability bed increases as the groundwater head falls.

Assuming that the plan area of the aquifer unit is 200 km^2, the following *very approximate* water balance can be deduced:

(i) runoff from hills, 1.7 m^3/d/m for a   68 Ml/d
    length of 40 km
(ii) flow from shallow aquifer at margins   10 Ml/d
    of low *K* bed (say)
(iii) flow through overlying low *K* bed,   30 Ml/d
    0.15 mm/d × 200 km^2

(iv) fall in water table, 2 m per year for   27 Ml/d
    50 km^2 with $S_Y = 0.1$
    total   ***135 Ml/d***
    estimated abstraction   225 Ml/d

Although the above quantities are approximate, they are unlikely to be over-estimates; consequently a further source of water must be identified.

10.4.3 Input of deep thermal water

A detailed study of the quality and temperature of waters from deeper aquifers is presented by Carrillo-Rivera (1992); by examining the major and minor ions and the temperature of water samples, the probable origin of the water can be identified. In this discussion, reference will be restricted to the fluoride content and the temperature of the deep aquifer water.

To assist in the analysis, the aquifer waters are divided into three groups:

Group 1: regional flow water from volcanic rocks, thermal water 36–41°C,
Group 2: water mainly originating from the shallower granular undifferentiated strata, relatively cold water 23–27°C,
Group 3: mixed flow from Groups 1 and 2, slightly to moderately thermal (27–35°C); this water is from both granular and volcanic material.

Three figures are selected to illustrate the findings:

- Figure 10.18 shows the relationship between temperature and fluoride for samples from the deeper aquifer,
- Figure 10.19 is a plan of the aquifer system showing the temperatures of the deeper aquifer water; areas are demarcated where the temperatures are 30 to 36°C or 36 to 41°C,
- Figure 10.20 records changes in temperature and fluoride content during a pumping test in the deeper granular aquifer.

An examination of Figure 10.18 shows that there is a reasonable correlation between temperature and

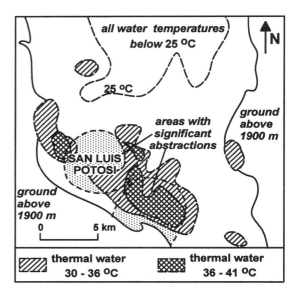

Figure 10.19   Temperature distribution in the deeper granular aquifer

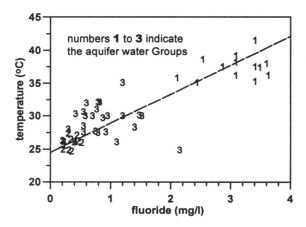

Figure 10.18   Relationship between temperature and fluoride concentration for the deeper aquifer system

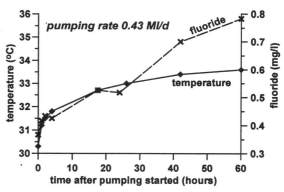

Figure 10.20   Change in temperature and fluoride concentration during a pumping test in the deeper granular aquifer

fluoride with Group 1 water having a higher temperature and fluoride content while Group 2 water has a far lower temperature and fluoride content. Group 3 samples, the mixed water, lie between Groups 1 and 2; for some of the mixed samples the correlation is poor. Fluoride is considered to originate from tertiary volcanic rocks.

The distribution of temperature in the deeper aquifer, Figure 10.19, provides a clear indication that thermal water enters the deep aquifer system. Group 1 water, with temperatures in the range 36 to 41°C, are found in regions with significant abstraction. It is only in the north that the temperature of the water is below 25°C; elsewhere much of the water is mixed, suggesting there may be a natural flow of thermal water from deep within the aquifer system mixing with water which enters on the margins of the aquifer system or through the overlying clay.

The changing temperature and fluoride content during a pumping test, Figure 10.20, supports the concept that the upward movement of thermal water is increased due to pumping. Within two days from the start of the test, the temperature of the pumped water increased by more than 3°C while the fluoride concentration more than doubled.

All this evidence supports an upward movement of thermal water into the deeper granular aquifer as indicated in the conceptual model of Figure 10.18. However, because the origin of this thermal water is unknown, with all the samples to some extent mixed samples, it is not possible to calculate the magnitude of the inflow of thermal water. However, the input of thermal water can explain the lack of balance of the approximate water balance of Section 10.4.2; the lack of balance is 90 Ml/d compared to a total abstraction of 225 Ml/d.

10.4.4 Further considerations

This brief summary of data gathered for the San Luis Potosi aquifer shows how the information can be interpreted to develop a conceptual model. This is supported by a tentative water balance calculation. Two questions need to be addressed.

First, is there presently sufficient information to prepare a mathematical model, or could a mathematical model be developed following further fieldwork? There is little quantified information about the aquifer system. In terms of aquifer properties, the vertical hydraulic conductivity of the low permeability bed is unknown, although estimates have been made of the transmissivities of the shallow and deeper aquifers.

Furthermore, the magnitudes of the inflows to the aquifer system are largely unknown. Estimates of recharge to the shallow aquifer could be made, but extensive fieldwork would be required to estimate the runoff from the hills and the proportion of that runoff which enters the aquifer system. Nor is there an adequate understanding of the source and inflow mechanisms of the thermal water. No doubt some form of mathematical equation could be devised to represent the inflow of thermal water and thereby develop a 'calibrated model' for current conditions in the granular aquifer system. However, the model is unlikely to represent the true physical processes of the release of thermal water. The lack of extensive historical data means that it is not possible to check the adequacy of any numerical model. Therefore any predictions of future aquifer response using a mathematical model would be open to question.

Second, is there sufficient field evidence to plan for the future exploitation of the aquifer? There are warning signs, such as a decline in the piezometric head and increasing temperature of the water. It is probable that the mechanism of vertical flow through low permeability layers (similar to that identified in the Mehsana aquifer; Figure 10.6b) is occurring for both downward flow from the shallow aquifer and upward flow of thermal water; increased abstraction rates will lead to more water being drawn from these sources. Continuing monitoring of groundwater heads, temperature and water chemistry is required. A reduction in abstraction from the deeper aquifer to achieve stable conditions is advisable.

10.5 BROMSGROVE SANDSTONE AQUIFER, UK

Objectives of the study: determine whether the abstraction from the aquifer system is sustainable; consider the apparently anomalous features of a steep groundwater gradient yet moderate to high yield of boreholes.

Key issues: insights from historical conditions when a large number of springs existed; importance of different responses of shallow and deep observation boreholes.

In the Bromsgrove Sandstone aquifer the water table gradient is sometimes steeper than 1 : 50 with a moderate recharge averaging about 1 mm/d. This steep gradient would normally indicate that the aquifer has a low transmissivity of less than 100 m^2/d and that borehole yields would be poor. However, the aquifer supports boreholes yields of 4 to 8 Ml/d. These apparently contradictory conditions are illustrated in Figure 10.21. As indicated on the cross-section, the sandstone aquifer has a saturated thickness varying from 100 to 600 m; the boreholes often penetrate 300 m to achieve these high yields. In some locations the water table intersects the ground surface, forming springs.

To understand the behaviour of this aquifer system, information concerning the geology, recharge, abstraction and groundwater heads together with spring and river flows is assembled, collated and interpreted to develop a conceptual model. The numerical model derived from the conceptual model confirms that bands of low hydraulic conductivity within the sandstone are important in understanding the aquifer response. Further information about the aquifer system can be found in Rushton and Salmon (1993).

10.5.1 Summary of field information

The Bromsgrove sandstone aquifer, which is about 20 km south of Birmingham, the UK, covers an area of 51.5 km² ; it is defined by a number of major faults. Ground elevations vary from more than 250 m to below 100 m; Figure 10.22. The main streams draining the aquifer are the Spadesbourne, Battlefield Brook and Tenni Brook. In the early 1900s, prior to major exploitation of the aquifer, there were many springs along the valleys; the perennial flows were sufficiently reliable for the operation of a number of water mills. These water mills were vital to the development of small industries in the area; as indicated in Figure 10.22 there were fifteen water mills operating in 1900.

The Triassic Sandstone of the study area is of fluvial origin; the dip is between 1° and 5° to the south. Major faults isolate the sandstone from adjacent lower permeability strata and also cause substantial changes in the sandstone aquifer thickness; Figure 10.21. Three sandstone strata have been identified in the study area. Borehole cores indicate that there are extensive mudstone and marl bands within the sandstone. Typical these bands are 1 m thick but some of the low permeability zones have a thickness of several metres.

Abstraction from the aquifer increased steadily

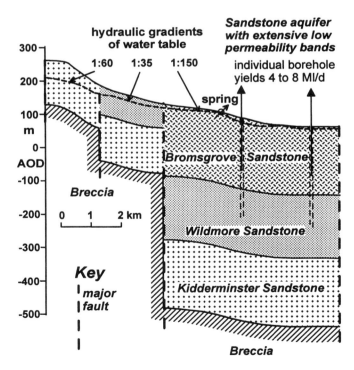

Figure 10.21   Illustration of the apparently inconsistent conditions of steep water table gradients but significant abstraction from production boreholes in Bromsgrove aquifer

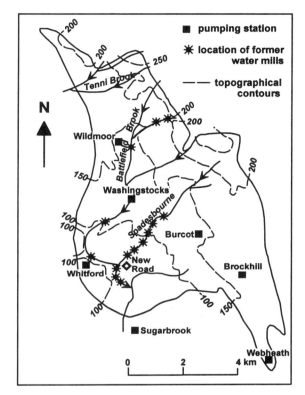

Figure 10.22 Map of the outcrop of the Bromsgrove sandstone aquifer. Reprinted from *Journal of Hydrology* **152**, Rushton and Salmon, significance of vertical flow through low-conductivity zones in Bromsgrove sandstone aquifer, pp. 131–52, copyright (1993) with permission from Elsevier Science

during the twentieth century. Table 10.5 lists the abstraction from the seven pumping stations during the period 1965–91; the locations of the pumping stations are shown in Figure 10.22. Most of the pumping stations have multiple boreholes, although many of the boreholes are for backup purposes. Two new pumping stations were commissioned, one in the 1970s and the other in the 1980s. Even though these new pumping stations are located on the margins of the aquifer, they provide acceptable yields.

An examination of the pumped water levels in Table 10.5 shows steady declines apart from at Wildmoor Pumping Station where the relatively stable pumping levels occur because Wildmoor is located in the northern part of the aquifer with no other operational boreholes. However, at Washingstocks there has been a substantial decline in pumping levels. This is probably due to the proximity of a fault; a trial borehole failed to reach the water table within 35 m of ground level. Significant falls in pumping levels have occurred at the two new boreholes at Whitford and Webheath. Does this indicate that over-exploitation is occurring? This issue will be considered further towards the end of this discussion.

Recharge from rainfall is estimated using the soil moisture balance technique of Sections 3.3 and 3.4. Due to appreciable differences in ground elevation, the area is divided into five rainfall zones; the rainfall at Sugarbrook is multiplied by the following factors:

Elevation	Multiplying factor
<100	1.02
100–150	1.10
150–200	1.18
200–250	1.27
250+	1.39

Table 10.5 Abstraction rates and pumping water levels for the Bromsgrove Sandstone aquifer. Reprinted from *Journal of Hydrology* **152**, Rushton and Salmon, significance of vertical flow through low-conductivity zones in Bromsgrove Sandstone aquifer, 131–52, copyright (1993) with permission from Elsevier Science

Pumping station	No. of boreholes	Abstraction Ml/d				*Pumping levels (m AOD)*			
		1965	1974	1983	1991	*1965*	*1974*	*1983*	*1991*
Brockhill	2	6.0	6.7	6.4	7.1	*75.9*	*75.3*	*75.0*	*59.2*
Burcot	4	5.5	5.0	4.9	4.4	*81.2*	*83.0*	*74.5*	*66.6*
Sugarbrook	4	2.7	11.1	9.0	11.5	*17.8*	*−1.2*	*−13.5*	*−28.3*
Washingstocks	2	5.4	4.3	4.1	4.2	*61.7*	*52.4*	*44.8*	*39.7*
Wildmoor	2	7.8	7.1	7.4	7.5	*72.8*	*78.7*	*78.4*	*74.5*
Webheath	3	0.0	5.0	6.8	7.6	*–*	*58.2*	*42.2*	*26.1*
Whitford	1	0.0	0.0	4.4	4.3	*–*	*–*	*40.1*	*33.2*

These factors are based on other rain-gauge stations which have a shorter record than Sugarbrook.

When these rainfall figures are included in water balance calculations, the average estimated recharge over the high ground is 1.7 times that for the lower ground. In urban areas in the south of the study area, recharge also occurs due to the provision of soakaways for domestic property, leaking water mains and losses from sewers. This leads to an increase of 6 Ml/d in the estimated recharge; this is added to the precipitation recharge, which averages 40 Ml/d.

Although a number of groundwater head hydrographs are available in the study area, there was no location at which both the shallow and deep responses have been monitored. Therefore a shallow and a deep piezometer were drilled at New Road which is located towards the centre of a group of four major pumping stations; see the open diamond in Figure 10.22. The hydrographs are included in Figure 2.34 and discussed in Section 2.8.2. Reference to Figure 2.34 shows that the shallow piezometer, which monitors the water table fluctuations, generally recovers by 0.5 to 1.0 m during the winter recharge season of December to April; this is followed by a slow decline. Low rainfall occurred during the years 1990–92; this is reflected by a decline lasting several years. The deeper piezometer is strongly influenced by the four surrounding pumping stations; decreases in total abstraction from Sugarbrook result in a recovery in the deep piezometer (for example, December 1989). The deep piezometer is also influenced by the long-term balance of inflows and outflows.

10.5.2 Conceptual model

Figure 10.23 is a representative vertical section which illustrates the conceptual response of the Bromsgrove sandstone aquifer system. Important features of the conceptual model include:

* rainfall recharge, which is indicated as vertical arrows to the water table,
* springs which occur where the water table intersects the ground surface,
* abstraction boreholes; water is taken out from the relevant layers,
* the three sandstone strata shown in Figure 10.21 are represented by five continuous layers; the top layer represents the Kidderminster Sandstone, then the Wildmoor Sandstone and finally the Bromsgrove Sandstone,
* although not explicitly shown in Figure 10.21, there

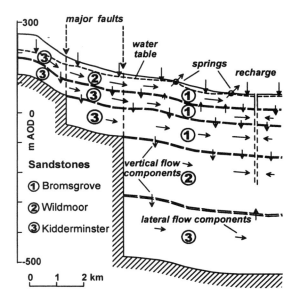

Figure 10.23   Cross-section conceptual model of the Bromsgrove aquifer system

are low permeability bands within the sandstone, they are included in the calculations of the equivalent lateral and vertical hydraulic conductivities,
* possible flow paths are shown on Figure 10.23; they are represented as lateral and vertical components. Although the magnitude and direction of these flows is unknown, they must be included in the conceptual model (note that this conceptual cross-section does not represent flows in the third dimension).

10.5.3 Mathematical model

Decisions have to be made about the form of the grid used for the mathematical model of the Bromsgrove aquifer. Unlike the Mehsana aquifer models, which use orthogonal grids, the layers are designed to follow the dip of the strata, as shown in Figure 10.24. A square grid of sides 0.5 km is used in plan. This is a pseudo-three-dimensional model (Javandel and Witherspoon 1969). The manner in which the different sandstone strata are represented as model layers is indicated in Figure 10.23.

The lateral and vertical equivalent hydraulic conductivities are calculated using the approach of Figure 10.7. For example, if in a vertical mesh interval of 50 m there are two layers of low hydraulic conductivity

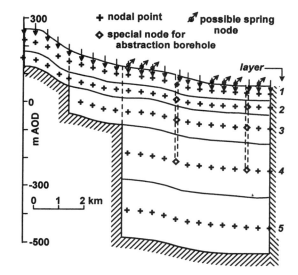

Figure 10.24 Features of the mathematical model of the Bromsgrove aquifer system including the representation of the different layers. Reprinted from *Journal of Hydrology* **152**, Rushton and Salmon, significance of vertical flow through low-conductivity zones in Bromsgrove sandstone aquifer, pp. 131–52, copyright (1993) with permission from Elsevier Science

sented as discharges from the appropriate layers (layers 2 to 4 for the boreholes represented in Figure 10.24) with the further condition that all the groundwater heads in a single borehole take the same value. This is achieved by setting a high vertical hydraulic conductivity between nodes on the vertical line representing the pumped borehole; see Section 12.3.

Starting conditions for the simulation are that the water table heads (groundwater heads in layer 1) are initially set to equal the estimated field values. Groundwater heads in the remaining layers are then calculated. This process is slow to converge since there is no abstraction from the aquifer at the start of the simulation. The achievement of consistent lateral and vertical flows throughout the full depth of the aquifer system involves a delicate water balance. Next the groundwater heads at the water table are released and allowed to converge to values consistent with the recharge, deeper flows and spring outflows. The achievement of starting conditions with consistent flows is a prerequisite of a reliable simulation. In this particular problem, successive over-relaxation was used to solve the simultaneous equations; other suitable iterative procedures can be used.

10.5.4 Presentation of model outputs

Once adequate starting conditions have been achieved, the simulation proceeds with larger time steps up to 1950 then with time steps of 10 days for the period when more detailed results are required. Even though this is an aquifer with high unconfined storage coefficients, rapid changes occur in the number and location of flowing springs, hence the need for a time step of 10 days. Comparisons with the available groundwater head hydrographs and stream flow records shows adequate agreement, but matching this limited field data does not provide a rigorous test of the validity of the simulation. Therefore the model outputs are presented in a different way to explore the validity of the aquifer response.

In Figure 10.25, groundwater contours are plotted for September 1900 and September 1984; abstractions in 1984 are indicated by filled squares. Component flows within the aquifer system for 1900 and 1984 on a roughly north–south cross-section are presented in Figure 10.26. From these diagrams, an understanding of the flow processes can be developed.

- In Figure 10.25a and b, the half-filled circles represent flowing springs; in 1900 ninety-five springs are

(K = 0.001 m/d), one 2 m thick and the other 4 m thick, with the vertical permeability of the more permeable strata equal to 1.0 m/d, the effective vertical hydraulic conductivity between the nodal points is 0.0083 m/d. In the model, the specific yield is set at 0.10 with the confined storage of each layer equal to 0.0001. The simulation covers the period 1900 to 1998; up to 1950 the time step is 100 days, it is then reduced to 10 days.

To the north, east and west the boundaries of the study area are faults which are represented as no-flow boundaries. To the south the aquifer is overlain by Mercia Mudstone. As indicated in Figure 10.22, Sugarbrook pumping station is to the south of the sandstone outcrop within the region overlain by Mercia Mudstone. Further south the groundwater is of increasing salinity which indicates little fresh groundwater flow; therefore the model is extended for 1 km south of Sugarbrook PS where a zero-flow condition is enforced.

Since there were a large number of springs before the start of exploitation of the aquifer, all low-lying nodes in layer 1 act as springs whenever the groundwater head reaches the ground elevation. Boreholes are repre-

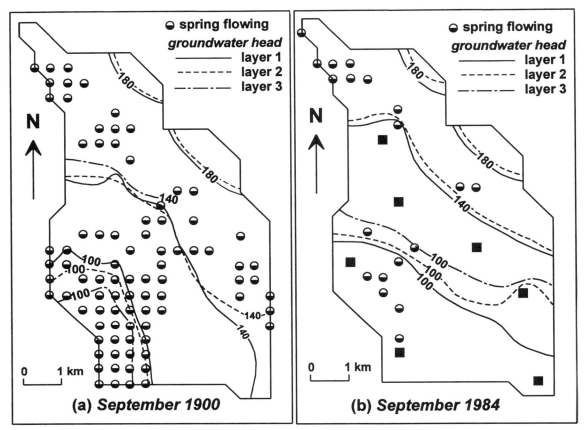

Figure 10.25   Results from the mathematical model simulation of the Bromsgrove aquifer system; all contours are in metres above Ordnance Datum. Reprinted from *Journal of Hydrology* **152**, Rushton and Salmon, significance of vertical flow through low-conductivity zones in Bromsgrove sandstone aquifer, pp. 131–52, copyright (1993) with permission from Elsevier Science

simulated as flowing whereas in 1984 there are only twenty-one flowing springs. Both of these results are consistent with field information.

- In Figure 10.25 the unbroken line represents the groundwater head in uppermost layer, layer 1, and the broken line refers to layer 2; when layer 3 is present the groundwater heads are indicated by a chain dotted line. In Figure 10.25a for September 1900 the 180 m contour for layer 2 (broken line) is to the north of that of layer 1 (unbroken line); this indicates a vertically downward flow. For the 100 m contour, groundwater heads in layers 2 and 3 are to the south of layer 1, consequently flows are upwards. The 140 m contours do not show a consistent pattern indicating that in some locations flows

are predominantly upward while in other locations they are mainly downward.

- The representative vertical section of Figure 10.26a confirms that in 1900 the predominant flow mechanism was downward in the north due to recharge with upward flows to springs in the south. Note also that water moves from the high ground in the north, through low permeability zones to depths of several hundred metres below ground level before moving upward to be released from springs in the centre and south.
- Comparisons between Figures 10.25a and b show that changes occur in positions of the contours. There is little lateral movement in the 180-m contours although the increased distance between the unbro-

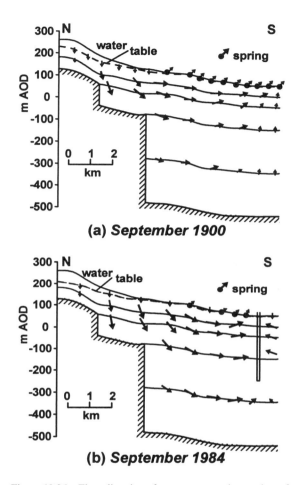

Figure 10.26 Flow directions for a representative section of the Bromsgrove aquifer system. Reprinted from *Journal of Hydrology* **152**, Rushton and Salmon, significance of vertical flow through low-conductivity zones in Bromsgrove sandstone aquifer, pp. 131–52, copyright (1993) with permission from Elsevier Science

ence of boreholes results in larger flows within the aquifer system (the size of the arrows is a rough indication of the magnitude of the flow). Whereas in 1900 the flows in the deeper parts of the aquifer were small, in 1984 the large pumped drawdowns, typically of more than 50 m, draw water from deep within the aquifer system.

Returning to the question posed earlier about the apparent inconsistency between high groundwater gradients and the high pumping rates of the production boreholes, the key is the substantial energy (and hence groundwater head) required to move water through the low permeability bands. Before the development of the aquifer, there were substantial vertical head gradients as the water moved downward and then upward across the low permeability bands; Figure 10.26a. As abstraction from the aquifer increased, pumped drawdowns of more than 50 m were required to draw water downwards and laterally to the abstraction boreholes; see Table 10.5. However, abstraction from the aquifer has not resulted in a cessation of all the spring flow; in some locations groundwater still moves upwards to leave the aquifer through springs.

10.5.5 Management issues

The purpose of this study is an assessment of groundwater resources. The total abstraction from the Bromsgrove aquifer increased from 27.4 Ml/d in 1965 to 46.6 Ml/d in 1991, an average increase of 2 per cent per year. Does this suggest over-exploitation? The outflows from the springs have reduced to an average of around 15 Ml/d, with the upper reaches of many of the streams becoming dry.

If the criterion for over-exploitation is that streams should continue to flow as they did before pumping from the aquifer commenced, then the current pumping would be classified as over-exploitation. However, if the definition of over-exploitation is mining of groundwater, the abstraction from the Bromsgrove sandstone aquifer is just acceptable, with abstraction equal to 90–95 per cent of recharge.

The continuing decline in most pumping water levels does not necessarily indicate over-exploitation. In Section 12.6.2 an expression is quoted for the time following a change in conditions before equilibrium is approached,

$$t_{eq} = 1.5 S L^2 / T \tag{12.8}$$

ken and broken lines indicates that the vertical flows are higher. It is the 100 m contours which show the greatest movement. Some of the contours have moved several kilometres, especially in the southeast. Furthermore, the 100-m contours for layers 2 and 3 are always to the north of those for layer 1, indicating a predominantly downward movement of water towards the pumped boreholes.

• Consideration of Figure 10.26b shows that the pres-

where L is the length of a typical flowline, S is the appropriate storage coefficient and T is the transmissivity. For the Bromsgrove aquifer using parameter values from Rushton and Salmon (1993),

$T = 290$ m^2/d, $S = 0.10$ and assuming a typical length of flowline of 5 km then

$t_{eq} = 1.5 \times (5000)^2 \times 0.10/290 = 12\,931$ days ≈ 35 years.

This is only an approximate calculation but it does suggest that the declining pumped water levels recorded in Table 10.5, which relate to a period of 26 years, are primarily due to the time taken for new equilibrium conditions to be reached following the commissioning of pumping stations.

10.6 FURTHER EXAMPLES WHERE VERTICAL COMPONENTS OF FLOW ARE SIGNIFICANT

10.6.1 Madras aquifer

The Minjur-Panjetty aquifer to the north of Madras City, southern India, is an important source for public water supply and irrigation. The alluvial aquifer can be idealised as an upper aquifer of sand and silt overlying an aquitard consisting mainly of clay and silt; beneath this aquitard is the lower aquifer consisting of coarse to medium sand (Elango and Manickam 1987).

Initially the upper aquifer was exploited using dug wells, but subsequently tubewells were drilled into the lower aquifer which has a thickness of about 17 m. In the lower aquifer, well fields have been constructed for industries, for augmentation of the Madras city water supply and for the irrigation of rice. The upper aquifer receives recharge from rainfall, losses from ricefields (see Section 4.5) and inflow from major rivers.

The exploitation of the lower aquifer has led to an overall decline in the water table in the upper aquifer (Figure 10.27) (from Krishnasamy and Sakthivadivel 1986). Recovery in the water table occurs during the north-east monsoon from October to January. In the lower aquifer, recovery of the piezometric heads also occurs during the monsoon period but this is primarily due to reduced pumping when rainfall is available for the rice crop. There is also a long-term overall decline in the piezometric head of about one metre per year. During the last two years recorded in Figure 10.27, the piezometric head in the lower aquifer fell below the base of the clay layer; this has serious implications. First, the groundwater head in the lower aquifer is below sea level thereby increasing the risk of saline intrusion. Second, the lower aquifer has become disconnected from the overlying clay layer and the upper aquifer. Flow through the clay layer decreases due to a reduced hydraulic gradient; furthermore cones of depression form in the lower aquifer around the pumped tubewells. Krishnasamy and Sakthivadivel

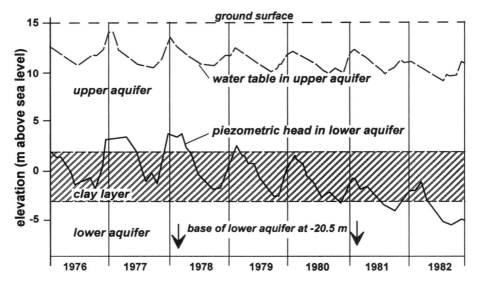

Figure 10.27 Groundwater heads in the upper and lower aquifers in Madras

(1986) developed a numerical model to represent this situation and showed that dewatering of the lower aquifer is a continuing process requiring further deeper replacement tubewells.

10.6.2 Waterlogging in Riyadh, Saudi Arabia

High water tables leading to waterlogging were not anticipated in Riyadh, Saudi Arabia, since the rainfall is low and the city is in the middle of a desert. There are two main causes of the rising groundwater levels in Riyadh. First, recharge to the aquifer system is high with recharge components due to losses from water mains, leaking domestic underground water storage tanks, the use of septic tanks and the watering of public parks, roadside verges and gardens. The smallest recharge intensities are estimated to be 2.5 mm/d with recharge of more than 10 mm/d where there is intense watering of parks and gardens (Rushton and Al-Othman 1994). The second cause of waterlogging is the low vertical permeability of the Arab Aquitard which is present over large areas of the city. The Arab Aquitard has vertical hydraulic conductivities in the range 0.0002 to 0.003 m/d.

Controlling waterlogging by lowering the water table

has proved to be difficult. Horizontal drainage (see Section 4.4) is not successful due to the low hydraulic conductivity of the Arab Aquitard. Water table control is especially important where a three-lane dual carriageway through the centre of Riyadh is located in a cutting which is mainly in the Arab Aquitard (Figure 10.28). During road construction, initial attempts at dewatering the Arab Aquitard using boreholes penetrating 20 m into the aquitard proved to be unsuccessful due to their small radius of influence of about 10 m. It was by constructing boreholes into the underlying Arriyadh aquifer, at 15 m centre on either side of the cutting, that slow dewatering was achieved which permitted excavation of the cutting and construction of the road.

Large quantities of water were pumped from these dewatering boreholes which penetrated to the Arriyadh aquifer; the maximum total discharge reached 75 Ml/d. This high pumping rate was necessary because dewatering required a substantial hydraulic gradient across the Arab Aquitard to cause the vertical downwards flow through the aquitard. The effect of pumping extends for more than 1 km on either side of the cutting due to the transmissivity of the Arriyadh aquifer of about 1000 m²/d. Fortuitously, this results in

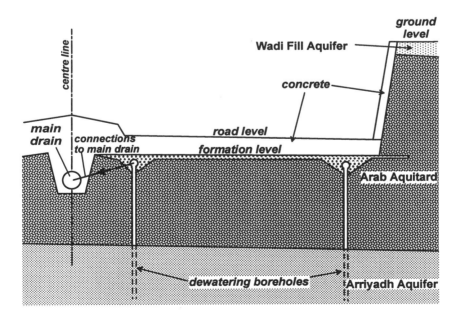

Figure 10.28   Cross-section through King Fahd Road, Riyadh, showing the dewatering wells into the Arriyadh aquifer which maintain a depressed water table

groundwater control over a considerable area in Central Riyadh.

Long-term pumping to control the water levels is not an acceptable strategy for the road cutting, yet if there is no withdrawal of water, the concrete base and cutting sides act as a 'boat' which will float if the water table is allowed to returns to its original level. Groundwater control is achieved using the existing dewatering boreholes with additional boreholes into the Arriyadh aquifer. Water from these boreholes is transferred to the main drain (one connection is shown in Figure 10.28) from which water is pumped. As this system was commissioned, the quantity pumped from the main drain slowly reduced to 20 Ml/d.

10.6.3 SCARPS: Saline Control and Reclamation Projects in Pakistan

A report by the Geological Survey of the United States (Greenman *et al.* 1967) highlighted the serious waterlogging problems in Pakistan: 'Rising water tables and the salinisation of land as a result of canal irrigation threaten the agricultural economy of the Punjab.' The report explains how canal irrigation, which commenced around 1850, is the principal cause of rising water levels and constitutes the major component of groundwater recharge in the Punjab. Most of the recharge originates from water lost from the main canals, less from the smaller canals and field channels. Reference should be made to Section 4.2 for information about the mechanisms of losses from canals. The report also describes a long-range programme for reclaiming the irrigated lands of the Punjab: 'The essential feature of the programme is a proposed network of tubewells located with an average density of one per square mile (2.6 km^2). Where the groundwater is of acceptable quality, the wells discharge into the canal system with the pumping rate of each well determined by the supplemental irrigation requirements of the surrounding land. Thus the groundwater withdrawals serve the dual purpose of satisfying irrigation requirements and providing subsurface drainage.' Reference is also made to areas where the quality of groundwater is unsatisfactory; the water is 'to be disposed in drainage ditches', but there is no mention of what would ultimately happen to this saline water.

Did this approach to saline control and reclamation succeed? Reference has already been made to the successful use of *horizontal drains* in Mardan SCARP (Saline Control and Reclamation Project): see Section 4.4. To examine the viability of water table control

using *vertical drainage*, reference will be made to SCARP 1. Initially more than 1300 tubewells were installed in the alluvial aquifer along the main canals. However, due to recirculation of water, these tubewells were not successful in controlling waterlogging. In the full SCARP 1 project, which covers an area of 5000 km^2, 2044 tubewells were installed in the interfluve areas of the alluvial aquifer above the flood plains between the Ravi and Chenab Rivers. Capacities of the tubewells were between 4.9 and 12.2 Ml/d; the total design discharge of the tubewells equalled 14 600 Ml/d compared to the canal supply of 6400 Ml/d. The effectiveness of the tubewells in SCARP 1 can be gauged from two diagrams:

• Figure 10.29a shows the average abstraction from SCARP 1 tubewells. It was during 1960–63 that construction of the tubewells occurred. At the end of this period the annual average abstraction was 9270 Ml/d compared to the design discharge of

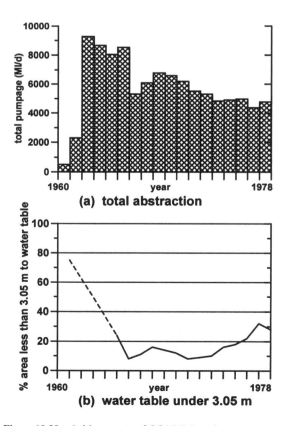

Figure 10.29   Achievements of SCARP 1 project

14 600 Ml/d. This rate was almost maintained for a few years; subsequently there was a decline so that from 1973 to 1978 the average discharge was consistently less than 5000 Ml/d.

- An alternative method of assessing the success of the SCARP tubewells is presented in Figure 10.29b which shows the percentage area with the water table below 10 feet (3.05 m) during June immediately before the monsoon season. By 1966, only 8 per cent of the total area had the water table within 3.05 m of the ground surface but the area at risk gradually increased until in 1977 the area totalled 32 per cent.

There are several reasons for the reductions in groundwater discharge and the increasing area with the water table within 3.05 m of the ground surface.

1. Closure of tubewells due to failure of the borehole (screen, gravel pack etc.), failure of the pump and pumping brackish water.
2. Declining yield of tubewells to typically 70 per cent of design capacity; as explained in the study of the Mehsana aquifer in Section 10.2, interference between adjacent boreholes is a common reason for declining yields.
3. Loss of operation time caused by electrical or mechanical faults or conveyance channel faults such as breaching of the water-course channels.
4. Increased losses from canals as the water table is lowered (see Section 4.2); losses from canals can double as the vertical hydraulic gradient from canal water level to the underlying groundwater head increases.
5. Upconing of poor quality water from depth (see Section 2.9.4).

Awan (1985) considered the limited success of the SCARP 1 project and suggested that the original objective of reducing water tables to pre-irrigation levels is not feasible. He adds that 'the problem of waterlogging and salinity did not diminish in the same degree as the amount of huge investment made'. A further significant item of information from Awan's review is that the utilisation factor for public tubewells was 35 per cent compared to the 84 per cent envisaged in the original plans. A study of tubewell design in Pakistan and proposals for achieving long-term reliable yields is presented by Bakiewicz *et al.* (1985).

Due to the unreliability of the SCARP tubewells, a number of farmers constructed their own tubewells so that they could guarantee irrigation water even though the SCARP tubewells were not operating. By 1978

there were almost 4000 private tubewells which are typically 30 m deep with yields of 1.2–2.4 Ml/d. The number of private tubewells continued to increase; by the early 1990s there were ten private tubewells for every SCARP tubewell, with about 75 per cent of the total abstraction from the private tubewells. This resulted in a limited decrease in the area with the water table within 3.05 m of the ground surface. However, in areas where the pumped water is brackish or saline, water table control is not successful. Most government tubewells have now been handed over to the private sector or replaced by subsidised private tubewells. Initial results suggest that water table control is being achieved. Nevertheless, locations with unsatisfactory groundwater quality still present serious problems. One solution to the disposal of saline water is provided by the Left Bank Outfall Drain project (McCready 1987) in which a spinal drain about 250 km long is used to transport saline water to the Arabian Sea.

10.6.4 Fylde aquifer, UK

An integrated, multi-functional approach required for the investigation of water resources is demonstrated by the Fylde aquifer project in central Lancashire, England (Seymour *et al.* 1998). This discussion will focus on the representation of recharge through drift to the underlying sandstone aquifer.

The Sherwood Sandstone of the Fylde aquifer is bounded in the east and underlain by Carboniferous strata comprising inter-bedded mudstones, shales, sandstones and limestones (Figure 10.30). Permo-Triassic sediments were deposited on the Carboniferous; the Sherwood Sandstone Group is the main aquifer. From its contact with the Carboniferous in the east, the thickness increases to over 500 m; it is then downthrown by up to 600 m beneath the siltstones and mudstones of the Mercia Mudstone Group. The Sherwood Sandstone Group is predominantly a fine-to-medium-grained sandstone with occasional inter-bedded mudstone (marl) beds. The sandstone aquifer is almost entirely covered by drift deposits, which are mainly inter-bedded boulder clay (till) with sands and gravels of glacial origin. The thickness of the drift varies from 5 to 30 m.

Since the 1970s, the Fylde aquifer has been part of the Lancashire Conjunctive Use Scheme in which upland reservoirs, river abstractions, river transfers and borehole sources are used conjunctively to utilise the cheaper and more abundant surface water when available, but relying on groundwater to meet shortfalls especially in times of drought (Walsh 1976). Since the

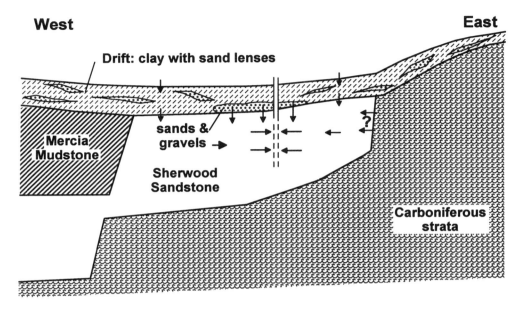

Figure 10.30 Schematic cross-section through the Fylde aquifer

groundwater resources are only called upon in drought years, skill is required in utilising these resources. Abstractions are licensed according to three-year 'rolling' totals; abstraction is spread over the whole aquifer by using groups of boreholes. To avoid the risk of saline intrusion, groundwater gradients must be positive towards the boundaries. River augmentation is used to lessen the impact of groundwater abstraction with 'hands-off' conditions enforced at certain observation boreholes to prevent derogation of other groundwater sources.

Due to the extensive low-permeability drift cover, the *original* conceptual model of the Fylde aquifer was that the vertical flow through the drift would be negligible but that pumping would draw water from the Carboniferous strata to the east; this is indicated on Figure 10.30 by arrows with a question mark. A mathematical model was derived in which the Carboniferous strata provided most of the water pumped from the sandstone aquifer.

This initial analysis proved to be unreliable. A careful study of the Carboniferous strata shows that, due to faulting and the low permeability of many of the strata, little water can be drawn into the sandstone aquifer from the Carboniferous. Furthermore, it is possible to draw water downward through the low permeability drift in a similar manner to the Mehsana aquifer

(Section 10.2). In a recent study (Seymour *et al.* 1998), conceptual and mathematical models of the Sherwood sandstone aquifer are developed with an upper layer representing the drift. As abstraction occurs from the sandstone aquifer, water is drawn down through the drift. These flows are represented in Figure 10.30 as small arrows at the top and bottom of the drift.

However, certain observation borehole responses at some distance from the major abstraction sites were not reproduced adequately by the model. When a single layer is used to represent the vertical leakage through the drift, the storage properties of the drift are ignored. As shown in Section 7.5.4, failure to represent the storage properties of the drift can lead to incorrect simulations. A more detailed investigation showed that towards the bottom of the drift there are extensive sand and gravel deposits, as indicated Figure 10.30. When substantial abstraction occurs from the sandstone aquifer, the source of water most easily accessible is from the sand and gravel deposits towards the bottom of the drift. This means that it is easier to attract water than if it has to be drawn through the full thickness of the low permeability drift layer. Local dewatering of these sand and gravel deposits has subsequently occurred. Consequently, larger pumped drawdowns are now required to draw water through the drift into the sandstone aquifer and then into the pumped boreholes.

When the true nature of the drift is represented in the numerical groundwater model, improved simulations are achieved.

10.7 CONCLUDING REMARKS

Vertical flow components are crucially important in understanding flow processes in many regional groundwater flow problems. The four major case studies in this chapter illustrate issues and techniques involved in developing appropriate conceptual models. For two of the case studies, numerical models are developed. For the other two case studies quantified conceptual models provide sufficient information to make management decisions.

The first case study relates to an alluvial aquifer in India. An earlier inadequate understanding of the aquifer system resulted in unrealistically high anticipated yields from production boreholes constructed into the deeper aquifer units. Rapidly falling pumped water levels and the failure of some of the earlier tubewells led to a reappraisal of the conceptual model. Information from individual piezometers in the shallow and deeper aquifers indicates that vertical flow components through the low permeability zones could be significant. The finding from a mathematical model that more than 90 per cent of the abstracted water originates from the overlying aquifer was unexpected. Two numerical models are described, a vertical-slice model and a time-variant three-dimensional model of a restricted area.

The Vanathavillu aquifer system in Sri Lanka is selected to illustrate that a failure to quantify the aquifer parameters correctly can lead to over-optimistic assessments of the aquifer resources. Leaky aquifer pumping test analysis was used to estimate the vertical hydraulic conductivity of the low permeability strata and the transmissivity of the underlying limestone aquifer. Because both aquitard storage and the substantial vertical gradient across the aquitard prior to the test were ignored, the vertical hydraulic conductivity and the transmissivity were both over-estimated by a factor of about five. This explains why the anticipated high yields from the limestone aquifer were not sustained.

Regional groundwater models for the aquifer system of San Luis Potosi in Mexico were not prepared, but careful and imaginative fieldwork allowed the development of realistic conceptual models. Contributions from deep thermal water have been identified from the chemical composition and the temperature of the groundwater. A pumping test in which both the fluoride concentration and the temperature increased with time confirms the significance of the deep thermal water.

When vertical flows in the Bromsgrove sandstone aquifer in the UK are ignored it is not possible to reconcile the steep water table gradient of about 1 : 50 with the high yields of boreholes. However, when account is taken of the head losses across the low permeability marl bands in a multi-layered numerical model, satisfactory simulations can be achieved. One important piece of information was crucial to the achievement of adequate starting conditions, namely that extensive spring flows supported several water mills in the early 1900s.

Four additional case studies provide further insights into the response of multi-layered aquifer systems. For each of the field examples, identifying the important flow processes and preparing quantified conceptual models led to the successful completion of the study. In fact, for most of the examples described in Chapter 10, the initial conceptual models were incorrect. By re-examining and reinterpreting the field information and data, the correct conceptual models were derived.

Chapters 9 and 10 consider situations where the aquifer properties do not change significantly with time. In Chapter 11, attention is turned to limestone and chalk aquifers where fluctuating water tables lead to changes in aquifer saturated thickness and perhaps the dewatering of zones of high hydraulic conductivity; this causes significant seasonal variations in transmissivity.

Regional Groundwater Flow with Hydraulic Conductivity Varying with Saturated Thickness

11.1 INTRODUCTION

This chapter refers to unconfined aquifer systems where the hydraulic conductivity varies with the saturated thickness. Frequently an important characteristic of these aquifers is a marked difference in the yield of wells and boreholes between high and low water tables. Provided that there are no significant extensive continuous low-permeability layers, the analysis can be carried out using a single-layer approximation with the transmissivities varying with changing water table elevations. The case studies refer to limestone and chalk aquifer systems although the approach is valid for other aquifer systems with a small to moderate saturated thickness. Development of valid conceptual and computational models involves an *approximate* representation of flow through complex systems of fissures and fractures.

The discussion in this chapter concentrates on limestone and chalk aquifers in which groundwater flows are mainly through macro and micro fractures, although there may also be some rapid flows through major fractures or conduits. Typical examples of limestone aquifers include Middle Jurassic Limestones such as the Lincolnshire Limestone in UK. Carboniferous Limestones, which are often dominated by karst behaviour, are not considered in this presentation although Teutsch (1993) demonstrates that in karstified limestone it is possible to use standard porous media models or dual porosity models where the karstification is more advanced. Furthermore, he considers that typically only about 25 per cent of the groundwater passes through fast-flowing conduit type systems.

Chalk is the principal aquifer in the UK; it is a soft white limestone consisting of shell debris and foraminifera together with minute calcareous shells of plankton and their disintegration products. The primary permeability of the Chalk is small but the secondary permeability, due to fissures and joints and the enlargement of these fissures by percolating water, provides pathways through which groundwater can flow. Flow occurs in the Chalk through numerous interconnected small cracks and fissures; the flow is normally laminar apart from in the vicinity of pumped boreholes. A valuable paper by MacDonald *et al.* (1998) considers rapid flow and karst-type behaviour in the Chalk of southern England. Examples are given of some locations where high velocities can be identified from pumping tests, tracer tests, the presence of bacteria and the success of adits in collecting water. In almost all of the examples quoted there is also evidence that the greater proportion of the flow occurs through a network of smaller fractures. In regional groundwater flow studies it is acceptable to use equivalent hydraulic conductivities and specific yields which vary with the saturated depth. However, the equivalent transmissivity does not represent in detail the rapid flow pathways which are potential sources of aquifer contamination. Extensive information about the hydrogeology of Chalk and its occurrence and importance in north-west Europe can be found in Downing *et al.* (1993).

Four major case studies are selected in this chapter to illustrate the conceptual and numerical models required for limestone and chalk aquifer systems. Three further field examples introduce additional features which explain the distinctive responses observed in the field.

Groundwater Hydrology: Conceptual and Computational Models. K.R. Rushton
© 2003 John Wiley & Sons Ltd ISBN: 0-470-85004-3

11.2 CHALK AQUIFER OF THE BERKSHIRE DOWNS

Objectives of the study: investigate the suitability of a chalk aquifer to provide augmentation flows to the River Thames in times of drought. *Key issues*: pump testing and pilot schemes are used to identify and quantify the resources of the aquifer under low water levels; developing a mathematical model which represents the significant reduction in transmissivity with decreasing saturated thickness.

Spring flows from the chalk aquifer of the Berkshire Downs provides a contribution to the baseflow of the River Thames; water abstracted from the Thames is used for public supply in many towns including London. The basis of the Thames Groundwater Scheme is to pump water from aquifers to augment river flows during severe droughts when spring flows are insufficient. This presentation concentrates on the Lambourn catchment which feeds into the River Kennet and then to the River Thames; subsequently the whole Kennet catchment was studied. Details of the catchment including a north–south cross-section are presented in Figure 11.1.

The chalk aquifer, together with the underlying Upper Greensand, lies above the Gault Clay; the Chalk is overlain over part of the study area by the London Clay and Reading Beds. The north–south cross-section of Figure 11.1b shows the Chalk outcropping in the north but confined beneath the Tertiary strata of the Reading Beds and London Clay in the south.

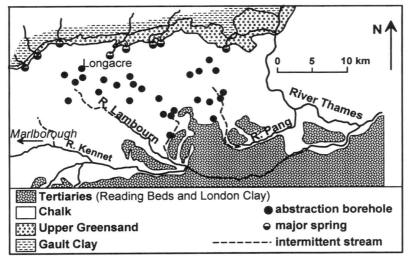

(a) Lambourn catchment and augmentation boreholes

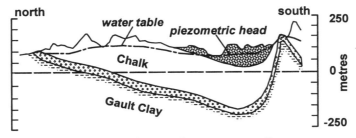

(b) representative north-south cross-section

Figure 11.1 Plan and cross-section of the Lambourn sub-catchment of the River Thames © Oxford University Press, 1981. Reprinted from *Case studies in Groundwater Resources Evaluation* edited by J.W. Lloyd (1981) by permission of Oxford University Press

11.2.1 Changing estimates of yields of the Lambourn Valley catchment

Investigations of the Lambourn Valley catchment covered a period of more than twenty years. As field tests were carried out and the understanding of the aquifer system developed, estimates of the yield of the scheme were modified. The yield of the scheme is defined as the additional water that can be pumped from the aquifer for a period of six months during a severe drought. Table 11.1 summarises the changed assessments of the catchment yield at different stages of the project; wide-ranging information about the Thames Groundwater Scheme can be found in Thames Water (1978) and Owen (1981).

It is instructive to consider how the estimated yield of the scheme was reduced on a number of occasions as the understanding of the aquifer system developed.

1. Based on a desk study using accepted values for the transmissivity and specific yield of the Chalk, the initial estimate of increased groundwater discharge for Stage 1 (Lambourn catchment) from 38 boreholes was 345 Ml/d for six months during a drought year.
2. When it was recognised that there are factors which cause the yield of individual boreholes to fall towards the end of the six month drought, the predicted yield was reduced to 250 Ml/d.
3. As drilling and pump testing of the boreholes progressed, certain boreholes failed to produce the anticipated yield; the estimated total yield was then reduced to 200 Ml/d.
4. By 1975, the scheme had been approved, with the locations of the boreholes as shown in Figure 11.1. Since 1975 was a dry year, a pumping test was carried out on a group of five boreholes which had already been constructed in the Upper Lambourn Valley; the test continued from 1 September to 1

December. Information about the test is contained in Figure 11.2. Drawdowns on 1 December are shown in Figure 11.2a with a cross-section showing the water tables on 1 September and 1 December in Figure 11.2b. The total yield fell from 25 to 15 Ml/d; the net gain (increase in river flow divided by the quantity of water abstracted) was estimated to be 95 per cent. A careful examination of Figure 11.2 provides insights as to why there is difficulty in maintaining high borehole yields.

- The effect of the pumping spreads several kilometres from the well field. However, to the north and north-west the effective depth of the aquifer decreases and becomes zero at the junction with the Lower Greensand.
- At the start of the test, the water table was between 25 and 40 m above the effective base of the chalk aquifer. As the test proceeded, the saturated thickness of the aquifer decreased by six to eight metres in the vicinity of some boreholes (see Figure 11.2a), with the result that certain zones with higher hydraulic conductivity were dewatered.
- At the pumped boreholes with drawdowns of 20 to 35 m, the saturated thickness decreased so much that it became difficult to draw water into the boreholes.

  The reduction in yield during the test provided a warning that yields of boreholes were likely to reduce during operation of the scheme in drought conditions. Therefore the anticipated yield of the scheme was reduced to 113 Ml/d.

When the full scheme commenced operation in 1976, the initial yield of the boreholes was 88 Ml/d; due to stoppages the actual yield fell to 79 Ml/d. If heavy rain had not occurred during the autumn of that year it is likely that the yield would have fallen to about 60 Ml/d after six months.

Table 11.1 Summary of anticipated yields during planning and development of the Lambourn Valley Stage of the Thames Groundwater Scheme; the objective is to augment flows in the River Thames for six months during a drought period

Date	Anticipated Yield	Comments
1956	345 Ml/d	Assumes that T and S of chalk remain constant, anticipated yield reduced to 250 Ml/d to allow for losses
1969	200 Ml/d	Following test of nine boreholes in Winterbourne and Lambourn valleys, impact of reduced borehole yields
1975	113 Ml/d	Upper Lambourn group test showed severe impact of reductions in T and S due to falling water tables
1976	80 Ml/d	Taking note of actual yield of boreholes and stoppages
1976	(60 Ml/d)	Probable actual yield if the scheme had operated for 6 months; autumn 1976 was very wet

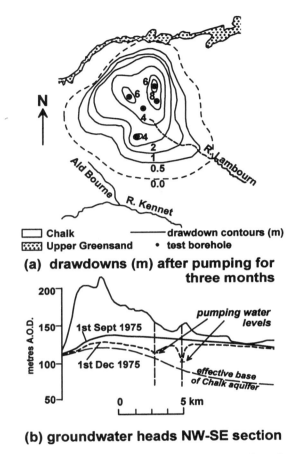

Chalk ⬚   **drawdown contours (m)** ———

Upper Greensand ▥   **test borehole** •

(a) drawdowns (m) after pumping for three months

(b) groundwater heads NW-SE section

Figure 11.2 Results of long-term pumping test from five boreholes in upper Lambourn catchment

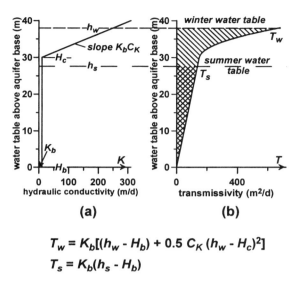

$$T_w = K_b[(h_w - H_b) + 0.5\, C_K\, (h_w - H_c)^2]$$
$$T_s = K_b(h_s - H_b)$$

Figure 11.3 Variable hydraulic conductivity with saturated thickness: (a) hydraulic conductivity variation, (b) corresponding transmissivity variation

a numerical model indicates that there is an enhanced hydraulic conductivity in the zone of water table fluctuation, but beneath this zone an approximately constant hydraulic conductivity applies. These findings are similar to the hydraulic conductivity variation shown in Figure 11.3a; equations used to calculate the transmissivities are quoted in the figure. The corresponding changes in transmissivity (Figure 11.3b), reflect the differences between winter and summer conditions. For the particular example of Figure 11.3 the transmissivity for the winter water table elevation is $T_w = 655\,\text{m}^2/\text{d}$, whereas for the summer water table the transmissivity reduces to $T_s = 140\,\text{m}^2/\text{d}$.

There are also areal changes in transmissivity. From about thirty pumping tests, a distribution of transmissivities has been inferred as shown in Figure 11.4 (Owen 1981). Transmissivities are highest in the valleys and lowest in the interfluves. The transmissivity distribution is based on classical methods of pumping test analysis using the Theis theory. However, several of the assumptions of the Theis theory, such as the transmissivity remaining constant throughout the test, are violated. Consequently the transmissivities of Figure 11.4 are used as relative, not absolute values. Insights into the behaviour and properties of Chalk in both the unsaturated and saturated zones can be found in Price *et al.* (1993).

The regional groundwater model covers the whole of the Kennet catchment. Important features of the

This description demonstrates that there are many practical difficulties in developing a successful river augmentation scheme for drought periods due to decreases in the saturated thickness of the aquifer. Field studies, pumping tests, pilot schemes testing small groups of boreholes and mathematical modelling are all needed in planning and implementing these schemes.

11.2.2 Conceptual and mathematical modelling

As indicated above, there are substantial reductions in borehole yields when water tables are lower; this is due to reducing transmissivities. Crucial insights into the change in transmissivity with saturated depth are obtained from the analysis of two pumping tests carried out in the same borehole at different rest water levels (see Section 8.7.2). Analysis of these tests using

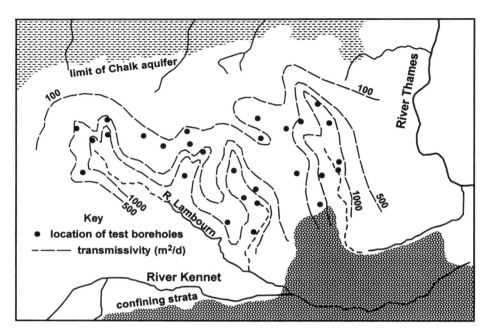

Figure 11.4   Contours of estimated transmissivity © Oxford University Press, 1981. Reprinted from *Case Studies in Groundwater Resources Evaluation* edited by J.W. Lloyd (1981) by permission of Oxford University Press

groundwater model include the representation of the variable hydraulic conductivity with depth and the simulation of intermittent streams and rivers (see Section 12.2.1). The range of transmissivity values is indicated in Figure 11.4; the specific yield averages 0.01, with values ranging between 0.004 and 0.02. The simulation covers a period of twelve years which include sequences of wet and dry years; the representation of both groundwater heads and river flows are generally good (Rushton *et al.* 1989).

The importance of including the variation of transmissivity with saturated depth is demonstrated by the hydrographs in Figure 11.5 which refer to water years 1975 to 1979 (a water year starts in October of the previous year). Figure 11.5a compares field measurements and model simulations of river flows at Marlborough, which is on the River Kennet about 30 km upstream of the confluence with the River Lambourn. Field results are indicated by filled circles. The unbroken line corresponds to flows calculated from the mathematical model of the whole Kennet catchment when the hydraulic conductivity varies with the saturated depth in the manner shown in Figure 11.3. The broken line, which refers to a simulation in which the hydraulic conductivity remains constant, shows significant

differences from the field readings. Since low flows are critical to this study, the adequacy of the simulation is assessed by focussing on low flows. During periods of low flow, especially 1976 and 1978, the simulation with the hydraulic conductivity varying with depth provides a reliable representation of the flows, whereas the constant hydraulic conductivity simulation over-estimates low river flows.

Comparisons at Longacre observation borehole between field results and modelled groundwater heads, based either on variable or constant hydraulic conductivities, are presented in Figure 11.5b. With a constant hydraulic conductivity, groundwater heads are over-estimated and groundwater head fluctuations are underestimated. Improved agreement is achieved with hydraulic conductivites varying with depth as shown by the unbroken line in Figure 11.5b. However, there are clear differences during the drought of water year 1976; most mathematical models of chalk and limestone aquifers in England provide a poor representation of field groundwater heads for the period immediately following the drought. Further information about the conceptual and mathematical models of the Berkshire Downs chalk aquifer system can be found in Rushton *et al.* (1989).

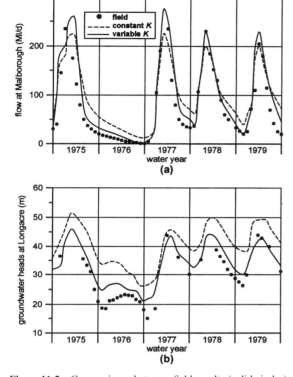

Figure 11.5 Comparisons between field results (solid circles) and groundwater model results either based on a constant hydraulic conductivity (broken line) or with hydraulic conductivity as a function of saturated thickness (unbroken line): (a) river flows at Marlborough, (b) groundwater heads at Longacre observation borehole. Reproduced by permission of Geological Society, London, from Rushton *et al.* (1989)

11.3 SOUTHERN LINCOLNSHIRE LIMESTONE

Objectives of the study: quantify the various recharge mechanisms and identify acceptable abstraction patterns.

Key issues: wide variety of recharge processes with significant runoff-recharge; importance of limestone inliers in the confined region; reduced transmissivity with decreasing saturated thickness; presenting river–aquifer interaction in diagrammatic form.

Studies of the Southern Lincolnshire Limestone catchment have been underway for more than three decades; as the understanding of aquifer systems developed, improvements have been made to conceptual and

numerical models. The focus of this presentation is the development and refinement of conceptual models for interacting surface water and groundwater flow processes, quantifying these conceptual models and then transferring this information to mathematical models. Interpretation of the model outputs has also been central to the resolution of issues such as the effect of abstraction on stream and river flows. The processes of particular importance in this aquifer system include precipitation and runoff-recharge, substantial variations in transmissivity with water table fluctuations, complex interaction between groundwater and surface water and uncontrolled overflowing (wild) boreholes.

11.3.1 General description of Southern Lincolnshire Limestone catchment

Detailed information concerning the Southern Lincolnshire Limestone catchment in eastern England is presented in Figures 11.6 and 11.7. To the west of the study area the Limestone crops out on a scarp slope at elevations of about 120 m, to the east the topography becomes flat fenland with elevations close to sea level. Surface drainage is provided by the West and East Glen rivers which flow in incised valleys in a roughly SSE direction until they turn in a north-easterly direction to join fenland water-courses. In addition the River Gwash crosses the limestone outcrop in the south-west.

The Lincolnshire Limestone aquifer is of the Middle Jurassic; to the south it is bounded by the Marholm–Tinwell Fault but continues for about 120 km to the north to the Humber Estuary. This discussion is focussed on the catchments of the West and East Glen Rivers; Figure 11.6. The Limestone is exposed only in limited areas; elsewhere, boulder clay or a sequence of low permeability strata, which are described by the term 'Overlying Beds', limit direct recharge to the limestone aquifer. The Overlying Beds, which are unshaded in Figure 11.6, can include the Upper Estuarine Series, Great Oolite Limestone, Great Oolite Clay, Cornbrash and Kellaway Beds. A detailed review of the groundwater hydrology of the Lincolnshire Limestone can be found in Downing and Williams (1969).

Originally the Marholm–Tinwell Fault (shown to the south in Figure 11.6) was assumed to act as an impermeable boundary; quarries constructed to the south of the fault were used for waste disposal. However, deterioration in the quality of water from a pumping station to the north of the fault indicated that contaminated water does cross the fault. A thorough field investigation has shown that a plume of contaminated

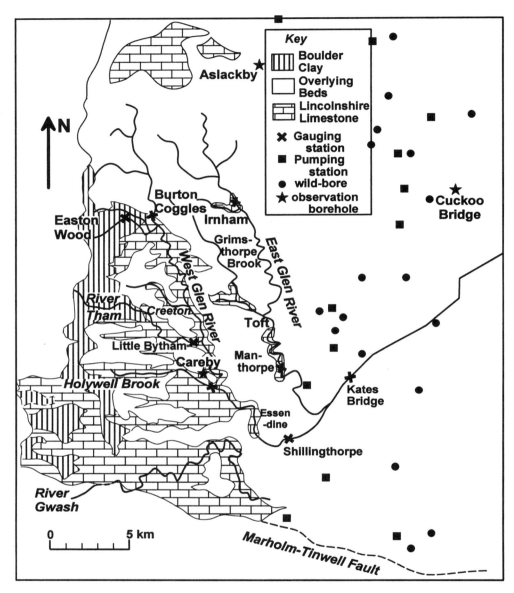

Figure 11.6   Southern Lincolnshire limestone aquifer system

water has formed in the Lincolnshire Limestone to the north of the fault. Furthermore, a detailed study of the Marholm–Tinwell Fault revealed that, rather than being a single fault line which isolates the Lincolnshire Limestone, there are a number of smaller faults which provide continuity of permeable strata from south to north across the fault.

11.3.2 Recharge including runoff-recharge

Recharge to the Southern Lincolnshire Limestone aquifer has many components. Each component is introduced with reference to the *schematic* diagram of Figure 11.8. The lower-case letters on the figure refer to different recharge pathways.

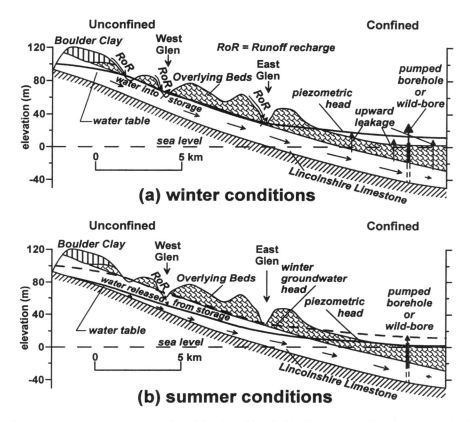

Figure 11.7 Representative east–west cross-section of Southern Lincolnshire limestone aquifer; the cross-section is located to the south of Easton Wood and Burton Coggles

Component (a) – conventional rainfall recharge on the limestone outcrop: (a) is estimated from a soil moisture balance following the approach of Section 3.4; the detailed recharge estimation of Figure 3.18 refers to the Southern Lincolnshire Limestone catchment.

Component (b) – runoff-recharge from boulder clay: since boulder clay contains deposits of more permeable material which act as minor aquifers, much of the potential recharge from the soil zone moves into the boulder clay and is stored; the processes are illustrated in Figure 11.9. From the boulder clay store, a small quantity of water moves vertically downwards to become recharge to the limestone aquifer; this is estimated to be 0.05 mm/d. Water is also released from boulder clay storage to move laterally through permeable zones in the boulder clay to springs, ditches or drains; an exponential decay relationship is used to quantify this delayed runoff (see the detailed discussion in Section 3.6). When this delayed runoff crosses onto the

limestone outcrop it is available to enter the aquifer as runoff-recharge through swallow holes. Six locations are identified by (b) in Figure 11.8; in practice there are more locations on the margins of the limestone outcrop where runoff-recharge occurs.

Component (c) – runoff-recharge from isolated areas of Overlying Beds: component (c) is similar to (b) but a smaller delay coefficient is used to reflect the slower release of water from Overlying Beds than from boulder clay.

Component (d) – runoff-recharge in the West Glen at Burton Coggles: the West Glen catchment above Burton Coggles covers an area of approximately 31 km². An examination of Figures 11.6 and 11.8 shows that most of the upper catchment is covered by Overlying Beds, hence the flows due to direct runoff and delayed runoff are substantial. Calculations of the direct and delayed runoff are based on the conceptual model of Figure 11.9. When runoff in the West Glen crosses onto the limestone outcrop south of Burton Coggles it can enter

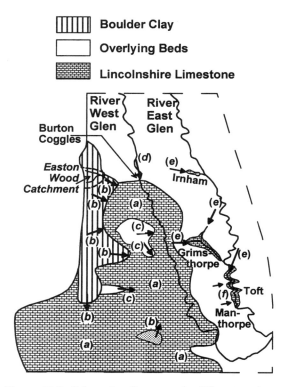

Figure 11.8 Schematic diagram of different recharge processes in the Southern Lincolnshire limestone. *Journal of Hydrology* 211, reprinted from Bradbury and Rushton, Estimating runoff-recharge in the Southern Lincolnshire limestone catchment, UK, pp. 86–99, copyright (1998) with permission from Elsevier Science

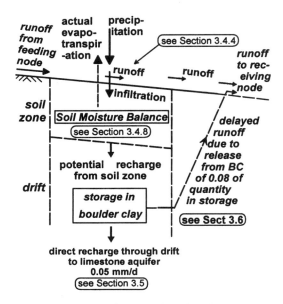

Figure 11.9 Computational model of recharge processes associated with boulder clay overlying the limestone aquifer. *Journal of Hydrology* 211, reprinted from Bradbury and Rushton, Estimating runoff-recharge in the Southern Lincolnshire limestone catchment, UK, pp. 86–99, copyright (1998) with permission from Elsevier Science

the aquifer system provided that the groundwater head is below river surface level. However, the quantity of water entering the aquifer at any node cannot exceed the river coefficient at that node. As groundwater heads rise at the river nodes, the quantity of water actually entering the aquifer decreases; this is considered further in Section 11.3.3.

Component (e) – runoff-recharge through limestone inliers in the East Glen: during initial studies of the Southern Lincolnshire Limestone aquifer, the East Glen river was ignored since it was considered to be effectively isolated from the limestone aquifer. The flow characteristics of the East Glen differ from those of the West Glen; following heavy rainfall there is substantial runoff but during the summer the East Glen often becomes dry. Nevertheless, there are limestone inliers of limited extent where the East Glen or its tributaries are in contact with the limestone; the locations of the

three main inliers can be identified in Figures 11.6 and 11.8. These inliers provide a route whereby water enters the limestone aquifer. Careful monitoring of stream flows above and below these inliers confirmed that they do provide pathways for flow from the East Glen into the aquifer system. As indicated in Figure 11.8, there are inliers on two of the tributary streams; at Irnham the inlier extends for about 2 km, on Grimsthorpe Brook there is a more extensive area of contact with the limestone aquifer. There is also a reach of the East Glen in the vicinity of Toft where there is contact with the limestone aquifer over a distance of more than 4 km.

There are two stages in estimating the flows entering the limestone aquifer through the inliers in the East Glen. First, flows due to runoff in the main East Glen and its tributaries are estimated from soil moisture balances and releases from subsurface stores using the computational model of Figure 11.9. Computed flows at individual nodes are combined working down the catchment. The accreted flows at Irnham, Grimsthorpe and Toft can enter the aquifer system through the inliers provided that there is storage space available in the aquifer. Consequently the actual quantity entering the aquifer can only be determined during the regional groundwater model simulation. When the groundwater heads at the inliers reach the river surface level, no more flow can enter; when the groundwater heads rise above the river surface level the inliers act as discharge locations (i.e. springs) with no runoff-recharge entering the limestone aquifer.

Component (f) – runoff from Overlying Beds to inlier at Toft: this component is estimated in a similar manner to (c).

Six alternative recharge components have been identified; the distribution of the recharge components and the directions in which runoff is routed across the low permeable strata to locations where the water can enter the limestone aquifer are shown in Figure 11.10. This figure is based on a square grid of sides 1 km; locations where runoff can enter the limestone aquifer are indicated by solid circles. Since runoff across the catchment

is not instantaneous, the routing is arranged so that it takes one day to transfer runoff between nodes. The runoff-recharge calculation is carried out on a daily basis; these daily values are combined for the regional groundwater model which uses 15-day stress periods. More detailed information about the estimation of the various recharge components can be found in Bradbury and Rushton (1998).

11.3.3 Surface water–groundwater interaction

Interaction between the aquifer and springs or rivers is an important aspect of the catchment response. Some of the springs and rivers stop flowing during dry summers yet carry high flows in wet winters. Section 4.3 describes different types of interaction between aquifers and springs or rivers; Figure 11.11a is a

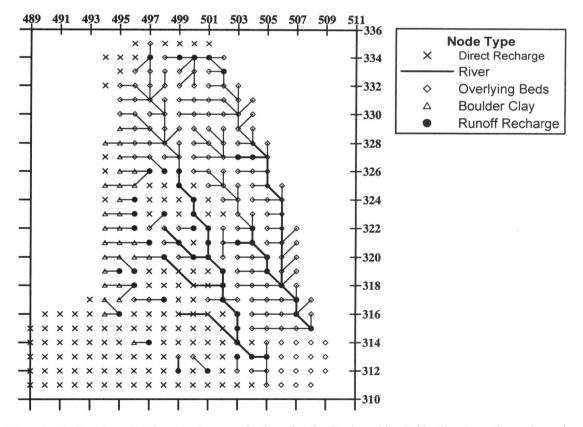

Figure 11.10 Location of different recharge mechanisms for the Southern Lincolnshire limestone; the mesh spacing is 1.0 km. *Journal of Hydrology* **211**, reprinted from Bradbury and Rushton, Estimating runoff-recharge in the Southern Lincolnshire limestone catchment, UK, pp. 86–99, copyright (1998) with permission from Elsevier Science

simplified form of Figure 4.11 which defines river–aquifer interaction. The vertical axis of Figure 11.11a represents the groundwater head at the river node relative to the river surface elevation; the horizontal axis indicates the flow from aquifer to river. When the groundwater head is above the river surface level, the flow from aquifer to river equals the river coefficient *RC* multiplied by the excess head; it is convenient to define *RC* for a 1-km reach of river. However, when the groundwater head is below river surface level the river loses water to the aquifer provided that there is sufficient water in the river. The loss from the river reaches a maximum value under unit head gradient (gravity) of $RC \times 1.0\,\mathrm{m^3/d/km}$; this constant loss applies where groundwater heads are more than 1.0 m below river surface level.

Figure 11.11 refers to three river reaches where river–aquifer interaction occurs. For each of the locations, the relevant area of the geological map of Figure 11.6 is reproduced; beneath the geological information there are sketches of the corresponding nodal points, river surface elevations and river coefficients.

Immediately below Burton Coggles (Figure 11.11b): the upper West Glen above Burton Coggles, includes a substantial area of Overlying Beds from which rapid runoff and delayed runoff occur. As water flows past Burton Coggles onto the Lincolnshire Limestone (see diagram of geology), there is the potential for flows from river to aquifer provided that the water table is below river level. Gaugings down the West Glen suggest that the loss can be as high as 6000 m³/d per kilometre for the first two kilometres, falling to 3000 m³/d further down the river. These values are the basis of the river coefficients *RC* which are 6000 m²/d/km for the first two nodes and 3000 m²/d/km for downstream nodes. River surface elevations are also included in Figure 11.11b.

Irnham, a tributary of the East Glen (Figure 11.11c): there is a small catchment of about 4 km² feeding onto the inlier at Irnham; flow gauging suggests that the maximum quantity of water which enters the aquifer through the inlier is 4000 m³/d. Since the Irnham inlier is represented by two nodes spaced 1 km apart, the river coefficient for each node is 2000 m²/d/km; the gradient of the river surface is 1.0 m per km.

Headwaters of the Holywell Brook (Figure 11.11d): Smith (1979) considers the nature of the springs which feed Holywell Brook; he suggests that springs are associated with fissures at different elevations; Figure 4.9c. The most important feature, which controls aquifer–river flows in Holywell Brook, is the rel-

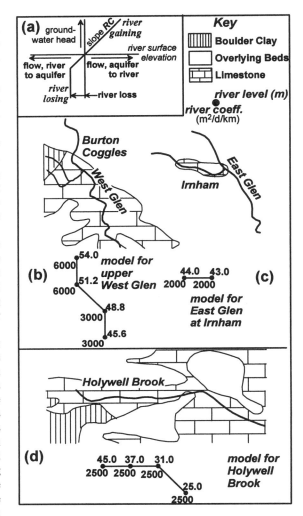

Figure 11.11 River–aquifer interaction: (a) diagram of algorithm used to describe river–aquifer interaction, (b), (c) and (d) geological information, equivalent nodal points, river surface elevations and river coefficients for West Glen below Burton Coggles, East Glen tributary at Irnham and Holywell Brook

atively steep gradient of the water surface. As indicated in Figure 11.11d the water surface elevation falls 8 m in the first kilometre. There is no direct way of estimating the river coefficient *RC*. However, since the river gains water at all nodes along Holywell Brook, the flows are not sensitive to the river coefficients. Therefore a value of 2500 m²/d/km is selected to be consistent with other locations.

11.3.4 Wild boreholes

In the nineteenth century, artesian boreholes for individual farms were drilled to depths of up to 60 m in the fenland to the east of the East Glen; these bores passed through Oxford Clay and Overlying Beds into the Lincolnshire Limestone. When many small farms amalgamated, boreholes were abandoned so that uncontrolled artesian discharges occurred, hence the expression *wild boreholes*. These boreholes respond in a similar manner to springs with flows depending on the difference between the groundwater head h and the elevation of the outlet z_0. However, the equation for flow from a spring quoted in Figure 4.9e,

$$Q = SC(h - z_0) \qquad (11.1)$$

is not appropriate since it fails to represent the characteristics of flow through the borehole pipe. In the spring equation, the spring coefficient SC is assumed to be independent of the velocity of the flow. However, from classical pipe flow theory, the equation for the head loss in a pipe is

$$head\ loss = \frac{flu^2}{2gd} + \frac{ku^2}{2g} \qquad (11.2)$$

where f = friction factor determined from the Moody graph (Daugherty *et al.* 1989)
l = length of pipe
u = velocity of water in the pipe
d = diameter of pipe
k = head loss coefficient for entry and exit losses

The pipe discharge Q is the product of the velocity of water in the pipe, which is proportional to $\sqrt{(h - z_0)}$, and the cross-sectional area of the pipe, hence

$$Q = PC(h - z_0)^{1/2} \qquad (11.3)$$

in which PC is the pipe coefficient which depends on the parameters f, l, d, k and the cross-sectional area of the pipe. Estimates of the pipe coefficient can be made from the diameter, length and frictional properties of the pipe. However, in the study area, considerable deterioration occurred in most wild-bore pipes, hence conventional parameters do not apply. Consequently, the coefficients of individual wild-bores are estimated from flow measurements and knowledge of the excess head. The estimated coefficients lie in the range 60 to 1800 $m^{5/2}$/d (Johnson *et al.* 1999).

Taking a typical pipe coefficient of 250 $m^{5/2}$/d, calculations based on Eq. (11.3) provide the following estimates of flows from a wild-bore:

$(h - z_0)$ m	0.0	1.0	2.0	4.0	8.0	16.0
wild-bore flow m^3/d	0.0	250	354	500	707	1000

For an increase in head difference $h - z_0$ from zero to 8.0 m the flow increases by 707 m^3/d; for a further 8 m increase in head difference the flow only increases by 293 m^3/d. This demonstrates that flows from wild-bores do not increase proportionally under high winter water levels.

Most of the wild-bores have been controlled or sealed (Barton and Perkins 1994); this is represented in the model by reducing the pipe coefficient or setting it to zero.

11.3.5 Variable hydraulic conductivity with depth

The Lincolnshire Limestone is of marine origin consisting of shell beds, oolites, pisolites, cementstones, coral knobs and sandy limestone. Due to the passage of water, especially on the margins of the drift, and due to geological faulting and folding, there are extensive fissures and fractures. Consequently, significant variations occur in the hydraulic conductivity within the limestone. Direct measurements of the effective hydraulic conductivity on a scale appropriate to regional groundwater models is not possible; instead indirect information has to be used to devise credible variations in hydraulic conductivity at different depths within a vertical section.

As an example of indirect evidence, consider the two groundwater head hydrographs plotted in Figure 11.12; the upper hydrograph refers to Careby in the unconfined region of the Glen catchment, the lower hydrograph is for Cuckoo Bridge in the confined region (for locations see Figure 11.6). There is an unexpected difference; fluctuations in groundwater head in the confined region are greater than fluctuations in the unconfined region. Runoff-recharge type (e) of Figure 11.8 causes the fluctuations in the confined region. This is illustrated by the simplified conceptual model of Figure 11.13. The differing widths of the unconfined and confined 'reservoirs' reflect the differences between unconfined and confined storage coefficients. However, runoff-recharge entering directly into the confined region does not explain the differences between the hydrographs during the water year 1976 (October 1975 to September 1976). Due to low rainfall, there was only minor recharge during the water year 1976; this is reflected in the upper groundwater head hydrograph of Figure 11.12 where the decline at Careby in the unconfined region is less than one metre over a period of ten months. During the same period a decline of almost 10 m occurs at Cuckoo Bridge in the confined region

despite substantial reductions in abstraction in the later part of the water year.

Consequently a further feature needs to be introduced into the conceptual model of Figure 11.13. The '*pipe*' which connects the unconfined and confined '*reservoirs*' is equivalent to the transmissivity of the limestone aquifer. Water is withdrawn from the *pipe* in the confined region to simulate the abstraction and wild-bore flows. During 1976, when there was negligible inflow into the unconfined *reservoir*, water should have been drawn through the *pipe* from the unconfined *reservoir*; this would lead to a decline in the water level in the unconfined *reservoir* (groundwater heads in the unconfined region). The almost constant head at Careby during 1976 (Figure 11.12), shows that water was not drawn from the unconfined region despite the increasing head gradient. Consequently there must be some form of flow restriction in the *pipe* between the unconfined and confined *reservoirs* under low groundwater head conditions; see Figure 11.13. But what causes this restriction?

A careful examination of the schematic cross-section for summer conditions (Figure 11.7b), indicates that there is a substantial decrease in the saturated thickness in the limestone aquifer between the West and East Glens from the winter groundwater heads (shown by the broken line) to the summer groundwater heads (unbroken line). The impact is more severe than indicated by the decrease in saturated thickness, since major fissures in the upper part of the aquifer are dewatered due to the fall in groundwater head. There is

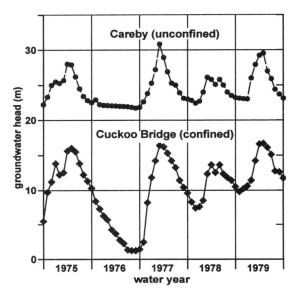

Figure 11.12 Groundwater head hydrographs in the unconfined and confined regions of the Southern Lincolnshire limestone system

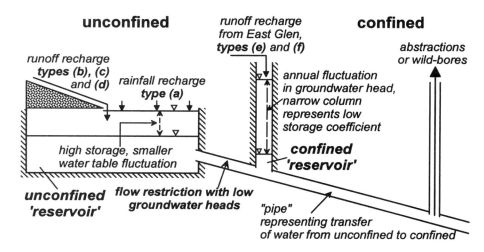

Figure 11.13 Simplified conceptual model of responses in unconfined and confined regions including restriction in flows under low groundwater heads

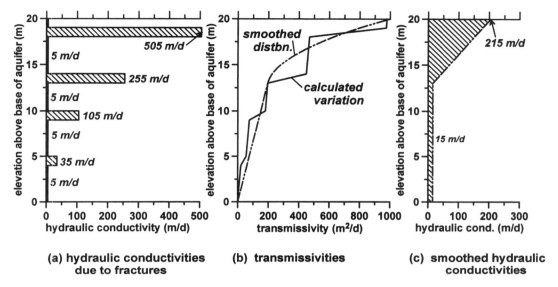

Figure 11.14   Assumed hydraulic conductivity variation with saturated depth and corresponding transmissivities and smoothed hydraulic conductivities

no precise field evidence for the actual distribution of the fissures, therefore a plausible distribution of fissure zones is selected as shown in Figure 11.14a. There are two fissure zones of higher hydraulic conductivity in the upper half of the section; these zones represent the effect of fissures over a 1-km mesh interval. Using this distribution, the variation in transmissivity with groundwater head elevation is indicated by the unbroken line of Figure 11.14b. If a smoothed hydraulic conductivity variation is chosen as shown in Figure 11.14c, this leads to a transmissivity distribution (the chain-dotted line in Figure 11.14b) which approximates to the transmissivity characteristics of the four distinct fissures zones. When the aquifer is fully saturated the transmissivity is 1000 m²/d; when only half the section is saturated the transmissivity is 150 m²/d. A smoothed distribution of hydraulic conductivity is used because:

* when incorporated in a numerical model, the hydraulic conductivity distribution between nodal points (or in cells) must represent the overall effect of a number different systems of fissures,
* for stability and convergence of numerical model solutions, it is preferable to avoid abrupt changes in hydraulic conductivity,
* sensitivity analyses are performed during the model refinement to explore appropriate hydraulic conduc-

tivity variations with saturated thickness; a distribution such as that of Figure 11.14c is more convenient for sensitivity analyses.

The smoothed hydraulic conductivity distribution of Figure 11.14c is basically the same as that of the Berkshire Downs Chalk aquifer (Figure 11.3); see also Rushton *et al.* (1982).

11.3.6 Resumé of conceptual models

The flow processes in the Southern Lincolnshire Limestone catchment have been presented using a number of quantified conceptual diagrams.

* The cross-sections of Figure 11.7 which illustrate differing conditions of runoff-recharge in winter and summer, interaction between groundwater and surface water, water taken into or taken from storage at the water table, lateral flow through the aquifer with the saturated thickness varying between winter and summer, abstraction from pumped boreholes or outflows from wild-bores and upward leakage through the Overlying Beds. Figure 11.13 is an idealised representation of these processes.
* Runoff-recharge with a schematic diagram indicating the types of runoff-recharge; Figure 11.8; representation of storage in drift and delayed runoff from

the drift, Figure 11.9, and director arrays which indicate how water is transferred across the drift, Figure 11.10.

- A conceptual model which permits the estimation of flows between river and aquifer, Figure 11.11a; river coefficients and elevations for three river reaches are included in Figure 11.11b–d.
- Behaviour of wild-bores and how conditions change when they are sealed, Eq. (11.3) and Section 11.3.4.
- Representation of changes in transmissivity due to varying saturated thickness and the consequent dewatering of fissures; Figure 11.14.
- One additional feature is upward flow from the Limestone in the confined region through the Overlying Beds; these flows are described as upward leakage in Figure 11.7a. The upward leakage depends on the difference between the groundwater head in the confined region and the land surface which is extensively drained. Flows are proportional to the vertical hydraulic conductivity of the Overlying Beds of 0.0002 m/d and inversely proportional to the thickness of the Overlying Beds which increase from 18 to 70 m. Upward leakage during a winter month can total 20 Ml/d.

11.3.7 Brief description of the total catchment models

Important features of the Southern Lincolnshire Limestone Catchment models are presented in Table 11.2; it is a major task to collect and collate all the data

Table 11.2 Summary of features in Southern Lincolnshire Limestone Catchment models

Feature	Information
Duration of simulation	1970–2000
Finite difference mesh	Square mesh, 1 km by 1 km
Discrete time steps	15 days and to the end of month
Rainfall recharge	Estimated using techniques of Section 3.4 for each nodal point for each day; averages are then calculated for each time step
Runoff-recharge	Estimated as described in Section 11.3.2 then averaged for each time step; runoff-recharge can occur at 48 nodes
Northern boundary	Actual model extends 25 km north of study area to a groundwater divide
Western boundary	Scarp slope, no flow
Southern boundary	Marholm–Tinwell Fault; flow across the fault depends on difference in groundwater head between aquifers on either side of the fault
Eastern boundary	Transmissivity becomes very low and saline water present, no flow boundary 20 km to east of major confined abstraction sites
Transmissivities	Unconfined region, transmissivities for the initial conditions are in the range 30 to 800 m²/d, they vary with water table elevation; in the confined region with substantial fissure development, transmissivities are 1000–3000 m²/d but fall to the east
Storage coefficients	Unconfined 0.01 to 0.05, confined 0.0001 to 0.0009
Leakage – confined region	Vertical head gradient multiplied by vertical hydraulic conductivities and divided by thickness of overlying strata
Abstractions	Monthly average abstractions
Wild-bores	Flows depend on square root of vertical head gradients (see Section 11.3.4)
Springs	Outflows depend on groundwater head minus spring elevation
Rivers	River coefficients in range 800 to 6000 m³/d/km (Section 11.3.3)
Initial conditions	Time-instant conditions with recharge for a typical year and abstraction for 1970; further information in Section 12.5.3
Data for checking models	Groundwater head hydrographs at 23 locations, seven river gauging sites, extensive accretion profiles from flow gauging throughout the year, hydrochemical information
Direct outputs	Recharge, runoff-recharge and groundwater head variations with time, river–aquifer flows, variation of transmissivity
Derived outputs	Diagrams of river–aquifer interaction including accretions, water balances for different regions, flow vectors for assessing whether saline water is being drawn into confined zone or for impacts of pollution incidents

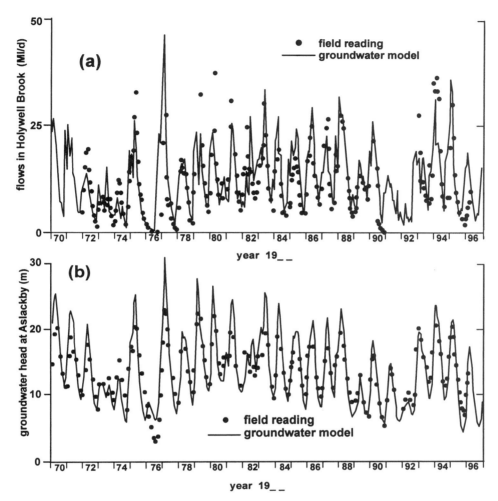

Figure 11.15   Comparisons between field and model results: (a) flows at gauging station on Holywell Brook, (b) groundwater heads at Aslackby in the confined region. Reproduced by permission of Geological Society, London from Rushton and Tomlinson (1995)

and other information. When there is uncertainty about the parameter values, they can be tested during sensitivity analyses of the mathematical models.

11.3.8 Selected results and insights from the numerical model

Comparison of field and model results

To illustrate the agreement between the groundwater model simulation and field readings, comparisons are made in Figure 11.15 between (a) flows at the gauging station on Holywell Brook and (b) groundwater heads at Aslackby in the northern confined region (locations

are shown in Figure 11.6); the comparison cover water years 1970–96. Generally agreement between the modelled and field results is satisfactory. Differences are certain to occur when account is taken of the complex physical nature of the aquifer systems and the idealisations introduced in the conceptual models. The following findings are of particular importance:

- low flows in Holywell Brook, Figure 11.15a, are represented adequately by the model; since low river flows are the result of many interacting processes, the agreement suggests that all the important features are included in the simulation,

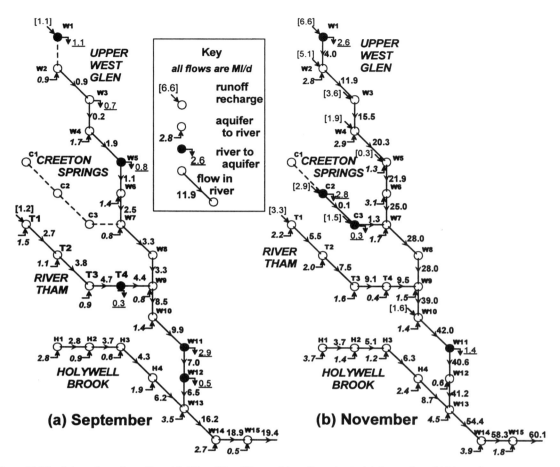

Figure 11.16   Interaction of aquifer with River West Glen and its tributaries in (a) September, (b) November

- the groundwater model also reproduces approximately high flows in Holywell Brook; differences are mainly due to the unreliability of the gauging station under high flow conditions,
- representation by the model of groundwater heads at Aslackby is also satisfactory with acceptable agreement for years with both large and small seasonal fluctuations,
- in the drought year of 1976 there are differences between field and modelled results for both flows in Holywell Brook and groundwater heads at the Aslackby observation borehole; the recovery in flows and heads are both too rapid probably due to inaccurate recharge estimates.

Insight into river–aquifer interaction in the West Glen catchment

Interaction between rivers and the aquifer and the ability of the aquifer to accept runoff-recharge are important facets of the West Glen catchment. Groundwater model results are used to explore how these components interact. Figures 11.16a and b refer to conditions in September and November 1984 for the West Glen and its tributaries. The gradient of the river bed of the West Glen is equivalent to about 2.5 m per kilometre in a generally north–south direction, which means that the river is almost perpendicular to the west–east direction of groundwater flow. This leads to sensitive river–aquifer interaction especially at times of low groundwater heads.

As indicated in the Key for Figure 11.16, there are three possible flow conditions at each nodal point namely, inflow due to runoff-recharge, inflow to the river from the aquifer when the groundwater head in the aquifer is above the river surface level, and flow from river to aquifer when the groundwater head is below river surface level. These three components, when added to the river flow from the upstream node, equal the river flow towards the downstream node.

In September (Figure 11.16a), groundwater heads are low. Flow from the Overlying Beds at Burton Coggles, node **W1**, is 1.1 Ml/d; all of this flow immediately enters the aquifer so that there is no river flow between nodes **W1** and **W2**. All along the West Glen, conditions change from node to node, with water lost from river to aquifer at some nodes (the filled circles) but with small flows from aquifer to river at other nodes. There is no flow from Creeton Springs. At the upper node of the River Tham there is a small runoff-recharge contribution; at most of the nodes on the river there is an inflow from the aquifer to the river. Holywell Brook is also groundwater-fed. Inflows from the River Tham and Holywell Brook are significant; they contribute 10.6 Ml/d of the total flow at the bottom of the West Glen of 19.4 Ml/d.

Considerable changes in flow conditions occur by November 1984; Figure 11.16b. Flow in the West Glen at Burton Coggles at node **W1** is 6.6 Ml/d; 2.6 Ml/d enters the aquifer and 4.0 Ml/d flows to node **W2**. At node **W2** there is a runoff-recharge of 5.1 Ml/d but the groundwater head is nearly 0.5 m above the river level so that there is a flow of 2.8 Ml/d *from* aquifer to river with the result that none of the runoff-recharge enters at node **W2**. At node **C2** of Creeton Springs, much of the runoff-recharge enters the aquifer; at node **C3** less of the water enters hence there is a small flow to the West Glen. Contributions to the West Glen from the River Tham and Holywell Brook are substantial; these contributions result mainly from aquifer to river flows.

Later in the year the groundwater heads rise further; at all locations the aquifer feeds the rivers so that no runoff-recharge enter the aquifer. Flows in the West Glen often exceed 150 Ml/d.

Impact of changed abstractions in confined region on river flows

Abstractions from the confined region undoubtedly reduce flows in the streams and rivers on the outcrop, but would reductions in abstractions improve low flow conditions? This issue was explored following the controlling and sealing of many of the wild-bores (Barton and Perkins 1994, Johnson *et al.* 1999). Since much of the water escaping from the wild-bores is not used productively but enters the surface drainage system, work was carried out in 1991–92 to cap the wild-bores; for some of the capped bores provision is made for limited compensation flows.

Using the mathematical model, comparisons are made between two predictive simulations extending over 25 years. In one simulation there is no control of wild-bores, the second represents sealing or control for the whole simulation period. Due to the capping of the wild-bores, the average saving is estimated to be 15.4 Ml/d. This leads to increases in natural groundwater outflows which are listed below as long-term averages:

increased upward leakage through Overlying Beds	5.5 Ml/d
increased groundwater flows to West Glen	2.3 Ml/d
increased groundwater flows to East Glen	2.7 Ml/d
increased groundwater flows to other rivers	4.9 Ml/d

The main purpose of the control of wild-bores is to improve flows in the River West Glen due to its high amenity value, but the increase in flows in the West Glen is only 16 per cent of the reduction in wild-bore flows. Furthermore, an examination of the hydrograph for flow increases in the West Glen shows that in drought periods the increase in low river flow is less than 0.5 Ml/d. Gains in flows in the West Glen are small because there are many routes by which water can exit from the confined region of the aquifer. The route which is distributed over the largest area is upwards leakage through the Overlying Beds, hence the greatest increase is in Overlying Beds flows. Rarely will reductions in the quantity of water pumped from an aquifer system result in an equivalent increase in river flows (Rushton 2002).

Source protection zones

The groundwater model has been used to identify source protection zones. Flow vector diagrams are prepared from the numerical model output (see Figure 2.23) to identify the origin of water moving towards nodal points which represent boreholes. Due to the substantial changes in flow patterns between winter and summer, there is a relatively small zone supplying a borehole during winter when high transmissivities occur, but in the summer the zones supplying each source increase in area by a factor of three or more.

11.4 MILIOLITE LIMESTONE AQUIFER IN WESTERN INDIA

Objectives of the study: understand the very different yields from wells between post-monsoon high water levels and pre-monsoon low water levels; also to determine whether artificial recharge could flush out saline water in the vicinity of the coast.

Key issues: differing transmissivities between high and low water tables; difficulty in retaining artificial recharge water within the aquifer system.

Exploitation from the Miliolite Limestone aquifer in western Gujarat, India, has resulted in the ingress of saline water for distances of several kilometres; Figure 11.17. The coastal area has extensive cultivated land used mainly for cash crops; groundwater is used for an irrigated second crop. However, due to the deterioration of groundwater quality in the coastal tract, there has been a fall in agricultural production. The salinity of water in the coastal tract is indicated by the electrical conductivity contours of Figure 11.18.

There are two principal geological formations, the Gaj beds, which consist of limestones, marls, clays, grits and sandstones, and the overlying Miliolite oolitic limestone which is karstified in some locations. The Gaj

formation has a thickness in excess of 30 m but has a transmissivity of less than 30 m²/d. However, the Miliolite Limestone, with a maximum thickness of 30 m, has transmissivities in the range 200–>1000 m²/d with higher values under high water table conditions. The discussion in Section 6.6.3 and Figure 6.14, concerning a specific borehole in the Miliolite Limestone, shows that the transmissivity, and hence the yield of wells, decreases appreciably as the water table falls.

Recharge, which is estimated using the soil moisture balance techniques of Chapter 3, occurs principally during the monsoon rainfall months of July to September. Due to the spread of canal irrigation in the north of the study area, recharge also occurs due to losses from irrigation canals and water-courses. Groundwater abstraction from a very large number of individual wells averages about 70 per cent of the rainfall recharge; other outflows are to the coast (equalling about 10 per cent of the rainfall recharge), from intermittent rivers to the east and west of the study area and from springs such as those indicated in Figure 11.18.

A regional groundwater model of the study area was prepared and tested; detailed information about the model is presented in Rushton and Raghava Rao (1988). An important feature of the model is the representation of the variation of transmissivity with saturated thickness; a relationship similar to that of Figure 11.14 is used. In the groundwater model, the ratios of minimum transmissivities immediately pre-monsoon and maximum transmissivities at the end of

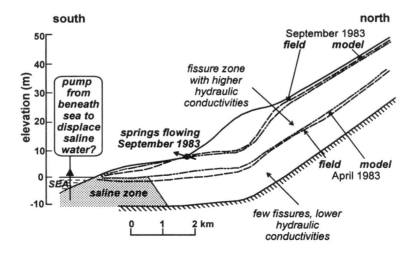

Figure 11.17   Section through Miliolite limestone aquifer system in western India; dashed line, field, and dash-dotted line, modelled groundwater heads. Reproduced by permission of IAHS Press from Rushton and Raghava Rao (1988)

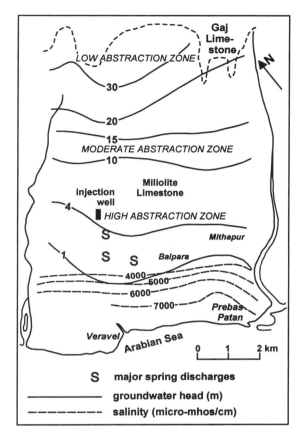

Figure 11.18 Groundwater heads and salinity measurements in Miliolite aquifer. Reproduced by permission of IAHS Press from Rushton and Raghava Rao (1988)

the recharge period lie between 1 to 2 and 1 to 6. Field and modelled groundwater heads are compared on the representative cross-section of Figure 11.17. Note that at the end of the monsoon period, September 1983, there were significant flows from the springs. Detailed comparisons of groundwater head hydrograph, reported by Rushton and Raghava Rao (1988), show differences of no more than 1.0 m compared to seasonal fluctuations of 10 to 15 m.

Artificial recharge to the Miliolite Limestone aquifer through injection wells presents no practical difficulties due to the high transmissivities; the location of the pilot injection well test site is shown in Figure 11.18. After confirming that the groundwater model adequately represents the pilot recharge experiment, the model was used to explore the likely response of more extensive artificial recharge during or soon after the

end of the monsoon period when surplus surface water is available. In the simulation, the outflows at springs increased, indicating the limited ability of the aquifer to store more water under high water-level conditions. Since the artificially recharged water passes through the more permeable regions of the aquifer, it has virtually no impact on the saline water trapped in parts of the aquifer below the minimum water table where little fissure development has occurred. The only means of displacing the saline zone is to pump the saline water from the aquifer in the region beyond the coastline (Figure 11.17); this approach has not yet been attempted.

11.5 GIPPING CHALK CATCHMENT, EASTERN ENGLAND

> *Objectives of the study*: identify the recharge processes in the chalk aquifer system where much of the chalk aquifer is overlain by low permeability drift.
> *Key issues*: interpreting differing responses in observation boreholes; developing a conceptual model of the surface water and groundwater processes.

The chalk aquifer in the catchment of the River Gipping in eastern England is covered by extensive boulder clay deposits; the outcrop of the Chalk (shown unshaded in Figure 11.19) covers a small area. However, rainfall recharge can move directly through sand and gravel deposits in the valleys to the chalk aquifer. Preliminary calculations suggest that no more than 25 per cent of the groundwater outflows from the Chalk aquifer (groundwater abstraction and flows from aquifer to river) is due to direct rainfall recharge on the chalk outcrop or where the Chalk in the valleys is overlain by permeable sands and gravels. Consequently this discussion will focus on the various sources of recharge in the study area.

11.5.1 Conceptual model

The conceptual model of Figure 11.20a is developed from the following information.

- *Geology*: the aquifer consists of the Chalk (shown unshaded in Figure 11.19) together with overlying sands and gravels. In the valley bottoms the transmissivity of the Chalk is high at about 1500 m²/d. However, on the interfluves, little development of the chalk has occurred so that the transmissivity is only 20 m²/d. The relative magnitudes of the chalk

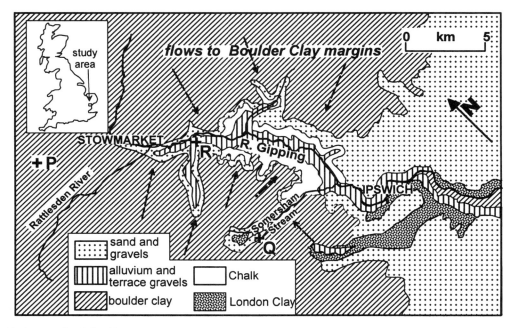

Figure 11.19   Details of the Gipping catchment. *Journal of Hydrology* **92**, Reprinted from Jackson and Rushton, Assessment of recharge components for a chalk aquifer unit, pp. 1–15, copyright (1987) with permission from Elsevier Science

transmissivities are indicated in the conceptual diagram of Figure 11.20 by the spacing of the horizontal and vertical lines of the chalk symbol. Under the interfluves the transmissivity is low (indicated by the widely spaced lines) but on the boulder clay margins and in the valleys, the Chalk transmissivity is higher. Sands and gravels lie above the Chalk and beneath the boulder clay; lateral flows occur through the sands and gravels (Lloyd *et al.* 1981).

The boulder clay, which is of glacial origin, contains extensive sand lenses. Consequently water can infiltrate into the boulder clay. Some of this water is subsequently released as a form of delayed runoff while a further outlet is by downward movement through the boulder clay to the underlying sands and gravels with a small proportion entering the Chalk.

* *Groundwater head hydrographs*: reference is made to three representative groundwater head hydrographs in the chalk aquifer; the locations are indicated on Figure 11.19.
  — **location P**, Figure 11.20b, beneath the boulder clay in a chalk interfluve; annual groundwater head fluctuations are in the range 0.1 to 0.3 m. There is no clear response to high recharge years

and there is no distinctive annual cycle; this would suggest a small, steady recharge to the low-transmissivity chalk.
  — **location Q**, Figure 11.20c, on the margins of the boulder clay; the annual fluctuations of groundwater heads in the Chalk can be as high as 8.0 m with distinctive responses to high and low recharge years.
  — **location R**, Figure 11.20d, within the chalk outcrop; annual fluctuations are in the range 0.5 to 1.2 m, depending on whether years have high or low recharge. Due to the proximity of the river and terrace gravels which have higher storage coefficients, there is a damping effect. Nevertheless, there is a clear response to rainfall recharge.

* *Field examination of flows in drainage channels, streams and rivers on boulder clay*: Another important source of information is stream flows on the boulder clay. By walking along small tributary streams the seepage and spring flows from the boulder clay, which maintain flows in the streams, can be observed. These flows continue well into the summer and in some instances there is flow throughout the year.

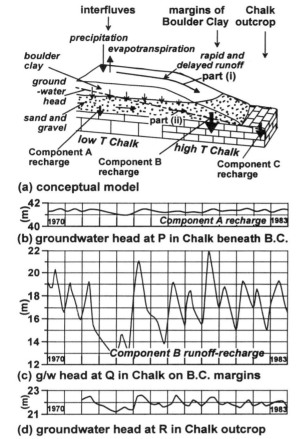

(a) conceptual model

(b) groundwater head at P in Chalk beneath B.C.

(c) g/w head at Q in Chalk on B.C. margins

(d) groundwater head at R in Chalk outcrop

Figure 11.20 Conceptual model for Gipping chalk aquifer system and hydrographs used to develop quantified conceptual model. *Journal of Hydrology* **92**, Reprinted from Jackson and Rushton, Assessment of recharge components for a chalk aquifer unit, pp. 1–15, copyright (1987) with permission from Elsevier Science

However, these streams lose water and often become dry at the boulder clay margins; this indicates that water enters the more permeable Chalk.

• *Hydrochemistry*: detailed hydrochemical studies were carried out (Heathcote and Lloyd 1984); three types of water were identified:

Type IA water which is indicative of recent recharge with no influence of boulder clay; this water occurs in places such as Location R.

Type IB water which is recent water from boulder clay covered areas; Location Q is a typical site.

Type IIB water which is indicative of long residence times in the Chalk; Location P on the interfluve is a representative setting.

From the above field information, the conceptual model of Figure 11.20a can be deduced. Some of the infiltration to the boulder clay moves vertically downwards until it reaches the underlying sand and gravels. Only a small proportion of this vertical flow enters the Chalk, due to the low transmissivity of the Chalk; this is termed Component A recharge. The fraction of vertical flow, which cannot enter the chalk, moves laterally through the sands and gravels until it reaches Chalk with a higher transmissivity where it infiltrates; this is *part (ii)* of Component B recharge. Field evidence shows extensive seepages and spring flows occur from the boulder clay for many weeks following heavy rainfall; these flows continue for much of the year. When they reach the boulder clay margins, where the chalk transmissivities are higher, they become *part (i)* of Component B recharge. The extent of the rapid and delayed runoff across the boulder clay to the boulder clay margins is indicated in Figure 11.19 by the broken lines with arrows. Over the outcrop of the Chalk, Component C recharge can be estimated from a daily soil moisture balance; see Chapter 3.

11.5.2 Quantified conceptual model

Many of the features of the quantified conceptual model are similar to the Berkshire Downs chalk aquifer discussed in Section 11.2; therefore, attention is directed to the quantification of the recharge. Estimation of the recharge components was based initially on physical water balances but subsequently refined using a regional groundwater model.

Location P is on an interfluve between valleys; the transmissivity is low and the fluctuations in the groundwater head are primarily a reflection of the groundwater head fluctuations down-gradient. Therefore it is reasonable to represent Component A as a constant recharge; from a sensitivity analysis using the groundwater model, a value of 24mm/yr was selected. Table 11.3 lists the monthly recharge for each of the components for a water year; Component A is included in the second column (note that the recharge is expressed as mm/month). In the vicinity of Component B recharge on the margins of the boulder clay, the chalk groundwater head hydrographs show much larger fluctuations; Figure 11.20c. Component B recharge consists of rapid and delayed surface runoff over the boulder clay plus water moving vertically through the boulder clay on the

Table 11.3 Recharge for each of the recharge components of the Gipping Catchment for a representative year

Month	Component A	Component B (mm/month)	Component C
Oct	2.1	26.7	2.5
Nov	2.0	20.7	6.3
Dec	2.1	18.6	16.1
Jan	2.1	26.7	29.1
Feb	1.9	36.1	27.4
Mar	2.1	53.3	13.5
Apr	2.0	45.0	5.3
May	2.1	40.0	0.0
Jun	2.0	32.1	0.0
Jul	2.1	26.7	0.0
Aug	2.1	21.4	0.0
Sep	2.0	18.0	0.0

interfluves and then moving horizontally through sands and gravels (Lloyd *et al.* 1981). A computational model similar to Figure 11.9 is used to estimate the magnitude of the different components; the large fluctuations of the hydrograph in Figure 11.20c helped in deducing suitable values for the infiltration and delay coefficients. Numerical values of Component B recharge are included in the third column of Table 11.3. Component C recharge for the chalk outcrop and areas where the Chalk is covered by permeable strata is estimated using the procedures of Chapter 3; numerical values are included in the final column of Table 11.3.

A comparison of the numerical values in Table 11.3 for the three recharge components illustrates the importance of recharge on the boulder clay margins (Component B). Not only is this component numerically greater than the other two components but it also continues into the summer. However, Component B recharge only applies on the boulder clay margins for an area of $23.3 \, \text{km}^2$; consequently it represents about 30 per cent of the total recharge (Jackson and Rushton 1987). In fact the greatest proportion of the recharge is from Component A, the slow vertical movement through the extensive boulder clay deposits into the Chalk.

11.6 FURTHER EXAMPLES

11.6.1 South Humberside Chalk

The inability of a groundwater model to reproduce field responses often leads to a reappraisal of the conceptual model and the identification of crucial features which were not recognized during the initial study. The chalk aquifer system in South Humberside (north-

eastern England) provides a good example. Initially the primary flow process in the aquifer system was assumed to be recharge to the outcrop Chalk with outflows of groundwater from springs where the Chalk becomes confined under boulder clay or to major abstraction sites in the confined region (Figure 11.21a). A groundwater model was developed based on this concept but it failed to reproduce the groundwater head responses during the drought period of 1975–76. According to the original groundwater model (the dashed line in Figure 11.21b), the groundwater head in the Grimsby area would fall to 1.1 m AOD whereas the minimum observed head (indicated by the unbroken line) was 6.3 m.

A careful review of the available information showed that the sands and gravels, which lie on top of the Chalk, could have a significant effect on the aquifer response. The plan view of Figure 11.21c shows the Chalk outcrop and the coverage of the sands and gravels which extend beneath the boulder clay, Figure 11.21d. The importance of the sands and gravels should have been identified earlier because of the historical existence of springs to the west of Grimsby, some of which were converted to pumping stations. Dating using C^{14} showed modern water in the vicinity of the sand and gravel deposits compared to older water, of more than 1000 years, away from the sands and gravels.

The sands and gravels influence the response of the aquifer system in two ways. First, they provide pathways for water to be transferred laterally; the original springs in the vicinity of Grimsby are a good example of their effect. Second, they provide an enhanced specific yield. The schematic diagram of Figure 11.21d shows that as the water table falls from 15 to 5 m, the width of sands and gravels at the water table increases from 350 m to 730 m. Furthermore, the specific yield of the sands and gravels is about 0.15 compared to a value of 0.01 for Chalk. Therefore an increasing volume of water is released from storage as the water table falls during a drought period.

When the transmissivities and specific yields of the sands and gravels are incorporated in the refined groundwater model, the results are as indicated by the chain-dotted line of Figure 11.21b; agreement with the field readings is greatly improved. Consequently, it is necessary to consider the aquifer system as a whole; that is, the Chalk together with sands and gravels.

11.6.2 Candover Augmentation Scheme

The Candover Chalk sub-catchment of the River Itchen in Southern England was selected for a river augmentation scheme (Southern Water Authority

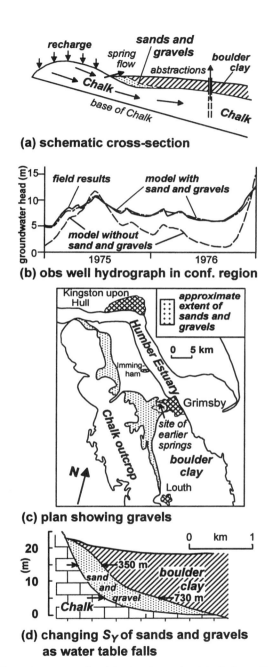

(a) schematic cross-section

(b) obs well hydrograph in conf. region

(c) plan showing gravels

(d) changing S_Y of sands and gravels as water table falls

Figure 11.21 Details of the South Humberside chalk aquifer system

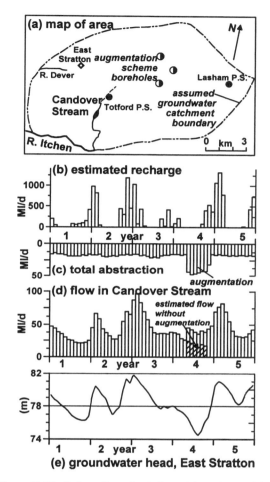

(a) map of area

(b) estimated recharge

(c) total abstraction

(d) flow in Candover Stream

(e) groundwater head, East Stratton

Figure 11.22 Information about the catchment associated with the Candover Stream (reprinted from Rushton and Rathod (1981), *Journal of Hydrology*, Copyright (1981) with permission of Elsevier Science)

1979, Headworth *et al.* 1982). Three pairs of augmentation boreholes were constructed with diameters of 550 mm and depths of 74, 100 and 117 m; the locations of the borehole sites are shown in Figure 11.22a. Since these boreholes are in the upper part of the catchment, water is taken by pipeline with a minor outfall just above and a major outfall just below the perennial head of the Candover stream. The purpose of the scheme is to augment flows into the River Itchen for abstraction further down the river. A requirement of the scheme is that there should be no detrimental effect on fishing interests in the catchment. Further information about the catchment in Figure 11.22b to e includes the estimated sub-catchment recharge, groundwater abstraction from the catchment, total flow in the Candover Stream about 500 m above its confluence with the River Itchen (note the different vertical scales) and the groundwater head at East Stratton.

Test pumping of the augmentation boreholes

was carried out in calendar year 4 which followed low recharge during the winter of year 3–4. The pumping rate from the six boreholes of 27 Ml/d was close to the design yield; there was no reduction in borehole yields with time. The period of augmentation, which can be identified in Figure 11.22c, results in the increase in flow in the Candover Stream shown in Figure 11.22d; the estimated stream flow without augmentation is shown hatched. Due to the poor recharge in the preceding winter and the increased abstraction, there is a noticeable decline in water table elevation at the observation borehole at East Stratton which is more than 6 km from the augmentation boreholes.

There were sufficient observation boreholes during the augmentation test to plot water table contours; from these contours it is possible to estimate the additional volume of aquifer dewatered. When the volume pumped during each time period plus the flow to the streams is divided by the additional volume dewatered, an estimate can be made of the overall effective specific yield *for that time period*. Table 11.4 shows the calculations for eight time increments. The most significant result is that the effective specific yield increases with time from below 0.01 to more than 0.025. Although the estimates are only approximate due to uncertainties in drawing the contours, there is nevertheless a clear indication that as the cone of depression due to the pumping spreads to include larger areas with smaller drawdowns, the higher specific yield in the zone of water table fluctuation becomes more significant. This assumption of a higher specific yield is supported by information from Figure 11.22e that annual fluctuations in groundwater heads are less than 4 m (not including year 4 which is due to the augmentation pumping) compared with the more usual fluctuations of 10–15 m for outcrop Chalk in this region.

To explore the significance of these apparently high values of specific yield, a lumped model was devised (Southern Water Authority 1979, Keating 1982). Details of the model are presented in Figure 11.23a; to ensure consistency of units the length of the aquifer block parallel to the stream is L, with B the dimension perpendicular to the stream. A specified inflow to the model is the recharge q (units L/T) with abstraction Q_P as a specified output. A flow Q_W occurs to the winterbourne (non-perennial) stream when the representative groundwater head h is above the elevation of the winterbourne stream z_W. In addition, there are outflows to the perennial part of the Candover Stream Q_C and to the River Itchen Q_I at elevations z_C and z_I. Since this is a lumped model, single *representative* values of groundwater head and stream elevation are used.

Zones of higher transmissivity and storage coefficient coincident with the zone of water table fluctuation are introduced, these parameters change with the elevation of the water table as shown on the right of Figure 11.23a. When the water table is below the winterbourne outlet the transmissivity and storage coefficient are T_b and S_b; when the water table is above the outlet the values are T_W and S_W. A transition between the two values is needed for computational stability. Outflows to the streams and rivers are calculated as follows. Flow only occurs to the winterbourne part of the Candover Stream when the groundwater head is above stream elevation:

$$Q_W = 2T_W B \frac{h - z_W}{L} \quad \text{for } h \geq z_W \tag{11.4a}$$

$$Q_C = 2T_b B \frac{h - z_C}{L} \tag{11.4b}$$

Table 11.4 Estimation of incremental specific yield of Candover Catchment for different time periods calculated from volume pumped, flow to streams and the estimated change in volume dewatered due to the fall in water table

Duration	Days pumping	Volume pumped (Mm³)	Flow to streams (Mm³)	Change in vol. dewatered (Mm³)	Specific yield for time period
5 May–2 July	30	0.843	0.155	99.4	0.007
3 June–18 July	15	0.364	0.070	26.9	0.011
19 June–3 July	15	0.404	0.080	40.7	0.008
4 July–17 July	14	0.366	0.075	29.0	0.010
18 July–2 Aug	16	0.388	0.080	21.0	0.015
3 Aug–14 Aug	12	0.365	0.065	14.6	0.021
15 Aug–27 Aug	13	0.349	0.065	9.0	0.031
28 Aug–11 Sept	15	0.354	0.080	10.4	0.026

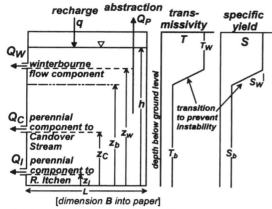

(a) lumped model of Keating

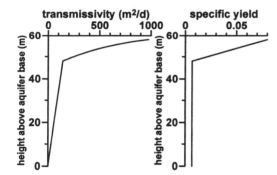

(b) variation of parameters in regional model

Figure 11.23 Variation of aquifer parameters with groundwater elevation: (a) lumped model, (b) regional groundwater model

$$Q_I = 2T_b B \frac{h - z_I}{L} \qquad (11.4c)$$

The equation of continuity for the aquifer block when the groundwater head is above the winterbourne outlet is

$$q(L \times B) - S_W \frac{\Delta h}{\Delta t} = Q_W + Q_C + Q_I + Q_P \qquad (11.5)$$

where $\Delta h/\Delta t$ is the rate of change of the groundwater head and S_W the specific yield when the winterbourne is flowing. When the groundwater head is below the winterbourne outlet Q_W is omitted, the storage coefficient may also change.

The above equation can be used to calculate the groundwater head at one time step from conditions at the previous time step. A monthly time step is used in the computation with values of the transmissivity and storage coefficient adjusted to obtain a good match between modelled and field values of stream flow and groundwater head; Figure 11.22d and e. Differences between field and modelled flows in the Candover Stream are in the range of 5 to 15 Ml/d; differences in groundwater head can be as high as 6 m during dry years although for most of the time the agreement is within 2 m (Keating 1982). Numerical values of the parameters in the lumped model are $T_W = 10\,000\,\mathrm{m^2/d}$, $T_b = 1000\,\mathrm{m^2/d}$, $S_W = 0.05$ and $S_b = 0.01$. The numerical values of the transmissivity do not have a direct physical meaning, although the relative magnitudes are significant. Numerical values of the specific yield can be used directly.

This lumped model analysis was supplemented by a regional model study (Rushton and Rathod 1981). The variation of transmissivity is of the same form as that introduced in Section 11.2; numerical values selected for the Candover study are shown in Figure 11.23b. The variation of specific yield with height of the water table above the base of the aquifer is important; the variation is shown on the right-hand diagram of Figure 11.23b. Agreement between the regional model and field results is good (Rushton and Rathod 1981) with the differences less than those for the lumped model. Nevertheless, the lumped model is invaluable in identifying the significance of the variation of specific yield with water table elevation.

11.6.3 Hesbaye aquifer, Belgium

The Hesbaye aquifer in Belgium has a Chalk outcrop of $350\,\mathrm{km^2}$; it is located to the north of the River Meuse near Liège (Dassargues and Monjoie 1993). The groundwater gradient in the Chalk is about 5 m in each km towards the River Geer. Information about strata associated with the chalk aquifer system is presented in Table 11.5.

Groundwater is withdrawn from the Campanian Chalk using a system of about 45 km of galleries; these galleries have a cross-section of 2.0 to $2.5\,\mathrm{m^2}$. Water which enters these galleries flows by gravity to supply Liège and its suburbs. A fourteen-year average annual water balance expressed as mm/annum is given below:

rainfall	740 mm
evapotranspiration	525 mm
effective infiltration (175 to 275 mm)	215 mm

Table 11.5 Geological information and aquifer properties of Hesbaye aquifer

Geological sequence	Thickness	Hyd. cond. (m/d)	Specific yield
Recent alluvial and colluvial deposits	up to 5 m	–	–
Sands and loess	2–20 m	0.0001 to 0.02	
Residual conglomerate	2–15 m	1.0 to 800	0.05 to 0.10
Maastrichtian chalk	10–15 m	20 to 500	0.07 to 0.15
Hardened upper Campanian chalk	>1 m	–	–
Campanian chalk	20–40 m	1 to 50	0.05 to 0.15
Hardened calcareous clay	≈10 m	effectively impermeable	

flow to River Geer (52 Mm^3 equivalent to)	120 mm
groundwater abstraction (60 Ml/d)	65 mm
storage release	15 mm
groundwater flow out of catchment area	15 mm

Values of aquifer properties deduced from about 150 pumping tests are listed in the final two columns of Table 11.5. Because they represent both matrix and major fracture properties, the hydraulic conductivity and specific yield are higher than those for Chalk in the UK quoted in Sections 11.2, 11.5 and 11.6.2. The main drainage axes of the aquifer are below dry valleys where there are both fractures and karstic features.

A groundwater model of the aquifer has been developed using a finite element approach. Of particular importance is the representation of the heterogeneous nature of the aquifer system. Special elements are used to represent the porous medium and infiltration; in addition, one-dimensional 'pipe' elements have been developed to simulate the galleries. The model has been calibrated against historical data and tested against variations in hydraulic conductivity and storage coefficient.

11.7 CONCLUDING REMARKS

Limestone and chalk aquifers exhibit substantial changes in transmissivity between high and low water table conditions. Provided that there are no substantial extensive low permeability layers within the aquifer, a single-layer model approach, similar to that discussed in Chapter 9, can be used. Specific issues for limestone and chalk aquifers include the identification of variation in hydraulic conductivity with depth, the representation of intermittent springs and streams and the estimation of recharge when drift covers part of the aquifer system.

For the chalk aquifer of the Berkshire Downs, two pumping tests conducted in the same borehole with different rest water levels provide data from which the variation of hydraulic conductivity with depth can be estimated. Significant regional variations in transmissivity also occur. Comparisons between field hydrographs for stream flows and groundwater heads, and values obtained from numerical models with variations of hydraulic conductivity with depth, show improved agreement compared to results when the variation of hydraulic conductivity with depth is ignored.

Studies of the Southern Lincolnshire Limestone aquifer system over several decades have resulted in a number of advances in groundwater hydrology. Of particular significance is the variation of hydraulic conductivity with depth, the contribution of runoff-recharge from less permeable strata and the substantial inflows and outflows when streams and rivers cross permeable limestone inliers. A further important feature is the representation of uncontrolled overflowing artesian boreholes.

Similar numerical models are used for a Miliolite limestone aquifer in western India. However, the conceptual model has different features. Recharge occurs during the monsoon season over most of the aquifer; during the dry season there is extensive abstraction of groundwater for irrigated crops. This irrigation pumping has resulted in further ingress of saline water in the coastal plain. To move the saline water in the coastal plain back towards the coast, artificial recharge of the aquifer system was developed. Although there is no difficulty in inputting the artificial recharge water into the aquifer system, this additional water usually leaves the aquifer system from high level springs rather than moving down-gradient to flush out the saline water.

Most of the Chalk of the Gipping catchment is covered by drift. Little water moves through the boulder clay on the interfluves to recharge the aquifer; this is partly due to the low chalk transmissivities of the interfluves. However, water is discharged from the boulder clay as delayed runoff. It joins water which moves through the sands and gravels above the chalk aquifer to enter the Chalk at the boulder clay margins where the transmissivity increases significantly. Runoff from the boulder clay is a major source of aquifer recharge.

A common theme of each of these case studies is the rapid lateral movement of groundwater through the aquifer system during high water table conditions following significant recharge. The storage properties of limestone and chalk aquifers are limited so that much of the recharge is lost during the rainy season/winter. Consequently stream flows and water available for abstraction is significantly reduced during the non-recharge period.

12

Numerical Modelling Insights

12.1 INTRODUCTION

A number of books consider in detail the method-ologies and techniques associated with the development and testing of numerical models for groundwater flow (e.g. Anderson and Woessner 1992, Kresic 1997, Pinder 2002). Handbooks associated with various groundwater modelling packages are also available. In earlier chapters of this book there are descriptions of many important features of numerical models. In the introduction to Part III: Regional Groundwater Models, lists are presented of different aspects of numerical models which are addressed in the case studies.

The purpose of this chapter is to consider additional topics which are central to the development of reliable numerical models. The title of this chapter, *Numerical Modelling Insights*, indicates that the aim is to identify certain strategic features associated with numerical models. For example, rivers may be intermittent, or the depth and width of the river may change significantly during a year; methods of defining and simulating these conditions are described. Boreholes pumping water from multi-level aquifers must be represented consistently. Steady-state solutions are often used for starting conditions but real steady-state conditions rarely occur in practice; instead a *time-instant* concept can be used in which the *equivalent recharge* is a func-tion of the actual recharge and the release of water from storage. As demonstrated in Chapters 7 and 8, the time-instant approach is particularly useful for radial flow problems. This chapter also includes a discussion concerning both the areal extent of the aquifer system and the time period which must be included in numer-ical groundwater models. The selection of suitable boundary conditions and aquifer parameters, and real-istic targets for agreement between field data and mod-elled results, are also considered. Strategies for model refinement are proposed. Sensitivity analyses and pre-dictive solutions are reviewed.

Due to the rapid and continuing development of new computer hardware and software (including new lan-guages), this material on numerical models is not linked directly to existing codes or packages. Instead the dis-cussion concentrates on the principles and practicali-ties of preparing numerical models from quantified conceptual models.

This chapter concludes with an *Evaluation of Conceptual and Computational Models*. Topics include approaches to groundwater modelling and a review of monitoring, recharge, model calibration and refine-ment. Sustainability, social implications and climate change are also examined. Finally, selected substantive issues requiring further investigation are identified.

12.2 REPRESENTATION OF RIVERS

This section considers two issues. First, the inclusion of intermittent rivers in regional groundwater models and, second, the representation of rivers which change in size and hence river coefficient during high flows.

12.2.1 Intermittent rivers

In many aquifers which respond quickly to seasonal changes, streams or rivers become dry over certain reaches. This can occur due to changes in seasonal con-ditions, the effect of groundwater pumping or due to the discharge of water to the stream above the perennial head. This discussion relates to the approach used in the MODFLOW Streamflow-Routing Package, STR1

Groundwater Hydrology: Conceptual and Computational Models. K.R. Rushton
© 2003 John Wiley & Sons Ltd ISBN: 0-470-85004-3

(Prudic 1989, Anderson and Woessner 1992) which includes the effect of intermittent streams. Note that in this discussion the flow *from* the aquifer *to* the river Q_r is of the opposite sign to $QRIV$ in MODFLOW which is defined as the leakage from river to aquifer. The technique of modelling intermittent rivers is introduced with reference to Figure 12.1 in which Figure 12.1a, taken from Figure 4.11b, shows the relationship between aquifer to river flow and the difference between groundwater head and river surface elevation. Figure 12.1c shows conditions for a river reach; Figure 12.1b contains a flow chart of the calculations.

Three alternative conditions are identified in Figure 12.1. Conditions A and B are the standard aquifer–river interaction relationships, but they only apply if the inflow to the reach Q_{in} is greater than any loss from the river calculated from Conditions A or B. Condition C applies when there is insufficient water entering the river reach to meet the calculated loss; the loss from aquifer to river is then set to Q_{in}.

The procedure is illustrated by an example, Table 12.1. For river reaches 1 and 2 the groundwater heads are below stream level, hence there is no flow in the river so that Condition C of Figure 12.1 applies. However between reaches 2 and 3 there is an inflow from a treatment works of 2500 m³/d in the non-perennial part of the river. For reach 3 the groundwater head is below bed level (Condition B) hence the aquifer to river flow is 2400 (124.7 − 125.2) = −1200 m³/d/km. Consequently there is a river flow of 2500 − 1200 = 1300 m³/d entering reach 4. The groundwater head in reach 4 is between the bed level z_b and the stream surface level z_r, hence Condition A applies with an aquifer to river flow of −1253 m³/d/km; this leaves an inflow into reach 5 of only 47 m³/d. In reach 5 the calculated aquifer–river flow is −565 m³/d/km; this is greater than the inflow into the reach hence Condition C applies and the quantity of water lost from this reach is 47 m³/d. There is no river inflow into reach 6. However for reach 6 and the following reaches the groundwater head is above river surface level; the resultant groundwater inflows ensure a steady increase in accumulated river flows downstream.

Table 12.1 quotes the groundwater heads along the river reach; these groundwater heads are determined within the iterative procedure of solving the finite difference equations. When there are several interacting non-perennial rivers, the time step must be small (preferably less than 10 days) with a suitable iterative method of solution of the finite difference equations selected. The preconditioned conjugate gradient method, the strongly implicit procedure and successive over-relaxation have all proved to be suitable.

12.2.2 Rivers with significant seasonal changes in stage and wetted perimeter

In Section 4.3.4 the discussion of river–aquifer interaction assumes that the cross-section of the river and the river surface elevation do not change appreciably. Therefore the equation for flow from the aquifer to the interconnecting river is

$$Q_r = RC(h - z_r) \quad \text{for } h \geq z_b \tag{12.1}$$

The river stage (elevation of the river surface) z_r and the coefficient defining the interaction RC are assumed to be constants. However, for major rivers in semi-arid areas the water surface elevation and the wetted perimeter do change significantly. Certain regional groundwater flow codes and the original MODFLOW code permits different values of the river stage and river coefficient for each stress period, but many of the graphical interfaces do not include this option.

If varying coefficients are to be included, how are they determined? The following example refers to the River Yobe in north-eastern Nigeria (for further information see Section 12.6.7 and Figure 12.11). River flows are gauged at Gashua and Geidam which are about 100 km apart. The river has a width of 25 to 40 m, the riverbed is about 5 m below the floodplain. Headwaters of the river system are on the Jos Plateau in central Nigeria, the river flows some 800 km northeast to Lake Chad; Gashua is approximately halfway down the catchment with the catchment area above the gauge of 30 000 km². The rainy season on the Joss Plateau commences about two months earlier than in the study area. Consequently, runoff from the basement rocks of the upper catchment arrives in the study area before the onset of the rainy season. Figure 12.2a shows averaged monthly flows for the two gauging stations.

There are four distinct time periods with different flow patterns between the upstream and downstream gauges.

(a) By June there is a measurable flow at the upstream gauge, while the flow at the downstream gauge is lower, indicating that water is lost between the two gauges; most of this water flows from the river to recharge the floodplain aquifer between the two gauging stations.

(b) During August, September and October flows at the upstream gauge increase appreciably; overbank flow occurs, leading to flooding extending up to 1 km from the river bank. This flooding explains

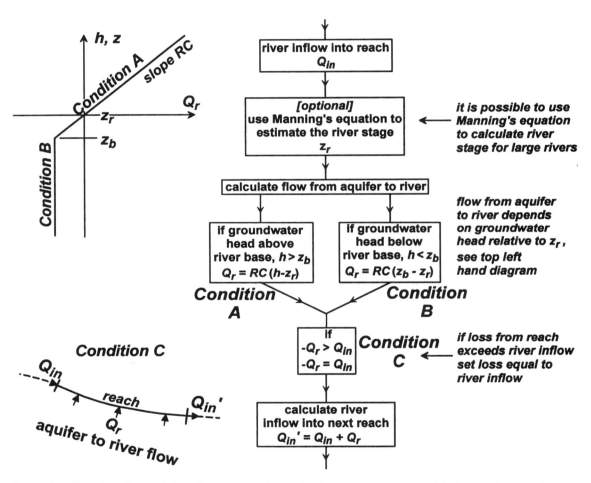

Figure 12.1   Flow chart for simulation of aquifer–river interaction for a river reach which might become dry; note that Q_r is the flow from aquifer to river

Table 12.1 Typical calculations of aquifer–river interaction with inflow of 2500 m³/d between reaches 2 and 3; each river reach has a length of 1.0 km

Reach no.	River level z_r (m)	Bed level z_b (m)	River coefficient RC (m³/d/m/km)	Accumulated flow Q_{in} (m³/d)	Groundwater head h (m)	Condition	Aquifer to river flow Q_r (m³/d)
				0			
1	130.0	129.5	1200		126.456	C	0
				0			
2	127.6	127.1	1800		125.142	C	0
				0 + 2500			
3	125.2	124.7	2400*		124.123	B	−1200
				1300			
4	122.8	122.3	3000		122.382	A	−1253
				47			
5	120.4	119.9	3600		120.243	C	−47
				0			
6	118.0	117.5	4200		118.114	A	479
				479			
7	116.0	115.5	4800		116.117	A	562
				1041			
8	114.0	113.5	5400		114.147	A	794
				1835			

Notes: Condition A: groundwater head above streambed level, both +ve and −ve flows possible (Figure 12.1).
Condition B: groundwater head below streambed level, stream loses water to aquifer.
Condition C: either Condition A or B with loss from stream to aquifer but insufficient water in stream to meet the loss; consequently loss equals inflow to reach.
* Maximum loss from stream reach = 1200 m³/d.

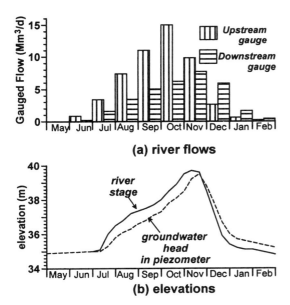

Figure 12.2 Information for River Yobe: (a) averaged measured flows at upstream and downstream gauges, (b) river stage and groundwater head in nearby piezometer

part of the loss between upstream and downstream gauges; the remainder of the loss is caused by river water entering the aquifer through the bed and sides of the river channel.

(c) For November and December there is a decrease in total flow at the upstream gauge. In November the flow at the upstream gauge is still higher than at the downstream gauge but by December the flow at the downstream gauge is higher than at the upstream gauge, indicating that the aquifer supplies water to the river.

(d) In January and February the aquifer continues to supply water to the river as evidenced by the downstream flows remaining higher than the upstream flows; however, the contribution from the aquifer in February is small.

In Figure 12.2b there is a plot of the river stage and the groundwater head in a piezometer in the vicinity of the river. From this information it is possible to identify whether the river feeds the aquifer or the aquifer feeds the river. The river stage is higher than the groundwater head from June to November, indicating that the

Table 12.2   Estimation of river coefficient for four different river stages (see also Figure 12.3)

	River Stage (m)	Head in obs well (m)	Aquifer to river flow Ml/d/km	River coeff Ml/d/m/km	Comments
(a)	35.8	35.2	−9	15	River has just started flowing, river stage is above aquifer heads, only part of the river is filled, small river coefficient
(b)	39.5	38.5	−66	66	River channel is full, very significant flows from river to aquifer, aquifer heads rise due to inflow from river, high river coefficient
(c)	37.5	38.2	36	51	River channel emptying due to reduction in flow from upstream, groundwater heads remain high, river coefficient still high
(d)	35.5	36.2	12	17	Small flow in river, mainly groundwater-fed, aquifer head still above river level

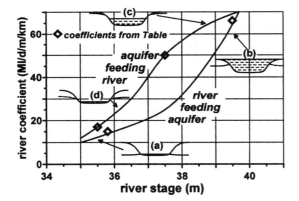

Figure 12.3   Diagram showing how the river coefficient of the Yobe River changes with river stage

river feeds the aquifer; this corresponds to the river flow periods (a), (b) and the first month of (c). The groundwater head is higher than the river stage from December to February; that is for the second half of period (c) and period (d).

With information about changes in flows and differences between the river stage and groundwater head, it is possible to estimate how the river coefficient varies with stage. Four specific times are identified in Table 12.2; they lie within the four time periods (a) to (d) above. The table shows how values of the river coefficient are calculated from the river stage and the groundwater head in an observation piezometer; the

flow from river to aquifer is calculated from the difference between upstream and downstream gauges. Equation (12.1) is used to estimate the river coefficient; the flow from aquifer to river and the river coefficient are quoted for a typical 1-km reach of river. Smoothing and adjustments to certain field readings are necessary due to the difficulty in obtaining accurate flow measurements, especially when the river is in flood. Results from calculations, such as those of Table 12.2, are used to construct Figure 12.3 which shows how the river coefficient for rising river stages varies in a different manner to the river coefficient for falling river stages. The actual numerical values of the river coefficient from Table 12.2 are indicated in Figure 12.3 by the open diamond symbols. Section 12.6.7 describes a simulation when the varying river coefficients are included in a groundwater model.

12.3 REPRESENTING BOREHOLES PUMPING WATER FROM MULTI-LAYERED AQUIFERS

When a single pumped borehole penetrates a multi-layered aquifer, the distribution of the discharge between the layers is unknown. The selection of appropriate conditions is important (Ruud and Kabala 1997). Bennett *et al.* (1982) represent this situation by modifying the finite difference equations at each node in each layer which is open to the borehole. Normally a node in a finite difference three-dimensional layered system is connected to six nodes, four in the same layer, one above and one below. However, for a node representing the borehole there is an additional connection

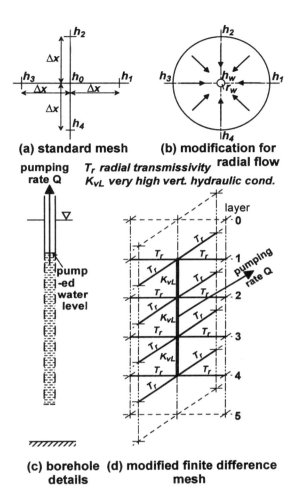

(a) standard mesh

pumping rate Q

T_r *radial transmissivity*
K_{vL} *very high vert. hydraulic cond.*

(b) modification for radial flow

(c) borehole details   **(d) modified finite difference mesh**

Figure 12.4   Representation of boreholes in multi-layered aquifers

$$\frac{T}{\Delta x^2}(h_1 + h_2 + h_3 + h_4 - 4h_0) = -q = \frac{Q}{\Delta x^2} \qquad (12.2)$$

in which the final term is the abstraction Q (of opposite sign to the recharge) divided by the mesh area to give an equivalent recharge. For a well at node 0 with well radius r_w and head h_w, the Thiem equation (Eq. (5.3)) can be rewritten in terms of the nodal heads to describe the radial flow

$$(h_1 + h_2 + h_3 + h_4 - 4h_0) = \frac{2Q}{\pi T}\ln\left(\frac{\Delta x}{r_w}\right) \qquad (12.3)$$

A modified transmissivity for radial flow T_r is introduced for the mesh lines connected to the pumped borehole as shown in Figure 12.4d, hence Eq. (12.2) becomes

$$\frac{T_r}{\Delta x^2}(h_1 + h_2 + h_3 + h_4 - 4h_0) = -q = \frac{Q}{\Delta x^2} \qquad (12.2a)$$

Equations (12.2a) and (12.3) are identical if

$$T_r = \frac{\pi T}{2\ln(\Delta x/r_w)} \qquad (12.4)$$

Therefore if the transmissivity is modified at all nodes connected to the open section of the borehole as shown in Figure 12.4d, radial flow is represented. For example, with a borehole of 0.3 m radius and a mesh spacing of 500 m, the modified transmissivity is 0.212 of the original value. If there is an increased resistance to flow approaching the borehole due to clogging or the presence of a seepage face (see Eq. (5.5) and Figure 5.9b), the modified transmissivity will be lower than that calculated from Eq. (12.4).

Representing vertical flow in the open borehole: an open borehole has no resistance to vertical flow. An alternative explanation is that groundwater heads on the periphery of the borehole are the same over the full depth of the open or screened section of the borehole, but the magnitude of this groundwater head is unknown. In the numerical model this feature is represented by setting a very high vertical hydraulic conductivity between the nodes on the section of the borehole open to the aquifer as shown by the thick vertical line in Figure 12.4d. The discharge from the borehole can be taken from any location on this vertical line in a similar manner to the freedom to locate a pump at any depth in the water column. In the iterative solution of the finite difference equations, the groundwater head

to all the nodes on the open section of the borehole. In an alternative approach introduced below, the radial flow towards the borehole through each of the layers is simulated by a modified transmissivity representing radial flow; in addition, a high vertical conductance is introduced on the line of nodes which represent the open borehole. This approach is illustrated in Figure 12.4.

Revised transmissivity to represent radial flow: the standard finite difference mesh is shown in plan in Figure 12.4a; a square mesh is selected for convenience. The finite difference equation for steady-state flow with recharge, based on the methodology of Section 2.7, is

in the well converges to a value so that the sum of the quantities taken from each of the modified transmissivities T_r equals the pumping rate Q.

Figure 12.4c shows details of a typical partially penetrating borehole while Figure 12.4d is the corresponding finite difference mesh. The borehole has solid casing through layer 0, layer 5 is below the penetration of the borehole; no changes are necessary to the transmissivities in these layers. For layers 1 to 4, modifications are required. For all the horizontal mesh lines connected to the borehole the transmissivity is modified using Eq. (12.4). Also, the vertical hydraulic conductivity on the line of the well connecting the modified transmissivities is set to a high value; a multiplying factor of 1000 is usually adequate. The total pumped discharge can be taken from any one of the borehole nodes.

12.4 TIME-INSTANT CONDITIONS

12.4.1 Introduction

Before carrying out time-variant numerical simulations, preliminary 'steady state' solutions are often recommended. What does a steady-state simulation mean? Is an aquifer ever in a steady-state condition? This question is explored using a representative example. Figure 12.5 refers to a one-dimensional aquifer, 6.0 km long, which feeds into a large lake in which the ground-water head remains at 0.0 m. There is a repetitive cycle of a lateral inflow of $1.0 \, \mathrm{m^3/d/m}$ for six months followed by zero inflow for six months. The aquifer is unconfined with a constant transmissivity of $200 \, \mathrm{m^2/d}$ and a specific yield of 0.04. Groundwater head profiles are plotted for four different times and compared with the linear head distribution which results from a steady average inflow of $0.5 \, \mathrm{m^3/d/m}$.

- Soon after recharge commences, line (a) in Figure 12.5, the head distribution is highly curved, with heads above the average value to 1 km but below the average for the remainder of the aquifer.
- Towards the end of the recharge period, line (b), the groundwater heads are near to a linear distribution with heads approaching double the average values.
- Soon after recharge ceases, line (c), the heads are again highly curved.
- Towards the end of the recovery period, line (d), the groundwater heads approach zero.

From these results it is clear that on no occasion is the time-variant groundwater head distribution close to the average value. If the transmissivity was lower, the specific yield higher, or the length of the aquifer greater, the differences between the average and time-variant values would be less. Nevertheless, in many practical situations, time-variant groundwater heads never coincide with the steady-state heads resulting from average

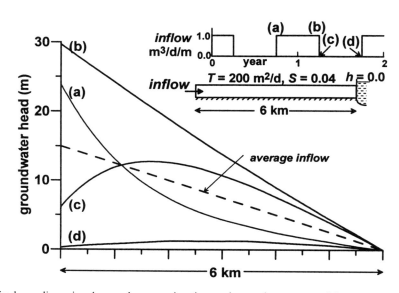

Figure 12.5 Idealised one-dimensional example comparing time-variant and average conditions

inflows and outflows. However a *pseudo-steady state* does have a physical meaning when the concept of *time-instant* simulations is examined.

12.4.2 Basis of *time-instant* approach

The differential equation describing time-variant regional groundwater flow in an unconfined aquifer ignoring confined storage effects is:

$$\left(\frac{\partial}{\partial x}T_x\frac{\partial h}{\partial x}\right)+\left(\frac{\partial}{\partial y}T_y\frac{\partial h}{\partial y}\right)=S_Y\frac{\partial h}{\partial t}-q \qquad (2.54c)$$

If the right-hand side (with the sign reversed) is written as

$$equivalent\ recharge = q - S_Y\frac{\partial h}{\partial t} \qquad (12.5)$$

with the equivalent recharge at a specific time estimated using field information, a pseudo-steady state equation is obtained:

$$\left(\frac{\partial}{\partial x}T_x\frac{\partial h}{\partial x}\right)+\left(\frac{\partial}{\partial y}T_y\frac{\partial h}{\partial y}\right)=-\ equivalent\ recharge$$
$$(12.6)$$

This is the basis of the *time-instant* approach.

During a period with high recharge, the equivalent recharge equals the actual recharge minus the quantity of water taken into storage. During a period with zero recharge, the equivalent recharge, which supplies the quantity of water drawn through the aquifer due to the transmissivity and the head gradients, is provided solely by the water released from storage. In deep sandstone aquifers, the equivalent recharge remains approximately constant from summer to winter because the transmissivity hardly changes and the head gradient also remains relatively constant. However, in chalk or limestone aquifers where there are significant variations in transmissivity between high and low groundwater heads (see Chapter 11), the equivalent recharge is likely to show substantial variations.

12.4.3 Examples of sandstone and limestone aquifers

The concept of equivalent recharge is explored further by considering monthly values of actual recharge to the water table and water released from storage. Years with highly variable rainfall and recharge are selected to illustrate the concept of equivalent recharge. The values quoted in the following tables are taken from numerical solutions; an example of the estimation of equivalent recharge from field information can be found in Section 12.5.2.

Considering first an unconfined sandstone aquifer, estimates of monthly recharge to the water table, storage changes and the resultant equivalent recharge are quoted in Table 12.3. Note that for this particular example, the recharge to the water table reaches a maximum in early summer; this occurs because of the substantial delays as the effect of the recharge moves from beneath the soil zone, through the unsaturated zone, to the water table (see Section 3.5.4). The water table tends to rise in early summer as water is taken into storage, indicated by the negative signs in the column headed Storage release. Some seasonal changes in equivalent recharge do occur, but they are small. With annual water table fluctuations of less than 1.0 m, the groundwater gradients remain relatively constant from summer to winter and the transmissivity is effectively constant. The groundwater system has considerable 'momentum' due to the high specific yield; consequently changes cannot occur quickly. The average equivalent recharge of 17.4 mm/month, when multiplied by the recharge area, represents the quantity of water moving through the aquifer system; this supplies the abstraction demands and maintains the flows to springs and rivers.

Table 12.4 refers to a limestone aquifer with substantial recharge in November, January and April but moderate to low recharge for December, February and March. In November most of the recharge refills storage, hence the equivalent recharge is only moderate. The next high recharge month is January;

Table 12.3 Equivalent recharge for a sandstone aquifer (all units are mm/month)

Month	Actual recharge	Storage release	Equivalent recharge
October	+3.7	+14.0	17.7
November	+5.3	+12.1	17.4
December	+2.5	+14.6	17.1
January	+2.2	+14.3	16.5
February	+6.7	+9.5	16.2
March	+12.3	+3.9	16.2
April	+33.3	−16.4	16.9
May	+32.7	−15.6	17.1
June	+50.0	−31.6	18.4
July	+33.6	−14.8	18.8
August	+19.4	−0.8	18.6
September	+11.6	+6.7	18.3

Table 12.4 Equivalent recharge for a limestone aquifer (all units are mm/month)

Month	Actual recharge	Storage release	Equivalent recharge
October	+0.2	+8.4	8.6
November	+50.3	−39.8	10.5
December	+10.6	+1.7	12.3
January	+58.2	−36.3	21.9
February	+5.5	+18.9	24.4
March	+20.2	+4.5	24.7
April	+38.4	−14.1	24.3
May	+4.9	+17.5	22.4
June	+9.8	+10.2	20.0
July	+0.1	+14.8	14.9
August	+0.7	+10.8	11.5
September	+0.0	+9.4	9.4

due to increases in water table elevations and the resultant higher transmissivities, more water passes through the aquifer system, hence the equivalent recharge is higher and the quantity taken into storage is proportionally smaller. In February, the actual recharge is small. However, because the water table elevation has increased following January's high rainfall, the transmissivities are higher. Consequently the quantity of water drawn through the aquifer is high. This requires an increased equivalent recharge which results in considerable volume of water taken from storage.

The highest equivalent recharge in this limestone aquifer is in March; it is almost three times the lowest equivalent recharge recorded in Table 12.4. This contrasts with the example of the sandstone aquifer, Table 12.3, where the highest equivalent recharge is 1.16 times the lowest value.

12.4.4 Time-instant solutions

The *time-instant* approach has many applications in groundwater resource estimation; the application of the time-instant technique in radial flow problems is discussed in Section 7.3. As demonstrated in the examples of the sandstone and limestone aquifers, the time-instant concept provides insights as to how the aquifer system works and in particular whether recharge is stored in the aquifer or moves quickly to streams and rivers. The concept also indicates why water is available to flow through the aquifer during a dry summer with no recharge. Precise field information may not be available about the magnitude of the recharge or the quan-

tity of water taken into or from storage. Nevertheless, with approximate estimates of the equivalent recharge, the time-instant approach provides a methodology for quantifying the impact of changes.

Steady-state solutions are often obtained during the early stages of groundwater model development; the time-instant approach provides a more logical and meaningful explanation of these solutions.

12.5 INITIAL CONDITIONS

When defining a time-variant problem in mathematical terms, initial conditions must be specified. For analytical solutions, such as the Theis theory (Section 5.4), the initial condition is that the drawdown is zero everywhere at zero time. For regional groundwater studies, initial conditions are far more complex. Consequently a crucial aspect is the selection of appropriate initial conditions which realistically represent the historical response of the aquifer and also provide appropriate flow conditions at the start of the simulation.

The discussion on initial conditions is divided into three parts. The first issue is whether specified heads or specified flows should be used for initial conditions. Second, initial conditions for high storage aquifers are considered; sandstone aquifers are used to illustrate the approach. Finally, a more challenging situation is addressed; initial conditions for aquifers with significant seasonal variations in transmissivity.

12.5.1 Specified heads or specified flows

Two alternatives are available when specifying initial conditions for numerical solutions. Either the simulation is started from specified groundwater heads or the simulation starts with specified flows (although when specified flows are imposed, groundwater heads must be defined at some locations to achieve a stable solution). It is not possible for both groundwater head and vertical inflow to be specified at the same location, although certain groundwater model codes appear to allow both to be enforced. If a vertical inflow is applied at each nodal point (or cell) together with the transmissivity and boundary conditions, the groundwater head distribution can be determined. However, if the groundwater head is specified together with the transmissivity and boundary conditions, this automatically defines the recharge. The recharge, which results from specified groundwater heads, is often in error.

These two approaches are explored using a simple one-dimensional example as shown in Figure 12.6a. The aquifer is 4.0 km long, has a uniform transmissiv-

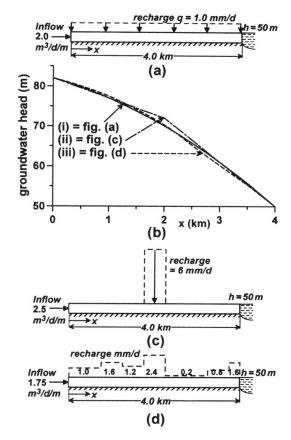

Figure 12.6 Comparison of inferred recharge for simulations with specified flows or specified heads: (a) problem with specified flows, (b) alternative head variations, (c) flows corresponding to two straight-line head approximation, (d) flows corresponding to smoothed curve for heads

$$q_0 = -mK(h_{-1} - 2h_0 + h_1)/\Delta x^2 \qquad (2.60a)$$

According to Eq. (2.60a) there is no recharge wherever the groundwater head has a linear variation. Therefore, recharge only occurs at the intersection of the two straight lines. At the intersection, the groundwater head is 72 m with values of 74.5 and 66.5 m for the nodes on either side, hence the recharge at the intersection point

$$q = -500(74.5 - 2 \times 72.0 + 66.5)/500^2 = 0.006 \, \text{m/d}$$

This recharge of 6.0 mm/d is shown in Figure 12.6c; the inflow on the left-hand boundary of 2.5 m³/d/m is calculated from Darcy's Law. Clearly these flows differ significantly from the inflow and recharge defined in Figure 12.6a.

As an improvement, the initial groundwater heads are approximated by the curved broken line (iii) in Figure 12.6b, which is almost indistinguishable from the groundwater heads calculated for the uniform recharge. The equivalent inflows, shown in Figure 12.6d, are a lateral flow of 1.75 m³/d/m and recharge at nodal points varying from 0.2 to 2.4 mm/d. Even with smoothed initial groundwater heads, the resultant recharge distribution is somewhat erratic. Therefore, when specified heads are used as initial conditions, the inflows and outflows may be far from the correct values. Furthermore, it may take several years for the effect of incorrect initial flows to dissipate.

12.5.2 Initial conditions for sandstone and other high storage aquifers

For sandstone aquifers, especially if they are heavily exploited, initial conditions should preferably represent the situation before significant abstraction occurred. A typical example is the Permo-Triassic sandstone aquifer of the Mersey Basin in north-west England. (Figure 12.7 and Section 9.4). Intense industrial and public supply abstraction has changed the groundwater conditions significantly with many springs drying up and substantial quantities of water drawn from the Mersey Estuary. In the Lower Mersey Area, there is little field information about water table elevations for the mid-1800s when industrial abstraction commenced. Therefore initial conditions in 1847 are based on the long-term average recharge and estimated abstractions for the mid-1800s. Significant discharges occurred from major springs which were identified from old maps (indicated by the diamond symbols on Figure 12.7). With these initial conditions, reasonable agreement is

ity of 500 m²/d and feeds into a large lake at an elevation of 50.0 m. There is a lateral inflow on the left-hand side of 2.0 m³/d per metre width of aquifer. With a uniform recharge $q = 1.0$ mm/d, the resultant groundwater head distribution, determined from a finite difference solution of Eq. (2.60) with a mesh interval of 500 m, is plotted as the unbroken line in Figure 12.6b and labelled (i).

This parabolic initial groundwater head distribution can be approximated by the two straight chain-dotted lines (ii) in Figure 12.6b. If these two straight lines are used as the initial condition, the equivalent inflows can be derived by rewriting the finite difference equation Eq. (2.60) (see also Figure 2.28) as

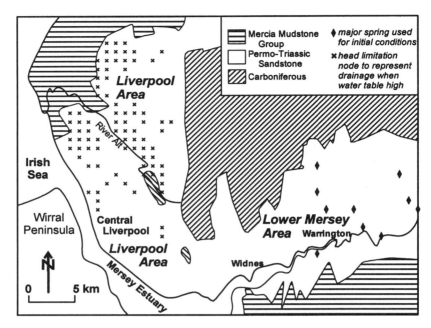

Figure 12.7   Details of the Mersey Permo-Triassic sandstone aquifer system

achieved with available field information. The springs operating on high ground and the connection with the Mersey Estuary exert a strong constraint on the groundwater heads.

For the westerly part of the aquifer, the Liverpool Area (Figure 12.7) abstraction patterns were more complex, with little reliable historical information. By 1955, groundwater abstractions had reached a maximum, subsequently abstractions have shown a steady decline (Brassington and Rushton 1987). Due to the limited field information before 1947, *time-instant* initial conditions are used. For 1948 there is sufficient field information to make a reasonable estimate of the quantity of water taken from storage. The average annual *fall* of the water table in 1948 was 0.25 m (Rushton *et al.* 1988) hence the storage release is equal to the daily *fall* (0.25/365) multiplied by the specific yield of 0.14, which equals 0.000096 m/d or 0.096 mm/d. The estimated actual recharge from rainfall and leaking sewers and water mains averages 0.50 mm/d, hence

$$equivalent\ recharge = q - S_Y\ \frac{\partial h}{\partial t}$$

$$= 0.5 - (-0.096) = 0.596\ mm/d.$$

This value was used for time-instant initial conditions; the groundwater model then continued with a time-variant solution and normal monthly time steps. If the release of water from storage is ignored with actual recharge used rather than equivalent recharge, groundwater heads for the initial conditions are up to six metres too low.

A further important feature of the Liverpool Area is that artificial drainage has been provided in many areas to bring the water table below ground level. These drainage areas are described as *head limitation nodes* whereby water is drained away if the groundwater head exceeds the specified drainage elevation. The locations are shown in Figure 12.7 by crosses. Only a limited number of the head limitation nodes operated during the historical simulation; more now operate following decreases in abstraction from the aquifer.

12.5.3 Initial conditions for aquifers with seasonal changes in transmissivity

For aquifers in which the hydraulic conductivity varies appreciably over the saturated depth, so that the transmissivities in winter are several times the transmissivities in summer, the achievement of acceptable initial conditions is more difficult. Since average conditions may not coincide with any specific time, it is preferable to use a *time-instant* approach. There are a number of important considerations.

- selecting months for which the equivalent recharge can be estimated for time-instant solutions,
- devising a suitable methodology for including the base of the aquifer and the variation of hydraulic conductivity with saturated depth so that a stable initial solution can be obtained. Difficulties occur due to the non-linear variation of the transmissivity (see Chapter 11) and the non-perennial nature of many springs and streams (Section 4.3.4); these features interact and may result in instability or non-convergence in achieving initial conditions,
- ensuring that parameter values can be modified during the initial model refinement.

In the early stages of model development, it is helpful to obtain at least two preliminary time-instant solutions, one during the recharge season and the second towards the end of a period with no recharge. For these preliminary solutions, transmissivity values are selected to represent typical high and low groundwater head distributions. The equivalent recharge for each situation must also be defined; if there is little information about water table fluctuations and hence the storage release, an equivalent recharge of, say, 110 per cent of the average annual recharge can be used for the recharge season, with 50 per cent of annual recharge for the low water table condition. These equivalent recharge values are consistent with Table 12.4. For the Berkshire Downs (Section 11.2), where groundwater flows to streams are so rapid that some of the water moves from recharge to stream discharge in a time less than the model time step, the equivalent recharge values were set at 80 per cent and 35 per cent of the average recharge. Results from time-instant solutions should be compared with groundwater heads and also with the accretion of stream flows. Stream flow comparisons are especially important with a network of non-perennial streams. This procedure of attempting to reproduce flows and groundwater heads under both high and low water table conditions is crucial to the development of an understanding of the flow processes. Only when acceptable time-instant simulations have been obtained should full time-variant solutions be attempted.

There are four *stages* in achieving initial conditions and moving towards a time-variant solution; the individual stages are shown in Figure 12.8 which

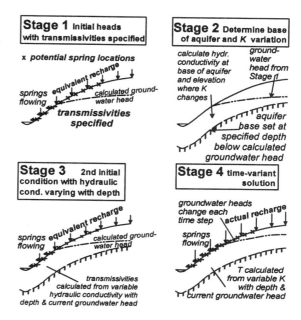

Figure 12.8 Schematic diagram of the four stages for initial conditions when starting from a time-instant solution; suitable for aquifers with significant changes in hydraulic conductivity with saturated depth

uses a schematic cross-section to illustrate the purpose of each stage.

Stage 1: Time-instant numerical solution (first initial condition) with an equivalent recharge, a specified transmissivity distribution plus river or spring coefficients and their elevations; normally a time-instant simulation for the recharge season is selected.

Stage 2: Using the groundwater heads determined during Stage 1, calculate the base hydraulic conductivity, the elevation of the effective base of the aquifer and the elevation at which the hydraulic conductivity ceases to be constant (see the equations in Figure 11.3); this calculation refers to locations between nodal points (or between cells).

Stage 3: Obtain a second initial condition solution using the equivalent recharge, with the inter-nodal transmissivities calculated from groundwater heads, hydraulic conductivity distribution and the base of the aquifer; the calculated groundwater heads should be identical to those obtained during Stage 1.

Stage 4: Move to a time-variant solution with the actual recharge input to the water table; it is often helpful to have two cycles with a typical year before moving on to the full time-variant simulation.

Taylor *et al.* (2001) provide information about the inclusion of this technique in MODFLOW.

12.6 DIMENSIONS AND DETAIL OF NUMERICAL MODELS

Earlier sections of this book have stressed the skills and experience required to develop conceptual models which represent the essential features of the physical flow processes. Similar skills are needed in transferring conceptual models into numerical models. With increasingly powerful and flexible computer packages and without the historical restrictions of limited computing power, there is a tendency to make numerical models too complex. This often occurs when there is a lack of careful planning during the early stages of model development.

This section considers the selection of the area to be represented by a numerical model (both in plan and vertically) and also the time period required to test whether or not the model does represent the historical behaviour of the aquifer system. Often it is not feasible to represent the whole aquifer system in a numerical model. If a reduced area is selected, a judicious choice of conditions on the boundaries of the modelled area is essential. Appropriate parameter values must be specified over the whole of the modelled area; with limited information, this stage of the model development must be carried out with care. A further crucial issue is the selection of demanding yet realistic criteria for the target accuracy between field and modelled results for both groundwater heads and flows. For further information about these issues the reader should consult texts such as Anderson and Woessner (1992), ASTM (1993), Environment Agency (2002), Pinder (2002) or recent conferences on groundwater modelling.

12.6.1 Identification of the area to be represented by a numerical model

Numerical model simulations may be unreliable unless the area included is appropriate and the design of the model grid allows the representation of important detail. Although the area of particular concern may be relatively small, the numerical model must include all

parts of the aquifer system which influence this area of concern. The different scales introduced by Tóth (1963) are helpful; he identified three types of flow system: (1) *local flow systems* where water flows to a nearby discharge area, (2) *regional flow systems* where groundwater travels greater distances, discharging into major rivers, large lakes or the sea and (3) *intermediate flow systems* with one or more topographical highs and a number of discharge locations.

The following questions may be helpful in identifying the required extent of the numerical model; where appropriate, reference is made to case studies where these conditions arise.

1. Are the effects of groundwater abstractions felt over large areas? If so the numerical model should include the whole of the area influenced by the pumping. For both the Nottinghamshire Sherwood Sandstone aquifer system (Section 9.2) and the Southern Lincolnshire Limestone (Section 11.3) it was necessary to extend the numerical model further north than was anticipated initially.

2. Does the aquifer system cover a large area without clearly identifiable boundaries either with known flows or effectively constant groundwater head conditions? When this situation occurs, it is necessary to make an intelligent choice of boundary conditions with the suitability of the assumed boundaries tested using sensitivity analyses.

3. Are there significant regional groundwater gradients? If it is not possible to include the whole of the aquifer system in the numerical model, boundaries should be selected with specified flows enforced; these flows are likely to change with the season. For computational reasons it may be advisable to set one of the boundaries with specified groundwater heads or head-dependent flows (again with seasonal variations); a careful check must be made on the flows entering or leaving this boundary. Sensitivity analyses are essential to examine the impact of artificial boundaries. The western boundary of the Nottinghamshire Sherwood Sandstone aquifer (Section 9.2) and the eastern boundary of the Fylde aquifer (Section 10.6.4) are typical examples.

4. Is the regional groundwater gradient relatively flat with most of the discharge from rivers or boreholes originating from recharge or storage release in the vicinity of these discharges? This is a local system in Tóth's classification. The extent of the modelled area is unlikely to be critical and the opportunity should be taken to have small mesh intervals in areas of special interest. Modelled areas often cover

thousands of square kilometres; if mesh spacings of 5 km or more are selected, detailed representation of features such as canals, rivers or individual boreholes is not possible. The detailed model of the Mehsana area (Section 10.2.7) is an example of a local system model.

5. Have long-term changes in groundwater heads or groundwater flows been noted? Under such conditions the mathematical model should extend to the real physical boundaries. For the Lower Mersey aquifer system (Sections 9.4 and 12.5.2) valid physical boundaries together with a simulation starting in the mid-nineteenth century are required.

The selection of too small a modelled area is likely to result in a numerical model which may be unreliable when used for predictions. The area covered by a model can often be extended using a graded mesh with a larger mesh spacing for the extended area even though the data coverage is of a lower standard.

The representation of the *vertical dimension* of the aquifer system must also be considered carefully. With the powerful modelling packages now available it is possible to represent a large number of geological layers, but is this necessary? Points to bear in mind when selecting the vertical extent of the model include:

1. Vertical flow components (i.e. in the z-direction) in relatively homogeneous aquifers are automatically represented by analyses in the x-y plane (see Section 2.6).
2. If the vertical hydraulic conductivities of extensive low permeability zones in the aquifer system are less than 1 per cent of the horizontal hydraulic conductivities, a multi-layer approach must be used (Section 2.6.3).
3. If the regional groundwater model is used subsequently for contaminant transport studies, a detailed representation of vertical flow components is essential.
4. In some practical situations, three model layers are *just* sufficient, with the upper layer representing the water table, the middle layer representing the main zone of abstraction and the lower layer representing aquifer zones below the main flow horizons.
5. An adequate representation of geological faults is also important. Some faults prevent lateral flow; other faults restrict lateral flow and can be represented by reduced lateral hydraulic conductivities. Faults may provide connection between strata which are not normally laterally connected. Whenever

faults occur, all available information must be examined including drilling records, groundwater head hydrographs and hydrochemical information on either side of the fault together with pumping tests in the vicinity of the fault. This information should be used to develop plausible conceptual models; the mathematical model must be equivalent to these conceptual models.

An approach called *telescopic mesh refinement* has proved to be useful when the overall numerical model must cover an extensive region but detailed representation is required in specific locations. An important example of this approach for a hazardous waste site in the Miami River valley near Dayton, Ohio, USA is described by Ward *et al.* (1987) and discussed by Anderson and Woessner (1992).

A similar approach has been used for the Nottinghamshire Sherwood sandstone aquifer in which the initial analysis included the whole aquifer system using a single-layer regular grid model; this was followed by a detailed multi-layered study of a smaller area. The analysis of the aquifer system using a single-layer approach with a 1-km square grid is described in Section 9.2. This single-layer model, which is contained within an area of 58 km by 50 km (Figure 9.1), provided a good representation of the regional groundwater flow processes and was crucial in the identification of the various sources of recharge and the importance of river–aquifer interaction. However, the single-layer model is not acceptable for the study of conditions at a pumping station with four boreholes located a few hundred metres north of the River Poulter. A detailed study of the response of the individual boreholes required a multi-layered variable grid model covering an area of 5 km by 7 km with conditions on the boundary of this detailed model taken from the regional model.

The extent of the detailed model area is shown in Figure 12.9; the 1 km grid of the original model is indicated by the broken lines. In the plan view, horizontal mesh intervals decrease from 500 m on the boundaries to 60 m in the vicinity of the boreholes. The vertical spacings, which are illustrated in Figure 12.9b, vary between 20 and 60 m. The top layer represents the water table and includes river–aquifer interactions; lower layers simulate the pumped boreholes which are connected to the relevant layers using the technique described in Section 12.3. On the boundaries of the multi-layered variable grid model, boundary flows are derived from the single-layer model and distributed according to the width and depth of the model cells. In

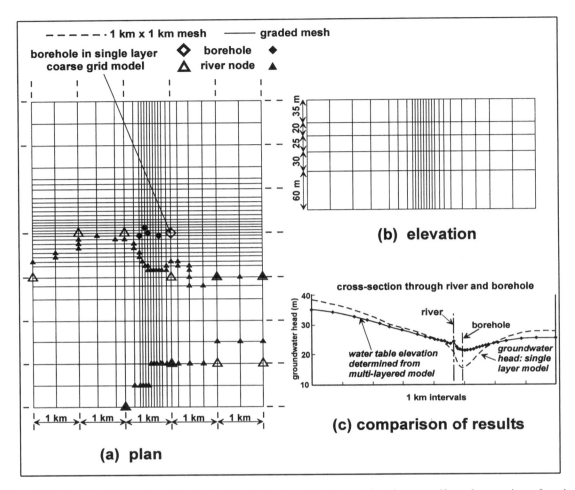

Figure 12.9   Details of fine mesh model of part of the Nottinghamshire Sherwood sandstone aquifer and comparison of results between coarse grid and fine grid models; open diamonds and triangles represent coarse grid model, filled symbols the fine grid model

the variable grid model the locations of the four boreholes are indicated by the filled diamonds; on the 1 km single-layer model abstraction is from one node, as indicated by the open diamond. The rivers are represented as specified river elevations shown in Figure 12.9a with open triangles indicating the locations on the 1 km grid and filled triangles for the variable grid model.

The importance of using a fine grid multi-layered model to represent the river and boreholes is apparent from Figure 12.9c which shows groundwater head distributions on a cross-section through the river and a borehole. The broken line refers to results taken from

the single layer model; the diamonds joined by the continuous line refer to the uppermost layer (i.e. the water table) of the fine grid multi-layer model.

12.6.2 Identification of the duration of a numerical model simulation

Each aquifer system has a natural equilibrium time period t_{eq} which represents the time taken for any change in condition to spread throughout the aquifer. This natural time period is defined by considering a simple one-dimensional problem in which a fluctuating discharge is taken from an aquifer which is originally at rest. The one-dimensional aquifer, of length L, transmissivity T and storage coefficients S, is illustrated in Figure 12.10b. On the right-hand side of the aquifer there is a large lake which ensures that the groundwater head remains constant. At time $t = 0$ all the groundwater heads are zero. On the left-hand boundary for $t > 0$ there is a sinusoidal variable discharge

$$Q = \overline{Q}\left(1 + \cos\left(\frac{2\pi t}{p}\right)\right) \tag{12.7}$$

in which $\overline{Q}$ is the mean discharge, t is the time since the start of the discharge and p is the period of the variable discharge.

An analytical solution is used to calculate the drawdowns on the left-hand boundary which are plotted in Figure 12.10c; gradually the drawdowns converge to the broken line which represents the long-term cyclic pattern of drawdowns. The time t_{eq} for this drawdown to become within 1 per cent of the cyclic pattern is

$$t_{eq} = \frac{1.5SL^2}{T} \tag{12.8}$$

This expression is independent of the period of the sinusoidal discharge (Rushton and Wedderburn 1973).

Time periods to equilibrium for certain of the case studies discussed in Chapters 9 to 11 are presented in Table 12.5. Since these regional groundwater problems involve two-dimensional flow, the distance L is taken as the length of a flow-path *to* an outlet at a body of water (such as a river, lake or the coast) *from* aquifer locations which are most distant from these water bodies.

For the three sandstone aquifers considered in Chapters 9 and 10, the time periods to equilibrium for the unconfined regions are forty years or more; the numerical model simulations all covered several

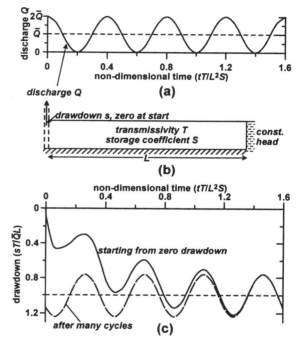

Figure 12.10 One-dimensional example of time taken for convergence to cyclic pattern: (a) cyclic discharge with a mean of $\overline{Q}$, (b) one-dimensional constant transmissivity aquifer with discharge at left hand boundary, (c) increasing drawdown from initial value of zero, broken line represents final cyclic pattern

Table 12.5 Time period to equilibrium for selected regional aquifer studies

Case study	Reported in Sect.	T (m²/d)	S	L (km)	t_{eq} (years)	Comment
Nottingham Sst	9.2	250	0.15	4.5	50	Unconfined: length of flow-path to rivers 4.5 km
		300	0.001	15	3.1	Confined
L. Mersy Sst	9.4	350	0.14	10	164	Simulation started in 1847
Bromsgrove Sst	10.5	250	0.10	5	41	Due to layering of aquifer with low K_z, t_{eq} is > than 41 yrs
Berks Downs Chalk	11.2	400	0.01	8	6.6	Unconfined: flow path to perennial rivers ≈ 8 km
S. Lincs Lmst	11.3	500	0.05	5	10.3	Unconfined: possibly a smaller t_{eq} due to high T in winter
		2000	0.0005	10	0.1	Confined

decades. Within the confined region of a sandstone aquifer the time period is much smaller. For the chalk and limestone aquifers of Chapter 11, time periods are no more than a decade; shorter times than those quoted may be appropriate due to the higher transmissivities under winter conditions.

The time periods to equilibrium quoted above are an indication of the required duration of numerical model simulations. If a model is calibrated over a shorter time period it should be used with caution for predictive purposes since insufficient years have been used to test the adequacy of the model. In such circumstances it is advisable to increase the duration of the simulation even though there may be little reliable data during the early stages of the historical simulation.

12.6.3 External boundary conditions

In transferring a conceptual model to a regional groundwater model, the boundaries of the area to be modelled must be defined with suitable conditions identified on each boundary. Often the actual aquifer system extends beyond the boundary of the study area; great care is then required to select appropriate conditions on the external boundaries of the model. An incorrect choice of boundary condition can lead to an unsafe model. When there are uncertainties, the boundary condition should be included in sensitivity analyses of both historical and predictive simulations.

There are three possible external boundary conditions:

1. *Specified groundwater head* is a boundary condition in which, at each node on the boundary, either constant or time-variant groundwater heads are enforced. This condition is defensible when the aquifer is in direct contact with the sea or a large lake; it is unlikely to be justifiable along the line of a river unless the river penetrates to at least 50 per cent of the aquifer depth.
2. *Specified flow* is a boundary condition which is appropriate either when there is a physical boundary across which the flow is zero, or when the cross-boundary flow is small and can be estimated using Darcy's Law from the properties and groundwater head gradient of the adjoining strata. This approach is used both for the western boundary of the Nottinghamshire Sherwood sandstone aquifer, where there is contact with Permian strata (Section 9.2), and for the boundary between the Lower Mersey sandstone aquifer and the Carboniferous strata (see Figures 9.13 and 9.14).
3. *Head-dependent groundwater flow* is used as a boundary condition when there is a flow across the boundary but the magnitude of the flows cannot be predetermined because conditions in both the modelled area and the area beyond the boundary are changing. This situation is illustrated by boundary conditions to the north of the Doncaster study area; Figure 9.8. The northern boundary of the Doncaster model is located between the Doncaster and Selby well-fields which both draw water from the Sandstone aquifer. When abstraction from the Selby well-field increases, there is a decrease in groundwater heads in the Selby area, resulting in a change in flows across the northern model boundary. The method of simulating a head-dependent flow can be visualised by supposing that the aquifer model is extended to the north by n mesh intervals for a distance $n\Delta y$ (typically 8 km) with the same transmissivities T as the northern part of the Doncaster model (i.e. about 500 m^2/d for the main part of the aquifer falling to 40 m^2/d beneath the Mercia Mudstone). Groundwater heads h_S at the northern limit of this aquifer extension vary to represent changing groundwater heads in the Selby area. Consequently the groundwater flow Q_N across each node on the northern boundary of the Doncaster model can be calculated from the expression,

$$Q_N = T\Delta x \frac{h_N - h_S}{n\Delta y} \tag{12.9}$$

where h_N is the unknown groundwater head on the northern boundary of the Doncaster model. For a square finite difference grid, Eq. (12.9) becomes

$$Q_N = T(h_N - h_S)/n \tag{12.10}$$

In the numerical model, rather than extending the model as visualised above, the cross-boundary flows Q_N are calculated from Eq. (12.10) within the iterative solution of the numerical model. This approach is similar to the General Head Boundary Package in MODFLOW (Anderson and Woessner 1992).

A specified head boundary is often inappropriate. In an early study of the Fylde aquifer (Section 10.6.4 and Figure 10.30) it was noted that only small changes in head occurred in the Carboniferous strata to the east. Consequently, in the initial model, the Carboniferous strata were represented as specified constant heads on the eastern edge of the Sandstone. The groundwater model predicted that increased abstractions from the Sandstone would draw most of the water from the Carboniferous strata. Subsequent studies showed that this is equivalent to the unrealistic assumption that the Carboniferous can provide large volumes of water. In reality, the actual quantities of water which can be drawn from the Carboniferous strata are limited. In current models the boundary is represented as a small specified flow.

12.6.4 Estimating parameter values for a numerical model

Perhaps the most demanding task in preparing a numerical model from a conceptual model is the selection of appropriate values of the aquifer parameters. Inevitably there is insufficient information. Even if additional fieldwork is carried out, the selection of suitable parameter values requires skill, experience and judgement. Furthermore, the selection of parameter values is a time-consuming task. Unless it is carried out systematically and thoroughly, a great deal of rethinking may be required in later stages of model refinement. For each parameter it is advisable to quote the best estimate and a range which represents the uncertainty.

The issue of *assigning parameter values* is examined in detail in Section 3.4 of Anderson and Woessner (1992); they discuss the estimation of aquifer parameters and transferring this information to the model grid. In this present discussion certain important issues are highlighted.

- How should the transmissivity estimates, which are derived from pumping tests, be incorporated in a regional groundwater model? The accuracy of the transmissivity estimates may be uncertain due to differences between actual field conditions and the assumptions of the methods of pumping test analysis (see Part II, especially chapters 5 and 7). Furthermore, pumping tests or other field estimates such as slug tests (Bouwer 1989, Kruseman and de Ridder 1990) or packer tests (Section 8.3) reflect the particular conditions in the vicinity of the test borehole.
- In certain aquifers (especially chalk, limestone and alluvial aquifers) the individual transmissivity estimates (and hence the horizontal hydraulic conductivities) may vary by half an order of magnitude compared to neighbouring test boreholes; a geometric mean of several test boreholes is usually appropriate.

- Techniques such as Kriging (a statistical interpolation method) can be used to devise a transmissivity or hydraulic conductivity variation (de Marsily 1986). However, the resultant distribution may be an over-estimate since it is usually the more productive boreholes which are tested, with fewer transmissivity estimates from the less productive boreholes.
- The recommended approach for including areal hydraulic conductivity variations is to identify a number of zones in which similar hydraulic conductivity values apply. These zones are based on hydrogeological information, the historical development of the aquifer system and experience of exploitation of the aquifer system. Typically about seven distinctive zones should be identified with aquifer parameters defined for each zone; in more extensive aquifers more zones may be appropriate. The selection of zones permits quick changes to be made to aquifer parameters during model refinement.
- Vertical hydraulic conductivity is a very difficult parameter to estimate; probably the best guide can be gained from previous studies in similar aquifer systems. However, when using information from other modelling studies it is necessary to determine whether the quoted vertical hydraulic conductivity refers to a mesh interval or the actual low conductivity layer (see Figure 10.7).
- The specific yield appropriate to long-term regional aquifer responses is frequently underestimated in pumping test analysis (Section 5.17). Estimates of the drainable porosity are more likely to be appropriate; for an informative discussion on the estimation of specific yield, see Healy and Cook (2002).
- In multi-layered aquifer systems, care is essential when assigning parameter values to the grid. In the initial stages it is helpful to draw a cross-section to a large scale, superimpose the model grid and calculate some of the model coefficients by hand. Not only does this approach clarify procedures but it also provides numerical values for checking the coefficients calculated by the package.
- Many regional groundwater flow packages have graphical interfaces which assist in the input of aquifer parameters. However, this software may contain assumptions of which the user is unaware. For instance, many packages do not allow for river coefficients varying between stress periods, as required for the simulation of the River Yobe (Figure 12.3). Therefore it is important to check that the parameters which are actually used in the computation are the parameters which the user intends.

Examples of numerical groundwater models, which were derived from quantified conceptual models, are presented in Chapters 9, 10 and 11. For instance, the distribution of transmissivity in the Berkshire Downs is shown in Figure 11.4. Detailed information about input parameters for the Southern Lincolnshire Limestone aquifer system, including river locations and coefficients, are recorded in Figures 11.11 and 11.16, recharge types and locations are presented in Figure 11.10, with a summary of many features of the numerical model in Table 11.2. For the Sherwood Sandstone aquifer in Doncaster a diagram of the locations of river spring nodes, different types of recharge nodes and leakage nodes, is included in Figure 9.8.

12.6.5 Refinement of numerical groundwater models

When the model code is run for the first time there are likely to be many differences between field and modelled values of groundwater heads and flows. First of all there will be differences because of mistakes in preparing the input data; very careful checks must be made to ensure that the model boundaries, parameter values, inflows and outflows are correctly included in the numerical model.

Precise agreement between field groundwater heads and flows and modelled values will never occur in practical situations. Due to the approximations inherent in regional groundwater models, it is not possible to represent all the detailed flow processes. Also, there are usually deficiencies in field information due to limited accuracy of field measurements, poor spatial coverage, low frequency of readings, etc. These issues must be considered when setting realistic targets for agreement between field and modelled results.

Differences between field and modelled values also occur because of limitations due to assumptions in conceptual models including flow mechanisms, boundaries and parameter values. The process of developing a reliable model is often called *calibration*. Anderson and Woessner (1992) explain that, 'Calibration of a flow model refers to a demonstration that the model is capable of producing field-measured heads and flows which are the calibration values. Calibration is accomplished by finding a set of parameters, boundary conditions and stresses that produce simulated heads and fluxes that match field-measured values within a prescribed range of error. Finding this set of values amounts to solving what is known as the inverse problem.'

This author prefers to use the phrase *model refinement* to describe this process. In model refinement, time-instant or time-variant simulation are used.

Physically realistic modifications are made to flow processes, aquifer parameters and inflows and outflows until an acceptable match is achieved between field and modelled groundwater heads and flows at sufficient locations covering time periods appropriate to the study (see Section 12.6.2). During model refinement it is essential to have a selection of field information against which comparisons can be made. This should include hydrographs of flows in rivers or springs, a selection of groundwater head hydrographs from different parts of the aquifer system, and a limited number of cross-sections roughly following groundwater flow paths on which high and low groundwater heads are plotted. Additional comparisons can be made during later stages of the model refinement. A record should be kept of the steps in model refinement, including comparisons between field and modelled results.

The initial task in model refinement is to determine whether any crucial flow processes have been omitted in the quantified conceptual model and hence in the numerical model. The model refinement must be carried out with frequent reference to the conceptual model. It is the experience of the author that in each modelling study *at least* one fundamental feature was not identified in the initial conceptual model.

- In early studies of the Southern Lincolnshire Limestone (Section 11.3) neither the variation of hydraulic conductivity with saturated depth nor runoff-recharge were included.
- When preparing models of the Nottinghamshire Sherwood Sandstone aquifer, no account was taken of recharge through the Colwick Formation (Section 9.2.1).
- In the sandstone aquifer of the Liverpool area, the initial conditions used average recharge rather than equivalent recharge (Section 12.5.2).
- For the Lower Mersey study area (Section 9.4.1), the significance of springs, which no longer operate, was overlooked.
- In the study of the Mehsana alluvial aquifer (Section 10.2) there was a failure to recognise that most of the water pumped from deeper aquifers originated from vertical flow through overlying low permeability strata.

The following list includes features which may need to be reviewed during model refinement.

- Are the starting conditions appropriate; does the numerical model run for a sufficiently long time for any inadequacies in the model to be identified?

- Do the locations and conditions on the boundaries of the numerical model represent the field conditions?
- Have all the recharge processes been included; is there a possibility that a source of recharge through lower permeability strata has not been considered?
- Does the river–aquifer interaction reflect field behaviour; are all possible spring locations included in the model?
- Are there too many or too few layers in the mathematical model?
- Have any features, such as the variation of hydraulic conductivity with saturated depth or changes between unconfined and confined conditions, been omitted?

After these questions have been addressed, adjustments to values of aquifer parameters should be explored. A possible strategy for model refinement is outlined below.

- To test whether the overall flow balance is adequate, comparisons should be made between field and modelled accreted river flows and long-term trends in groundwater head hydrographs. If these comparisons are unsatisfactory, all the inflow and outflow components and specific yield values should be reviewed.
- An examination of the groundwater head distributions on cross-sections (and later by comparing field and modelled groundwater head contour plots) will indicate whether the hydraulic conductivity values are appropriate; the effect of changing individual zones or groups of zones should be explored.
- Groundwater head fluctuations away from rivers and other water bodies provide insights into the selection of the specific yield or the delay in recharge reaching the water table (often it is difficult to distinguish between these two effects).
- A comparison of field and modelled groundwater flows and groundwater heads for average years, low recharge years and high recharge years verifies the detailed capability of the model. If, for example, the study is concerned with the alleviation of low flows, parameter adjustment to improve this aspect of the simulation should be explored.
- A detailed examination of streamflow accretion and spring flows will indicate whether refinement of the river or spring coefficients and their elevations is necessary.

In this text there are several examples of acceptable agreement between field and modelled groundwater flows and heads.

- Comparisons of field and modelled groundwater flows at gauging stations (these flows are an integrated response), examples include the River Kennet at Marlborough, Figure 11.5a, and Holywell Brook in the Glen Catchment, Figure 11.15a. Comparisons between total river flows are also important but the agreement may only be approximate; Figures 3.24 and 9.5.
- Indirect comparisons which relate to flows such as the ingress of saline water from the Mersey Estuary, Figure 9.19, or the number of flowing springs in the Bromsgrove aquifer in 1900 and 1984, Figure 10.25.
- There are numerous comparisons of field and modelled groundwater head hydrographs; see for example Figures 9.4, 9.10, 10.10, 11.5b and 11.15b, which refer to observation boreholes, and Figure 9.18 which refers to a pumped borehole. Figure 11.17 compares high and low groundwater heads on a cross-section.

A helpful and detailed explanation of model refinement for a particular aquifer system is described by Gerber and Howard (2000).

12.6.6 Sensitivity analysis

When a numerical model has been developed which is consistent with available field information, there is a need to carry out a sensitivity analysis, since many of the mechanisms and parameters included in the model are approximate estimates. Sensitivity analyses are required both for a refined model and for predictive runs. In a sensitivity analysis, comparisons are made between outputs from the refined model and the corresponding model results when a condition or parameter is changed. Diagrams should be prepared of the output of the refined model and sensitivity runs; it is helpful if field information is also recorded on these diagrams. Table 12.6 is presented to provide guidance as to how conditions and parameter values could be modified in a sensitivity analysis.

12.6.7 Preparing exploratory groundwater models with limited field information

Published descriptions of numerical groundwater model studies are usually based on detailed field studies, with aquifer properties estimated from several pumping tests and extensive databases of groundwater head and flow hydrographs covering many years. Due to the high data demands of these studies, it may appear that numerical model studies should not be carried out when field information is limited, especially in terms of groundwater head hydrographs. However,

Table 12.6   Conditions and parameters to be modified in a sensitivity analysis

Feature	Suggested modifications
Recharge	Consider rainfall, crop factors etc. changing by ±5%; convert to changes in recharge using methodology of Chapter 3. Consider other possible sources of recharge through less permeable strata; does runoff recharge occur? Actual recharge may be less than potential recharge, delays may occur as recharge moves through overlying drift
River coefficient	Initially try doubling and halving the coefficients, also adjust the river level; consider especially the effect on losing rivers
Boundaries	Change boundary flows or coefficients by ±30%
Specific yield	Change specific yield by ±50%
Vertical K	Increase or decrease vertical hydraulic conductivity by half an order of magnitude
Horizontal K	Change horizontal hydraulic conductivity ±20%

useful insights can be gained from exploratory groundwater models; a case study from north-eastern Nigeria is used to illustrate the approach.

The study is concerned with the floodplain aquifer of the River Yobe between gauging stations at Gashua and Geidam, Figure 12.11. Since flows in the River Yobe originate from the Jos Plateau, which is about 450 km to the west, river flows increase from late June when rainfall occurs over the Plateau, whereas rainfall in the study area does not commence until July, with little recharge before August. The purpose of the study is to estimate how much river water enters the floodplain aquifers. Figure 12.2a shows that the flow at the upstream gauge, Gashua, exceeds the flow at the downstream gauge at Geidam from June to November but from December onwards conditions are reversed. Flows from river to aquifer occur when the river stage (river surface elevation) is above groundwater heads in the aquifer as shown in Figure 12.2b but from December onwards the groundwater heads are above the river stage, indicating that the aquifer feeds the river. Further information from field studies includes:

- exploratory boreholes and groundwater head monitoring to the north and to the south of the river,
- estimates of hydraulic conductivity from the exploratory drilling and pumping tests,
- river coefficients which vary with the river stage (derived in Section 12.2.2),
- recharge estimates based on a soil moisture balance (see Figure 3.19 which refers to a nearby area),
- areas which are flooded when the river flows outside its banks; the flooding is sufficient for rice to be grown.

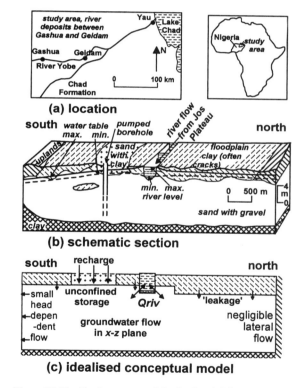

(a) location

(b) schematic section

(c) idealised conceptual model

Figure 12.11   Exploratory model of a floodplain aquifer

Using this limited information, the schematic cross-section of Figure 12.11b is developed. Confined conditions apply to the north whereas unconfined conditions occur over much of the area to the south. Recharge occurs through the sandy deposits in the south, there is also a small 'leakage' through the confining clay deposits. From this schematic cross-section the idealised conceptual model of Figure 12.11c is derived. Note that the northern boundary is represented as having a negligible cross-boundary flow, whereas the southern boundary has a small head-dependent outflow.

The numerical model simulation continues for three years with recharge and river stage variations remaining the same for each year. The *plausibility* (Carter *et al.* 2002) of the simulation is confirmed by the following observations.

- the general form of the incomplete field groundwater head hydrograph is reproduced by the model,
- modelled flows from the river to the aquifer are consistent with the decrease in flow between Gashua and Geidam,
- The model reproduces flows from aquifer to river from December onwards. From March to June, the model predicts a small flow from the aquifer into the river channel; this is confirmed by the availability of water in the river bed.

This vertical section model is used to explore the consequences of pumping from the aquifer during the dry season; a typical location of a pumped borehole is indicated in Figure 12.11b. The simulation shows that pumping would not draw much more water from the river into the aquifer; dewatering of the aquifer in the vicinity of the pumped boreholes is more likely to limit the quantity of water that can be abstracted.

12.7 PREDICTIVE SIMULATIONS

Groundwater models are usually prepared for two reasons, firstly to assist in developing an understanding of the groundwater system and secondly to predict the future behaviour of the aquifer system under changed conditions.

Anderson and Woessner (1992) suggest that there are two major pitfalls involved in making predictions: uncertainty in the calibrated model and uncertainty about future hydrologic stresses. The use of separate sensitivity analyses to explore each of these sources of uncertainty is emphasised. In a discussion of post-audits of earlier modelling studies, Anderson and Woessner show that inaccurate predictions were partly caused by errors in the conceptual models. They quote an example where one of the sources of recharge was not included. Another consideration is that some early conceptual models were influenced by the limited computing power available at the time when the original study was carried out. Unreliable predictions have also occurred due to the failure to use appropriate values for assumed future stresses. Predictive simulations should be planned and assessed by a group of people with a *range* of expertise to ensure that the best estimates of future stresses are included.

12.7.1 Issues to be considered

For satisfactory predictive simulations, the following issues must be addressed.

- *Define the purpose of the prediction*: possible objectives include the effect of increasing or decreasing abstraction rates, repositioning abstraction sites to improve conditions at wetlands or in rivers, the effect of prolonged droughts, the consequences of climate change, etc. When the objectives have been defined, scenarios should be developed to explore the proposed changes; these should be realistic scenarios.
- *Choice of suitable starting conditions* for the predictive simulation: usually it is advisable to start the predictive simulation from conditions at the end of a modelled historical period.
- *Select a baseline for comparison*: with each predictive simulation the purpose is to explore how aquifer conditions would alter (improve or deteriorate) due to modifications to the abstraction, recharge etc. The significance of the changes becomes clear when comparisons are made with a baseline predictive simulation which is usually derived from inputs and outputs similar to current conditions (Shepley and Taylor 2003).
- *Recharge*: a realistic recharge pattern is a crucial feature of a comprehensive predictive simulation. In the UK, the last three decades of the twentieth century provided a wide range of climatic conditions, consequently predictive simulations often use these historical recharge sequences. However, in areas where there is a clear change in climatic conditions, recharge for the predictive simulations should be devised to reflect those changes. Rather than devising recharge data, it is preferable to use plausible values of rainfall and evapotranspiration to calculate the recharge as described in Chapter 3.
- *Duration of historical simulation and prediction*: if the historical simulation and the predictive simulation do not cover a sufficiently long time period, the reliability of the predictions will be uncertain. For aquifers which respond quickly and refill most years (chalk and limestone aquifers often fall into this category) simulation periods of ten years may be adequate.

However for aquifers with a higher specific yield and hence slower response, twenty-five years is the minimum predictive period.

- *Checking whether conditions during the predictive simulation remain within those which have been tested during the historical simulation*: if the predictive simulation results in groundwater heads or river–aquifer flows which are outside the range covered during the historical simulation period, it is essential to check that the flow processes are realistic. Examples of conditions which may not be represented adequately include additional springs which form when abstraction is decreased, part of the aquifer becoming dewatered when abstraction is increased, or a reduction in actual recharge because the aquifer is full of water.
- *Clarity in presenting results*: numerical models provide excessive outputs. However, to be able to identify clearly the consequences of possible changes, a limited number of diagrams should be prepared to highlight the important features of the aquifer response.

12.7.2 Representative example

To illustrate different aspects of predictive simulations, a case study from the UK is considered. The simulation refers to the Nottinghamshire Sherwood Sandstone aquifer which is described in Section 9.2; see also Shepley and Taylor (2003). The schematic cross-section of Figure 9.2b shows that recharge occurs on the sandstone outcrop and through the lower permeability Colwick formation; water is also drawn through the low-permeability Mercia Mudstone. Water is pumped from boreholes in the sandstone outcrop beneath the Colwick Formation and beneath the Mercia Mudstone. There is also significant interaction between the aquifer and rivers.

The purpose of the predictive simulation described below is to consider the consequences of halving the abstraction from two boreholes in the sandstone outcrop. At borehole P, Figure 12.12a, the abstraction is reduced from 8.6 to 4.3 Ml/d, at borehole Q the reduction is from 11.8 to 5.9 Ml/d. For the baseline prediction, abstractions continue at constant values which generally coincide with the average abstraction for the last five years of the historical simulation. The results of the predictive simulation are summarised in four diagrams which are selected to indicate, in a concise form, the impact of the changed abstractions.

(a) Increases in groundwater heads at the end of the prediction period (i.e. after 27 years) *compared* to

the baseline simulation are shown in the contour plot of Figure 12.12a. In the vicinity of the boreholes where the abstraction is reduced, the groundwater heads rise more than 8 m. The diagram also contains contours for groundwater head rises of 0.1, 0.3, 0.5 m and 1.0, 2.0, 3.0 m, the contours for rises of 4.0 and 5.0 m are not labelled. Perhaps the most remarkable result is a rise within the confined region of 0.5 m at 20 km from the boreholes where the abstractions are modified.

(b) Figure 12.12b refers to the groundwater heads at an observation borehole about 1 km from borehole Q. Results are presented for field values, the historical simulation, the baseline prediction and the prediction with reduced abstractions. The historical simulation is consistently between 2 and 3 m below the field values, so the predicted results are likely to have a similar variance. For the predictive simulation the groundwater head, due to reduced abstraction, is 5 m higher than the baseline at the end of the prediction period. Even after five years the difference is more than 3.5 m, indicating that relatively rapid changes occur at 1 km from one of the boreholes with reduced abstraction.

(c) Perhaps the most informative graph is the diagram showing the difference between the flow balance components for the reduced abstraction and the baseline simulation; Figure 12.12c. The lower half of the diagram shows a reduction in abstraction of 10.2 Ml/d. The immediate response is an increase in the quantity of water stored in the unconfined region due to the rapid rise in groundwater head. However, after a decade, this component becomes less significant because the rise in groundwater head leads to a reduction in the quantity of water being drawn through the Colwick Formation to recharge the aquifer. At the end of the predictive period, this reduction in recharge totals 4.5 Ml/d. The increase in baseflow to rivers averages about 3 Ml/d but varies between years of high or low recharge. The increase in confined and unconfined storage averages about 2 Ml/d at the end of the predictive period; this indicates that a new equilibrium has not yet been reached (see Section 12.6.2).

(d) For the final diagram, flow duration curves are presented for flows from the sandstone aquifer to the River Maun, Figure 12.12d. From this figure it is possible to assess the improvement in flows in the river closest to the boreholes at which the abstraction is reduced. The flow duration curves show the percentage of time during the last five years of the simulation that a certain flow is exceeded. Note

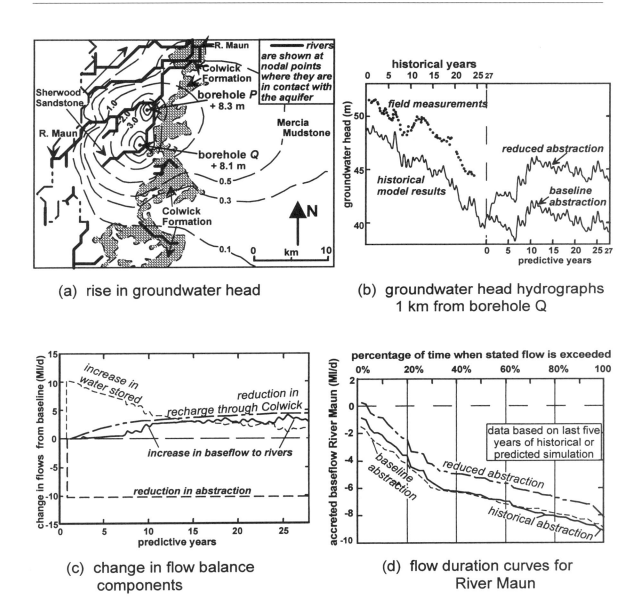

(a) rise in groundwater head

(b) groundwater head hydrographs
1 km from borehole Q

(c) change in flow balance
components

(d) flow duration curves for
River Maun

Figure 12.12 Predictive simulations; four diagrams illustrating the predicted consequences of reduced abstractions

that the accreted baseflows are usually negative, indicating that abstractions from the sandstone aquifer result in a net loss of water from the river to the aquifer even during periods of high recharge. Reducing abstractions leads to a smaller loss from the River Maun of about 1.2 Ml/d compared to the baseline scenario; this should be compared with the reduction in abstraction of 10.2 Ml/d.

The above example shows how four diagrams provide sufficient information to *understand* the consequences of changes in abstraction. For other aquifer systems with predictive simulations which have different objectives, alternative diagrams will be required. However, the aim should always be to present the important consequences of the changes in no more than four diagrams; this will allow the non-expert to be involved in management decisions.

12.8 EVALUATION OF CONCEPTUAL AND COMPUTATIONAL MODELS

The discussion in this section draws together some of the experiences and wider issues identified during the development of conceptual and computational models. Comments reflect the scientific challenge of groundwater studies, the need to be pragmatic about groundwater investigations due to limitations in time and budget, and some of the social and management issues encountered during the projects. Investigations involving a large number of practical field problems have influenced the author's approach to conceptual and computational modelling.

12.8.1 Approach to groundwater modelling

Throughout this book the need to develop conceptual models before moving on to computational models has been stressed. For those with extensive experience of numerical groundwater models, there is a tendency for existing modelling software to become the framework on which a groundwater study is built. Similarly, when analysing pumping tests the approach frequently followed is to look at the drawdown plots and suggest that the data looks as though it could fit a Theis curve, a leaky aquifer curve, etc., rather than considering the field setting. The groundwater scientist must resist this approach. Especially with the analysis of pumping test results, the geology and topography must be examined and a conceptual model developed.

Developing a conceptual model is not straightforward. It is necessary to examine all available data and other information, visit the area under different climatic conditions, talk to those who have used the aquifer (a farmer with a well, the operations engineer of a water company). Insights can be gained from case studies in similar areas, but there will always be new features to identify since every aquifer system has unique features.

The data and information used to test the adequacy of the computational model needs careful selection. Spatial coverage is important. Frequently there is a scarcity of data in some locations but there is often some valid information, such as the fall of water level causing a farmer's well to become dry, intermittent springs or the occurrence of waterlogging, which can help to confirm the adequacy of the computational model.

Another important aspect of model development is identifying which features *must* be included in the computational model. This will determine the type of model to be used and the required duration of the simulation. However, numerical models should not have unnecessary complexity in terms of numbers of layers, number of mesh divisions and size of time steps. Another issue which requires careful attention is that there are some features which are not conveniently represented in certain numerical model codes. As an example, difficulties can occur in multi-layered systems when a model layer is confined over part of the area and unconfined elsewhere. Great care is required in model design when a geological fault causes cross-connections between layers, or when recharge is dependent on the current groundwater heads.

Model outputs must be checked very carefully; frequently mistakes are made when inputting data. Again, the graphical interface used for the input of data may not carry out tasks as anticipated. Further checks should explore whether initial conditions are appropriate. Expect to find mistakes in your initial models; do not be satisfied until you have found several mistakes.

An intelligent selection of outputs assists the modeller in discovering how the model is responding; it also allows another person to check progress. It is easy to be swamped with a large number of plots; too many lines on single graph often lead to confusion. Preparing hydrographs covering both a few years and the full duration of the simulation are helpful.

These suggestions may appear unnecessary but time spent in improving outputs is likely to reduce the time spent in converging to a satisfactory final solution.

12.8.2 Monitoring

During a study there is always a wish for more data. However, when a study covers a historical period of twenty years or more, the data set is likely to have limitations. Nevertheless, there is usually further information, which may not be sufficiently accurate to be classified as data, which can be useful in preparing conceptual models and checking the adequacy of numerical models.

When planning new monitoring, it is important to think carefully about what needs to be monitored, including the duration and frequency of readings.

Often existing river and stream gauges are not in ideal locations for groundwater monitoring; whenever possible they should be sited where a river passes onto or leaves an aquifer. Monitoring should also include frequent site visits to record information such as when springs start and stop flowing, the size of culverts and whether they can always cope with flows. Precision in the measurement of flows is important because groundwater components are usually small compared to overall flows. Examples of satisfactory flow monitoring to identify groundwater components include a water balance for an operating canal in Pakistan (Section 4.2.6) and river–aquifer interaction in the River Meden (Section 4.3.3).

When monitoring groundwater heads it is preferable to have a few locations with several piezometers at different depths; this applies to radial flow and regional groundwater flow studies. The frequency of monitoring is important. In fast-response aquifers such as limestone and chalk, at least two readings should be taken each month; longer intervals are generally adequate for sandstone aquifers.

12.8.3 Recharge

The presentation in Chapter 3 on the estimation of groundwater recharge concentrates on the soil moisture balance technique. It was used in most of the case studies discussed in Part III. Alternative methods of recharge estimation, which are summarised in Section 3.2, provide important insights into recharge processes in a variety of situations. Furthermore, the magnitude of the recharge estimated from these alternative methods serves as a check on soil moisture balance calculations. However, these alternative methods are not suitable for routine recharge estimation.

There has often been justifiable criticism of the soil moisture balance approach, especially when there has been no clear conceptual model. In some soil moisture balance calculations a realistic method of estimating actual evapotranspiration from potential evapotranspiration is not used. Also, many of the earlier soil moisture balance calculations failed to represent soil and crop properties or ignored runoff. The three field examples quoted in Section 3.4 demonstrate that the technique can be used for a variety of climatic conditions.

It is only recently that recharge estimation using the soil moisture balance method has been developed to include explicitly soil properties, crop properties and bare soil conditions. The method of representing runoff as a function of rainfall intensity and soil moisture deficit is helpful, but needs refinement. Further information should be obtained from agriculturalists and soil scientists about soils and crops in the study area. Instrumented catchment studies are also needed to learn more about runoff and its dependence on rainfall intensity and soil moisture deficit.

For many aquifers there is some form of drift cover beneath the soil zone and above the saturated aquifer; much of this drift has a low permeability so that it restricts recharge. With low permeability drift there is an increased runoff. When the low permeability drift contains permeable lenses or layers, they act as minor aquifers, storing and then slowly releasing the stored water. This delayed runoff can support stream and river flows and may provide runoff recharge where the streams pass onto more permeable areas of the aquifer system. Examples of computational models for runoff-recharge are presented, but there is a need for information from appropriately instrumented sub-catchments to refine runoff-recharge models.

12.8.4 Model calibration and refinement

Detailed information is available in the groundwater literature concerning *model calibration*; see, for example, Anderson and Woessner (1992) and the conference *MODFLOW 2001 and Other Modeling Odysseys* (Hill et al. 2001); some papers from that MODFLOW conference are reprinted in issue no. 2 of *Ground Water* (see Hill et al., 2003). Model calibration can be achieved either using manual trial-and-error adjustments or automatic parameter estimation. Automatic parameter estimation enforces discipline and ensures that a record is kept of the process; a similar record should be kept when trial-and-error procedures are followed. As explained by Anderson and Woessner (1992), 'Solving the inverse problem by manual trial-and-error adjustments of parameters does not give information on the degree of uncertainty in the final parameter selection, nor does it guarantee the statistically best solution. An automatic statistically based solution of the inverse problem quantifies the uncertainty in parameter estimates and gives the statistically most appropriate solution for the given input parameters provided it is based on an appropriate statistical model of errors.' Some authors suggest that the model produced using automated techniques is not always superior to one based on manual trial-and-error. Anderson and Woessner (1992) also present a brief introduction to automated calibration procedures. The developments in automatic calibration are illustrated by many of the papers in Hill *et al.* (2001).

Manual trial-and-error methods were used for the case studies described in this book; the phrase *model refinement* is used to indicate the procedure used. For most of the case studies, the available groundwater head data are not of a sufficient precision, nor are there enough data points for automated methods. Groundwater flow data is of a lower precision and is often sporadic, yet information about groundwater flows is often crucial in developing an adequate model. The greatest challenge in model refinement is usually to identify critical features which were not included in the original conceptual models. For each of the case studies described in Part III, certain important features were not included in the initial conceptual models.

The trial-and-error procedure followed in most of the case studies involved aiming to improve the model hydrographs showing poorest agreement with field values. If modelled river flows are very different from field values, this suggests errors in flows, possibly the recharge. If modelled groundwater head hydrographs are out of phase with field information, this suggests incorrect storage values. The need to adjust river coefficients is indicated by the inability of the model to represent flow accretions in streams and rivers. When a series of adjustments are made to the aquifer parameters, yet there are still serious differences between model and field values, this indicates the need for a review of the conceptual model, including the boundaries. Once an adequate numerical model has been obtained, a sensitivity analysis must be carried out; this may lead to further improvements in the aquifer parameters. Throughout all these modifications it is essential to keep a record of progress, with the effect of changes recorded on key data plots.

12.8.5 Sustainability, legislation and social implications

Many of the case studies refer to sustainability issues, for example agrowells (Section 6.6.2, limited cropping areas), weathered-fractured aquifers (8.5.2, weathered zone dewatered), Lower Mersey (9.4, ingress of saline water), Mehsana (10.2, falling water table and pumping levels), Berkshire Downs (11.2, limited yield in drought periods). When planning groundwater development it is often difficult to identify the sustainability of a groundwater system; pilot test pumping may be in a non-critical year, the method of pumping test analysis may not be correct, a groundwater model may be based on aquifer response before heavy exploitation has

occurred, with the risk of incorrect predictions. Sustainability must be assessed over a long period. Falling groundwater heads may result from local changes due to new abstractions rather than long-term declines due to overabstraction.

Often there are opposing views about sustainability when environmental issues are involved. All groundwater abstractions cause some deterioration in spring and stream flows, but do they result in a serious ecological risk? Will reducing existing groundwater abstractions result in a substantial ecological gain? When an aquifer is already heavily exploited, there is a need for a balance between the cost of reducing abstractions and ecological gain. Reductions in abstraction close to a river often do not result in the anticipated gains in flow in that river (Rushton 2002). When considering wetlands, regional groundwater models may not have the precision to predict what will happen at the wetland site, since a wetland is usually caused by unusual physical circumstances which may not be represented by the regional model.

When existing abstraction cannot be sustained, what action should be taken? Using the Mehsana region in western India (Section 10.2) as an example, severe over-exploitation was identified by the groundwater scientists, but they had to persuade the decision makers that abstraction could not continue at its present rate. Numerical model results did not convince the decision makers, but they were persuaded by extrapolations of the rapid decline in pumped water levels which would mean that many of the existing tubewells would fail. What sort of legislation should be introduced? To attempt to return to the situation when there were only shallow wells with water withdrawn using animal power is not realistic. Furthermore, it is difficult to frame legislation when, historically, there has been a free-for-all. The only acceptable legislation was to prevent the drilling of further tubewells or the deepening of existing tubewells.

When groundwater is used for irrigation during the dry season, problems are certain to arise. For an area with an annual rainfall of, say, 700 mm the recharge is unlikely to exceed 100 mm. With a crop water requirement of 500 mm, either the cropping intensity must be restricted to 20 per cent or over-exploitation will occur.

For the weathered-fractured aquifer in central India (Section 8.5.2), much of the shallow weathered zone was dewatered due to the deep tubewells. The grape gardens and the deep tubewells were abandoned but local farmers, who rely on the shallow aquifer, struggled to grow crops during the very slow recovery of the water table. Limited improvements in water table recovery were achieved by catching as much of the rainfall as possible. The provision of small dams in the drainage channels limited the runoff; assistance was also given by providing crops with a low water demand. Nevertheless, local farmers suffered hardship for more than a decade due to pumping from deep tubewells.

With agrowells in Sri Lanka (Section 6.6.2), limited finance meant that only one in five farmers were subsidised to dig their own well; they were also given loans to purchase pumps. Farmers were selected based on the size of their family and whether they had ever defaulted on loans. Farmers with agrowells were able to irrigate their land holding of about 1 hectare (100 m by 100 m); they quickly repaid the loan and their standard of living improved substantially. To the other farmers who did not have agrowells, this appeared to be unfair. Some of these farmers obtained finance for their own wells, but when they attempted to irrigate a crop over the whole of their land holding, interference occurred with the wells constructed earlier so that both the original owners and the new owners ran out of water before the end of the growing season. Over-exploitation can be prevented only by restricting wells to one farmer in five, but those with wells should share their water with surrounding farmers. Because farmers were initially encouraged to irrigate the whole of their land holding, it is now difficult to persuade them to grow irrigated crops over only 20 per cent of their land and share water with their neighbours.

12.8.6 Climate change

Regional groundwater models are ideally suited to climate change predictions. The impact of climate change on groundwater resources in the UK has been investigated using a number of the regional groundwater models described in Part III. The approach to Predictive Simulations described in Section 12.7 is the basis of this approach. New recharge estimates are calculated from the climate change scenarios for rainfall and potential evapotranspiration. For more detailed climate change scenarios, changes in crop type and cropping pattern can be included in the soil moisture balance model to provide new estimates of recharge; increased groundwater and surface water abstraction required for the increasing crop water demand can also be included.

Initial climate change studies in the Nottinghamshire Sherwood Sandstone aquifer and the Southern Lincolnshire Limestone aquifer system predict an increase in recharge during the wetter winters, based on the climate change scenarios. These initial assessments need further refinement to account for the likely changes in cropping patterns and the need to use groundwater to irrigate crops.

12.8.7 Substantive issues requiring further investigation

Of the many issues related to groundwater hydrology which require further investigation, four are highlighted below.

Assistance to practitioners in conducting groundwater studies: groundwater investigations are challenging, especially for practitioners with limited experience. Mistakes are frequently made in developing both conceptual and computational models. Yet much of the published literature is not intended for the practitioner; furthermore quoted case studies often relate to very well instrumented locations. There is a need for more research and wider dissemination of information about groundwater investigations based on limited data and limited finance.

Integrated groundwater and surface water conceptual and computational models: although software is available for integrated groundwater-surface water models, the surface water components are frequently subroutines added to groundwater models. A limited number of quantified conceptual models describing the whole catchment response have been developed, but their importance is not widely recognised. One difficulty is that surface water and groundwater investigations tend to use different approaches, primarily because of the different objectives and different timescales. A methodology for developing quantified conceptual models for combined groundwater–surface water systems is required, together with appropriate software. The software may not be based on MODFLOW.

Improved and imaginative fieldwork: developments of suitable integrated groundwater–surface water models require improved fieldwork. Detailed field investigations are often omitted or severely limited due to the increasing expense of drilling to identify the geology of the aquifer and the drift; also, monitoring of groundwater heads and spring, stream and river flows is constrained. Additional groundwater modelling is often seen as an alternative to field studies. Yet appropriate fieldwork is the key to groundwater investigations, especially when groundwater–surface water interaction is significant. Further research is required on the efficient planning and implementation of field studies.

Catchment-wide water quality studies: significant progress has been made in investigating contaminated aquifers in which pollution incidents have occurred: see, for example, Zheng and Bennett (2002). However, many catchment-wide studies of pollution risk are less reliable. When identifying source protection zones for boreholes, questionable idealisations are often made about aquifer behaviour. For example, in chalk and limestone aquifers, steady-state conditions are usually assumed to occur even though there are very different flow patterns between summer and winter and hence different source protection zones. Too often source protection zone studies are based on a specified software package, rather than preparing conceptual models and deciding which computational models should be used.

Appendix: Computer Program for Two-Zone Model

Detailed information about the two-zone radial flow model is provided in Section 7.4, with further examples of its application in Chapter 8. At the end of this Appendix there is a printed version of the computer code in FORTRAN for the two-zone model; this Appendix also includes a typical input data set and information about outputs.

Explanations of the numerical model and the meaning of the symbols can be gathered from Section 7.4 and associated figures, from comments within the computer code and from the explanation of the input data in Figure A.1. The data set for the problem in Figure 7.15 is listed in Figure A.1. Parameter values as specified in the data are transferred to array variables for all the nodes. However, individual parameter values can be overwritten as illustrated in line 6 of the data record. Alternatively, additional coding can be written to include parameter values varying with radius.

Figure A.2 is a copy of the first part of the output in EXAMPLE.OUT; information from this file can be used to confirm that the correct parameter values are used in the program. In addition, the file PLOT.OUT provides data for plotting. In its present form the output for plotting includes the variation with time of the quantity of water supplied by well storage plus selected drawdowns. Other results can be provided by changing the statement WRITE(6,6300).

In Figure A.3 samples of the standard output are presented; three groups of three time steps are included with the time in days written as T = 0.00040 etc. The first line of the output for an individual time step shows how the well discharge of 1500 m³/d is supplied from the Upper zone (QUP), the Lower zone (QLO) and from well storage; the final quantity is the total leakage through the Overlying layer. Note that the contribution from well storage is 570 m³/d at 0.0004 day (0.58

minutes) but by 0.0159 day (23 minutes) it has fallen to 3 m³/d. A greater proportion of the borehole discharge is drawn from the Lower zone. This occurs because the product of the thickness and permeability of the Lower zone is 245 m²/d compared to 90 m²/d for the Upper zone. Drawdowns for the Upper and Lower zones are quoted for radial distances varying from 0.2 to 5000 m.

A further section of the output file EXAMPLE.OUT is reproduced in Figure A.4; this provides information about drawdowns and component flows at a specified time, in this example at 1.0 day. The first four columns list the node number, the radial distance from the pumped borehole and the drawdowns in the Upper and Lower zones. At 2940 m the drawdowns are around 0.001 m (1 mm), indicating that the impact of pumping has begun to reach 3 km after one day. The remaining columns of figures provide information about the component flows (all in units of m³/d). To illustrate the significance of these flows, values for nodes 2 to 4 (0.2 m to 0.43 m) and nodes 20 to 23 (200 m to 632 m) are presented on a diagram which represents the Overlying and Middle layers and the Upper and Lower zones. The numbers with grey shading refer to the quantity of water per node released from storage as the drawdowns increase. A careful examination of all the results shows that the main source of water is the release of water from storage in the Lower zone; vertical flows through the Middle layer make a smaller contribution. At longer times, vertical flows through the Overlying layer QLEAKAGE become more significant.

Many more outputs from the two-zone model are presented in Chapters 7 and 8. The program in FORTRAN, the data file, the output file and an executable file of the program can be copied from http://members.lycos.co.uk/twznprog/

Groundwater Hydrology: Conceptual and Computational Models. K.R. Rushton
© 2003 John Wiley & Sons Ltd ISBN: 0-470-85004-3

```
      data                                  description
0.20, 5000.0                    RWELL, RMAX: well radius, max radius
0.0,25.0,50.0,65.0,90.0,125.0   RWL, TPOL, TPUP, BSUP, TPLO, BSLO: note 1
2                               JFIX: JFIX=1 recharge bound, JFIX=2 no-flow bound
0.02,0.00                       PERMVOL, RCH1: see note 1
    6.0, 1.2, 0.0004, 0.08        PERMRUP, PERMVUP, SCONUP, SUNCNUP: note 1
8, 6.0, 1.2, 0.0004, 0.08       I,PERMRUP, PERMVUP, SCONUP, SUNCNUP: note 2
0, 2.3, 0.7, 0.0001, 0.02       I,PERMRUP, PERMVUP, SCONUP, SUNCNUP: note 3
    0.005                         PERMVMD: see note 1
0, 0.003                        I,PERMVMD: see note 3
    7.0, 2.5, 0.0006              PERMRLO, PERMVLO, SCONLO: see note 1
0, 3.2, 1.1, 0.0001             I,PERMRLO, PERMVLO, SCONLO: see note 3
-2.0, 9.0, 6.0                  ALPHA, WLOSSUP, WLOSSLO: see note 4
3                               IWELL: see note 5
4                               NOBS: number of observation boreholes
10.0, 30.0, 100.0, 300.0        DOBS(): radial distances to observation boreholes
3                               LEVPRN: see note 6
0.01, 0.1, 1.0, 0.0, 0.0        TP(): times of detailed output if LEVPRN=3
1500.0, 2.50                    QPUMP, TSTOP: pumping rate until time TSTOP
    0.0, 3.50                   QPUMP, TSTOP:   QPUMP=0.0 indicates recovery
  -10.0, 4.50                   QPUMP, TSTOP:   -ve QPUMP signifies end of run
```

Note 1   These variables are defined in Section 7.4 and Figs 7.12 & 7.13. Initially values are read into a single variable but they are then transferred to arrays, PERMPUP() etc.

Note 2   This line can be used to change values for node 8

Note 3   I=0 terminates input of non-standard values, these parameter values are ignored

Note 4   ALPHA = delay index, negative value indicates no delayed yield
WLOSSUP and WLOSSLO are well loss factors for Upper and Lower zones (see Sect 5.5.4)

Note 5   IWELL = 1, abstraction from Lower zone only: IWELL = 2, abstraction from Upper zone only:
IWELL = 3, abstraction from both zones

Note 6   LEVPRN determines detail of output, LEVPRN=1 minimum output, LEVPRN=3 maximum output

Figure A.1   Input data for two-zone model for problem defined in Figure 7.15

```
RADIAL FLOW MODEL WITH VERTICAL COMPONENTS OF FLOW
  6   MESH INTERVALS PER DECADE
 10 TIME STEPS PER DECADE
RWELL=0.200   RMAX= 5000.00
INITIAL PIEZOMETRIC LEVEL =    0.00
TOP OF OVERLYING LEAKY ZONE =  25.00
TOP OF UPPER ZONE= 50.00  BASE OF UPPER ZONE= 65.00
TOP OF LOWER ZONE= 90.00  BASE OF LOWER ZONE=125.00
**IMPERMEABLE BOUNDARY AT RMAX**
*** NO DELAYED YIELD ***
PERMEABILITY OF LEAKY LAYER=0.200D-01
WELL LOSS FACTORS
FOR UPPER ZONE  9.000      FOR LOWER ZONE  6.000
**FULLY PENETRATING WELL, ABSTRACTION FROM BOTH THE ZONES**
```

Figure A.2   Initial output from program

```
PUMPING RATE= 1500.0 UNTIL 0.250D+01 DAYS

          R=   0.200      10.000      30.000     100.000     300.000    5000.000

   QUP=  226.8   QLO=   703.0   WELL STORAGE=   570.2  TOTAL LEAKAGE=     0.1
T=  0.00040   2.805       0.125       0.003       0.000       0.000       0.000
             2.805       0.230       0.015       0.000       0.000       0.000

   QUP=  252.9   QLO=   781.5   WELL STORAGE=   465.6  TOTAL LEAKAGE=     0.2
T=  0.00050   3.187       0.174       0.007       0.000       0.000       0.000
             3.187       0.304       0.027       0.000·      0.000       0.000

   QUP=  277.6   QLO=   855.6   WELL STORAGE=   366.8  TOTAL LEAKAGE=     0.2
T=  0.00063   3.566       0.234       0.013       0.000       0.000       0.000
             3.566       0.391       0.045       0.000       0.000       0.000


   QUP=  374.4   QLO=  1122.2   WELL STORAGE=     3.4  TOTAL LEAKAGE=    11.3
T=  0.01585   6.035       1.415       0.710       0.117       0.001       0.000
             6.035       1.786       0.998       0.252       0.005       0.000

   QUP=  375.1   QLO=  1122.3   WELL STORAGE=     2.7  TOTAL LEAKAGE=    14.3
T=  0.01995   6.122       1.493       0.783       0.155       0.001       0.000
             6.122       1.872       1.081       0.311       0.009       0.000

   QUP=  375.7   QLO=  1122.2   WELL STORAGE=     2.1  TOTAL LEAKAGE=    18.1
T=  0.02512   6.208       1.571       0.856       0.198       0.003       0.000
             6.208       1.958       1.165       0.373       0.016       0.000


   QUP=  396.1   QLO=  1103.9   WELL STORAGE=     0.0  TOTAL LEAKAGE=   605.3
T=  1.58489   7.510       2.620       1.854       1.042       0.410       0.000
             7.510       3.328       2.541       1.680       0.910       0.000

   QUP=  397.7   QLO=  1102.3   WELL STORAGE=     0.0  TOTAL LEAKAGE=   694.9
T=  1.99526   7.555       2.646       1.877       1.061       0.426       0.000
             7.555       3.380       2.593       1.733       0.963       0.000

   QUP=  399.1   QLO=  1100.9   WELL STORAGE=     0.0  TOTAL LEAKAGE=   790.3
T=  2.50000   7.595       2.668       1.896       1.077       0.440       0.000
             7.595       3.425       2.639       1.780       1.009       0.001
```

Figure A.3 Selection of detailed output for three groups of time steps

NO	RADIUS	UPPER D/D	LOWER D/D	QLEAKAGE	QST-UPPER	UPPER FLOW	VERT. FLOW	QST-LOWER	LOWER FLOW
2	0.200D+00	0.7405D+01	0.7405D+01	0.5714D-03	0.5492D-05	0.3927D+03	-.3573D+03	0.8238D-05	0.1107D+04
3	0.294D+00	0.5006D+01	0.5749D+01	0.8323D-03	0.1940D-04	0.3927D+03	-.3302D-04	0.3705D-04	0.1107D+04
4	0.431D+00	0.4740D+01	0.5472D+01	0.1698D-02	0.4077D-04	0.3927D+03	0.6534D-04	0.8039D-04	0.1107D+04
5	0.632D+00	0.4473D+01	0.5196D+01	0.3452D-02	0.8562D-04	0.3927D+03	0.1389D-03	0.1744D-03	0.1107D+04
6	0.928D+00	0.4207D+01	0.4920D+01	0.6994D-02	0.1797D-03	0.3927D+03	0.2953D-03	0.3783D-03	0.1107D+04
7	0.136D+01	0.3940D+01	0.4644D+01	0.1411D-01	0.3770D-03	0.3926D+03	0.6278D-03	0.8207D-03	0.1107D+04
8	0.200D+01	0.3674D+01	0.4368D+01	0.2835D-01	0.7902D-03	0.3926D+03	0.1334D-02	0.1780D-02	0.1107D+04
9	0.294D+01	0.3408D+01	0.4092D+01	0.5665D-01	0.1655D-02	0.3926D+03	0.2835D-02	0.3861D-02	0.1107D+04
10	0.431D+01	0.3141D+01	0.3816D+01	0.1125D+00	0.3463D-02	0.3925D+03	0.6021D-02	0.8375D-02	0.1107D+04
11	0.632D+01	0.2875D+01	0.3540D+01	0.2218D+00	0.7241D-02	0.3922D+03	0.1279D-01	0.1816D-01	0.1107D+04
12	0.928D+01	0.2609D+01	0.3264D+01	0.4336D+00	0.1513D-01	0.3918D+03	0.2714D-01	0.3939D-01	0.1107D+04
13	0.136D+02	0.2343D+01	0.2988D+01	0.8390D+00	0.3157D-01	0.3910D+03	0.5756D-01	0.8541D-01	0.1107D+04
14	0.200D+02	0.2077D+01	0.2712D+01	0.1603D+01	0.6580D-01	0.3895D+03	0.1220D+00	0.1852D+00	0.1107D+04
15	0.294D+02	0.1813D+01	0.2436D+01	0.3014D+01	0.1370D+00	0.3866D+03	0.2580D+00	0.4014D+00	0.1106D+04
16	0.431D+02	0.1551D+01	0.2161D+01	0.5554D+01	0.2848D+00	0.3813D+03	0.5439D+00	0.8697D+00	0.1105D+04
17	0.632D+02	0.1292D+01	0.1885D+01	0.9969D+01	0.5908D+00	0.3719D+03	0.1140D+01	0.1882D+01	0.1102D+04
18	0.928D+02	0.1040D+01	0.1611D+01	0.1728D+02	0.1222D+01	0.3557D+03	0.2363D+01	0.4067D+01	0.1095D+04
19	0.136D+03	0.7982D+00	0.1337D+01	0.2859D+02	0.2512D+01	0.3294D+03	0.4810D+01	0.8747D+01	0.1082D+04
20	0.200D+03	0.5747D+00	0.1068D+01	0.4434D+02	0.5101D+01	0.2895D+03	0.9476D+01	0.1865D+02	0.1054D+04
21	0.294D+03	0.3782D+00	0.8052D+00	0.6287D+02	0.1009D+02	0.2342D+03	0.1768D+02	0.3898D+02	0.9968D+03
22	0.431D+03	0.2193D+00	0.5567D+00	0.7854D+02	0.1882D+02	0.1669D+03	0.3009D+02	0.7819D+02	0.8886D+03
23	0.632D+03	0.1061D+00	0.3352D+00	0.8183D+02	0.3094D+02	0.9814D+02	0.4402D+02	0.1435D+03	0.7011D+03
24	0.928D+03	0.3945D-01	0.1604D+00	0.6558D+02	0.3935D+02	0.4327D+02	0.5006D+02	0.2176D+03	0.4334D+03
25	0.136D+04	0.1009D-01	0.5234D-01	0.3613D+02	0.3223D+02	0.1260D+02	0.3769D+02	0.2242D+03	0.1715D+03
26	0.200D+04	0.1535D-02	0.9593D-02	0.1184D+02	0.1416D+02	0.2084D+01	0.1548D+02	0.1209D+03	0.3507D+02
27	0.294D+04	0.1206D-03	0.8499D-03	0.2006D+01	0.2929D+01	0.1685D+00	0.3019D+01	0.2883D+02	0.3225D+01
28	0.431D+04	0.6263D-05	0.4601D-04	0.1400D+00	0.2322D+00	0.1757D-01	0.2213D+00	0.2425D+01	0.5781D+00
29	0.500D+04	0.2679D-05	0.2435D-04	0.2327D-01	0.4117D-01	0.3230D-15	0.4687D-01	0.5312D+00	0.2936D-14

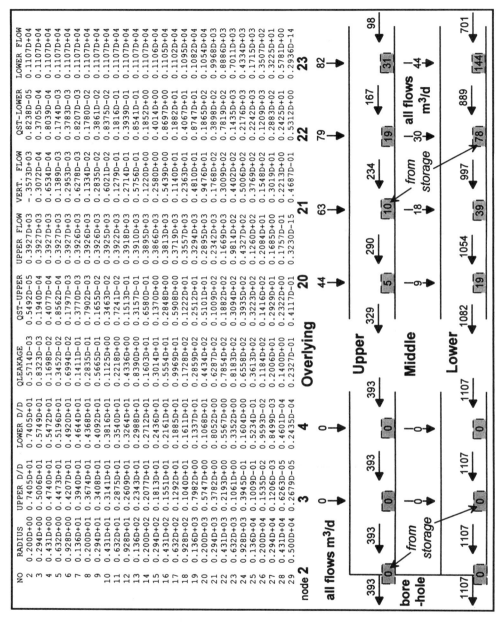

Figure A.4 Detailed output at 1.0 day with diagram showing selected flows FORTRAN code for two-zone radial flow model

Listing of Program in FORTRAN

```fortran
      PROGRAM TWZNPROG
C     ***************************************************
C     *NUMERICAL MODEL FOR PUMPING FROM TWO-ZONE LAYERED AQUIFER*
C     *     prepared by K. S. Rathod, Civil Engineering        *
C     *     UNIVERSITY, BIRMINGHAM. B15 2TT  (UNITED KINGDOM).  *
C     ***************************************************
      PARAMETER (NA=40,NB=5)
      IMPLICIT DOUBLE PRECISION (A-H, O-Z)
      COMMON R(NA),RR(NA),DLO(NA),DUP(NA),OLDDLO(NA),OLDDUP(NA),
     1 RECH(NA),PERMRUP(NA),PERMVUP(NA),SCONUP(NA),SUNCNUP(NA),
     2 PERMVMD(NA),TSUP( NA),PERMRLO(NA),PERMVLO(NA),SCONLO(NA),
     3 VL(NA),HUP(NA),HLO(NA),V(NA),TSLO(NA),AUP(NA),BUP(NA),
     4 CUP(NA),E(NA),FUP(NA),ALO(NA),BLO(NA),CLO(NA),FLO(NA),
     5 U1(NA),U2(NA),V1(NA),V2(NA),RJD(NA),RLN(NA),X(NA),Y(NA),
     6 DOBS(NB),DPAC(NB),ARAY(NB),ARRAY(NB),TP(NB),C1,RWL,TMB,
     7 TIMIN,NMAX,NOB(NB),NMONE,NOBS,OUT1,OUT2
      INTEGER OUT1,OUT2
      IN1=5
      OUT1=6
      OUT2=7
      OPEN (IN1, FILE='EXAMPLE.DAT')
      OPEN (OUT1,FILE='EXAMPLE.OUT')
      OPEN (OUT2,FILE='PLOT.OUT')
C     INPUT WELL RAD. DIST. TO OUTER BOUND. AND SET LOG RADIAL MESH
      READ(IN1,*) RWELL, RMAX
C     MM=6  !MM IS NO. MESH INTERVALS PER TENFOLD INCREASE IN RAD
      DO 320 N=1,40
      AN = FLOAT(N - 2) / FLOAT(MM)
      R(N) = RWELL * (1.0D+01 ** AN)
      IF (R(N).LT. RMAX ) GOTO 310
300   R(N) = RMAX
      RR(N) = RMAX * RMAX
      NMAX = N
      NMONE = N - 1
      GOTO 330
310   RR(N) = R(N) * R(N)
320   CONTINUE
330   AN = 1.0 / FLOAT(MM)        ! CALC DELA AND DELA SQUARED
      DELA = AN * DLOG(1.0D+01)
      DELA2 = DELA * DELA
C     **********  MAIN INPUT SECTION **********************
C     INPUT ELEVATIONS OF UPPER AND LOWER ZONES, DATUM IS RWL
C     T IS TOP, B IS BASE, UP IS UPPER AND LO IS LOWER ZONE
420   READ(IN1,*) RWL, TPOL, TFUP, BSUP, TPLO, BSLO
C     INPUT OUTER BOUND TYPE, JFIX=1 FOR RECHARGE, =2 FOR NO-FLOW
470   READ(IN1,*) JFIX
C     INPUT VERTICAL PERMEABILITY FOR THE OVERLYING ZONE AND RECHARGE
      READ(IN1,*) PERMVOL,RCH1
C     INPUT STANDARD PARAMETERS; FIRST FOR THE UPPER ZONE
      READ(IN1,*) PRADUP, PVERTUP, CONUPS, UNCNUPS
      DO 570 N = 1, NMAX
      PERMRUP(N)=PRADUP
      PERMVUP(N)=PVERTUP
      SCONUP(N) = CONUPS
      SUNCNUP(N) = UNCNUPS
570   CONTINUE
C NOW INPUT NON-STANDARD PARAMETERS FOR THIS ZONE
590   READ(IN1,*) I,PERMRUP(I),PERMVUP(I),SCONUP(I),SUNCNUP(I)
      IF (I.GT. 0) GOTO 590
C     INPUT VERTICAL PERMEABILITY BETWEEN THE UPPER AND LOWER ZONE
      READ(IN1,*) VMDS

      DO 650 N = 1, NMAX
      PERMVMD(N) = VMDS
650   CONTINUE
C NOW INPUT NON-STANDARD VERTICAL PERMEABILITY FOR THIS ZONE
670   READ(IN1,*) I, PERMVMD(I)
      IF (I .GT. 0) GOTO 670
C     INPUT STANDARD PARAMETERS FOR THE LOWER ZONE
      READ(IN1,*) PRADLO, PVERTLO, CONLOS
      DO 740 N = 1, NMAX
      PERMRLO(N)=PRADLO
      PERMVLO(N)=PVERTLO
      SCONLO(N) = CONLOS
C NOW INPUT NON-STANDARD PARAMETERS FOR THIS ZONE
760   READ(IN1,*) I, PERMRLO(I), PERMVLO(I), SCONLO(I)
      IF ( I .GT. 0) GOTO 760
C     INPUT FACS. FOR SPECIAL FEATURES DELAYED YIELD AND WELL LOSS
      READ(IN1,*) ALPHA, WLOSSUP, WLOSSLO
C     INPUT FACTORS FOR WELL CONSTRUCTION
C     IWELL=1, FULLY PENETRATING WELL, SOLID CASING IN UPPER ZONE,
C     IWELL=2, WELL PENETRATING ONLY THE UPPER ZONE,
C     IWELL=3, FULLY PENETRATING WELL WITHOUT ANY SOLID CASING.
      READ(IN1,*) IWELL
C     INPUT NO. OF OBSERVATION BOREHOLES & THEIR DISTANCES; MAX. 5 OBH
      READ(IN1,*) NOBS
      READ(IN1,*) (DOBS(J), J=1,NOBS)
C     INPUT LEVEL OF OUTPUT REQUIRED, 1,2 OR 3 (LOW, MEDIUM, HIGH)
      READ(IN1,*) LEVPRN
C     INPUT 5 TIMES FOR FULL PRINTING, 1,2 OR 5, READ THE REST AS ZERO
      READ(IN1,*) (TP(I), I=1,5)
C     ****************** END OF MAIN INPUT SECTION
C
C     INITIALISE VARIOUS VARIABLES AND ARRAYS
C     KT IS NUMBER OF TIME STEPS PER DECADE
      KT = 10
      AKT = -1 / FLOAT(KT)
      DT = (1.0D+01** AKT) - 1.0
      TMB=0.0
      IPRINT = 0
      ITP = 1        ! INITIALIZE COUNTERS FOR PRINTING TIME
      PACA = 0.0
      PACB = 0.0
      PACC = 0.0      !INITIALIZE DELAYED YIELD FACTOR
      PLEAK = 1.0     !LEAKAGE FACTOR
      THLEAK=TPUP-TPOL
      IF (THLEAK .LT. 1.0D-10 .OR. PERMVOL .LE. 0.0) PLEAK=0.0
      RCH=RCH1
      IF (TPUP.GT.RWL) RCH=0.0
      PI=4.0*DATAN(1.0D+00)
      C1=2.0*PI*DELA      !CONST. USED FOR CALCULATION OF FLOW
      DRAWMX = 0.9 * BSUP + 0.1 * TPUP
      IF (IWELL.EQ. 1) RATIO=1.0
      IF (IWELL.EQ. 2) RATIO=0.0
      IF (IWELL.EQ. 3) RATIO=0.5        ! INITIALIZE ARRAYS
      DO 2640 N = 1, NMAX
      RECH(N)=RCH
      DLO(N)=RWL
      DUP(N)=RWL
      VL(N)=0.0
      OLDDLO(N)=DLO(N)
      OLDDUP(N)=DUP(N)
2640  CONTINUE
C     **************  OUTPUT INITIAL INFORMATION  ****************
      WRITE(OUT1,3020)
3020  FORMAT(1X,'RADIAL FLOW MODEL WITH VERTICAL COMPONENTS OF FLOW')
      WRITE(OUT1,3030) MM
3030  FORMAT(1X,I2,' MESH INTERVALS PER DECADE')
```

```fortran
3040  WRITE(OUT1,3040) KT
      FORMAT(1X,I3,' TIME STEPS PER DECADE')
3050  WRITE(OUT1,3050) RWELL,RMAX
      FORMAT(1X,'RWELL=',F5.3,'   RMAX=',F8.2)
3060  WRITE(OUT1,3060) RWL
      FORMAT(1X,'INITIAL PIEZOMETRIC LEVEL = ',F6.2)
3070  WRITE(OUT1,3070) TPUP,BSUP
      FORMAT(1X,'TOP OF UPPER ZONE=',F6.2,' BASE OF UPPER ZONE=',F6.2)
3080  WRITE(OUT1,3080) TPLO,BSLO
      FORMAT(1X,'TOP OF LOWER ZONE=',F6.2,' BASE OF LOWER ZONE=',F6.2)
      IF (JFIX .EQ. 1) WRITE(OUT1,3090)
3090  FORMAT(1X,'**RECHARGE BOUNDARY AT RMAX***')
      IF (JFIX .EQ. 2) WRITE(OUT1,3100)
3100  FORMAT(1X,'**IMPERMEABLE BOUNDARY AT RMAX***')
      IF (ALPHA .LT. 0.0) WRITE(OUT1,3110)
3110  FORMAT(1X,'*** NO DELAYED YIELD ***')
      IF (ALPHA .GT. 0.0) WRITE(OUT1,3120)ALPHA
3120  FORMAT(1X,'DELAYED YIELD INDEX=',D10.4)
      IF (PERMVOL .LE. 0.0) GOTO 3145
      WRITE(OUT1,3140) PERMVOL
3140  FORMAT(1X,'PERMEABILITY OF LEAKY LAYER=',D9.3)
3145  IF (RCH .GT.0.0) WRITE(OUT1,3146)RCH
3146  FORMAT(1X,'UNIFORM RECHARGE=',F6.4,' M/DAY')
      IF (RCH .NE. 0.0) WRITE(OUT1,3149)
3149  FORMAT(1X,'NOTE: NO RECHARGE WHEN THERE IS OVERLYING ZONE')
      WRITE(OUT1,3150)
3150  FORMAT(1X,'WELL LOSS FACTORS ')
      WRITE(OUT1,3160) WLOSSUP,WLOSSLO
3160  FORMAT(1X,'FOR UPPER ZONE ',F5.3,'        FOR LOWER ZONE ',F5.3)
      IF (IWELL.EQ.1) WRITE(OUT1,3180)
3180  FORMAT(1X,'**FULLY PENETRATING WELL WITH UPPER ZONE CASED***')
      IF (IWELL .EQ. 2) WRITE(OUT1,3190)
3190  FORMAT(1X,'**WELL PENETRATING ONLY THE UPPER ZONE***')
      IF (IWELL .EQ. 1 .OR. IWELL .EQ. 2)  GOTO 3220
      WRITE(OUT1,3200)
3200  FORMAT(1X,'**FULLY PENETRATING WELL,',
     1 '  ABSTRACTION FROM BOTH THE ZONES***')
3220  IF (TP(1).LE.0.00001 .OR. LEVPRN .LT. 3) GOTO 3250
      WRITE(OUT1,3230) (TP(I),I=1,5)
3230  FORMAT(1X,'FLOWS WILL BE PRINTED OUT AT TIMES CLOSEST',
     1 ' TO -',5(1X,D10.3)//)
C ******** END OF OUTPUT OF INITIAL INFORMATION ***
3250  IF (LEVPRN .GE. 2) THEN
C******* PRINT AQUIFER PARAMETERS
      WRITE(OUT1,3520)
3520  FORMAT(1X,'NO. RADIUS        ',12HO'LYING KV   ,'  UPPER KH ',
     1 'UPPER KV   LOWER KV  LOWER SCON ')
      DO 3580 N=1, NMAX
      WRITE(OUT1,3550) N, R(N), PERMVOL, PERMVUP(N), PERMVMD(N),
     1 SUNCNUP(N), SCONUP(N), PERMVMD(N), PERMRLO(N), PERMVLO(N),
     2 SCONLO(N)
3550  FORMAT(1X,I2,1P,D10.3,1X,9D12.4)
3580  CONTINUE
      ENDIF
C FACTORS TO INTERPOLATE DRAWDOWNS AT OBHS.
      I=2
      DO 1170 J=1,NOBS
C THE NODE JUST BEFORE THE JTH OBH IS I-1
      DO WHILE (R(I) .LT. DOBS(J))
      I=I+1
      END DO
      NOB(J)=I-1
C FACTOR TO INTERPOLATE DRAWDOWN AT OBH
      DFAC(J)=(DLOG(DOBS(J))-DLOG(R(I-1))) / (DLOG(R(I))-DLOG(R(I-1)))
1170  CONTINUE

1250  READ(IN1,*) QPUMP, TSTOP
      IF (QPUMP .LT. 0.0) GOTO 2120
      WRITE(OUT1,1270) QPUMP.TSTOP
1270  FORMAT(1X/,1X,'PUMPING RATE=',F7.1,' UNTIL ',D10.3,' DAYS')
      QABST = QPUMP / C1  ! ABSTRACTION IS -VE RECHARGE AT NODE 1
      IND = 0
      TIMIN = 0.0
      DELI =1.0D-07
      DELT = DELI
      WRITE(OUT1,7500)R(2),(DOBS(J),J=1,NOBS),R(NMAX)
7500  FORMAT(1X//,10X,'R=  ',F7.3,6F11.3//)
1360  TME = TME + DELT
      IF( (TME+0.0025*DELT) .LT. TSTOP)   GOTO 1390
      DELT = TSTOP - TME + DELT
      TME = TSTOP
      IND = 100
1390  TIMIN = TIMIN + DELT
C ***** INCLUDE DELAYED YIELD    *************
      IF (ALPHA .LT. 0.0) GOTO 4040
      PA = ALPHA * DELT
      IF (PA .GT. 100.0) GOTO 4040
      PACA = DEXP(-PA)
      IF (QPUMP .LE. 0.0) PACA=0.0
4040  FACB = 1.0 - PACA
      FACC = FACB / (ALPHA * DELT)
      DO 4060 N = 1, NMAX
      X(N) = FACA * Y(N)
      RBCH(N) = ALPHA * SUNCNUP(N) * X(N) + RCH
4060  CONTINUE
C ******* ITERATIVE LOOP REQUIRED FOR UNCONFINED AQUIFERS   *******
      DO 1830 NUM = 1, 4
      DO 1620 N = 1, NMONE
      STORLO = SCONLO(N)
      VN1=0.0
      VN2=0.0
      VN3=0.0
      D = 0.5 * (DUP(N) + DUP(N + 1))            ! AV. FREE SURFACE DRAWDOWN
      IF (D .LT. TPUP) GOTO 1480
      STORUP = SUNCNUP(N) * FACB + SCONUP(N)     !UNCONFINED CONDITION
      SD = BSUP - D
      GOTO 1490
1480  STORUP = SCONUP(N)
      SD = BSUP - TPUP
1490  HUP(N) = DELA2 / (SD * PERMRUP(N))
      HLO(N) = DELA2 / ((BSLO - TPLO) * PERMRLO(N))
      VN1 = 0.5 * (BSLO - TPLO) / (PERMVLO(N))
      VN2 = SD / (PERMVUP(N))
      IF(DABS(TPLO-BSUP).LT.0.00001 .OR. PERMVMD(N).LE. 0.0) GOTO 1530
      VN3 = (TPLO - BSUP) / PERMVMD(N)
      V(N) = (VN1 + VN2 + VN3) / RR(N)
      TSUP(N) = DELT / (RR(N) * STORUP)
      TSLO(N) = DELT / (RR(N) * STORLO)
1530  CONTINUE
C CALCULATE LEAKAGE COEFF. ONLY IF LEAKY LAYER IS PRESENT AND
C LEAKAGE PERMEABILITY IS NON-ZERO POSITIVE VALUE.
      IF(FLBAK .LE. 0.00) GOTO 1620
      VL(N)=(PERMVOL*RR(N))/THLBAK
1620  CONTINUE
      HLO(1) = .00005 * HLO(1)
      HUP(1) = .00005 * HUP(1)
      HLO(2) = HLO(2) * WLOSSLO
      HUP(2) = HUP(2) * WLOSSUP
C    SETTING UP CONDITIONS FOR WELL CASING
      IF(IWELL .NE. 1) GOTO 1700
C ABSTRACTION FROM LOWER LAYER ONLY
      TSLO(1) = 2.0 * DELT * DELA / RR(2)
      TSLO(2) = 2.0 * TSLO(2)
```

```fortran
      TSUP(1) = TSUP(1) * 1.0D+10
      V(1) = V(1) * 1.0D+20
      GOTO 1760
 1700 IF (IWELL .NE. 2) GOTO 1740
C                   ABSTRACTION FROM UPPER LAYER ONLY
C
      TSUP(1) = 2.0 * DELT * RR(2)
      V(1) = V(1) * 1.0D+20
      GOTO 1760
C                   ABSTRACTION FROM BOTH THE LAYERS
 1740 TSLO(1) = 4.0 * DELT * DELA / RR(2)
      TSUP(1) = TSLO(1)
 1750 TSLO(2) = 2.0 * TSLO(2)
      V(1) = 1.0D-10
      V(2) = V(1)
 1760 TSUP(2) = 2.0 * TSUP(2)
C ******* MODIFY RESISTANCES AT THE OUTER BOUNDARY **************
      VN1=0.0
      VN2=0.0
      VN3=0.0
      HLO(NMAX) = 2.0D+10
      HUP(NMAX) = 2.0D+10
      DELAN = DLOG(R(NMAX) / R(NMONE))
      DELAN2 = DELAN * DELAN
      HUP(NMONE) = 2.0 * DELAN2 / (SD * PERMRUP(NMONE))
      HLO(NMONE) = DELAN2 / ((BSLO - TPLO) * PERMRLO(NMONE))
      ACONS = 2.0 * DELT * DELA
      RN = R(NMAX) * (R(NMAX) - R(NMAX - 2))
      TSLO(NMONE) = ACONS / (STORLO * RN)
      TSUP(NMONE) = ACONS / (STORUP * RN)
      RN1 = R(NMAX) * (R(NMAX) - R(NMONE))
      TSLO(NMAX) = ACONS / (STORLO * RN1)
      TSUP(NMAX) = ACONS / (STORUP * RN1)
      IF (JFIX .NE. 1) GOTO 4610
      TSUP(NMAX) = 1.0D-15 * TSUP(NMAX)
      TSLO(NMAX) = 2.0D-15 * TSLO(NMAX)
 4610 VN1 = 0.5 * (BSLO - TPLO) / (PERMVLO(NMONE))
      D = 0.5 * (DUP(NMONE) + DUP(NMAX))
      IF ( D .LT. TPUP) D = TPUP
      VN2 = (BSUP - D) / (PERMVUP(NMONE))
      IF (DABS(TPLO-BSUP).LT. 0.00001 .OR.
     1 PERMVMD(NMONE).LE.0.0) GOTO 4640
      VN3 = (TPLO - BSUP) / PERMVMD(NMONE)
 4640 V(NMONE) = (VN1 + VN2 + VN3) * 2.0 * DELA / RN
      VN1 = 0.5 * (BSLO - TPLO) / (PERMVLO(NMAX))
      D = DUP(NMAX)
      IF ( D .LT. TPUP) D = TPUP
      VN2 = (BSUP - D) / (PERMVUP(NMAX))
      IF(DABS(TPLO-BSUP).LT.0.00001 .OR. PERMVMD(NMAX).LE.0.0)GOTO 4680
      VN3 = (TPLO - BSUP) / PERMVMD(NMAX)
 4680 V(NMAX) = (VN1 + VN2 + VN3) * 2.0 * DELA / RN1
      IF (FLEAK .LE. 0.0) GOTO 4750
      VL(NMONE) = (PERMVOL * RN)/(THLEAK * 2.0 * DELA)
      VL(NMAX) = (PERMVOL * RN1)/(THLEAK * 2.0 * DELA)
 4750 IF (JFIX .EQ. 1) V(NMAX) = 1.0D-15 * V(NMAX)
C *** SOLVE EQNS. USING GAUSSIAN ELIMINATION *****
      ALO(1) = 0.0
      AUP(1) = 0.0
      B(1) = 1.0 / V(1)
      CUP(1) = 1.0 / HUP(1)
      BUP(1) = -1.0 / HUP(1) + 1.0 / TSUP(1) )
      FUP(1) = -OLDDUP(1) / TSUP(1) - (1.0 - RATIO) * QABST
      VL(NMAX) = (PERMVOL * RN1)/(THLEAK * 2.0 * DELA)
      CLO(1) = 1.0 / HLO(1)
      BLO(1) = -(1.0 / HLO(1) + 1.0 / TSLO(1))
      FLO(1) = -OLDDLO(1) / TSLO(1) - RATIO * QABST
      DO 5120 N = 2, NMAX
      AUP(N) = 1.0 / HUP(N - 1)

      CUP(N) = 1.0 / HUP(N)
      E(N) = 1.0 / V(N )
      BUP(N) = -(AUP(N) + CUP(N) + E(N) + 1.0 / TSUP(N) + VL(N))
      FUP(N) = -OLDDUP(N) / TSUP(N) + RR(N) * RECH(N) - VL(N) * RWL
      ALO(N) = 1.0 / HLO(N - 1)
      CLO(N) = 1.0 / HLO(N)
      BLO(N) = -(ALO(N) + CLO(N) + E(N) + 1.0 / TSLO(N))
      FLO(N) = -OLDDLO(N) / TSLO(N)
 5120 CONTINUE
      FUP(NMONE) = -OLDDUP(NMONE)/TSUP(NMONE) +
     1 0.5*RECH(NMONE)*RN/DELA - VL(NMONE) *RWL
      FUP(NMAX) = -OLDDUP(NMAX)/TSUP(NMAX)
     1 0.5*RECH(NMAX)*RN1/DELA - VL(NMAX)*RWL
C     GAUSSIAN ELIMINATION --        FORWARD SOLUTION
      U1(1) = BLO(1)
      V1(1) = FLO(1)
      U2(1) = BUP(1)
      V2(1) = FUP(1)
      RJD(1) = 0.0
      RLN(1) = 0.0
      DO 5250 N = 2, NMAX
      M = N - 1
      U1(N) = BLO(N) - CLO(M) / U1(M) * ALO(N)
     1 + RJD(N) * RLN(M) / U1(M) / U2(M) * ALO(N)
      RLN(N) = E(N) + CUP(N) / U1(M) * RLN(M) / U2(M) * ALO(N)
      V1(N) = FLO(N) - V1(M) / U1(M) * ALO(N)
     1 + V2(M) / U1(M) * RLN(M) / U2(M) * ALO(N)
      U2(N) = BUP(N) - CUP(M) / U2(M) * AUP(N) - RLN(N) / U2(M) * E(N)
      RJD(N) = -CLO(N) / U1(N) * E(N)
     1 + CLO(N) / U1(N) * RJD(M) / U2(M) * AUP(N)
      V2(N) = FUP(N) - AUP(N)/U2(M)*V2(M) - E(N)/U1(N)*V1(N)
     1 + RJD(M)/U2(M)*AUP(N)/U1(N)*V1(N)
 5250 CONTINUE
C          NOW BACK SUBSTITUTION
      DUP(NMAX) = V2(NMAX) / U2(NMAX)
      DLO(NMAX) = V1(NMAX)/U1(NMAX) - RLN(NMAX)/U1(NMAX)*DUP(NMAX)
      DO 5320 N = NMONE, 1, -1
      M = N + 1
      DUP(N) = V2(N) / U2(N) - RJD(N) / U2(N) * DLO(M)
     1 -CUP( N) / U2(N) * DUP(M)
      DLO(N) = V1(N) / U1(N) - RLN(N) / U1(N) * DUP(N)
     1 -CLO(N) / U1(N) * DLO(M)
 5320 CONTINUE
C
      IF (DUP(1) .LT. DRAWMX) GOTO 1830
      WRITE(OUT1,1800)
 1800 FORMAT(1X,'**EXCESSIVE DRAWDOWN IN THE WELL**')
      CALL OUTPT            ! FOR PRINTING HYD.RES,D/D, ETC.
      STOP            !*** STOP DUE TO EXCESSIVE DRAWDOWN IN THE WELL
 1830 CONTINUE            !*** END OF ITERATIVE LOOP FOR UNCONFINED AQUIFER
C         ADDITIONAL STATEMENTS FOR DELAYED YIELD
      DO 1860 N = 1, NMAX
      Y(N) = X(N) + FACC * (DUP(N) - OLDDUP(N))
 1860 CONTINUE
      IF(TIMIN .LT. 0.00035) GOTO 1900 ! SKIP PRINTING AT EARLIER TIMES
      CALL OUTFLD !TO CALC &PRINT FLOWS AND DRAWDOWNS AT OBHS.
 1900 IF(LEVPRN .LT. 3) GOTO 2050
      IF(DABS(TMB-TP(ITP)) .GT. 0.5*TIMIN*DT) GOTO 2050
      CALL OUTPT
      WRITE(OUT1,7500)R(2),(DOBS(J),J=1,NOBS),R(NMAX)
      ITP=ITP+1
 2050 DELT = TIMIN * DT
      IF (IND .EQ. 0) GOTO 2070        !NEW DELT
      CALL OUTPT        !ONE PHASE COMPLETED, PRODUCE SUMMARY PRINTOUT
 2070 DO 2075 N=1, NMAX
```

```fortran
          OLDDLO(N) = DLO(N)
          OLDDUP(N) = DUP(N)
2075      CONTINUE
          IF(IND .NE. 0) GOTO 1250
          GOTO 1360
C **********    END OF TIME STEP LOOP    *****************
C TO 1360 FOR NEXT TIME STEP OR IF THIS WAS THE LAST TIME STEP FOR
C THE PHASE, THEN TO 1250 TO READ Q AND TIME FOR THE NEXT PHASE
2120      CONTINUE
          WRITE(OUT1,2130)
2130      FORMAT(1X,'END OF RUN')
          STOP
          END
C
C ****** SUBROUTINE TO INTERPOLATE AND PRINT D/D AND FLOWS **
          SUBROUTINE OUTFLD
          PARAMETER (NA=40,NB=5)
          IMPLICIT DOUBLE PRECISION (A-H, O-Z)
          COMMON R(NA), RR(NA), DLO(NA), DUP(NA), OLDDLO(NA), OLDDUP(NA),
         1 RECH(NA), PERMRUP(NA), PERMVUP(NA), SCONUP(NA), SUNCNUP(NA),
         2 PERMVMD(NA), TSUP( NA), PERMRLO(NA), PERMVLO(NA), SCONLO(NA),
         3 VL(NA), HUP(NA), HLO(NA), V(NA), TSLO(NA), AUP(NA), BUP(NA),
         4 CUP(NA), E(NA), FUP(NA), ALO(NA), BLO(NA), CLO(NA), FLO(NA),
         5 U1(NA), U2(NA), V1(NA), V2(NA), RJD(NA), RLN(NA), X(NA), Y(NA),
         6 DOBS(NB), DPAC(NB), ARAY(NB), TP(NB), C1, RWL, TMB, TMIN,
         7 NMAX, NOB(NB), NMONE, NOBS, OUT1, OUT2
          INTEGER OUT1, OUT2
          DO 6030 I = 1, NOBS
          I1 = NOB(I)
          ARAY(I) = DLO(I1) + DFAC(I) * (DLO(I1 + 1) - DLO(I1))
          ARAY(I) = DUP(I1) + DFAC(I) * (DUP(I1 + 1) - DUP(I1))
6030      CONTINUE
          QUP = C1 * (DUP(2) - DUP(3)) / HUP(2)    ! UPPER RADIAL FLOW
          QLO = C1 * (DLO(2) - DLO(3)) / HLO(2)    ! LOWER RADIAL FLOW
C QST IS WELL STORAGE CONTRIBUTION TO HALFWAY BET'N NODE 2 AND 3
          QST1 = C1 * (DLO(2)-OLDDLO(2)) * (1.0/TSLO(1) + 1.0/TSLO(2))
          QST2 = C1 * (DUP(2)-OLDDUP(2)) * (1.0/TSUP(1) + 1.0/TSUP(2))
          QST = (QST1 + QST2)
          QL = 0.0                    ! CALCULATE TOTAL LEAKAGE QL
          DO 6110 N = 2, NMAX
          QL = QL + C1 * (DUP(N) - RWL) * VL(N)
6110      CONTINUE
          WRITE(OUT1,6120) QUP, QLO, QST, QL
6120      FORMAT(1X,'QUP= ',F6.1,'   QLO= ',F6.1,'   WELL STORAGE= ',
         1 F6.1,'  TOTAL LEAKAGE= ',F6.1)
          WRITE(OUT1,6160)TIMIN,DUP(1),(ARAY(I),I=1,NOBS),DUP(NMAX)
6160      FORMAT(1X,'T=',F9.5,F9.3, 7(2X,F9.3))
          WRITE(OUT1,6200)DLO(1),(ARAY(I),I=1,NOBS),DLO(NMAX)
6200      FORMAT(8X,F13.3,7F11.3)
          WRITE(OUT1,*)
C OPTIONAL OUTPUT FOR PLOTTING
          WRITE(OUT2,6300)TIMIN,QST,DUP(1),DLO(1),ARRAY(1),ARAY(1)
6300      FORMAT(F10.4,5F10.3)
          RETURN
          END
C
C *** PRINT DRAWDOWNS AND FLOWS FOR THE WHOLE AQUIFER **
          SUBROUTINE OUTPT
          PARAMETER (NA=40,NB=5)
          IMPLICIT DOUBLE PRECISION (A-H, O-Z)
          COMMON R(NA), RR(NA), DLO(NA), DUP(NA), OLDDLO(NA), OLDDUP(NA),
         1 RECH(NA), PERMRUP(NA), PERMVUP(NA), SCONUP(NA), SUNCNUP(NA),
         2 PERMVMD(NA), TSUP( NA), PERMRLO(NA), PERMVLO(NA), SCONLO(NA),
         3 VL(NA), HUP(NA), HLO(NA), V(NA), TSLO(NA), AUP(NA), BUP(NA),
         4 CUP(NA), E(NA), FUP(NA), ALO(NA), BLO(NA), CLO(NA), FLO(NA),
         5 U1(NA), U2(NA), V1(NA), V2(NA), RJD(NA), RLN(NA), X(NA), Y(NA),
         6 DOBS(NB), DPAC(NB), ARAY(NB), TP(NB), C1, RWL, TMB, TMIN,
         7 NMAX, NOB(NB), NMONE, NOBS, OUT1, OUT2
          INTEGER OUT1, OUT2
          WRITE(OUT1,7010)TME,TIMIN
7010      FORMAT(1X,'   SUMMARY AT TOTAL TIME', F8.5,'  DAYS',
         1 '  PHASE TIME  ',F8.5,'  DAYS'/)
          WRITE(OUT1,7040)
7040      FORMAT(1X,'NO    RADIUS   UPP. RHR   LOW. RHR   VERT. R',
         1 '   UPP. D/D   LOW. D/D    QLEAK     QST-UPP.',
         2 '   UPP.FLOW   VERT.FLOW QST-LOWER LOW. FLOW  '/)
          DO 7150 I = 1, NMAX
          QUP = C1 * (DUP(I) - DUP(I + 1)) / HUP(I)
          QLO = C1 * (DLO(I) - DLO(I + 1)) / HLO(I)
          QV = C1 * (DLO(I) - DUP(I)) / V(I)
          IF(I .EQ. 1) QV=0.0
          QL = C1 * (DUP(I) - RWL) * VL(I)
          QSTUP=C1*(DUP(I)-OLDDUP(I))/TSUP(I)
          QSTLO=C1*(DLO(I)-OLDDLO(I))/TSLO(I)
          WRITE(OUT1,7120) I, R(I),HUP(I),HLO(I),V(I),DUP(I),DLO(I),
         1 QL, QSTUP, QUP, QV, QSTLO, QLO
7120      FORMAT(1X,I2,F9.3,1P,1X,11(D9.2,1X))
7150      CONTINUE
          RETURN
          END
```

List of Symbols

Many of the commonly used symbols are listed below; symbols used in Chapter 3 are defined in Section 3.1.2.

a	mesh interval in radial finite difference formulation, $a = \ln(r)$	K_c	corrected hydraulic conductivity for radial flow [L/T]
a	specified radius [L]	K_C	hydraulic conductivity of underlying low conductivity layer [L/T]
A	cross sectional area [L²]		
A, B	constants of integration	K_L	hydraulic conductivity of canal lining [L/T]
b	original saturated aquifer thickness when considering upconing [L]	K_r	radial hydraulic conductivity [L/T]
B	leakage factor $(Kmm'/K)^{1/2}$ [L]	K_x, K_y, K_z	hydraulic conductivity in Cartesian co-ordinate directions [L/T]
B, C	well loss coefficients		
C	concentration of contaminant	l	distance along axis [L]
$CRIV$	streambed conductivity MODFLOW [L²/T for specified reach of river]	L	specified length [L]
		L	distance between drains [L]
d	depth of water in channel [L]	m	saturated thickness [L]
D	dispersion coefficient [L²/T]	m'	saturated thickness of aquitard [L]
D	height of drain above impermeable base [L]	m_0	reference saturated thickness [L]
		n	direction normal to boundary
D_e	equivalent value of D for modified ellipse approach [L]	N	porosity
		p	pressure [various units]
D_p	depth from base of channel to underlying high permeability layer [L]	p_f, p_s	pressures in fresh and in saline water
		pn	well penetration [L]
g	acceleration due to gravity [L/T²]	q	vertical recharge [L/T]
h	groundwater head [L]	q_L	loss from canal per unit plan area of water surface [L/T]
h_w	groundwater level in well [L]		
H	specified groundwater head, groundwater head on water table [L]	q_m	maximum value of recharge [L/T]
		Q	discharge [L³/T] or [L²/T] for vertical section formulation
In	infiltration from an intermittent stream [L³/T]		
		Q_a	well discharge component from aquifer [L³/T]
K	hydraulic conductivity (permeability) [L/T]		
		Q_C	loss from canal per unit length [L³/T/L]
K_0	Bessel function	Q_r	flow from aquifer to river [L³/T for specified reach of river]
K'	hydraulic conductivity of aquitard [L/T]		
K_A	hydraulic conductivity of aquifer beneath canal [L/T]	Q_s	discharge component from well storage [L³/T]

Groundwater Hydrology: Conceptual and Computational Models. K.R. Rushton
© 2003 John Wiley & Sons Ltd ISBN: 0-470-85004-3

Q_x	horizontal flow [L³/T]	W_b	width of base of channel [L]
$QRIV$	flow from river to aquifer, MODFLOW [L³/T for specified reach of river]	W_s	width of water surface in canal [L]
		$W(u)$	well function
r	radial co-ordinate [L]	x, y	co-ordinates in horizontal directions [L]
r_d	radius of drain [L]		
r_w	radius of well [L]	z	vertical co-ordinate [L]
R	radius to outer boundary [L]	z_b	elevation of base of river [L]
RC	river coefficient [L²/T for specified reach of river]	z_r	river surface elevation [L]
		$\bar{z}$	vertical distance measured downwards to interface [L]
s	drawdown [L]		
s_w	drawdown in well [L]	α	coefficient, angle below horizontal
S_C	confined storage coefficient	α	delayed yield coefficient [T⁻¹]
S_S	specific storage [L⁻¹]	δH	vertical rise in water table during time step δt [L]
S_Y	specific yield		
SC	spring coefficient [L²/T]	Δa	radial mesh increment [L]
t	time [T]	$\Delta x, \Delta y, \Delta z$	mesh intervals of finite difference grids [L]
t_{eq}	time to reach equilibrium [T]		
t_L	thickness of canal lining [L]	Δt	time interval in numerical solution [T]
T	transmissivity [L²/T]	Δh	incremental vertical rise in water table [L]
T_C	thickness of underlying low conductivity layer [L]		
		η	elevation of interface between fresh and saline water [L]
u	$r^2 S/4Tt$		
v	Darcy velocity [L/T]	ρ	density [M/L³]
v_s	seepage velocity [L/T]	ρ_f, ρ_s	density of fresh and saline water [M/L³]
v_x, v_y, v_z	component Darcy velocities [L/T]		

References

Abdel-Gawad, S.M., Iman, E.H. and Ali, H.M. (1993) Experimental investigation into the effect of imperfect canal linings on seepage losses. *ICID 15th Congress, The Hague* **Vol 1-C**: 1249–61.

Abu-Zied, M.A. and Scott, V.H. (1963) Non-steady flow for wells with decreasing discharge. *J. Hydraul. Div. ASCE* **89**: 119–32.

Abu-Zreig, M., Attom, M. and Hamasha, N. (2000) Rainfall harvesting using sand ditches in Jordan. *Agric. Water Man.* **46**: 183–92.

Ahmad, N. (1974) *Groundwater Resources of Pakistan.* Ripon Printing Press, Lahore.

Ahmadi, M.Z. (1999) Use of piezometers to find the depth to impermeable layer in the design of drainage systems. *Hydrol. Sci. J.* **44**: 25–31.

Alam, M.S. (1993) Stratigraphical and paleo-climatic studies of the quaternary deposits in north-western Bangladesh. DSc thesis, Free University of Brussels.

Al-Niami, A.N.S. and Rushton, K.R. (1978) Radial dispersion to an abstraction well. *J. Hydrol.* **39**: 287–300.

Al-Othman, A.A.R. (1991) Investigation of groundwater flow in complex aquifer systems in Saudi Arabia using numerical models. PhD thesis, University of Birmingham, UK.

Anderson, M.P. and Woessner, W.W. (1992) *Applied Groundwater Modeling: Simulation of Flow and Advective Transport.* Academic Press, San Diego.

Asad-uz-Zaman, M. and Rushton, K.R. (1998) Barind: predicted a failure, proved a success. In: H.S. Wheater and C. Kirby (eds) *Hydrology in a Changing Environment*, Vol. II, Wiley, Chichester, pp. 195–204.

Asano, T. (ed.) (1985) *Artificial Recharge of Groundwater.* Butterworth, Boston.

ASTM (1993) *Guide for the Application of a Ground-water Flow Model to a Site-specific Problem.* American Society for Testing and Materials, Philadelphia.

Athavale, R.N., Singh, V.S. and Subrahmanyam, K. (1983) A constant discharge device for aquifer tests on large diameter wells. *Ground Water* **21**: 752–5.

Awan, N.M. (1985) Water resources development and management strategy for irrigated agriculture in Pakistan. In: T.H.Y. Tebbutt (ed.) *Advances in Water Engineering*, Elsevier Applied Science, New York, pp. 73–83.

Azuhan, M. (1997) The exploitation of shallow and limited aquifer systems. PhD thesis, University of Birmingham, UK.

Bakiewicz, W., Milne, D.M. and Pattle, A.D. (1985) Development of public tubewell design in Pakistan. *Quart. J. Eng. Geol.* **18**: 63–77.

Ball, D.F. and Herbert, R. (1992) The use and performance of collector wells within the regolith aquifer of Sri Lanka. *Ground Water* **30**: 683–9.

Banerji, A.K. (1983) Importance of evolving a management plan for groundwater development in the Calcutta Region of the Bengal Basin. *IAHS Publ. No.* **142**: 45–54.

Barker, J.A. (1981) A formula for estimating fissure transmissivities from steady-state injection test data. *J. Hydrol.* **52**: 337–46.

Barker, J.A. and Herbert, R. (1982) Pumping tests in patchy aquifer. *Ground Water* **20**: 150–5.

Barton, B.M.J. and Perkins, M.A. (1994) Controlling the artesian boreholes of the South Lincolnshire Limestone. *J. Inst. Water & Env. Man.* **8**: 83–96.

Batu, V. (1998) *Aquifer Hydraulics.* Wiley, Chichester.

Bear, J. (1972) *Dynamics of Fluids in Porous Media.* Elsevier, New York.

Bear, J. (1979) *Hydraulics of Groundwater.* McGraw-Hill, New York.

Bear, J. and Dagan, G. (1964) Moving interface in coastal aquifers. *J. Hydraul. Div. ASCE* **90**: 193–216.

Bear, J. and Verruijt, A. (1987) *Modelling Groundwater Flow and Pollution.* D Reidel, Dordrecht.

Bear, J. and Corapcioglu, M.Y. (eds) (1991) *Transport Processes in Porous Media.* Kluwer Academic, Dordrecht.

Bedient, P.B., Rifai, H.S. and Newell, C.J. (1994) *Ground Water Contamination: Transport and Remediation.* Prentice-Hall, New Jersey.

Bennett, G.D., Mundorff, M.J. and Hussain, S.A. (1968) Electrical-analog studies of brine coning beneath freshwater wells in the Punjab Region, West Pakistan. *USGS Water Supply Paper* **1608-J**: 31pp.

Bennett, G.D., Kontis, A.L. and Larson, S.P. (1982) Representation of multi-aquifer well effects in three-dimensional ground-water flow simulation. *Ground Water* **20**: 334–41.

Berakovic, M. (1998) Depression of groundwater levels in the Zagreb area. In: H.S. Wheater and C. Kirby (eds) *Hydrology in a Changing Environment*, Vol. II, Wiley, Chichester, pp. 137–43.

Bhagavantam, D. (1975) An experimental study of seepage through partially lined earthen channels. *Civil Eng. Trans., Inst. Eng. (Australia)* 77–9.

Bhutta, M.N., Javed, I. and Dukker, P. (1997) Losses from canals in Pakistan. In: I. Simmers (ed.) *Recharge of Aquifers in (Semi-) Arid Areas*. International Contributions to Hydrogeology, AA Balkema, Rotterdam, **19**: 259–69.

Birtles, A.B. (1978) Identification and separation of major base flow components from a stream hydrograph. *Water Resour. Res.* **14**: 791–803.

Bliss, J.C. and Rushton, K.R. (1984) The reliability of packer tests for estimating the hydraulic conductivity of aquifers. *Quart. J. Eng. Geol.* **17**: 81–91.

Boak, R.A. and Rushton, K.R. (1993) Assessing aquifer exploitation from observation wells and piezometers. *British Hydrological Society, 4th National Symposium, Cardiff*, 1–72.

Bos, M.G. and Nugteren, J. (1982) *On Irrigation Efficiencies*. Publ No. 19, International Institute for Land Reclamation and Improvement, Wageningen, The Netherlands.

Boulton, N.S. (1954) The drawdown of the water table under non-steady conditions near a pumped well in an unconfined formation. *Proc. Inst. Civ. Eng. (London)* **3**: 564–79.

Boulton, N.S. (1963) Analysis of data from non-equilibrium pumping tests allowing for delayed yield from storage. *Proc. Inst. Civ. Eng. (London)* **26**: 469–82.

Bouwer, H. (1969) Theory of seepage from open channels. In: V.T. Chow (ed.) *Advances in Hydrosciences* **5**, Academic Press, New York, pp. 121–72.

Bouwer, H. (1978) *Groundwater Hydrology*. McGraw-Hill, New York.

Bouwer, H. (1979) Soil water hysteresis as a cause of delayed yield from unconfined aquifers. *Water Resour. Res.* **15**: 965–66.

Bouwer, H. (1989) The Bouwer and Rice slug test – an update. *Ground Water* **27**: 304–9.

Bouwer, H. (2002) Artificial recharge of groundwater: hydrogeology and engineering. *Hydrogeol. J.* **10**: 121–42.

Bradbury, C.G. and Rushton, K.R. (1998) Estimating runoff-recharge in the Southern Lincolnshire Limestone catchment, UK. *J. Hydrol.* **211**: 86–99.

Bradley, E. and Phadtare, P.N. (1989) Paleohydrology affecting recharge to over-exploited semi-confined aquifers in Mehsana area, Gujarat State, India. *J. Hydrol.* **108**: 309–22.

Brassington, F.C. (1988) *Field Hydrogeology*. Open University Press, Milton Keynes, UK.

Brassington, F.C. (1992) Measurement of head variations within observation boreholes and their implication for groundwater monitoring. *J. Inst. Water & Env. Man.* **6**: 91–100.

Brassington, F.C. and Walthall, S. (1985) Field techniques using borehole packers in hydrogeological investigations.

Quart. J. Eng. Geol. **18**: 181–93.

Brassington, F.C. and Rushton, K.R. (1987) A rising water table in Central Liverpool. *Quart. J. Eng. Geol.* **20**: 151–8.

Brown, I.T. (1997) Numerical modelling of pumping tests in unconfined aquifers. PhD thesis, University of Birmingham, England.

Bush, P.W. (1977) Connector well experiment to recharge the Floridan aquifer, East Orange County, Florida. *USGS Water Supply Paper No.* **2210**: 25pp.

Butler, J.J. (1988) Pumping tests in non-uniform aquifers: the radially symmetric case. *J. Hydrol.* **101**: 15–30.

Canter, L.W., Knox, R.C. and Fairchild, D.M. (1987) *Ground Water Quality Protection*. Lewis Publishers, Michigan.

Carrillo-Rivera, J.J. (1992) The hydrogeology of the San Luis Potosi area, Mexico. PhD thesis, University College, University of London, England.

Carrillo-Rivera, J.J., Cardona, A. and Moss, D. (1996) Importance of vertical components of groundwater flow: a hydrogeochemical approach in the valley of San Luis Potosi, Mexico. *J. Hydrol.* **185**: 23–44.

Carter, R.C., Rushton, K.R., Eilers, V.H.M. and Hassan, M. (2002) Modelling with limited data: 'Plausibility' as a measure of model reliability. *ModelCARE Conference*, Prague, pp. 328–30.

Cartwright, K., Hunt, C.S., Hughes, G.M. and Brower, R.D. (1979) Hydraulic potential in Lake Michigan bottom sediments. *J. Hydrol.* **43**: 67–78.

Cedergren, H.R. (1967) *Seepage, Drainage and Flow Nets*. Wiley, New York.

Chan, Y.K. (1976) Improved image well technique for aquifer analysis. *J. Hydrol.* **29**: 149–64.

Chandler, R.L. and McWhorter, D.B. (1975) Upconing of salt-water – fresh-water interface beneath a pumping well. *Ground Water* **13**: 354–9.

Charni, I.A. (1951) A rigorous derivation of Dupuit's formula for unconfined seepage with seepage surface. *Dokl. Akad. Nauk. USSR* **79**: 6.

Chen, K. and Ayers, J.F. (1998) Aquifer properties determined from two analytical solutions. *Ground Water* **36**: 783–91.

Cheng, X. and Anderson, M.P. (1993) Numerical simulation of groundwater interaction with lakes allowing for fluctuating lake levels. *Ground Water* **31**: 929–33.

Childs, E.C. (1943) The water table, equipotentials and streamlines in drained land. *Soil Science* **56**: 317–30.

CIRIA (1989) *The Engineering Implications of Rising Groundwater Levels in the Deep Aquifer beneath London*. CIRIA, Inst. Civ. Eng. (London), Special Publication **69**.

Clarke, L. (1977) The analysis and planning of step-drawdown tests. *Quart. J. Eng. Geol.* **10**: 125–43.

Connorton, B.J. (1985) Does the regional groundwater-flow equation model vertical flow? *J. Hydrol.* **79**: 279–99.

Connorton, B.J. and Reed, R.N. (1978) A numerical model for the prediction of long-term well yield in an unconfined chalk aquifer. *Quart. J. Eng. Geol.* **11**: 127–38.

Cookey, E.S., Rathod, K.S. and Rushton, K.R. (1987) Pumping from an unconfined aquifer containing layers of different hydraulic conductivity. *Hydrol. Sci. J.* **32**: 43–57.

Cooper, H.H. and Jacob, C.E. (1946) A generalized graphical method for evaluating formation constants and summarizing well field history. *Trans. Am. Geophys. Union* **27**: 526–34.

Cox, D.W. (1994) The effects of changing groundwater levels on construction in the City of London. In: W.B. Wilkinson (ed.), *Groundwater Problems in Urban Areas*, Thomas Telford, London, pp. 263–77.

Crompton, D.M. and Heathcote, J. (1994) Cardiff Bay barrage and groundwater rise. In: W.B. Wilkinson (ed.), *Groundwater Problems in Urban Areas*, Thomas Telford, London, pp. 172–96.

Cross, G.A., Rushton, K.R. and Tomlinson, L.M. (1995) The East Kent Chalk aquifer during the 1988–92 drought. *J. Inst. Water & Env. Man.* **9**: 37–48.

Dassargues, A. and Monjoie, A. (1993) The hydrogeology of the Chalk of north-west Europe – Belgium, In: R.A. Downing, M. Price and G.P. Jones (eds) *The Hydrogeology of the Chalk of North-West Europe*, Oxford Science Publications, Clarendon Press, Oxford, pp. 153–69.

Daugherty, R.L., Franzini, J.B. and Finnemore, E.J. (1989) *Fluid Mechanics with Engineering Applications*. McGraw-Hill, New York.

de Marsily, G. (1986) *Quantative Hydrogeology*. Academic Press, London.

de Silva, C.S. and Rushton, K.R. (1996) Interpretation of the behaviour of agrowell systems in Sri Lanka using radial flow models. *Hydrol. Sci. J.* **41**: 825–35.

Desai, B.I., Gupta, S.K., Mistry, J.F., Shahdadpuri, H.A., Sharma, S.C., Shar, C.R. and Char, B.V. (1978) Artificial recharge experiments for underground storage of water based on siphon principle. *J. Indian Water Works Assn* **10**: 273–8.

Dierickx, W. (1993) Research and developments in selecting drainage materials. *J. Irrig. Drainage Systems* **6**: 291–310.

Dierickx, W., Bhutta, M.N., Rehman, S. and Rafiq, M. (1995) Laboratory evaluation of synthetic drain envelopes. *IWASRI Publ No.* **145**, Water and Power Development Authority, Lahore, Pakistan.

Domenico, P.A. and Schwartz, F.W. (1997) *Physical and Chemical Hydrogeology*, 2nd edn, Wiley, New York.

Downing, R.A. and Williams, B.P.J. (1969) *The Groundwater Hydrology of the Lincolnshire Limestone*. Water Resources Board, Reading, England, Publ No. **9**.

Downing, R.A. and Wilkinson, W.B. (eds) (1991) *Applied Groundwater Hydrology: a British Perspective*. Clarendon Press, Oxford.

Downing, R.A., Ashford, P.L., Headworth, H.G., Owen, M. and Skinner, A.C. (1981) The use of groundwater for river augmentation. In *A Survey of British Hydrogeology*, The Royal Society, London, pp. 153–72.

Downing, R.A., Price, M. and Jones, G.P. (1993) *The Hydrogeology of the Chalk of North-West Europe*. Oxford Science Publications, Clarendon Press, Oxford.

Driscoll, F.G. (1986) *Groundwater and Wells*, 2nd edn, Johnson Filtration Systems Inc, St. Paul, Minnesota.

Ehlig, C. and Halepaska, J.C. (1976) A numerical study of confined-unconfined aquifers including effects of delayed yield and leakage. *Water Resour. Res.* **12**: 1175–83.

Elango, L. and Manickam, S. (1987) Hydrogeochemistry in the Madras aquifer, India: spatial and temporal variation in chemical quality of ground water. In: *The Role of Geology in Urban Development*, Geological Society of Hong Kong, Bulletin **3**: 525–34.

Elci, A., Molz, F.J. and Waldrop, W.R. (2001) Implications of observed and simulated ambient flow in monitoring wells. *Ground Water* **39**: 853–62.

Engelen, G.B. and Jones, G.P. (1986) Developments in the analysis of groundwater flow systems. *IAHS Publ No.* **165**.

Environment Agency (2002) *Groundwater Resource Modelling: Guidance Notes and Template Project Brief* (Version 1), Environment Agency, Bristol, BS12 4UD, UK.

Evenari, M., Shanan, L. and Tadmore, N. (1971) *The Negev, the Challenge of a Desert*. Harvard University Press.

FAO (1998) *Crop Evapotranspiration; Guidelines for Computing Crop Water Requirements*. FAO Irrigation and Drainage Paper **56**, Rome.

Fetter, C.W. (1999) *Contaminant Hydrogeology*, 2nd edn, Prentice Hall, New Jersey.

Fetter, C.W. (2001) *Applied Hydrogeology*, 4th edn, Prentice Hall, New Jersey.

Forster, C. and Smith, L. (1988) Groundwater flow systems in mountainous terrain, 1 Numerical modeling technique, 2 Controlling factors. *Water Resour. Res.* **24**: 999–1023.

Foster, S.S.D. and Milton, V.A. (1974) The permeability and storage of an unconfined chalk aquifer. *Hydrol. Sci. Bull.* **19**: 485–500.

Foster, S.S.D. and Crease, R.I. (1975) Hydraulic behaviour of the chalk aquifer in the Yorkshire Wolds. *Proc. Inst. Civ. Eng. (London)* **59**: 181–8.

Freeze, R.A. and Cherry, J.A. (1979) *Groundwater*. Prentice-Hall, New York.

Gambolati, G. (1976) Transient free surface flow to a well: an analysis of theoretical solutions. *Water Resour. Res.* **12**: 27–39.

Gardener, W.R. (1958) Some steady-state solutions of the unsaturated moisture flow equations with applications to evaporation from the water table. *Soil Sci.* **85**: 228–32.

Gerber, R.E. and Howard, K.W.F. (2000) Recharge through a regional till aquitard: three-dimensional flow water balance approach. *Ground Water* **38**: 410–22.

Gerber, R.E., Boyce, J.I. and Howard, K.W.F. (2001) Evaluation of heterogeneity and field-scale groundwater flow regime in a leaky till aquitard. *Hydrogeol. J.* **9**: 60–78.

Glover, R.E. (1959) The pattern of fresh water flow in a coastal aquifer. *J. Geophys. Res.* **64**: 457–9.

Gonzalez, M. and Rushton, K.R. (1981) Deviations from classical behaviour in pumping test analysis. *Ground Water* **19**: 510–16.

Greenman, D.W., Swarzenski, W.V. and Bennett, G.D. (1967) Groundwater hydrology of the Punjab, West Pakistan with emphasis on problems caused by canal irrigation. *USGS Water Supply Paper* **1608-H**.

Grout, M.W. and Rushton, K.R. (1990) Identifying causes for large pumped drawdowns in an alluvial wadi aquifer. In P. Howsam (ed.) *Water Wells, Monitoring, Maintenance and Rehabilitation*. E & F N Spon, London, pp. 130–47.

Gureghian, A.B. and Youngs, E.G. (1975) The calculation of steady-state water table heights in drained soils by means of the finite element method. *J. Hydrol.* **27**: 15–32.

Hall, P. (1996) *Water Well and Aquifer Test Analysis*. Water Resources Publications, Colorado.

Hantush, M.S. (1956) Analysis of data from pumping tests in leaky aquifers. *Am. Geophys. Union Trans.* **37**: 702–14.

Hantush, M.S. (1960) Modification of the theory of leaky aquifers. *J. Geophys. Res.* **65**: 3713–25.

Hantush, M.S. (1964) Drawdown around wells of variable discharge. *J. Geophys. Res.* **69**: 4221–35.

Hantush, M.S. and Papadopulos, I.S. (1962) Flow of groundwater to collector wells. *J. Hydraul. Div. ASCE* **88 HY5**: 221–4.

Hardcastle, B.J. (1978) From concept to commissioning. In: *Thames Groundwater Scheme*, Inst. Civ. Eng. (London).

Harr, M.E. (1962) *Groundwater and Seepage*. McGraw-Hill, New York.

Harvey, F.E. and Sibray, S.S. (2001) Delineating groundwater recharge from leaking irrigation canals using water chemistry and isotopes. *Ground Water* **39**: 408–21.

Hawker, P.J., Harries, K.J. and Osborne, N.F. (1993) Investigations of river flow failure at Sleaford. *J. Inst. Water & Env. Man.* **7**: 235–43.

Headworth, H.G.H., Keating, T. and Packman, M.J. (1982) Evidence for a shallow highly permeable zone in the Chalk of Hampshire. *J. Hydrol.* **55**: 93–112.

Healy, R.W. and Cook, P.G. (2002) Using groundwater levels to estimate recharge. *Hydrogeol. J.* **10**: 91–109.

Heathcote, J.A. and Lloyd, J.W. (1984) Groundwater chemistry in southeast Suffolk (UK) and its relation to Quaternary geology. *J. Hydrol.* **75**: 143–65.

Hendrickx, J.M.H. and Walker, G.R. (1997) Recharge from precipitation. In I. Simmers (ed.) *Recharge of Aquifers in (Semi-) Arid Areas*. International Contributions to Hydrogeology, AA Balkema **19**: 19–111.

Herschy, R.W. (ed.) (1978) *Hydrometry, Principles and Practices*. Wiley, New York.

Herschy, R.W. (1985) *Streamflow Measurement*. Elsevier, London.

Hill, M.C., Poeter, E., Zheng, C. and Doherty, J. (eds) (2001) *MODLOW 2001 and Other Modeling Odysseys*. Colorado School of Mines, Golden, Colorado.

Hill, M.C., Poeter, E., Zheng, C. and Doherty, J. (eds) (2003) Selected papers from MODFLOW 2001 and other modeling odysseys. *Ground Water* **41**: 119–283.

Hillel, D. (1982) *Introduction to Soil Physics*. Academic Press, London.

Hillel, D. (1998) *Environmental Soil Physics*. Academic Press, London.

Holmes, D.W., Wooldridge, R., Weller, J.A. and Gunston, H. (1981) Water management study at Kadulla irrigation scheme Sri Lanka, III, Interim report covering seasons Yala 1978 to Maha 1980/81. *Hydraulics Research Station, Wallingford, UK*, Report no. 38.

Howland, A.F., Rushton, K.R., Sutton, S.E. and Tomlinson, L.M. (1994) A hydrogeological model for London Docklands. In W.B. Wilkinson (ed.) *Groundwater Problems in Urban Areas*, Thomas Telford, London, pp. 76–92.

Huisman, L. and Olsthoorn, T.N. (1983) *Artificial Groundwater Recharge*. Pitman, London.

Hunt, B. (1999) Unsteady stream depletion from ground water pumping. *Ground Water* **37**: 98–102.

Hunt, B., Weir, J. and Clausen, B. (2001) A stream depletion field experiment. *Ground Water* **39**: 283–9.

Huyakorn, P.S., Andersen, P.F., Mercer, J.W. and White, H.O. (1987) Saltwater intrusion in aquifers: development and testing of a three-dimensional finite element model. *Water Resour. Res.* **23**: 293–312.

Hvorslev, M.J. (1951) *Time Lag and Soil Permeability in Groundwater Observations*. Bulletin 36, Waterways Experimental Station, US Army Corps of Engineers, Vicksburg, Mississippi.

Inouchi, K., Kishi, Y. and Kakinuma, T. (1985) The regional unsteady interface between fresh water and salt water in a confined coastal aquifer. *J. Hydrol.* **77**: 307–31.

IOH (1992) *Low River Flows in the UK*. Report No. **108**, Institute of Hydrology, Wallingford, England.

Isaacs, L.T. and Hunt, B. (1986) A simple approximation for a moving interface in a coastal aquifer. *J. Hydrol.* **83**: 29–43.

Jackson, D. and Rushton, K.R. (1987) Assessment of recharge components for a chalk aquifer unit. *J. Hydrol.* **92**: 1–15.

Jacob, C.E. (1946) Radial flow in a leaky artesian aquifer. *Trans. Am. Geophys. Union* **27**: 198–208.

Jacob, C.E. (1947) Drawdown tests to determine effective radius of artesian well. *Trans. Am. Soc. Civ. Eng.* **112**: 1047–64.

Jacob, C.E. and Lohman, S.W. (1952) Nonsteady flow to a well of constant drawdown in an extensive aquifer. *Trans. Am. Geophys. Union* **33**: 559–69.

Javandel, I. and Witherspoon, P.A. (1969) A method of analysing transient fluid flow in multi-layered aquifers. *Water Resour. Res.* **5**: 856–69.

Jiao, J.J. and Zheng, C. (1997) The different characteristics of aquifer parameters and their implications on pumping-test analysis. *Ground Water* **35**: 25–9.

Johnson, D., Rushton, K.R. and Tomlinson, L.M. (1999) The effect of borehole sealing on the Southern Lincolnshire limestone catchment. *J. C. Inst. Water & Env. Man.* **13**: 37–46.

Jousma, G., Bear, J., Haimes, Y.Y. and Walter, F. (eds) (1989) *Groundwater Contamination: Use of Models in Decision-Making*. Kluwer Academic, Dordrecht.

Kavalanekar, N.B., Sharma, S.C. and Rushton, K.R. (1992) Over-exploitation of an alluvial aquifer in Gujarat, India. *Hydrol. Sci. J.* **37**: 329–46.

Kawecki, M.W. (2000) Transient flow to a horizontal water well. *Ground Water* **38**: 842–50.

Kazda, I. (1978) Free surface seepage flow in earth dams. In: *Finite Elements in Water Resources*, 2nd International Conference, Pentech Press, London 1.103–1.119.

Keating, T. (1982) A lumped parameter model of a chalk

aquifer–stream system in Hampshire, United Kingdom. *Ground Water* **20**: 430–6.

Kellaway, G.A. (ed.) (1991) *Hot Springs of Bath*. Bath City Council.

Khan, S. and Rushton, K.R. (1996) Reappraisal of flow to tile drains; I Steady state response, II Time variant response, III Drains with limited flow capacity. *J. Hydrol.* **183**: 351–66, 367–82, 383–95.

Kirkham, D. and Gaskell, R.E. (1951) The falling water table in tile and ditch drainage. *Soil Sci. Soc. Am. Proc.* **15**: 37–42.

Kitching, R., Shearer, T.R. and Shedlock, S.L. (1977) Recharge to Bunter Sandstone determined from lysimeters. *J. Hydrol.* **33**: 217–32.

Kitching, R., Edmunds, W.M., Shearer, T.R., Walton, N.G.R. and Jacovides, J. (1980) Assessment of recharge to aquifers. *Hydrol. Sci. Bul.* **25**: 217–35.

Kresic, N. (1997) *Quantative Solutions in Hydrogeology and Groundwater Modeling*. CRC Press, Florida.

Krishnasamy, K.V. and Sakthivadivel, R. (1986) Regional modelling of non-linear flows in a multi-aquifer system. *UNESCO Regional Workshop on Groundwater Modelling*, Roorkee, India, pp. 85–105.

Kruseman, G.P. (1997) Recharge from intermittent flow. In I. Simmers (ed.) *Recharge of Aquifers in (Semi-) Arid Areas*. International Contributions to Hydrogeology, AA Balkema, **19**: 145–84.

Kruseman, G.P. and de Ridder, N.A. (1990) *Analysis and evaluation of pumping test data*, 2nd edn, Publ No. 47, International Institute for Land Reclamation and Improvement, Wageningen, The Netherlands.

Kuusisto, E.E. (1985) Lakes: their physical aspects. In J.C. Rodda (ed.) *Facets of Hydrology II*, Wiley, Chichester, pp. 153–81.

Lai, R.Y., Karadi, G.M. and Williams, R.A. (1973) Drawdown at time-dependent flow rate. *Water Resour. Bul.* **9**: 892–900.

Lawrence, A.R. and Dharmagunwardena, H.A. (1983) Vertical recharge to a confined limestone in northwest Sri Lanka. *J. Hydrol.* **63**: 287–97.

Lennon, G.P., Liu, P. and Liggett, J.A. (1979) Boundary integral equation solution to axisymmetric potential flows: 1. Basic formulation, 2. Recharge and well problems in porous media. *Water Resour. Res.* **15**: 1102–15.

Lerner, D.N. and Teutsch, G. (1995) Recommendations for level-determined sampling in wells. *J. Hydrol.* **171**: 355–77.

Lerner, D.N., Issar, A.S. and Simmers, I. (1990) *Groundwater Recharge. A Guide to Understanding and Estimating Natural Recharge*. International Contributions to Hydrogeology, Verlag Heinz Heise **8**.

Liggett, J.A. and Liu, P.L.-F. (1983) *The Boundary Integral Equation Method for Porous Media*. George Allen & Unwin, London.

Lloyd, J.W. and Heathcote, J.A. (1985) *Natural Inorganic Hydrochemistry in Relation to Groundwater*. Clarendon Press, Oxford.

Lloyd, J.W., Harker, D. and Baxendale, R.A. (1981) Recharge mechanisms and groundwater flow in the chalk and drift deposits of East Anglia. *Quart. J. Eng. Geol.* **14**: 87–96.

Lovell, C.J. and Youngs, E.G. (1984) A comparison of steady-state land drainage equations. *Agric. Water Man.* **9**: 1–21.

Lowe, T.C. and Rushton, K.R. (1990) Water losses from canals and ricefields. *Irrigation and Water Resources Symposium, Agricultural Engineering Society of Sri Lanka*, 107–22.

Luther, K. and Haitjema, H.M. (1999) An analytic element solution to unconfined flow near partially penetrating wells. *J. Hydrol.* **226**: 197–203.

MacDonald, A.M., Brewerton, L.J. and Allen, D.J. (1998) Evidence for rapid groundwater flow and karst-type behaviour in the Chalk of southern England. In N.S. Robbins (ed.) *Groundwater Pollution, Aquifer Recharge and Vulnerability*. Geol. Soc. (London) Special Publications **130**: 95–106.

Mace, R.E. (1999) Estimation of hydraulic conductivity in large-diameter hand-dug wells using slug methods. *J. Hydrol.* **219**: 34–45.

Mailvaganam, Y., Ramili, M.Z., Rushton, K.R. and Ong, B.Y. (1993) Groundwater exploitation of a shallow coastal sand aquifer in Sarawak, Malaysia. *IAHS Publ. No. 216*: 451–62.

Mander, R.G. and Greenfield, B.J. (1978) Hydrograph generation and augmented flow analysis. *Thames Groundwater Scheme*, Inst. Civ. Eng. (London) 47–59.

Marino, M.A. and Luthin, J.N. (1982) *Seepage and Groundwater*. Elsevier, New York.

McCready, W. (1987) Left bank outfall drain in Pakistan. *ICID Bulletin* **36**: 15–19.

McWhorter, D.B. and Sunada, D.K. (1977) *Groundwater Hydrology and Hydraulics*. Water Resources Publications, Littleton, Colorado.

Mehta, M. and Jain, S.K. (1997) Efficiency of artificial recharge from percolation tanks. In I. Simmers (ed.) *Recharge of Aquifers in (Semi-) Arid Areas*. International Contributions to Hydrogeology, AA Balkema, **19**: 271–7.

Miah, M.M. and Rushton, K.R. (1997) Exploitation of alluvial aquifers having an overlying zone of low permeability: examples from Bangladesh. *Hydrol. Sci. J.* **42**: 67–79.

Miles, J.C. and Rushton, K.R. (1983) A coupled surface water and groundwater model. *J. Hydrol.* **62**: 159–77.

Miles, J.C. and Kitmitto, K. (1989) New drain spacing formula. *J. Irrig. Drain Div. ASCE* **115**: 215–30.

Milojevic, M. (1963) Radial collector wells adjacent to river banks. *J. Hydraul. Div. ASCE* **89**: 133–51.

Mishra, G.C. and Chachadi, A.G. (1985) Analysis of flow to a large diameter well during recovery period. *Ground Water* **23**: 646–51.

Moench, A.F. (1995) Combining the Neuman and Boulton models for flow to a well in an unconfined aquifer. *Ground Water* **33**: 378–84.

Moench, A.F. and Prickett, T.A. (1972) Radial flow in an infinite aquifer undergoing conversion from artesian to water table conditions. *Water Resour. Res.* **8**: 494–9.

Moody, W.T. (1966) Non-linear differential equations of drain spacing. *J. Irrig. Drain Div. ASCE* **92**: 1–9.

Motz, L.H. (1992) Salt-water upconing in an aquifer overlain by a leaky confining bed. *Ground Water* **30**: 192–8.

Narashimha Reddy, T. and Prakasam, P. (1989) A study in changes in cropping pattern with the intensification of groundwater development in hard rock areas of Andhra Pradesh. *Bhujal News, Central Ground Water Board, New Delhi*, 16–19.

National Research Council (1984) *Groundwater Contamination*. National Research Council (US) Geophysics Study Committee.

Neilson, D.M. (1991) *Practical Handbook of Groundwater Monitoring*. Lewis, Michigan.

Neuman, S.P. (1972) Theory of flow in unconfined aquifers considering delayed response of the water table. *Water Resour. Res.* **8**: 1031–45.

Neuman, S.P. (1974) Effect of partial penetration on flow in unconfined aquifers considering delayed gravity response. *Water Resour. Res.* **10**: 303–12.

Neuman, S.P. (1975) Analysis of pumping test data from anisotropic unconfined aquifers considering delayed gravity response. *Water Resour. Res.* **11**: 329–42.

Neuman, S.P. (1979) Perspective on 'delayed yield'. *Water Resour. Res.* **15**: 899–908.

Neuman, S.P. and Witherspoon, P.A. (1969a) Theory of flow in a confined two-aquifer system. *Water Resour. Res.* **5**: 803–16.

Neuman, S.P. and Witherspoon, P.A. (1969b) Applicability of current theories of flow in leaky aquifers. *Water Resour. Res.* **5**: 817–29.

Neuman, S.P. and Witherspoon, P.A. (1969c) *Transient flow of ground water to wells in multiple-aquifer systems*. Geotechnical Engineering Report, University of Berkeley, California.

Neuman, S.P. and Witherspoon, P.A. (1970) Finite element method of analysing steady seepage with a free surface. *Water Resour. Res.* **6**: 889–97.

Neuman, S.P. and Witherspoon, P.A. (1971) Analysis of non-steady flow with a free surface using the finite element method. *Water Resour. Res.* **7**: 611–23.

Neuman, S.P. and Witherspoon, P.A. (1972) Field determination of the hydraulic properties of leaky multiple aquifer systems. *Water Resour. Res.* **8**: 1284–98.

Nutbrown, D.A. and Downing, R.A. (1976) Normal mode analysis of the structure of baseflow recession curves. *J. Hydrol.* **30**: 327–40.

Oakes, D.B. and Wilkinson, B.J. (1972) *Modelling groundwater and surface water systems. I Theoretical base flow*. Water Resources Board, Reading, UK.

O'Shea, M.J., Baxter, K.M. and Charalambous, A.N. (1995) The hydrogeology of the Enfield-Haringey artificial recharge scheme, north London. *Quart. J. Eng. Geol.* **28**: 115–29.

Otto, C.J. (1998) Monitoring tools and type of recording. In H.A.J. van Lanen (ed.) *Monitoring for Groundwater Management in (Semi-) Arid Regions*. Studies and reports in hydrology, UNESCO **57**: 65–89.

Owen, M. (1981) The Thames Groundwater Scheme. In J.W.

Lloyd (ed.) *Case Studies in Groundwater Resources Evaluation*, Clarendon Press, Oxford, pp. 186–202.

Panu, U.S. and Filice, A. (1992) Techniques of flow rates into draintubes with circular perforations. *J. Hydrol.* **137**: 57–72.

Papadopulos, I.S. (1967) Drawdown distribution around a large diameter well. *Symp. on Groundwater Hydrology*, San Fransisco, Am. Water Resour. Assoc., Proc No. **4**: 157–68.

Papadopulos, I.S. and Cooper, H.H. (1967) Drawdown in a well of large diameter. *Water Resour. Res.* **3**: 241–4.

Patel, S.C. and Mishra, G.C. (1983) Analysis of flow to a large diameter well by a discrete kernel approach. *Ground Water* **21**: 573–6.

Pinder, G.F. (2002) *Groundwater Modelling Using Geographical Information Systems*. Wiley, New York.

Pinder, G.F. and Gray, W.G. (1977) *Finite Element Simulation in Surface and Sub-Surface Flow*. Academic Press, New York.

Polubarinova-Kochina, P.Y. (1962) *Theory of Groundwater Movement*. Princeton University Press, New Jersey.

Pontin, J.M.A., Abed Laila and Weller, J.A. (1979) *Prediction of seepage loss from the enlarged Ismailia Canal*. Report No. O.D. **28**, Hydraulics Research Station, Wallingford, UK.

Powell, S., Rushton, K.R. and Dev Burman, G.K. (1983) Groundwater resources of low yielding aquifers. *IAHS Publ. No.* **142**: 461–70.

Price, M. and Williams, A. (1993) The influence of unlined boreholes on groundwater chemistry: a comparative study using pore-water extraction and packer sampling. *J. Inst. Water & Env. Man.* **7**: 651–9.

Price, M., Morris, B. and Robertson, A.S. (1982) Chalk permeability – a study of intergranular and fissure permeability in chalk and Permian aquifers using double packer injection tests. *J. Hydrol.* **54**: 401–23.

Price, M., Downing, R.A. and Edmonds, W.M. (1993) The Chalk as an aquifer. In R.A. Downing, M. Price and G.P. Jones (eds) *The Hydrogeology of the Chalk of North-West Europe*, Clarendon Press, Oxford, pp. 35–58.

Prudic, D.E. (1989) Documentation of a computer program to simulate stream–aquifer relations using a modular finite difference groundwater flow model. USGS, Open File Report, 88–729.

Raghunath, H.M. (1987) *Ground Water*, 2nd edn, Wiley Eastern Ltd, New Delhi.

Rai, S.N. and Singh, R.N. (1995) Two-dimensional modelling of water table fluctuations in response to local transient recharge. *J. Hydrol.* **167**: 167–74.

Rai, S.N. and Singh, R.N. (1996) Analytical modelling of unconfined flow induced by time varying recharge. *Proceedings of Indian National Science Academy* **62A**: 253–92.

Ramnarong, V. (1983) Environmental impacts of heavy groundwater development in Bangkok, Thailand. *International Conference on Groundwater and Man*, Australian Government Publishing Service, Canberra **2**: 345–50.

Rao, N.H. and Sarma, P.B.S. (1981) Ground–water recharge from rectangular areas. *Ground Water* **19**: 271–4.

Rasmussen, T.C. and Crawford, L.A. (1997) Identifying and

removing barometric pressure effects in confined and unconfined aquifers. *Ground Water* 35: 502–11.

Rastogi, A.K. and Prasad, B. (1992) FEM modelling to investigate seepage losses from the lined Nadiad branch canal, India. *J. Hydrol.* 138: 153–68.

Rathod, K.S. and Rushton, K.R. (1984) Numerical method of pumping test analysis using microcomputers. *Ground Water* 22: 602–8.

Rathod, K.S. and Rushton, K.R. (1991) Interpretation of pumping from a two-zone layered aquifer using a numerical model. *Ground Water* 29: 499–509.

Reilly, T.E. and Goodman, A.S. (1986) Analysis of salt-water upconing beneath a pumping well. *J. Hydrol.* 89: 169–204.

Reilly, T.E., Frimpter, M.H., LeBlank, D.R. and Goodman, A.S. (1987) Analysis of steady-state salt-water upconing with applications at Truro well field, Cape Cod, Massachusetts. *Ground Water* 25: 194–206.

Replogle, J.A. and Bos, M.G. (1982) Flow measurement flumes: application to irrigation water management. In D. Hillel (ed.) *Advances in Irrigation* 1, Academic Press, New York, pp. 148–217.

Restrepo, J.I., Montoya, A.M. and Obeysekera, J. (1998) A wetland simulation model for the MODFLOW groundwater model. *Ground Water* 36: 764–70.

Rushton, K.R. (1980) Differing positions of saline interfaces in aquifers and observation boreholes. *J. Hydrol.* 48: 185–9.

Rushton, K.R. (1985) Interference due to neighbouring wells during pumping tests. *Ground Water* 23: 361–6.

Rushton, K.R. (1986a) Groundwater Models. In T.W. Brandon (ed.) *Groundwater: Occurrence Development and Protection*, Institution of Water Engineers and Scientists, London, pp. 189–228.

Rushton, K.R. (1986b) Vertical flow in heavily exploited hard rock and alluvial aquifers. *Ground Water* 24: 601–8.

Rushton, K.R. (1999) Discussion on unsteady stream depletion from ground water pumping. *Ground Water* 37: 805.

Rushton, K.R. (2002) Will reductions in groundwater abstractions improve low river flows? In K.M. Hiscock, M.O. Rivett and R.M. Davison (eds) *Sustainable Groundwater Development*. Geological Society, London, Special Publications 193: 193–204.

Rushton, K.R. and Wedderburn, L.A. (1973) Starting conditions for aquifer simulations. *Ground Water* 11: 37–42.

Rushton, K.R. and Booth, S.J. (1976) Pumping test analysis using a discrete time-discrete space numerical method. *J. Hydrol.* 28: 13–27.

Rushton, K.R. and Chan, Y.K. (1976) Pumping test analysis when parameters vary with depth. *Ground Water* 14: 82–7.

Rushton, K.R. and Redshaw, S.C. (1979) *Seepage and Groundwater Flow*. Wiley, Chichester.

Rushton, K.R. and Tomlinson, L.M. (1979) Possible mechanisms for leakage between aquifers and rivers. *J. Hydrol.* 40: 49–65.

Rushton, K.R. and Ward, C. (1979) The estimation of groundwater recharge. *J. Hydrol.* 41: 345–61.

Rushton, K.R. and Rathod, K.S. (1980) Overflow tests analysed by theoretical and numerical methods. *Ground Water* 18: 61–9.

Rushton, K.R. and Holt, S.M. (1981) Estimating aquifer parameters for large diameter wells. *Ground Water* 19: 505–9.

Rushton, K.R. and Rathod, K.S. (1981) Aquifer response due to zones of higher permeability and storage coefficient. *J. Hydrol.* 50: 299–316.

Rushton, K.R. and Howard, K.W.F. (1982) The unreliability of open observation boreholes in unconfined aquifer pumping tests. *Ground Water* 20: 546–50.

Rushton, K.R. and Senarath, D.C.H. (1983) A mathematical model study of an aquifer with significant dewatering. *J. Hydrol.* 62: 143–58.

Rushton, K.R. and Singh, V.S. (1983) Drawdowns in large-diameter wells due to decreasing abstraction rates. *Ground Water* 21: 670–7.

Rushton, K.R. and Rathod, K.S. (1985) Horizontal and vertical flow components deduced from groundwater heads. *J. Hydrol.* 79: 261–78.

Rushton, K.R. and Weller, J. (1985) Response to pumping of a weathered-fractured granitic aquifer. *J. Hydrol.* 80: 299–309.

Rushton, K.R. and Singh, V.S. (1987) Pumping test analysis in large diameter wells with a seepage face by kernel function technique. *Ground Water* 25: 81–90.

Rushton, K.R. and Raghava Rao, S.V. (1988) Groundwater flow through a Miliolite limestone aquifer. *Hydrol. Sci. J.* 33: 449–64.

Rushton, K.R. and Rathod, K.S. (1988) Causes of non-linear step pumping test response. *Quart. J. Eng. Geol.* 21: 147–58.

Rushton, K.R. and Srivastava, N.K. (1988) Interpreting injection well tests in an alluvial aquifer. *J. Hydrol.* 99: 49–60.

Rushton, K.R. and Phadtare, P.N. (1989) Artificial recharge pilot projects in Gujarat, India. *IAHS Publ. No. 188*: 533–45.

Rushton, K.R. and Tiwari, S.C. (1989) Mathematical modelling of a multi-layered alluvial aquifer. *J. Inst. Eng. India* 70: CV2, 47–56.

Rushton, K.R. and Fawthrop, N.P. (1991) Groundwater support of stream flows in Cambridge Area, UK. *IAHS Publ. No. 202*: 367–76.

Rushton, K.R. and Salmon, S. (1993) Significance of vertical flow through low-conductivity zones in Bromsgrove sandstone aquifer. *J. Hydrol.* 152: 131–52.

Rushton, K.R. and Al-Othman, A.A.R. (1994) Control of rising groundwater levels in Riyadh, Saudi Arabia. In W.B. Wilkinson (ed.) *Groundwater Problems in Urban Areas*, Thomas Telford, London, pp. 299–309.

Rushton, K.R. and Tomlinson, L.M. (1995) Interaction between rivers and Nottinghamshire Sherwood sandstone aquifer system. *BHS Occasional Paper No. 6*: 101–16.

Rushton, K.R., Smith, E.J. and Tomlinson, L.M. (1982) An improved understanding of flow in a Limestone aquifer using field evidence and mathematical models. *J. Inst. Water Eng. & Sci.* 36: 369–87.

Rushton, K.R., Kawecki, M.W. and Brassington, F.C. (1988) Groundwater model of conditions in the Liverpool Sandstone Aquifer. *J. Inst. Water & Env. Man.* **2**: 67–84.

Rushton, K.R., Connorton, B.J. and Tomlinson, L.M. (1989) Estimation of the groundwater resources of the Berkshire Downs supported by mathematical modelling. *Quart. J. Eng. Geol.* **22**: 329–41.

Rushton, K.R., Owen, M. and Tomlinson, L.M. (1992) The water resources of the Great Oolite aquifer in the Thames Basin. *J. Hydrol.* **132**: 225–48.

Rushton, K.R., Fletcher, S.W. and Bishop, T.J. (1995) Surface water and groundwater components of the Nottinghamshire Sherwood sandstone aquifer system, *BHS 5th National Hydrology Symposium, Edinburgh*, 2.17–2.24.

Russac, D.A.V., Rushton, K.R. and Simpson, R.J. (1991) Insights into domestic demand from a metering trial. *J. Inst. Water & Env. Man.* **5**: 342–50.

Ruud, N.C. and Kabala, Z.J. (1997) Response of a partially penetrating well in a heterogeneous aquifer: integrating well-face flux vs. uniform well-face flux boundary conditions. *J. Hydrol.* **194**: 76–94.

Scanlon, B.R. and Cook, P.G. (eds) (2002) Theme issue: groundwater recharge. *Hydrogeol. J.* **10**.

Schilfgaarde, J.V. (ed.) (1974) *Drainage for Agriculture*. American Society of Agronomy, Madison, Wisconsin.

Senarath, D.C.H. and Rushton, K.R. (1984) A routing technique for estimating groundwater recharge. *Ground Water* **22**: 142–7.

Seymour, K.J., Wyness, A.J. and Rushton, K.R. (1998) The Fylde aquifer – a case study in assessing the sustainable use of groundwater resources. In H.S. Wheater and C. Kirby (eds) *Hydrology in a Changing Environment*, Vol. II, Wiley, Chichester, pp. 253–67.

Sharma, H.D. and Chawla, H.S. (1979) Canal seepage with boundary at finite depth. *J. Hydraul. Div. ASCE* **105**: 877–97.

Sharma, P. and Gupta, S.K. (1987) Isotope investigation of soil water movement: a case study in the Thar desert, western Rajasthan. *Hydrol. Sci. J.* **32**: 469–83.

Sharma, K.C., Chauhan, H.S. and Ram, S. (1985) Hydraulics of a well pumped with linearly decreasing discharge. *J. Hydrol.* **77**: 281–91.

Shaw, E.M. (1994) *Hydrology in Practice*, 3rd edn, Chapman & Hall, London.

Shepley, M.G. and Taylor, A. (2003) Exploration of aquifer management options using a groundwater model. *J. Inst. Water & Env. Man.* **17**: 176–80.

Simmers, I. (ed.) (1997) *Recharge of Phreatic Aquifers in (Semi-) Arid Areas*. International Contributions to Hydrogeology, AA Balkema, Rotterdam **19**.

Skaggs, R.W. (1978) Effect of drain tube openings on water-table drawdowns. *J. Irrig. Drain. Div. ASCE* **104**: 13–21.

Smith, D.B., Wearn, P.L., Richards, H.J. and Rowe, P.C. (1970) Water movement in the unsaturated zone of high and low permeability strata by measuring natural tritium. *Symp. on Use of Isotopes in Hydrol. IAEA Vienna* 73–87.

Smith, E.J. (1979) Spring discharge in relation to rapid fissure flow. *Ground Water* **17**: 346–50.

Smith, G.D. (1985) *Numerical Solution of Partial Differential Equations: Finite Difference Methods*, 3rd edn, Oxford University Press, Oxford.

Sophocleous, M. (2002) Interactions between groundwater and surface water: the state of the science. *Hydrogeol. J.* **10**: 52–67.

Sophocleous, M., Townsend, M.A., Vogler, L.D., McClain, T.J., Marks, E.T. and Coble, G.R. (1988) Experimental studies in stream–aquifer interaction along the Arkansas River in Central Kansas – field testing and analysis. *J. Hydrol.* **98**: 249–73.

Southern Water Authority (1979) *The Candover Pilot Scheme – Final Report*. Southern Water Authority, Worthing, UK.

Stone, D.B., Moomaw, C.L. and Davis, A. (2001) Estimating recharge distribution by incorporating runoff from mountainous areas in an alluvial basin in the Great Basin Region of the Southwestern United States. *Ground Water* **39**: 807–18.

Susanto, R.H. and Skaggs, R.W. (1992) Hydraulic head losses near agricultural drains during drainage and subirrigation. *Proc. Sixth Internat. Drain Symp. ASAE* 419–27.

Sutton, S.E. (1985) Assessment of layered aquifers with limited piezometry. In T.H.Y. Tebbutt (ed.) *Advances in Water Engineering*, Elsevier Applied Science, New York, pp. 116–23.

Talsma, T. and Haskew, H.C. (1959) Investigation of water table response to tile drains in comparison with theory. *J. Geophys. Res.* **64**: 1933–44.

Taylor, A., Hulme, P., Hughes, A. and Rushton, K.R. (2001) Representation of variable hydraulic conductivity with depth in MODFLOW. In H.S. Seo, E.P. Poeter, C. Zheng and O. Poeter (eds) *Proc MODFLOW 2001 and Other Modelling Odysseys*, IGWMC and Colorado School of Mines, 24–29.

Taylor, R.G. and Howard, K.W.F. (1996) Groundwater recharge in the Victoria Nile Basin of east Africa: support for the soil moisture balance approach using stable isotope tracers and flow modelling. *J. Hydrol.* **180**: 31–53.

Teutsch, G. (1993) An extended double porosity concept as a practical modelling approach to a karstified terrain. *IAHS Publ. No.* **207**: 281–92.

Thames Water (1978) *Thames Groundwater Scheme*. Institution of Civil Engineers, London.

Theis, C.V. (1935) The relation between the lowering of the piezometric surface and the rate and duration of discharge of a well using groundwater storage. *Trans. Am. Geophys. Union* **16**: 519–24.

Thiem, G. (1906) *Hydrologische Methode*. Gebhardt, Leipzig.

Todd, D.K. (1980) *Groundwater Hydrology*, 2nd edn, Wiley, New York.

Todsen, M. (1971) On solution of transient free-surface flow problem in porous media by finite-difference methods. *J. Hydrol.* **12**: 177–210.

Tóth, J. (1963) A theoretical analysis of groundwater flow in small drainage basins. *J. Geophys. Res.* **68**: 4795–812.

Tsao, Y.-S. (1987) Over-irrigation of paddy fields for the purpose of artificially recharging groundwater in the conjunctive use scheme of the Cho-Shui river basin. *IAHS Publ. No.* **169**: 99–109.

Tuong, T.P., Wopereis, M.C.S., Marquez, J.A. and Kropff, M.J. (1994) Mechanisms and control of percolation losses in irrigated puddled ricefields. *Soil. Sci. Soc. Am. J.* **58**: 1794–803.

USBR (1978) *Drainage Manual.* US Department of the Interior, Bureau of Reclamation, Government Printing Office, Washington.

van Lannen, H.A.J. (1998) *Monitoring for Groundwater Management in (Semi-) Arid Regions.* Studies and Reports in Hydrology, UNESCO **57**.

Volker, R.E. and Rushton, K.R. (1982) An assessment of the importance of some parameters for seawater intrusion in aquifers and a comparison of dispersive and sharp interface modelling approaches. *J. Hydrol.* **56**: 239–50.

Wachyan, E. and Rushton, K.R. (1987) Water losses from irrigation canals. *J. Hydrol.* **92**: 275–88.

Walker, S.H. and Rushton, K.R. (1984) Verification of lateral percolation losses from irrigated ricefields by a numerical model. *J. Hydrol.* **71**: 335–51.

Walker, S.H. and Rushton, K.R. (1986) Water losses through the bunds of irrigated ricefields interpreted through an analogue model. *Agric. Water Man.* **11**: 59–73.

Walsh, P. (1976) Investigation into the yield of an existing surface water reservoir and aquifer system. *J. Hydrol.* **28**: 215–28.

Walthall, S. and Ingram, J.A. (1984) The investigation of aquifer parameters using multiple piezometers. *Ground Water* **22**: 25–30.

Walton, W.C. (1962) Selected analytical methods for well and aquifer evaluation. *Illinois State Water Survey Bull.* **39**.

Walton, W.C. (1970) *Groundwater Resources Evaluation.* McGraw-Hill, New York.

Walton, W.C. (1987) *Groundwater Pumping Tests, Design and Analysis.* Lewis, Michigan.

Walton, W.C. (1991) *Principles of Groundwater Engineering.* Lewis, Michigan.

WAPDA (1984) *Mardan SCARP, Subsurface Drainage Design Analysis.* Water and Power Development Authority, Lahore, Pakistan.

Ward, D.S., Buss, D.R., Mercer, J.W. and Hughes, S.S. (1987) Evaluation of groundwater corrective action of the Chem-Dyne hazardous waste site using a telescopic mesh refinement modeling approach. *Water Resour. Res.* **23**: 603–17.

Watson, K.K. (1986) Numerical analysis of natural recharge to an unconfined aquifer. *Water Resour. Res.* **156**: 323–33.

Wellings, S.R. and Bell, J.P. (1980) Movement of water and nitrate in the unsaturated zone of Upper Chalk near Winchester, Hants, England. *J. Hydrol.* **48**: 119–36.

Wellings, S.R. and Bell, J.P. (1982) Physical controls of water movement in the unsaturated zone. *Quart. J. Eng. Geol.* **15**: 235–41.

Willardson, L.S. (1974) Envelope materials. In J.V. Schilfgaarde (ed.) *Drainage for Agriculture*, Am. Soc. Agronomy, Madison, Wisconsin, pp. 179–96.

Willardson, L.S., Boles, A.J. and Bouwer, H. (1971) Interceptor drain recovery of canal seepage. *Trans. ASAE* **14**: 738–41.

Wilsnack, J.I., Welter, D.E., Montoya, A.M., Restrepo, J.I. and Obeysekera, J. (2001) Simulating flows in regional wetlands with the MODFLOW wetlands package. *J. Am. Water Resour. Assoc.* **37**: 655–74.

Winter, T.C. (1999) Relations of streams, lakes and wetlands to groundwater flow. *Hydrogeol. J.* **7**: 28–45.

Winter, T.C. and Rosenberry, D.O. (1995) The interaction of groundwater with prairie pothole wetlands in Cottonwood Lake area, east-central North Dakota, 1979–1990. *Wetlands* **15**: 193–211.

WMO (1989) *Management of groundwater observation programmes.* Operational Hydrology Report **31**, Geneva.

Worstell, V. (1976) Estimating seepage losses from canal systems. *J. Irrig. Drain Div. ASCE* **102**: 137–47.

Wu. J., Xue, Y., Liu, P., Wang, J., Jiang, Q. and Shi, H. (1993) Sea-water intrusion in the coastal area of Laizhou Bay, China: 2. Sea-water intrusion monitoring. *Ground Water* **31**: 740–45.

Yitayew, M. (1998) Hydraulic design of drainage pipes: simplified solution. *Trans. ASAE* **41**: 1707–9.

Youngs, E.G. (1984) Developments in the physics of land drainage. *J. Agric. Eng. Res.* **29**: 167–75.

Yu, S.C. and Konyha, K.D. (1992) Boundary modelling of steady saturated flow to drain tubes. *Drainage and Water Table Control, Proc. 6th Internat. Drain Symp. ASCE*, pp. 305–13.

Zhang, B. and Lerner, D. (2000) Modelling of groundwater flow to adits. *Ground Water* **38**: 99–105.

Zhang, G. (1998) Groundwater overdraft and environmental impacts in the Huaihe River basin. In H.S. Wheater and C. Kirby (eds) *Hydrology in a Changing Environment*, Vol. II, Wiley, Chichester, pp. 69–75.

Zheng, C. and Bennett, G.B. (2002) *Applied Contaminant Transport Modeling*, 2nd edn, Wiley, New York.

Index

Groundwater Hydrology: Conceptual and Computational Models. K.R. Rushton
© 2003 John Wiley & Sons Ltd ISBN: 0-470-85004-3